Differential Equations for Beams

$$\frac{dv}{dx} = v' = \theta \tag{10.12}$$

$$\frac{d^2v}{dx^2} = v'' = \frac{M}{EI} \tag{10.9}$$

$$\frac{d}{dx}\left(EI\frac{d^2v}{dx^2}\right) = (EIv'')' = V \tag{10.13}$$

$$\frac{d^2}{dx^2}\left(EI\frac{d^2v}{dx^2}\right) = (EIv'')'' = w \tag{10.14}$$

Unit Load Method for Deflections

$$1 \cdot \Delta = \sum_{i=1}^{N} \frac{\mathbf{p}_i\mathbf{P}_iL_i}{A_iE_i} \tag{10.16}$$

$$1 \cdot \mathbf{v}_A = \int_0^L \frac{\mathbf{m}\cdot\mathbf{M}\,dx}{EI} \tag{10.17}$$

$$1 \cdot \phi = \int_0^L \frac{\mathbf{t}\cdot\mathbf{T}\,dx}{JG} \tag{10.19}$$

Euler Column Formula

$$P_{cr} = \pi^2 EI/L^2$$

Strain Energy

Axial Force

$$U = \frac{1}{2}\int_0^L \frac{P^2\,dx}{AE} \tag{13.3}$$

Transverse Bending

$$U = \frac{1}{2}\int_0^L \frac{M^2}{EI}\,dx \tag{13.6}$$

Transverse Shear for a Rectangular Beam

$$U = \frac{3}{5}\frac{V^2L}{AG}$$

Torsion of a Circular Rod

$$U = \frac{1}{2}\int_0^L \frac{T^2}{GJ}\,dx \tag{13.5}$$

Castigliano's Second Theorem for Beams

$$v_i = \frac{\partial U}{\partial P_i} = \int_0^L \frac{M}{EI}\frac{\partial M}{\partial P_i}\,dx \tag{13.18}$$

$$\theta_i = \frac{\partial U}{\partial C_i} = \int_0^L \frac{M}{EI}\frac{\partial M}{\partial C_i}\,dx \tag{13.19}$$

General Deflection Equation for Beam Structures

$$u\mathbf{i} + v\mathbf{j} + w\mathbf{k} = \int\left[\left(\frac{\partial M_x}{\partial P}\frac{M_x}{JG} + \frac{\partial M_y}{\partial P}\frac{M_y}{EI_y} + \frac{\partial M_z}{\partial P}\frac{M_z}{EI_z}\right)\mathbf{i} + \left(\frac{\partial M_x}{\partial Q}\frac{M_x}{JG} + \frac{\partial M_y}{\partial Q}\frac{M_y}{EI_y} + \frac{\partial M_z}{\partial Q}\frac{M_z}{EI_z}\right)\mathbf{j}\right.$$
$$\left. + \left(\frac{\partial M_x}{\partial R}\frac{M_x}{JG} + \frac{\partial M_y}{\partial R}\frac{M_y}{EI_y} + \frac{\partial M_z}{\partial R}\frac{M_z}{EI_z}\right)\mathbf{k}\right]dl \tag{A13.13}$$

Vector Transformation Matrix

$$[T] = \begin{bmatrix} \cos\theta & \sin\theta \\ -\sin\theta & \cos\theta \end{bmatrix} \tag{A8.10}$$

MECHANICS OF MATERIALS

George R. Buchanan

TENNESSEE TECHNOLOGICAL UNIVERSITY, COOKEVILLE, TENNESSEE

HOLT, RINEHART AND WINSTON, INC.
NEW YORK   CHICAGO   SAN FRANCISCO   PHILADELPHIA
MONTREAL   TORONTO   LONDON   SYDNEY   TOKYO

**To the students,
for without them,
this book would not have been possible.**

8 9 0 1   039   9 8 7 6 5 4 3 2 1

Library of Congress Cataloging in Publication

Buchanan, George R.
  Mechanics of materials.

  Includes bibliographies and index.
    1. Strength of materials.   2. Structures, Theory of.
I. Title.
TA405.B795 1988   620.1'12   87-21111
ISBN 0-03-07979-9

Holt, Rinehart and Winston, Inc.
The Dryden Press
Saunders College Publishing

CONTENTS

PREFACE

Mechanics of materials is a topic that is fundamental to engineering analysis and design. As an engineering course of study, it is a required course for civil and mechanical engineers. Additionally, other disciplines may require some knowledge of deformation and stress analysis for engineering materials. From a practical viewpoint many graduate engineers, such as chemical and electrical engineers, find that as their careers develop they delve into topics that fall into the category of mechanics of materials. With that in mind I have tried to present the material in this text at a level suitable for sophomores or juniors in engineering to begin the study of mechanics of materials with no prior knowledge other than vector statics and differential equations. On the other hand, I have tried to extend the basic coverage beyond the boundaries of a traditional undergraduate course and incorporate concepts and examples that will qualify the text as a significant reference source.

The preparation of this text was motivated by my experience with teaching combined stresses. Chapter 8 presents the analysis of combined stresses as a problem in vector analysis prior to its being a problem in mechanics of materials. During the past 20 years, as I have taught combined stresses, it appears that a major difficulty that students encounter is deciding where and how to get started solving the problem. I have merely emphasized that many diverse problems can be approached using an organized procedure and that vector analysis is the organizational tool. The chapters preceding Chapter 8 are traditional in coverage but are written to incorporate vector analysis so that the topic has been reviewed and the reader is prepared for Chapter 8.

As an instructor I certainly am aware that some faculty will not share my enthusiasm for incorporating vector analysis into a course of mechanics of materials. I have organized the material in each chapter so that a course could be structured in the traditional way by omitting the sections marked in the table of contents with a double asterisk (**), thereby omitting the majority of the material that involves the application of vector analysis. As the textbook evolved it seemed natural and correct to extend the use of vector analysis beyond the first half of the book and incorporate the formalities of the vector approach into the analysis of deflections. Again, I have kept the mechanics of materials in the forefront and relegated vector analysis to its use as a computational tool.

The topics that are considered to be absolutely necessary for a course in mechanics of materials have been established for years. When I compare my manuscript to the text that I first studied almost 30 years ago, it becomes obvious that the basic course has not changed. But there have been significant changes during that time in the basic presentation of the material and the depth of understanding that is expected of the student. Greater emphasis is given to first defining or assuming how a body deforms and how that deformation dictates a strain distribution. The assumed deformation is dependent upon the geometry and loading for the deformable structure. I have presented the relationship between geometry and loading, deformation assumptions, and the resulting strain distribution in detail in the first sections of Chapter 4 for a circular shaft subjected to a torque. The treatment of the relation between strain and stress, the constitutive relation, has remained the same in one respect, yet broadened in another sense. Isotropic and homogeneous material properties are assumed, and rightfully so for an introductory course; however, inelastic material behavior is often incorporated throughout a contemporary textbook. I have elected to concentrate on elastic material behavior and group the discussion of inelastic material behavior for all specific applications into one chapter. Contemporary design-oriented courses usually investigate inelastic material behavior and its applicability for analysis and design. A more important issue that I have discussed in Chapters 3 and 9 is the definition of engineering strain versus the definition of strain using the theory of elasticity. The introduction of vector analysis and vector transformations throughout the early chapters has allowed me to extend the standard treatment of stress transformation to illustrate the mathematical property of transformation using a transformation matrix. A study of the table of contents indicates that I have grouped all statically indeterminate problems in Chapter 11. Axially loaded indeterminate structures can be included in Chapters 2 and 4 with no loss of continuity. It is a matter of preference. Energy methods have enjoyed an increasing emphasis in modern textbooks and I have taken the liberty to exploit the use of vector analysis and develop a logical and organized discussion of these methods that is oriented toward the solution of practical determinate and indeterminate beam structures. The unit load method is included in Chapter 10 and an application to three-dimensional problems is illustrated. I included the unit load method in Chapter 10 in order to make Chapter 11 a more complete and thorough coverage of indeterminate structures. Special or advanced topics are marked with an asterisk (*).

The use of vector analysis necessitated the introduction of a vector sign convention. I made the comparison between the traditional mechanics of materials beam sign convention and the vector sign convention in Chapter 5. Numerous examples illustrate the nuances of the traditional sign convention in an attempt to draw the reader's attention to the importance of the proper use of sign conventions. I do not intend to propose to revolutionize mechanics of materials by suggesting that the traditional sign convention be abandoned. However, eventually someone more resourceful and clever than I may propose a totally consistent sign convention for mechanics of materials.

I have included several lengthy example problems and feel that an explanation is in order for both students and instructors. The two examples near the beginning of Chapter 7 are teaching examples—one to illustrate the physical behavior of shear stress and the other to illustrate the computations for shear flow. The students should continue through these problems until they develop a clear understanding of the physical phenomenon associated with shear caused by transverse loading. Example 8.9 is quite lengthy and is included as an application of combined stresses that illustrates, in one problem, most of the concepts that have been introduced up to that point in the text. It is my idea of a teaching example as well as a reference example for those engineers who will not continue formal study in structural mechanics or machine design. The use of vector analysis is demonstrated in Chapters 11 and 13 for statically indeterminate three-dimensional structures, and considerable detail is included in the examples. I suggest that those examples are not as lengthy and involved as they might at first appear.

There is sufficient material in the book to teach a three-credit course each term for an academic year. Unfortunately, an engineering curriculum rarely permits that much time for mechanics of materials. A four-credit semester course or two three-credit quarter courses is probably the average coverage given to mechanics of materials. That would give the instructor ample time to cover the majority of the first eight chapters and parts of Chapters 9 through 13.

This textbook certainly incorporates a new approach for the presentation of material that is considered to be mechanics of materials. The coverage is not new, but has been guided and influenced by the many excellent textbooks that have been published during the past decades. I am grateful and appreciative for the many discussions that I have had with colleagues concerning the writing of this textbook and teaching mechanics of materials. Numerous reviewers took an interest in the text and expended considerable effort with the formidable task of reading and studying the manuscript. My special thanks to Professors P. C. Chan of the New Jersey Institute of Technology, Elmer Payne of the University of Dayton, Nadim Hassoun of the South Dakota State University, Jeffrey P. Laible of the University of Vermont, Chuck Milne of Montana State University, Kevin Truman of Washington University, Alan Miller of the New Mexico Institute of Mining and Technology, Howard Conlon of Oklahoma State University, and David Bourell of the University of Texas, and to several reviewers whose names I do not know for their efforts in behalf of this textbook and the future readers of the text. I am and will always be impressed by the monumental task that the editors and production staff confront each time a new text becomes a reality. I could not have asked for better friends and co-workers than John Beck, Kiran Kimbell, and Cynthia Godby at Holt, Rinehart and Winston and Lila M. Gardner at Cobb/Dunlop Publisher Services. Several of my students and friends assisted in solving and checking homework problems. I wish to acknowledge C. G. Tseng, S. Foroudastan, M. Sallah, K. F. Fong, J. A. Forsyth, and B. R. Jones. The original manuscript was deciphered and typed by JoAnn Boling. I shall forever be indebted to my wife, Susan A. Buchanan, for meticulously typing, editing, and proofreading the final manuscript and encouraging me to finish this project.

George R. Buchanan

LIST OF SYMBOLS

UNITS

F	force
L	length
T	time
Temp	temperature

$A, \Delta A$	area, increment of area (L^2)
b	width, cross-section dimension (L)
c	distance from the neutral axis of a beam to the outermost surface of the beam (L)
C_1, C_2, etc.	constants of integration
C_c	see Eq. (12.6)
d	distance, see parallel axis theorem (L)
E	modulus of elasticity, Young's Modulus (F/L^2)
e	eccentricity, see shear center (L)
f	angular frequency (cycles/T), see Section 4.9
f	shape factor, M_p/M_y, see Chapter 14
$\mathbf{F}$	force vector

$F, \Delta F$	force, increment of force (F)
F_x, F_y, F_z	force components (F)
G	shear modulus (F/L^2)
g	acceleration due to gravity (L/T^2)
h	height, cross-section dimension (L)
I, I_x, I_y, I_z	second moment of the area (L^4)
I_{xy}	product second moment of the area (L^4)
I_1, I_2	principal second moment of the area (L^4)
i, j, k	unit vectors
J	polar moment of the area (L^4)
K	bulk modulus of elasticity (F/L^2)
k	spring constant (F/L), correction factor for equivalent column length
k, k_t, k_{ts}	stress concentration factor
L	length of a beam or column structure (L)
M	moment vector (F·L)
$M, \Delta M$	moment, increment of moment (F·L)
M_x, M_y, M_z	moment components (F·L)
M_t	torque (F·L)
M_y, M_{ty}	bending moment at yield, torque at yield (F·L), see Chapter 14
M_p, M_{tp}	plastic moment, plastic torque (F·L), see Chapter 14
n	unit vector
n	angular frequency (rev/T), see Section 4.9
P	load, force (F)
P_{cr}	critical column load (F)
p	pressure (F/L^2)
Q	force (L)
Q, Q_x, Q_y, Q_z	first moment of an area (L^3)
ΔQ	increment of heat energy (F·L)
q	shear flow (F/L)
R	force, reaction (F), radius (L)
r	radius (L), radius of gyration $(I/A)^{1/2}$ (L)
r	position vector (L)
r, θ	polar coordinates (L), (radians or degrees)
S	force (L), section modulus (I/c) (L^3)
s	distance (L)
T	torque (F·L)
ΔT	increment of temperature (Temp)
t	thickness of a thin-walled member (L), width used in shear stress equation, Eq. (7.12) (L)
$U, \Delta U$	strain energy, increment of strain energy (F·L)
u	displacement in the x direction (L)

$V, \Delta V$	shear, increment of shear (F)
V_x, V_y, V_z	shear components (F)
$V, \Delta V$	volume, increment of volume (L^3)
v	displacement in the y direction (L)
v', v'', etc.	dv/dx, d^2/dx^2, etc.
$W, \Delta W$	external work, increment of external (F·L)
W_P	potential energy
w	uniform load (F/L), displacement in the z direction (L)
x, y, z	coordinates (L)
$\bar{x}, \bar{y}, \bar{z}$	coordinates of the centroid (L)
$S_1, S_2, B_c, D_c, n_u, n_y$	aluminum column design parameters
α	angular measure
α	coefficient of thermal expansion (L/L·Temp)
$\gamma_{\text{subscript}}$	unit weight (F/L^3)
γ	shear strain
$\gamma_1, \gamma_2, \gamma_T$	shear strain, see Eq. (9.20)
$\gamma_{rx}, \gamma_{x\theta}, \gamma_{r\theta}$	shear strain in cylindrical coordinates, see Fig. 4.7
δ	deflection (L)
Δ	deflection (L)
ε	normal strain (L/L)
$\varepsilon_1, \varepsilon_2$	principal strain
θ	angular measure
θ_N	angle that defines the plane of principal stress
θ_S	angle that defines the plane of maximum shear stress
ρ	curvature, $(1/\rho)$ (1/L)
ρ	radius of curvature (L)
π	3.14159
v	Poisson's ratio (L/L)
σ	normal stress (F/L^2)
$\sigma_x, \sigma_y, \sigma_z$	normal stress (F/L^2)
σ_1, σ_2	principal stress (F/L^2)
$\sigma_r, \sigma_\theta, \sigma_z$	normal stress in cylindrical coordinates (F/L^2)
σ_l, σ_h	longitudinal and hoop stress for a cylindrical pressure vessel (F/L^2)
σ_s	stress for spherical pressure vessel (F/L^2)
σ_{yp}	yield stress (F/L^2)
σ_0	yield stress for uniaxial state of stress (F/L^2), see Chapter 14
τ	shear stress (F/L^2)
$\tau_{xy}, \tau_{xz}, \tau_{yz}$	shear stress (F/L^2), see Fig. 8.1
$\tau_{\max}$	maximum shear stress (F/L^2)

$\tau_{rx}, \tau_{x\theta}, \tau_{r\theta}$	shear stress in cylindrical coordinates (F/L^2), see Fig. 4.7
ϕ	angle of twist
ω	angular velocity (rad/T) see Section 4.9
$\boldsymbol{\sigma}$	stress at a point (F/L^2)
$\boldsymbol{\varepsilon}$	strain at a point (L/L)
$\mathbf{M}$	second moment of the area
$[M]$	second moment of the area matrix
$[T]$	transformation matrix
$[T]^T$	transpose of the transformation matrix
$[\varepsilon]$	strain matrix
$[\sigma]$	stress matrix

CHAPTER

MECHANICS OF MATERIALS

1.1 INTRODUCTION

"In all of natural philosophy, the most deeply and repeatedly studied part, next to pure geometry, is mechanics."[1] This statement is taken from a lecture by Clifford Truesdell. Natural philosophy includes—besides mechanics—thermodynamics, electromagnetism, relativity, and so forth. These are but names used to identify and classify, in a broad sense, topics of natural philosophy. I do not, in any way, intend to suggest that this textbook is a treatise upon natural philosophy or that the student who uses this textbook is a student of natural philosophy.

Natural philosophy, in the 17th century, was a topic of study and research for mathematicians. The ultimate translation of their researches into practical and useful concepts was eventually recorded in textbooks and made available to the engineer.

The 17th-century natural philosophers researched the basic problems of stress distribution in a beam and the mathematical description

---

[1] C. Truesdell, *Six Lectures on Modern Natural Philosophy*, Springer-Verlag, New York, 1966, p. 1.

1

of the shape of a deflected beam. It turns out that a major portion of this textbook has roots dating to the 17th century.

As time passed, natural philosophers became specialized and eventually the early concept of the natural philosopher lost much of its significance. Specialization probably occurred when the sheer mass of understandable science became too great for one mind to master in a lifetime. This textbook is a beginning toward understanding the mechanics of materials of the 17th and 18th centuries. Mechanics of materials of the 19th and 20th centuries is a worthy challenge for professors and their students.

The history of mechanics of materials dates to the work of Galileo in the early 17th century and followed later in that same era by the famous Bernoulli brothers, Jacob and John. Eighteenth-century mathematicians and scientists, such as Lagrange, Euler, Coulomb, Navier, Cauchy, and Poisson had their impact on the fundamental structure of the theory.

The subject of mechanics of materials encompasses methods for computing the strength and deformation properties of load-carrying members. Examples of load-carrying members include the structural components of a building, bridge, radar beacon, and robotic mechanism, among countless others.

We begin the study of mechanics of materials with Newtonian mechanics. Recall from statics and dynamics that Newton's fundamental axiom of conservation of momentum applied to both particles and rigid bodies was the central theme of the course of study. We shall not be concerned with dynamics but will consistently be dependent upon the basic concept of equilibrium of a rigid body. The analytical methods of statics are employed to determine external reactions for a body subject to externally applied loads. Mechanics of materials is divided into several major topics, the first is the study of the internal resistance of materials to their external loading. The internal resistance is dependent upon both the strength of the material and the deformation properties of the material. Certain material parameters must be defined and determined experimentally; however, we will merely discuss experimentally determined material parameters and not delve deeply into the experimental methods for their measurement. Additional major topics include analytical methods for computing the deflection of beam-type structures, stability of axially loaded column structures, and several specialized topics.

The analytical methods of vector analysis are employed throughout the textbook. The use of vector analysis follows logically after its introduction in statics; in fact, no new concepts of vector analysis are introduced, merely additional practice in the application of this powerful computational tool.

In the next section of this chapter we present several concepts, some as review and some as new ideas, that will orient the reader toward the fundamental objective of the study of mechanics of materials. Free-body diagrams, equivalent-force systems, and the definition of stress are of importance for conceptually visualizing the analytical sequence that must occur for successful problem solving.

1.2   EQUILIBRIUM AND STRESS

We will not attempt a thorough review of equilibrium methods of statics, since presumably the reader is familiar with analytical concepts that lead to computation of equivalent-force systems and reactions for load-carrying members. For now the discussion will be conceptual. Later, certain topics of statics will be reviewed as considered pertinent.

The beam-type structure of Fig. 1.1 is considered to be in equilibrium with six actions occurring at each end of the beam. There are six and only six possible actions that can occur: three forces and three couples, which are evaluated using the equilibrium equations of statics. A section passed through the beam anywhere along its axis can be viewed as shown in Fig. 1.2. Internal forces and couples act on the cut section as illustrated and have magnitudes necessary to produce equilibrium for the free body. Specifically, we classify the forces and couples as in Fig. 1.3. The force vector acting along the beam axis is shown in Fig. 1.3a and causes either tension or compression on the section,

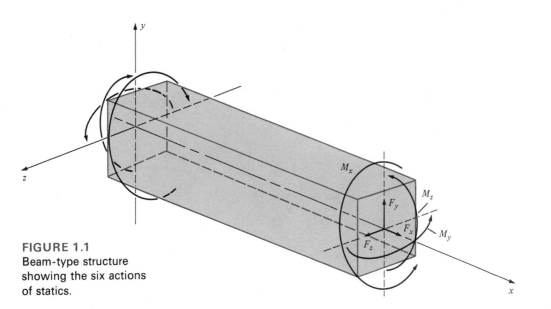

FIGURE 1.1
Beam-type structure showing the six actions of statics.

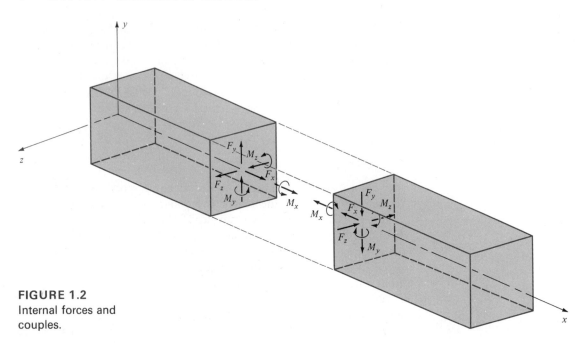

FIGURE 1.2
Internal forces and
couples.

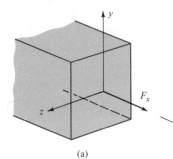

(a)

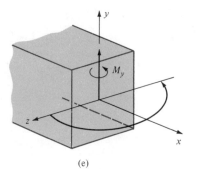

(b)

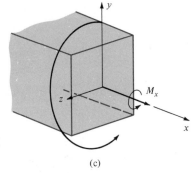

(c)

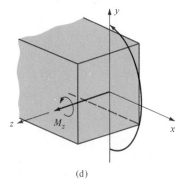

(d)

(e)

FIGURE 1.3
(a) Axial force; (b) shear
forces; (c) torque;
(d) bending about the
z axis; (e) bending about
the y axis.

depending upon its direction. The remaining two force vectors (Fig. 1.3*b*) produce shear loading on the cut section that is characterized by the forces acting tangent to the cut section as opposed to acting normal to the cut section as in the case of the axial force. The three couples of Fig. 1.2 are illustrated in Figs. 1.3*c–e* as vectors. The axial-couple vector of Fig. 1.3*c* represents a twisting couple whose direction is determined using the right-hand-screw rule. The twisting couple, referred to as torque, causes a shear action to occur on the cut cross section. The couples of Figs. 1.3*d* and *e* are referred to as bending moments, and the vector representation is interpreted as illustrated. It turns out that these couples cause a combination of tension and compression on the cut section.

The early chapters of this text are devoted to analytical methods for computing the magnitude and direction of these six actions and then the computation of the corresponding stresses. In particular, Chapter 2 deals with the action of Fig. 1.3*a*. Chapter 4 is devoted to the twisting couple of Fig. 1.3*c* applied to members of circular cross section. Analytical methods for computing shear forces and bending moments of Figs. 1.3*b*, *d*, and *e* are discussed in Chapter 5. Chapter 6 deals with the methods for computing stress caused by the bending moments of Figs. 1.3*d* and *e*, and Chapter 7 is devoted to the derivation of computational methods for the shear stress produced by the shear forces of Fig. 1.3*b*. Load-carrying members subject to various combinations of the forces and couples are analyzed in Chapter 8. This analysis is the culmination of the study of equivalent force systems and their associated stresses.

As we have stated, the six actions can combine to produce a combination of stresses. The question of exactly what is stress should be answered in a very simplistic manner that is also quite exact in a theoretical sense. We idealize a load-carrying body as shown in Fig. 1.4 and say that the body is a continuum. Essentially, a continuum is a

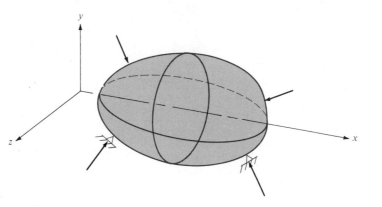

FIGURE 1.4
Idealized body.

collection of material particles, and its exact size is not important for this discussion. Materials are made up of clusters of molecules, and every material has a definite molecular structure. On a microscopic scale a material is composed of a space with atoms at specific locations. The continuum model is a collection of many molecules and is large enough that the individual molecular interactions for the material can be ignored and the total of all molecular interactions can be averaged and the continuum can be assigned some overall gross property to describe its behavior. The continuum is considered to be quite large compared to an atomistic model; however, it can still be imagined to be small enough to be of differential size. In other words, we can effect the mathematical concept of the limit of some quantity with respect to a length dimension. We assume that as we find the limiting value of a quantity as a length parameter approaches zero that we do not violate the material assumption of a continuum; the continuum still exists even though it may be of differential size.

Consider the continuum of Fig. 1.4 to be in statical equilibrium and imagine a slice taken through the continuum, as shown in Fig. 1.5. The continuum is still in equilibrium since internal forces and couples at the slice can be viewed as external balancing forces and couples. We are concerned with forces acting on the smooth, cut surface that are illustrated as acting on individual, small surface areas. On a small scale these forces are considered to be acting in a direction either normal or tangent to their respective areas. The forces originate from the molecular interactions at the microscopic level; however, as we have stated,

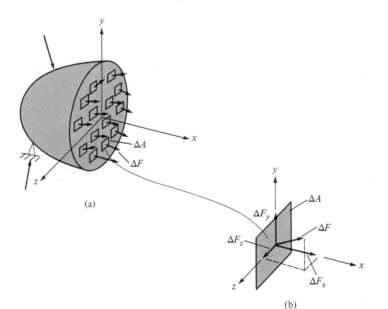

(a)

(b)

FIGURE 1.5

we assume that microscopic forces are averaged and their resultants act on the individual areas. We do not consider small couples acting on each element even though they may exist. It has been demonstrated, both experimentally and analytically, that small-body couples can be ignored for the general theory of mechanics of materials.

Every force depicted in Fig. 1.5 can be resolved into one normal component and two shear components. The definition of normal and shear arises naturally; normal forces act normally to their area and shear forces act tangentially to their area. Visualize one small area ΔA as shown in Fig. 1.5b. The force ΔF is shown in components ΔF_x, ΔF_y, and ΔF_z. Normal stress is denoted by σ (sigma) and is defined as the normal force per unit area; hence, the terminology normal stress. Normal stress is given as

$$\sigma_x = \frac{\Delta F_x}{\Delta A} \tag{1.1}$$

The Greek letter τ (tau) is usually used for shear stress and is the shear force divided by the area, or

$$\tau_y = \frac{\Delta F_y}{\Delta A} \tag{1.2}$$

and

$$\tau_z = \frac{\Delta F_z}{\Delta A} \tag{1.3}$$

A more complete description is given in Chapter 9 concerning shear and normal stresses that act at a point such as the small area of Fig. 1.5b.

Previously, it was noted that each action illustrated in Fig. 1.3 produced either a normal stress or shear stress. In every case we begin with a definition of stress, either normal or shear, as given by Eqs. (1.1), (1.2), and (1.3), and establish a theory that describes how a particular action causes a stress. The distribution of stress that occurs on the cross section of the member because of the action of forces and couples is dependent upon the way the load-carrying member deforms. In the case of the axial force, Fig. 1.3a, the member either elongates or shortens in the axial direction. If we assume a uniform deformation, meaning that every point on the cross section deforms an equal amount parallel to the beam axis, then we must investigate the limitation of the theory subject to that assumption.

It turns out that each of the six actions produces some corresponding deformation of the member, and prior to developing a theory for stress distribution we must establish the deformation characteristics of the member when subjected to a particular load. Chapters 2–9 are devoted to the study of concepts that have been briefly introduced in this section.

1.3 UNITS

Units of measure can be somewhat confusing, primarily because the universal standard is not yet well accepted. In October 1960, the Eleventh General (International) Conference on Weights and Measures redefined some original metric units and expanded the system to include other physical and engineering units. This was an attempt to bring some order to the confusion surrounding units of measure. The new system is called Le Système International d'Unités in French and abbreviated SI. In English this is the International System of Units. In the United States the Metric Conversion Act of 1975 committed the United States to a voluntary conversion to SI units.

Basic units are force F, length L, and time T. Table 1.1 illustrates the measures in four different systems. In the SI system there are seven base units, two supplementary units, and numerous derived units. The base units, supplementary units, and selected derived units are given in Table 1.2.

It is conventional to write SI units with prefixes as opposed to the accepted scientific notation used in the British System. In this text SI units are primarily written in prefix notation; however, the author prefers scientific notation simply because it is less confusing. The student needs practice in converting between the two systems and in dealing with prefix notation. Selected prefixes as used in this text are given in Table 1.3.

The conversion factors pertinent to mechanics of materials are given in Table 1.4. These conversion factors are taken from Marks' *Standard Handbook for Mechanical Engineers*, 8th ed., McGraw-Hill, 1978. That handbook, or a similar one, is recommended for additional information concerning units and conversion factors.

Mechanics of materials is usually limited to the study of deformable bodies at rest. The confusion that sometimes occurs when distin-

TABLE 1.1   Systems of Units

Name of Unit	Dimensions of Units in Terms of F, L, T	British "Gravitational" System, or "Foot-Pound-Second" System	Metric "Gravitational" System, or "Kilogram-Meter-Second" System	Metric "Absolute" System, or "cgs" System	SI (Newton, Meter, Second) System
Force	F	1 lb	1 kg	1 dyne	1 N
Length	L	1 ft	1 m	1 cm	1 m
Time	T	1 s	1 s	1 s	1 s

TABLE 1.2 SI Units

Quantity	Unit	SI Symbol	Formula
Base Units			
length	meter	m	
mass	kilogram	kg	
time	second	s	
electric current	ampere	A	
thermodynamic temperature	Kelvin	K	
amount of substance	mole	mol	
luminous intensity	candela	cd	
Supplementary Units			
plane angle	radian	rad	
solid angle	steradian	sr	
Derived Units			
area	square meter		m^2
density	kilogram per cubic meter		kg/m^3
energy	joule	J	$N \cdot m$
force	newton	N	$kg \cdot m/s^2$
frequency	hertz	Hz	$1/s$
power	watt	W	J/s
pressure	pascal	Pa	N/m^2
stress	pascal	Pa	N/m^2
thermal conductivity	watt per meter-kelvin		$W/m \cdot K$
volume	cubic meter		m^3
work	joule	J	$N \cdot m$

TABLE 1.3 SI Prefixes

Multiplication Factors	Prefix	SI Symbol
$1\,000\,000\,000 = 10^9$	giga	G
$1\,000\,000 = 10^6$	mega	M
$1\,000 = 10^3$	kilo	k
$0.001 = 10^{-3}$	milli	m
$0.000\,001 = 10^{-6}$	micro	μ

TABLE 1.4 Conversion Factors

To Convert From	To	Multiply by
pound-force (lbf avoirdupois)	newton (N)	4.448 222 E+00
pound-force-inch	newton meter (N·m)	1.129 848 E−01
pound-force-foot	newton meter (N·m)	1.355 818 E+00
pound-force/foot	newton/meter (N/m)	1.459 390 E+01
pound-force/foot^2	pascal (Pa)	4.788 026 E+01
pound-force/inch^2 (psi)	pascal (Pa)	6.894 757 E+03
inch	meter (m)	2.540 000 E−02
inch^2	meter^2 (m^2)	6.451 600 E−04
inch^3 (volume and section modulus)	meter^3 (m^3)	1.638 706 E−05
inch^3/minute	meter^3/second (m^3/s)	2.731 177 E−07
inch^4 (second moment of area)	meter^4 (m^4)	4.162 314 E−07
inch/second	meter/second (m/s)	2.540 000 E−02
kip (1000 lbf)	newton (N)	4.448 222 E+03
kip/inch^2 (ksi)	pascal (Pa)	6.894 757 E+06
foot	meter	3.048 000 E−01
foot^3/minute	meter^3/second (m^3/s)	4.719 474 E−04
foot^3 (volume and section modulus)	meter^3 (m^3)	2.831 685 E−02
foot^2	meter^2 (m^2)	9.290 304 E−02
foot-pound-force	joule (J)	1.355 818 E+00
foot-pound-force/hour	watt (W)	3.766 161 E−04
foot-pound-force/minute	watt (W)	2.259 697 E−02
foot-pound-force/second	watt (W)	1.355 818 E+00
degree Celsius	kelvin (K)	$t_K = t_{°C} + 273.15$
degree centigrade	kelvin (K)	$t_K = t_{°C} + 273.15$
degree Fahrenheit	degree Celsius	$t_{°C} = (t_{°F} - 32)/1.8$
degree Fahrenheit	kelvin (K)	$t_K = (t_{°F} + 459.67)/1.8$

guishing between force and mass will not occur. However, courses of study that build upon mechanics of materials, such as mechanical vibrations or dynamics of structures, involve the motion of deformable bodies.

Compare Table 1.1 and Table 1.2. In the SI system mass is a base unit with the kilogram as a unit of measure. Force is measured in newtons. Force measure is derived from mass measure using a basic equation of dynamics; that is, force equals mass times acceleration.

$$F = ma \tag{1.4}$$

or

$$1 \text{ N} = 1 \text{ kg·m/s}^2$$

A newton is the force required to accelerate one kilogram one meter per second squared. In the SI system there is never any confusion concerning units of force and mass.

The British gravitational system, defined in Table 1.1, uses a base unit of force given in pounds. Mass measure is derived using Eq. (1.4) as

$$\frac{1 \text{ lb}}{1 \text{ ft/s}^2} = 1 \text{ slug}$$

The slug is a derived unit of mass measure. Again, with care, there should be no confusion between force measure and mass measure.

The conversion from weight to mass is

$$m = \frac{w}{g} \qquad\qquad (1.5)$$

where

$$g = 32.2 \text{ ft/s}^2 \qquad\qquad (1.6)$$

and is commonly called the acceleration of gravity. In SI units,

$$g = 9.81 \text{ m/s}^2 \qquad\qquad (1.7)$$

The metric gravitational system can be confusing because both mass and force are measured using kilograms. It is necessary to designate kilograms mass or kilograms force. The absolute metric system is preferred.

For Further Study

Crowther, J. G., *Famous American Men of Science*, Norton, New York, 1937. (This book contains a short biography of J. Willard Gibbs, professor at Yale University, who was the first proponent for using vector analysis in science and engineering.)

Timoshenko, S. P., *History of Strength of Materials*, McGraw-Hill, New York, 1953.

Todhunter, I., and K. Pearson, *A History of the Theory of Elasticity and of the Strength of Materials*, Dover, New York, 1960.

CHAPTER

AXIALLY LOADED MEMBERS

2.1 INTRODUCTION

The concepts of stress and force along with the concepts of strain and deformation are introduced in this chapter. Stress is conceived as being of two types: normal stress and shear stress. Normal stress develops for situations where a force acts along a longitudinal axis of a structural member. Shear stress occurs when a sliding action can be imagined to exist between two adjacent parallel planes of a structural member. Strain is the nondimensional representation of the deformation of a material with respect to a given length dimension of that material.

Basic relations between stress and strain are referred to as mechanical properties of materials. Two important mechanical properties, Young's modulus and Poisson's ratio, are discussed because they are necessary for a complete representation of stress and strain in an axially loaded structural member. A more complete discussion of mechanical properties of materials is reserved for Chapter 3.

Methods of analysis developed in mechanics of materials represent fundamentally significant design concepts. From an academic viewpoint it is desirable to be exposed to mathematical and physical ideas that combine with engineering problems to make mechanics of materials an engineering course that bridges a gap between science and engineering.

The example problems and homework problems represent conceptually simple situations. The reader is expected to apply the equations of statics and use free-body analysis techniques when they are applicable.

2.2 NORMAL STRESS AND SHEAR STRESS

Once again we emphasize that there are two fundamental types of stress, and they are referred to as normal stress and shear stress. In this chapter we define normal stress in terms that are applicable to an axially loaded member such as the connecting rods of Fig. 2.1a. Two axially loaded rods are connected using a bolt or pin and forces of equal magnitude F are applied at each end. The rod assembly can be used to illustrate the idea of simple tension and simple shear. The free bodies of Fig. 2.1b show the two rods as axially loaded members with forces of equal magnitude and opposite direction applied at each end. An additional free body is obtained by cutting a section anywhere along

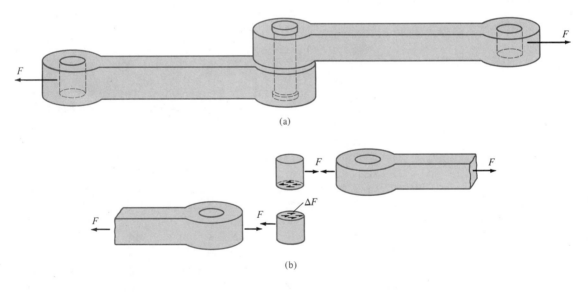

(a)

(b)

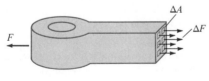

(c)

FIGURE 2.1
Axially loaded members illustrating the concept of normal stress and shear stress.

the rod normal to its axis as shown in Fig. 2.1c. Using the concept introduced in Chapter 1, imagine many small areas ΔA, each with a small force ΔF, as shown in the figure. We say that the overall internal reaction is the sum of all forces ΔF, and their sum must be equal in magnitude to the externally applied force F. We further assert that all forces ΔF act normally to their associated areas ΔA, and are uniformly distributed over the cross section. In essence, the assumption of uniform distribution of force on the cross section defines what is meant by simple tension. Our definition for normal stress was given by Eq. (1.1) and for the free-body section of Fig. 2.1c is

$$\sigma = \frac{\Delta F}{\Delta A} \tag{2.1}$$

In the limit we can obtain a differential relation for normal stress caused by simple tension

$$\sigma = \lim_{\Delta A \to 0} \frac{\Delta F}{\Delta A} = \frac{dF}{dA} \tag{2.2}$$

then

$$dF = \sigma \, dA \tag{2.3}$$

and we can evaluate the tensile force in terms of normal stress and area by integrating Eq. (2.3); that is,

$$F = \int_{\text{area}} \sigma \, dA = \sigma \int_{\text{area}} dA \tag{2.4}$$

The assumption that σ is constant and uniformly distributed over the cross section means that σ can be removed from the integral of Eq. (2.4), or is not a function of the coordinates that describe the area. Integrating Eq. (2.4) gives the fundamental relation between normal stress and simple tension.

$$F = \sigma A \qquad \text{or} \qquad \sigma = \frac{F}{A} \tag{2.5}$$

We shall return to Eq. (2.5) eventually and demonstrate mathematically a restriction on its use; basically, the axial force F applied to the load-carrying member must pass through the centroid of the cross section in order for Eq. (2.5) to be valid.

The assumption that normal stress is uniform and constant implies that the elongation of the rod is also uniform, or the cut section is a smooth plane normal to the axis of the rod before and after it is deformed. The assumptions could have been in the reverse order since we could assume uniform deformation and argue that uniform stress would be a logical result. Actually, for simple tension either argument is in order; however, for the remaining actions illustrated in Figs. 1.3b–e certain material parameters must be established that relate applied load and deformation prior to assuming the functional form of σ. Later in this chapter we shall discuss material parameters for axially loaded members.

Finally, it is reasonable to assert that Eq. (2.5) holds for the case of simple compression. The assumptions and limitations for simple tension would apply equally for simple compression except for one additional constraint. A slender compression member could deform by buckling, which would violate the assumption of uniform deformation of the cross section for a simple compression member. The topic of buckling of slender compression members is dealt with in Chapter 12 and for now we will assume that compression members qualify for simple compression and our assumption of uniform compressive stress distributed over the cross section will not be violated.

Several example problems should illustrate the use of Eq. (2.5), introduce the idea of force and stress diagrams, and serve as a review of the equilibrium equations of statics.

EXAMPLE 2.1

A circular bar with a diameter of 2 in. is subject to a tensile load of 200 kip acting at its centroid. Compute the average normal stress.

Solution:

$$\sigma = \frac{P}{A} = \frac{200\,\text{kip}}{\pi\,(1\,\text{in.})^2} = 63.66\ \text{ksi tension}$$

(*Note:* A kip represents 1000 pounds; 1 ksi is 1000 pounds per square inch.) In SI units the answer would be

$$\sigma = (63{,}660\ \text{psi})6.894(10^3)\ \text{Pa/psi}$$

$$\sigma = 4.388(10^8)\ \text{Pa} = 438.8\ \text{MPa}$$ ∎

EXAMPLE 2.2

A circular pipe is to be designed using a maximum compressive stress of 190 MPa. The outside diameter is 0.155 m and the inside diameter is 0.140 m. Determine the maximum compressive axial force, in newtons, that can be applied.

Solution:

$$\sigma = \frac{P}{A} = 190 \text{ MPa}$$

$$\text{Area} = A_{\text{outside}} - A_{\text{inside}}$$

$$\text{Area} = \pi(0.155/2)^2 - \pi(0.140/2)^2 = 3.475(10^{-3}) \text{ m}^2$$

$$P = \sigma A = (190 \text{ MPa})(3.475)(10^{-3}) \text{ m}^2 = 0.6603 \text{ MPa·m}^2$$

(*Note:* MPa is the same as MN/m^2.) Then

$$P = 0.6603 \text{ MN} = 660.3 \text{ kN}$$

The answer in pounds would be 14,845 lb.   ■

EXAMPLE 2.3

The rod of Fig. 2.2 can be visualized as supported at the upper end and subjected to forces P_1 and P_2 as illustrated. Neglect the weight of the rod and compute the axial-force distribution, compute the normal stress, and plot both the force and stress as a function of the axial coordinate. Let the area $= 0.26 \text{ in.}^2$, $P_1 = 120 \text{ lb}$, and $P_2 = 45 \text{ lb}$.

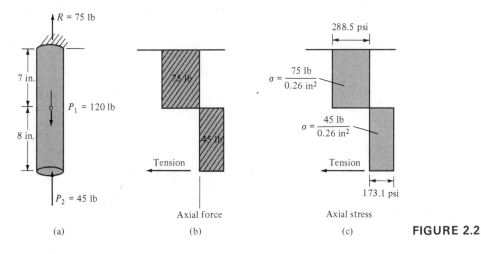

(a)   (b)   (c)   **FIGURE 2.2**

Solution:
The reaction is computed as $R = 120 - 45 = 75 \text{ lb}$ acting as shown in the figure. The upper portion of the rod carries an axial tension of 75 lb and is plotted as shown in Fig. 2.2b. The lower portion of the rod is in compression as indicated by the plot. The stress diagram of Fig. 2.2c is obtained by dividing the forces by the area of the rod.   ■

EXAMPLE 2.4

The circular bar illustrated in Fig. 2.3 is subject to the indicated loads. Construct an axial-force diagram and an axial-stress diagram.

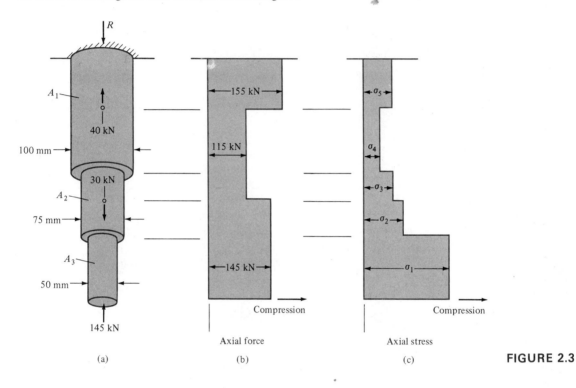

(a) (b) (c) **FIGURE 2.3**

Solution:

$$R = 145 - 30 + 40 = 155 \text{ kN}$$

The axial-force diagram is shown in Fig. 2.3b. The axial-stress diagram, Fig. 2.3c, changes with each change in axial force or area.

$$A_1 = \frac{\pi(100)^2}{4} = 7.853(10^3) \text{ mm}^2$$

$$A_2 = 4.418(10^3) \text{ mm}^2$$

$$A_3 = 1.964(10^3) \text{ mm}^2$$

$$\sigma_1 = \frac{145 \text{ kN}}{1.964(10^3) \text{ mm}^2} = 73.83(10^{-3}) \text{ kN/mm}^2$$

Multiply by $(10^3 \text{ N/kN})(1 \text{ mm}/10^{-3} \text{ m})^2$ to obtain N/m^2

$\sigma_1 = 73.83(10^3) \text{ Pa} = 73.83 \text{ MPa}$

$\sigma_2 = \dfrac{145}{4.418(10^3)} = 32.82 \text{ MPa}$

$\sigma_3 = \dfrac{115}{4.418(10^3)} = 26.03 \text{ MPa}$

$\sigma_4 = \dfrac{115}{7.853(10^3)} = 14.64 \text{ MPa}$

$\sigma_5 = \dfrac{155}{7.853(10^3)} = 19.74 \text{ MPa}$ ■

The foregoing examples illustrate the solution of some elementary mechanics of materials problems. The calculation of normal stress using the equation $\sigma = P/A$ may be referred to as simple stress, uniform stress, unit stress, or average stress. In order for the equation to represent the state of stress accurately, the axial force must pass through the centroid of the cross section of the loaded member. Consider the free body of Fig. 2.4; a differential force, dF, is acting on a differential area, dA, and produces a stress.

$$\sigma = \frac{dF}{dA} \tag{2.6}$$

Summing forces in the x direction gives

$$F = \int dF = \int \sigma \, dA \tag{2.7}$$

Also, summing moments about the z axis and y axis gives

$$M_z = \int \bar{y} \, dF = \int \bar{y}\sigma \, dA \quad \text{and} \quad M_y = \int \bar{z}\sigma \, dA \tag{2.8}$$

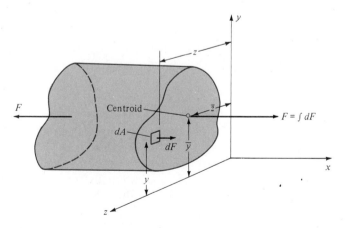

FIGURE 2.4

M_z and M_y would correspond to the couples that would balance the externally applied action of Figs. 1.3*d* and *e*. Assume that σ is uniform over the cross section independent of either *y* or *z*, then

$$F = \sigma \int dA = \sigma A \tag{2.9}$$

$$M_z = \sigma \int \bar{y}\, dA \tag{2.10}$$

$$M_y = \sigma \int \bar{z}\, dA \tag{2.11}$$

The requirement that only the axial force, as defined by Eq. (2.9), can act on the cut section dictates that both M_y and $M_z = 0$. For Eq. (2.10) to be zero there are two possibilities, $\sigma = 0$ or $\int \bar{y}\, dA = 0$. Of course, $\sigma = 0$ is not acceptable; hence, $\int \bar{y}\, dA = 0$. Therefore, if the force of Eq. (2.7) is applied at the centroid, the integrals of Eqs. (2.10) and (2.11) will be zero.

Simple shear is illustrated in Fig. 2.1*b*. The connecting pin for the rods transfers the force, *F*, from one rod to the other. The shear action is idealized as many forces, ΔF, that act adjacent or parallel to their corresponding area, ΔA. The assumption of "simple" implies that the forces are uniformly distributed over the area, and simple shear stress is given as

$$\tau = \frac{\Delta F}{\Delta A} \tag{2.12}$$

Once again we invoke the limiting process of differential calculus to obtain

$$\tau = \lim_{\Delta A \to 0} \frac{\Delta F}{\Delta A} = \frac{dF}{dA} \tag{2.13}$$

and

$$dF = \tau\, dA \tag{2.14}$$

It follows that τ is independent of the area associated with *dF*, and Eq. (2.14) can be integrated to give

$$F = \tau A \quad \text{or} \quad \tau = \frac{F}{A} \tag{2.15}$$

The concept of simple shear may be associated with rivets or bolts as illustrated in Fig. 2.5 where two structural members are bolted together. If a force is applied as shown in the figure, the system will be in equilibrium externally and the force *F* is transmitted from one member to the other. A free-body diagram of the structural joint is shown in Fig. 2.6 and indicates how the shear is transmitted between

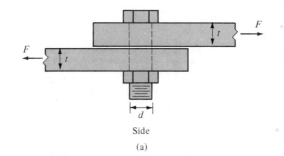

Side

(a)

FIGURE 2.5
The force *F* is transmitted between the plates and the bolt is subjected to single shear.

Top

(b)

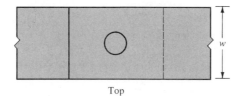

(a)

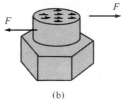

(b)

FIGURE 2.6
(a) A free body illustrating that *F* is in equilibrium with the distributed shear; (b) an element of the bolt showing how shear stress is assumed to be evenly distributed over the cross section of the bolt; (c) the geometry to be used for computing bearing stress.

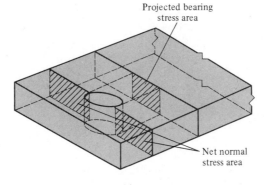

Projected bearing stress area

Net normal stress area

(c)

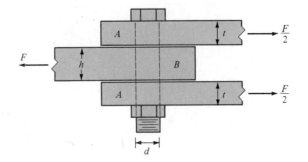

FIGURE 2.7
The bolt of this figure is in double shear as contrasted to single shear shown in Fig. 2.5.

the members. The force F is assumed to be uniformly distributed over the cross section of the bolt as shown in Fig. 2.6b. The assumed uniform distribution is force per unit area, or simple shear, computed as follows:

$$\tau = \frac{F}{A} = \frac{F}{\pi(d/2)^2} \tag{2.16}$$

where d is the diameter of the bolt.

Just as the shear illustrated in Fig. 2.5 is classified as single shear, double shear is illustrated in Fig. 2.7. The bolt in this figure is in double shear, which means that the total axial force is transmitted across two separate cross sections of the bolt.

Consider the following example to illustrate the application of the shear stress equation.

EXAMPLE 2.5

For the single-shear situation of Fig. 2.5, assume that $F = 950$ lb, $t = 0.25$ in., and the diameter of the bolt is 3/8 in. Compute the shear stress for the bolt. If $w = 2.5$ in., compute the critical net normal stress in the plates.

Solution:
Use Eq. (2.16).

$$\tau = \frac{F}{A} = \frac{950 \text{ lb}}{\pi[(3/8)/2]^2 \text{ in.}^2}$$

$$\tau = 8600 \text{ psi}$$

The greatest normal stress would occur where the area is the smallest or at the location of the bolt hole as shown in Fig. 2.6c. The normal stress is given by

$$\sigma = \frac{F}{(w - d)t}$$

$$\sigma = \frac{950}{(2.5 - 0.375)(0.25)} = 1788 \text{ psi}$$ ∎

Bearing stresses are a special case of compressive normal stresses. An important example is in soil mechanics where soils are characterized, among other parameters, according to bearing capacity. Consider the following example.

EXAMPLE 2.6

A particular soil will support 20 psi (meaning that each square inch of soil surface can support 20 lb). A column that is an integral part of a building is to transfer 250,000 lb to the ground through a foundation that is referred to as a footing. How large should the footing be that transmits the force to the ground?

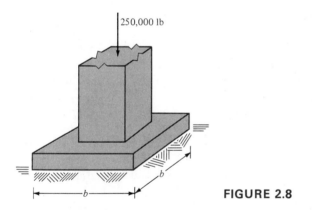

250,000 lb

FIGURE 2.8

Solution:
The problem is illustrated in Fig. 2.8. We assume a square footing and note that normal stress is given by $\sigma = P/A$. In this case $\sigma = 20$ psi and $P = 250,000$ lb. Then

$$A = \frac{P}{\sigma} = \frac{250,000}{20} = 12,500 \text{ in.}^2$$

$$A = b^2 = 12,500$$

and thus

$$b = 111.8 \text{ in.}$$

Only the dimensions of the footing have been determined. No information has been obtained concerning materials, footing thickness, or column design. However, all of these design concepts are based upon an understanding of the various topics to be presented in this textbook. It has been established that a square footing 111.8 by 111.8 in. will transmit 250,000 lb to a soil that will support 20 psi. It must be recognized that uniform bearing capacity for a soil is an assumption and that the actual pressure distribution may not be uniform; in fact, it may be very complicated.

∎

Bearing stresses also occur in the situation depicted in Fig. 2.5. We assume that the bolt or rivet bears against the curved surface of the plate. The variation in pressure between the bolt and the plate surface is not uniform; in fact, an exact description of the pressure variation is theoretically too advanced for this discussion. It is customary to use a simplified design procedure and assume that the force acting between the curved surface of the bolt and the plate is acting uniformly over an area equal to the projection of the curved surface upon a plane perpendicular to the direction of the force. The projected area is the diameter of the bolt times the thickness of the plate, as shown in Fig. 2.6c. For the single-shear situation of Fig. 2.5, the bearing stress would be $\sigma_{\text{bearing}} = F/td$ for either plate since both plates have a thickness t. The double-shear situation of Fig. 2.7 would result in a bearing stress of $F/2td$ for plate A and F/hd for plate B. The analysis of bearing stress at bolt and rivet holes using this simplified procedure is at best an idealized design formulation. It is a fact that some engineering design methods are inexact. The fact is recognized and very often experimental testing is employed in order to predict the stress at which failure of the material might occur. Hence, a design satisfying the bearing stress requirement would involve specifying an allowable bearing stress; either the bolt diameter or plate thickness, or both, must be adjusted to accommodate the allowable stress.

Three basic situations have been discussed thus far: simple normal stress, simple shear stress, and bearing stress. Before we conclude the discussion of bolts and rivets, one additional shearing stress situation should be mentioned. The bolt of Fig. 2.6 is illustrated in a top view in Fig. 2.9. As the bolt bears against the plate, a failure condition could occur along lines ab and cd that would be a shear-type failure. A section of the material, $abdc$, behind the bolt could slide out. A shear stress could be approximated for this situation as the force F divided by $2ht$, the area subject to shear. The analysis is somewhat crude for two reasons. The actual failure surface is more along the lines ae and cf. The distribution of shear stress along the failure surface is anything but uniform and quite complicated to describe analytically. To guard

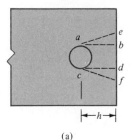

(a)

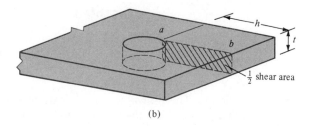

(b)

FIGURE 2.9
(a) A shear failure may occur along planes *ab* and *cd*; (b) one-half the shear area that is used for elementary computations.

against the pullout type of failure there are standard specifications that dictate the length h.

The discussion concerning rivets and bolts is not intended to cover the broad applications in engineering design of connections. Such a topic is more suitable as part of a course in the design of steel or aluminum structures. It is worthwhile to note that the rivet hole is predrilled or punched and is somewhat larger in diameter than the rivet. Rivets are sometimes inserted in a heated condition and allowed to expand and fill the rivet hole.

An additional source of pure shear that might be considered to occur in a simple way is the case of pure torque applied to a circular bar or shaft. Every time the reader has used a wrench to turn a bolt or a screwdriver to turn a screw, the physical situation has been one of pure shear due to a twisting motion. Torsional shear is given a thorough treatment in Chapter 4.

EXAMPLE 2.7

The structural joint of Fig. 2.10 is referred to as a lap joint, and it connects three structural members as shown. Assume the allowable normal stress is 22,000 psi and the allowable bearing stress is 16,000 psi for the plates. The rivets are 0.75 in. in diameter, and the force F is 25,000 lb. Compute the thickness of t_1 and t_2 of the plates based upon allowable normal and bearing stresses.

Not to scale

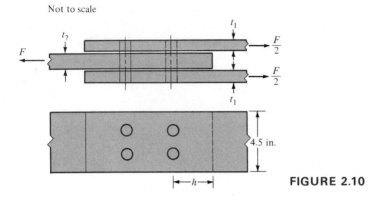

FIGURE 2.10

Solution:

The thickness t_1 will be computed first. The allowable normal stress must be developed over a net area that is the width of the plate less the two rivet holes times the thickness of the plate. Equation (2.5) gives the result

$$A = \frac{F/2}{\sigma}$$

$$t_1 [4.5 - 2(0.750)] = \frac{25,000/2}{22,000}$$

$$t_1 = 0.1894 \text{ in.}$$

The thickness t_1, using the allowable bearing stress, requires that we investigate the force acting on one rivet. Again, $A = (F/2)/\sigma_{bear}$, where A = rivet diameter times the plate thickness. Hence

$$0.75t_1 = \left(\frac{25,000 \text{ lb}}{2}\right)\left(\frac{1}{4}\right)\left(\frac{1}{16,000 \text{ psi}}\right)$$

(The quantity $1/4$ on the right-hand side divides the force, $F/2$, between the four rivets.)

$$t_1 = 0.2604 \text{ in.}$$

The thickness t_1 is governed by the allowable bearing stress and must be at least 0.2604 in.

Similar computations for the plate thickness t_2 give, for the normal stress,

$$t_2 [4.5 - 2(0.75)] = \frac{25,000}{22,000}$$

$$t_2 = 0.3788 \text{ in.}$$

For the bearing stress,

$$0.75t_2 = \left(\frac{25,000}{16,000}\right)\left(\frac{1}{4}\right)$$

$$t_2 = 0.5208 \text{ in.}$$

and again, bearing stress governs the design of the plate. The results for t_2 are exactly twice those for t_1, which is reasonable since the total force on plate t_2 is twice that acting on plate t_1. ■

EXAMPLE 2.8

The mechanism of Fig. 2.11 is pin-connected at joints A, B, and C and loaded as shown. The connection at A is a clevis pin as shown in the detail. Assume the pin connection at C is single shear. The bolts at all connections are 0.25 in. in diameter. Neglect the weight of the members and compute the following: *a.* shear stress on the bolt at A; *b.* bearing stresses at A; *c.* shear stress on the bolt at C.

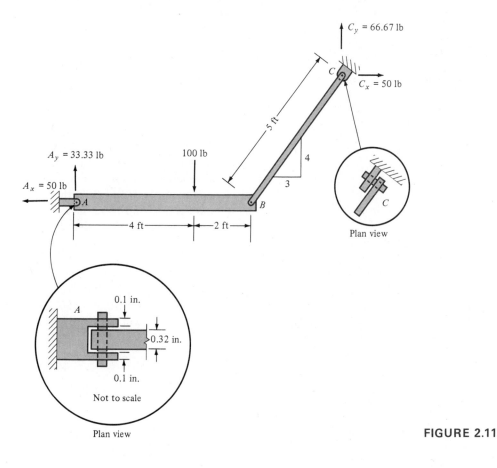

FIGURE 2.11

Solution:

The reactions at A and C are computed using the equations of statics and are shown in the figure. *Note:* The reactions A_x, A_y, C_x, and C_y are the actual forces acting on the pin connections. The resultant reaction is used for the computations.

a. The bolt at A is in double shear and the shear stress is given as one-half the resultant force divided by the area of the bolt.

$$\tau = \frac{60.1/2}{\pi(0.25/2)^2} = 612.17 \text{ psi}$$

b. The bearing stress for the clevis is computed using the total thickness that resists the resultant force.

$$\sigma_{bear} = \frac{\text{Force}}{(\text{Plate thickness})(\text{bolt diameter})}$$

$$\sigma_{bear} = \frac{60.1}{(2)(0.1)(0.25)} = 1202.0 \text{ psi}$$

The bearing stress for member AB is

$$\sigma_{bear} = \frac{60.1}{(0.32)(0.25)} = 751.25 \text{ psi}$$

c. The bolt at C is in single shear and the shear stress is computed as

$$\tau = \frac{83.33}{\pi(0.25/2)^2} = 16,980 \text{ psi} \qquad ■$$

Thus far, normal stress and shear stress have been treated as separate physical quantities. In general, one can always accompany the other. In the case of the axially loaded member of Fig. 2.1c, the stress acts in a direction normal to the cut section and there is no corresponding shear stress. The orientation of the cut section, normal to the applied load, causes the effect of pure simple tension. If the plane of the cut section was inclined with respect to the horizontal axis of the member, a different situation could be realized. Consider the axially loaded member of Fig. 2.12. A section cut normal to the axis of the member produces a normal stress, as illustrated in Figs. 2.12a and b. If the section is cut as shown in Fig. 2.12c, there will be a stress normal to the cut section, σ_N, and one that is tangent to the cut section, τ_T. These stresses are components of the stress of Fig. 2.12b; however, the magnitude is computed using the force of Fig. 2.12a, as shown in Fig. 2.12d. Note the representation of stress in Figs. 2.12b and c. The stress has been drawn as a distribution of forces in Fig. 2.12a. We now adopt in Figs. 2.12b and c a graphical representation of stress as a boldfaced arrow. Force is illustrated as a vector, and rightfully so, since force has

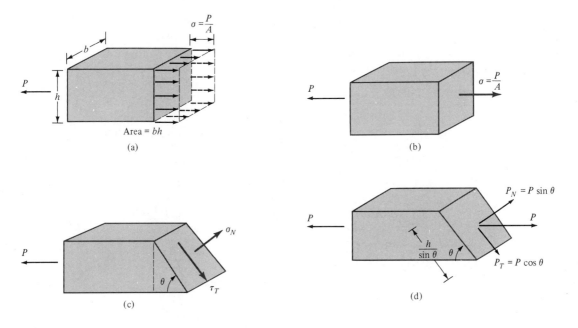

FIGURE 2.12
(a) and (b) Representation of normal stress on a plane; (c) normal and shear stress on an inclined plane; (d) normal and shear components of P on an inclined plane.

the mathematical properties of a vector; that is, magnitude and direction. In the mathematical sense stress is not a vector since it has magnitude and direction but is also a function of the area, force per unit area. One could say that stress has too many properties to be described as a vector; hence, we represent stress with a boldfaced arrow in order to call attention to the difference between stress and force. The axial force, P, of Fig. 2.12 is resolved into components P_N acting normal to the cut surface and P_T acting tangent to the cut surface. The force components are computed as

$$P_N = P \sin \theta \quad \text{and} \quad P_T = P \cos \theta \tag{2.17}$$

The area of the inclined cut surface is $bh/\sin \theta$. The normal and shear stresses acting on the inclined cut surface become

$$\sigma_N = \frac{P_N}{bh/\sin \theta} = \frac{P \sin^2 \theta}{A} \tag{2.18}$$

$$\tau_T = \frac{P_T}{bh/\sin \theta} = \frac{P \sin \theta \cos \theta}{A} \tag{2.19}$$

Equations (2.18) and (2.19) illustrate that the normal and tangential components are each less than the stress of Fig. 2.12b. However, there are physical situations when one of the components acting on the inclined surface might produce a more critical stress situation. The following example is conceptually elementary and yet illustrates a typical engineering analysis problem.

A short wooden compression member is shown in Fig. 2.13. The photograph of the actual test specimen (Fig. 2.13c) illustrates the failure mechanism. The failure surface is inclined approximately 58 degrees from the vertical, and a shear-type failure is readily observed as the top section appears to slide relative to the lower section. The material is weaker in shear than in compression and the failure will be explained using the foregoing analysis. Shear stress and normal stress are computed using Eqs. (2.18) and (2.19). Substituting the actual values of θ, b, and h gives

$$\sigma_N = \frac{P(\sin 58°)^2}{(50)^2} = 287.64P \text{ N/m}^2$$

$$\tau_T = \frac{P(\sin 58°)(\cos 58°)}{(50)^2} = 179.74P \text{ N/m}^2$$

The normal and shear stresses are shown idealized as they act on the cut section in Fig. 2.13d. The shear stress that is producing failure is shown in Figs. 2.13e and f.

For $\theta = 90$ degrees, a failure surface normal to the axis of the post, the stress would be

$$\sigma = \frac{P}{(50)^2} = 400P \text{ N/m}^2$$

But $\tau_T = 179.74$ and $\sigma_N = 287.64$ and the failure is a shear failure; hence, the failure stress is 45 percent of the axial normal stress. This result is due to actual material behavior that is an integral part of the study of mechanics of materials. The wood test specimen represents a material that is less resistant to failure in shear than in compression and the grain structure of wood lends itself to the failure mechanism of Fig. 2.13c. The calculation of stress is independent of material behavior. However, if stress is to be related to deformation, the mechanical properties of the material must be considered.

Equations (2.18) and (2.19) illustrate the earlier statement that stress is not a vector quantity. The force vectors of Fig. 2.12d have components described by a single trigonometric function. The stress of Fig. 2.12b, as illustrated in Fig. 2.12c, has components defined by the product of two trigonometric functions. We shall return to the problem of describing the state of stress on different planes in Chapter 9.

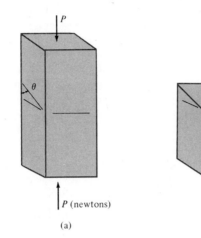

(a)

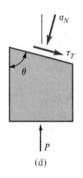

(b)

(c)

(d)

(e)

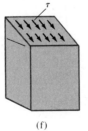

(f)

FIGURE 2.13
Failure of an actual wood test specimen.

2.3 NORMAL STRAIN AND SHEAR STRAIN

Stress can be thought of as a measure of the strength or load-carrying capacity of a structural element. Strain can be thought of as a measure of the deformation characteristics of a load-carrying member. Any structure will deform when subject to an external load; hence, strain is always associated with stress. The reverse is not always true—the notable exception being a structural element free to deform when subject to temperature increase or decrease.

An axially loaded member subject to a tensile load will suffer a simple elongation, which means that a cut section, viewed as a smooth plane surface, will deform parallel to itself with no rotation or warping of the cross section. Simple compression produces the same effect, except that the deformation is a uniform contraction of the member. Recall that simple deformation was assumed for the derivation of Eq. (2.5).

To be precise, normal strain or linear strain is a measure of deformation of a member per unit length of the member. Consider the rod of Fig. 2.14 and assume that it deforms as a result of the action of an axial force that passes through its centroid. Two points, x_1 and x_2, are defined at arbitrary points along the rod a distance Δx apart. An axial force applied to the rod will cause axial deformation that will be referred to as u. Material point x_1 will deform by an amount u and move to position x_1'. Similarly, point x_2 will deform to position x_2', thereby moving a distance $u + \Delta u$. The original element displaces an amount u as a rigid body and elongates an amount Δu. The definition of engineering strain, ε (epsilon), as postulated above, is

$$\varepsilon = \lim_{\Delta x \to 0} \frac{\Delta u}{\Delta x} = \frac{du}{dx} \tag{2.20}$$

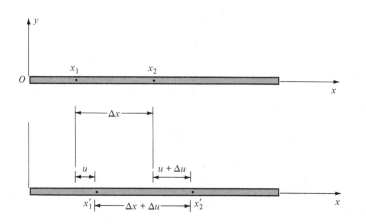

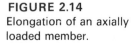

FIGURE 2.14
Elongation of an axially
loaded member.

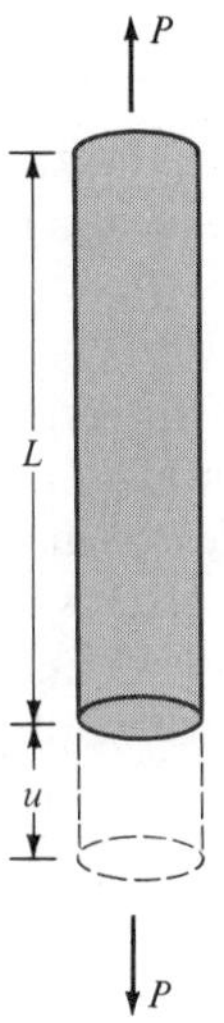

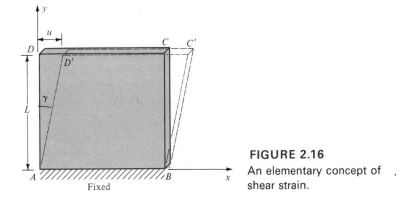

FIGURE 2.16
An elementary concept of shear strain.

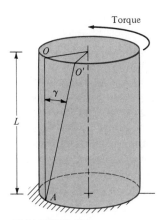

FIGURE 2.15

FIGURE 2.17
Shear strain caused by a torque acting on a cylinder.

The dimensional units of strain are inches per inch, or meters per meter, and therefore may be left dimensionless.

Consider a rod of length L, as shown in Fig. 2.15, with an axial force P applied along its centroidal axis. If the total deformation of the rod is u, the average strain for the rod, using Eq. (2.20), is

$$du = \varepsilon_{avg}\, dx \qquad \textbf{(2.21)}$$

then

$$\varepsilon_{avg} = \frac{u}{L} \qquad \textbf{(2.22)}$$

where L is the original length of the rod.

We will digress briefly from our discussion of axially loaded members and define shear strain, which may be visualized as shown in Fig. 2.16. A plate of material can be imagined to be held fixed and rigid at its base, AB, and a displacement applied at D. The deformation occurs as DC moves to the position $D'C'$. The shear strain γ (gamma) is the deformation divided by the length of the side, L. Hence

$$\gamma = \frac{u}{L} \qquad \textbf{(2.23)}$$

and shear strain is represented as the angle γ. The situation shown in Fig. 2.16 is very difficult to model experimentally. However, twisting shear can be modeled in the laboratory, and shear strain can be experimentally measured as shown in Fig. 2.17. The cylinder is fixed at the base and the line AO is vertical prior to the application of the torque. As twisting occurs, point O moves to O' and line AO rotates to line AO'. The strain, $\gamma = OO'/L$, can be measured experimentally and a relationship obtained between torque and the shear strain caused by the torque.

2.4   MECHANICAL PROPERTIES OF MATERIALS

Stress and strain have been introduced as independent topics. It must be emphasized that the relationship between stress and strain is fundamental to the study of mechanics of materials. Material properties are studied on both the microscopic scale and macroscopic scale. Topics such as metallurgy and material science are concerned with material properties on a small scale, such as the relation between individual atoms, crystal structure, and surface chemistry. In mechanics of materials, material properties are obtained for the continuum that was described in Chapter 1. The basic relation between axial stress and axial strain is termed a stress-strain relation and is shown graphically in Fig. 2.18. Young's modulus, or the modulus of elasticity, is the slope of the curve that results when stress is plotted versus strain. The standard method for obtaining the curve of Fig. 2.18 is termed a tension test. A rod of material is loaded in tension using a testing machine, and the deformation occurring in a gage length L is recorded and used to compute the strain. Young's modulus is defined for only the linear portion of the stress-strain curve, as illustrated in Fig. 2.18. Deformation beyond the linear range is discussed in Chapter 3. The general result, for our purpose, is that stress is related to strain using an experimental constant referred to as Young's modulus and symbolized by the letter E.

In a similar manner shear stress, τ, is related to shear strain, γ, such that

$$\sigma = E\varepsilon \tag{2.24}$$

$$\tau = G\gamma \tag{2.25}$$

Usually G is obtained from a torsion test and is referred to as the shear modulus.

It is easy to visualize that a structural member subject to a tensile loading would elongate. A rectangular rod subject to a tensile force is illustrated in Fig. 2.19. The rod elongates by some amount, u, giving a strain along the x axis as $\varepsilon_x = u/L_x$. The application of the tensile force causes a lateral contraction of the bar in the y and z directions. The deformation effect occurs in most materials, and, depending upon the type of material, one material may deform more than another. The deformation along the y and z axes will be referred to as v and w, respectively. Then the corresponding strain, referring to Fig. 2.19, would be $\varepsilon_y = v/L_y$ and $\varepsilon_z = w/L_z$.

The ratio of the lateral contractive strain to axial strain is a material property referred to as Poisson's ratio and will be denoted by v. According to the definition,

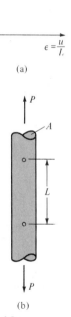

(a)

(b)

FIGURE 2.18
(a) Linear relation between normal stress and normal strain; (b) test specimen used to obtain a stress-strain relation.

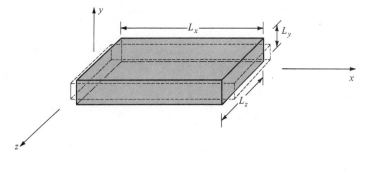

FIGURE 2.19
Deformation of a rectangular rod loaded in tension along the x axis.

$$v = \frac{-\,\text{Lateral strain}}{\text{Axial strain}} \tag{2.26}$$

Poisson's ratio is a constant equal to or less than 0.5 as long as the load deformation properties of the material remain linear—that is, as long as Eq. (2.24) remains valid. For the materials to be studied in this course, Poisson's ratio is assumed to be the same for either transverse coordinate direction. For the deformation of Fig. 2.19

$$v = |\varepsilon_y/\varepsilon_x| = |\varepsilon_z/\varepsilon_x| \tag{2.27}$$

and a tensile or compressive loading along any axis will produce a contraction or expansion at right angles to the applied force. The lateral contraction of Fig. 2.19 occurs without developing a corresponding lateral stress because the material is free to deform along the y and z axes.

2.5 DEFORMATION OF AXIALLY LOADED MEMBERS

The axial deformation of a rod caused by an axially applied force can be calculated using the equations that have been developed thus far. Beginning with Eq. (2.24), $\sigma = E\varepsilon$; then substituting Eq. (2.5), as $\sigma = P/A$, and Eq. (2.22), $\varepsilon = u/L$, to give $P/A = Eu/L$. Solving for u yields the desired result:

$$u = \frac{PL}{AE} \tag{2.28}$$

Equation (2.28) is valid as long as the following assumptions are satisfied.

1. The force P is applied through the centroid of the cross section.

2. The force P is constant with respect to the axial coordinate along the length L.

3. The parameters A and E are constant with respect to the axial coordinate along the length L.

Several additional example problems should serve to illustrate the use of the equations in this chapter.

EXAMPLE 2.9

A rod that is 2 in. in diameter is connected to a machinery component. The component will be suspended by the rod. If the machinery weighs 5000 lb and the rod is 8 ft long and made of steel, compute the axial deformation. Assume that $E = 30(10^6)$ psi.

Solution:
Use Eq. (2.28).

$$u = \frac{PL}{AE} = \frac{(5000 \text{ lb})(8 \text{ ft})(12 \text{ in./ft})}{\pi(1 \text{ in.})^2(30)(10^6) \text{ psi}} = 0.0051 \text{ in.} \qquad \blacksquare$$

EXAMPLE 2.10

A circular steel tube 3 ft long is to be used to support an axial force of 5000 lb. The thickness of the tube must be 0.12 in. Determine the inside diameter if the axial deformation is not to exceed 0.03 in. and $E = 30(10^6)$ psi.

Solution:
$A = PL/uE$; d_2 = outside diameter; d_1 = inside diameter.

$$\frac{\pi(d_2^2 - d_1^2)}{4} = \frac{(5000 \text{ lb})(36 \text{ in.})}{(0.03 \text{ in.})(30)(10^6) \text{ psi}}$$

$$d_2 = d_1 + t, \text{ where } t \text{ is the thickness}$$

$$(d_1 + t)^2 - d_1^2 = 0.2546 \text{ in.}^2$$

$$2td_1 + t^2 = 0.2546 \text{ in.}^2$$

$$d_1 = \frac{(0.2546) - (0.12)^2}{(2)(0.12)} = 1.00 \text{ in.} \qquad \blacksquare$$

EXAMPLE 2.11

A bar 0.25 m in length is subjected to an axial compression of 200 kN. The axial strain induced is 0.002.

a. For Young's modulus of 85 GPa, compute the stress.
b. Compute the cross-sectional area required to satisfy the criterion. Assume a square cross section.
c. Compute the lateral deformation if Poisson's ratio is 0.31.

Solution:
a. Use Eq. (2.24): $\sigma = E\varepsilon$.

$$\sigma = (85 \text{ GPa})(0.002) = 170 \text{ MPa}$$

b. Use Eq. (2.5): $A = P/\sigma$.

$$A = \frac{200 \text{ kN}}{170 \text{ MPa}} = 1177 \text{ mm}^2$$

For a square cross section,

$$d = \sqrt{1177} = 34.3 \text{ mm}$$

c. Using Eq. (2.27),

$$v = \frac{\varepsilon_y}{\varepsilon_x}$$

$$\varepsilon_y = v\varepsilon_x = (0.31)(0.002) = 6.2(10^{-4})$$

From Eq. (2.22),   ·

$$u = \varepsilon L = 6.2(10^{-4})(34.3 \text{ mm}) = 21.27 \ \mu\text{m}$$ ■

EXAMPLE 2.12

The square rod of Fig. 2.20 is loaded as illustrated. Plot the force, stress, and deformation diagrams if $E = 200$ GPa.

Solution:
The force diagram is shown in Fig. 2.20*b*. The computations for stress illustrated in Fig. 2.20*c* are as follows:

$$\frac{9 \text{ N}}{(75 \text{ mm})^2} = (1.6)(10^{-3}) \text{ N/mm}^2 = 1.6 \text{ kN/m}^2$$

$$\frac{9 \text{ N}}{(25 \text{ mm})^2} = 14.4 \text{ kN/m}^2$$

$$\frac{15 \text{ N}}{(25 \text{ mm})^2} = 24.0 \text{ kN/m}^2$$

Computation for deformation is illustrated in Fig. 2.20*d*. Begin at the point where $u = 0$ (in this problem, the left end).

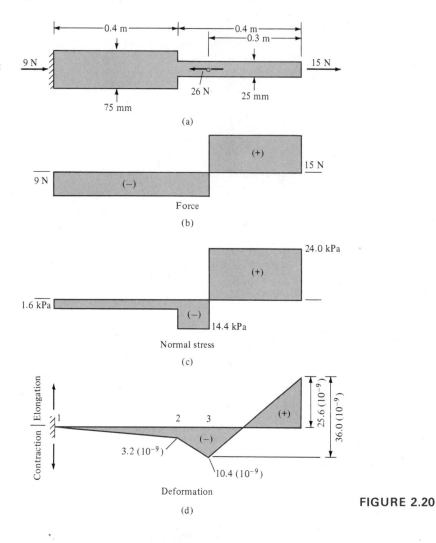

(a)

(b)

Force

(c)

Normal stress

(d)

Deformation

FIGURE 2.20

$$u_{1\text{-}2} = \frac{PL}{AE} = \frac{-(9\text{ N})(0.4\text{ m})}{(0.075\text{ m})^2(200)(10)^9\text{ N/m}}$$

$$= -3.2(10)^{-9}\text{ m} \quad \text{(contraction)}$$

$$u_{2\text{-}3} = \frac{-(9)(0.1)}{(0.025)^2(200)(10)^9} = 7.2(10^{-9})\text{ m} \quad \text{(contraction)}$$

$$u_{3\text{-}4} = \frac{(15)(0.3)}{(0.025)^2(200)(10)^9} = 36.0(10^{-9})\text{ m} \quad \text{(elongation)}$$

Note that the deformation, as given by Eq. (2.28), is linear in L, the axial co-ordinate. To calculate the deformation, Eq. (2.28) has been used three times, once for each constant section of axial load or area. The deformation actually varies as shown in Fig. 2.20—a straight line between each abrupt change in P or A. ∎

In practice of engineering analysis and design, Eq. (2.28) is usually satisfactory for calculating axial deformation. There are situations when the equation must be modified—for instance, if P or A is not a constant, but a function of the axial coordinate. The modulus of elasticity E could vary along the length of the member, but actually this concept is not very practical. Consider the rod of Fig. 2.21. The differential element of length dx can be viewed as in Fig. 2.21b. Summing the forces on the free body gives

$$P + dP - P - p\,dx = 0$$

or

$$dP = p\,dx \tag{2.29}$$

This equation can be integrated between appropriate limits as

$$\int_{P_1}^{P_2} dP = \int_{x_1}^{x_2} p\,dx$$

or

$$P_2 - P_1 = \int_{x_1}^{x_2} p\,dx \tag{2.30}$$

An interpretation for this equation will be given in Example 2.13.

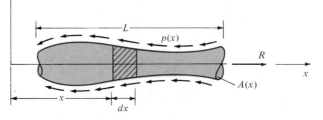

(a)

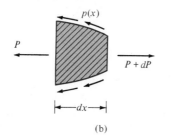

(b)

FIGURE 2.21
Axially loaded member with variable force and variable area.

The deformation of the element of Fig. 2.21 can be obtained using Eq. (2.20). Then $\varepsilon = du/dx$, where du represents the deformation of the differential length dx. Equations (2.5) and (2.24) hold even though P and A are assumed to be functions of the axial coordinate.

$$du = \varepsilon \, dx = \frac{\sigma \, dx}{E} = \frac{P(x)}{A(x)E} \, dx \tag{2.31}$$

The deformation between two points, x_1 and x_2 along the rod is, by integrating Eq. (2.31),

$$u_2 - u_1 = \int_{x_1}^{x_2} \frac{P(x) \, dx}{A(x)E} \tag{2.32}$$

If P and A are constant and the limits are $x_1 = 0$ and $x_2 = L$, Eq. (2.28) is the result.

EXAMPLE 2.13

Assume that the axial force for a rod varies as shown in Fig. 2.22 and compute the axial deformation in terms of P, A, L, and E. The rod is fastened to a support at the left end.

Solution:
The reaction R is equal to the total axial force; hence, using Eq. (2.30) with $P_1 = R$ and $P_2 = 0$ and Fig. 2.22e,

$$-R = \int_0^L p(x) \, dx = -\int_0^{0.8L} \frac{P_0 x}{0.8L} \, dx - \int_{0.8L}^L P_0 \, dx \tag{a}$$

$$R = \frac{P_0 x^2}{1.6L}\bigg|_0^{0.8L} + P_0 x \bigg|_{0.8L}^L = 0.6 P_0 L \tag{b}$$

R is in the direction indicated in the figure. Equation (2.30) can be interpreted as the change in axial force between two points and is equal in magnitude to the area of the axial load diagram between the same two points. Note that R is equal to the area of the applied load diagram. The axial force diagram can formally be obtained by writing an axial-force equation.

$$p(x) = R - \frac{P_0 x^2}{1.6L} \qquad 0 \le x \le 0.8L \tag{c}$$

The equation was obtained by considering a detail of Fig. 2.22b as shown in Fig. 2.22e and Eq. (2.30). $p(x)$ at any location x is the change in axial load between $x = 0$ and some value of x of the area of the detail section, which is the triangular area $(P_0 x/0.8L)(x)(1/2)$. Equation (c) can be written as follows by substituting Eq. (b).

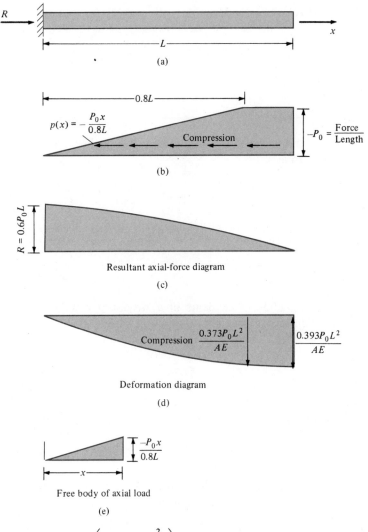

(a)

$p(x) = -\dfrac{P_0 x}{0.8L}$

Compression

$-P_0 = \dfrac{\text{Force}}{\text{Length}}$

(b)

$R = 0.6P_0 L$

Resultant axial-force diagram

(c)

Compression $\dfrac{0.373P_0 L^2}{AE}$ $\dfrac{0.393P_0 L^2}{AE}$

Deformation diagram

(d)

$\dfrac{-P_0 x}{0.8L}$

Free body of axial load

(e)

$$p(x) = -P_0 \left(0.6L - \frac{x^2}{1.6L} \right) \tag{d}$$

$$p(x) = -P_0 [L - x] \qquad 0.8L \le x \le L \tag{e}$$

The deformation is computed by substituting Eq. (d) into Eq. (2.32) with $u_1 = 0$ and integrating with limits $x = 0$ and $x = 0.8L$.

$$u_{0-0.8L} = \frac{-P_0}{AE} \int_0^{0.8L} \left(\frac{0.6L - x^2}{1.6L} \right) dx = \frac{-0.373P_0 L^2}{AE}$$

The deformation between $x = 0.8L$ and $x = L$ is found in a similar manner using Eq. (e).

$$u_{0.08L-L} = \frac{-P_0}{AE} \int_{0.8L}^{L} [L - x] \, dx = \frac{-0.02P_0L^2}{AE}$$

$$u_{\text{TOTAL}} = u_{0-0.8L} + u_{0.8L-L} = \frac{-0.393P_0L^2}{AE}$$ ■

2.6 SUMMARY

Normal stress and shear stress were introduced in this chapter. The equation for normal stress

$$\sigma = \frac{F_N}{A} \tag{2.5}$$

is applicable for axially loaded structural members when the axial force passes through the centroid of the cross section. Average shear stress

$$\tau = \frac{F_T}{A} \tag{2.15}$$

occurs when two adjacent elements of a structural member can be imagined to slide relative to each other along parallel planes.

Average strain was defined as axial deformation per unit length of a rod

$$\varepsilon_{\text{avg}} = \frac{u}{L} \tag{2.22}$$

or as an angular quantity corresponding to shear deformation.

$$\gamma = \frac{u}{L} \tag{2.23}$$

Young's modulus and Poisson's ratio are the material constants that linearly relate stress and strain for axially loaded rods.

$$\sigma = E\varepsilon \tag{2.24}$$

$$v = \frac{-\text{Lateral strain}}{\text{Axial strain}} \tag{2.26}$$

The corresponding linear elastic relation for shear stress and shear strain is

$$\tau = G\gamma \tag{2.25}$$

The deformation of an axially loaded member was shown to be

$$u = \frac{PL}{AE} \tag{2.28}$$

2.7 PROBLEMS

2.1 A steel bar of rectangular cross section, 20 mm by 50 mm, is subject to a compressive force of 100 kN acting at its centroid. Compute the average normal stress. Convert your answer to pounds per square inch.

2.2 A circular pipe with outside diameter of 4 in. and wall thickness of 1/2 in. is subject to an axial tensile force of 35,000 lb. Compute the average normal stress in pounds per square inch (psi) and convert your answer to SI units.

2.3 A circular pipe of outside radius 70 mm is subject to an axial compressive force of 400 kN. The average normal stress should not exceed 200 MPa. Compute the required wall thickness of the pipe.

2.4 The truss of Fig. P2.4 is loaded as illustrated. The support at A is pin-connected and has both horizontal and vertical reactions. The connection at B is a mech-

anism that will sustain only a horizontal reaction. The reactions are given, but could be calculated using the equation of statics. The area of all truss members is 2500 mm² (0.0025 m²). Compute the axial force in each member of the truss and calculate the axial stress.

2.5 The rod of Fig. P2.5 is fixed at the left end and subjected to the forces as illustrated. Compute the axial-force distribution. Calculate the normal stress and plot both the force and the stress as a function of length.

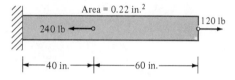

FIGURE P2.5

2.6 A machine element is connected at the left end and is subjected to axial loads as illustrated in Fig. P2.6. Compute and plot the axial force and axial stress in the bar. The area of the bar is $1.125(10^{-3})$ m².

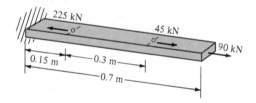

FIGURE P2.6

2.7 The square rod of Fig. P2.7 is fixed at the left end. Axial forces P_1 and P_2 are applied as shown. Compute the distribution of axial force and normal stress. Plot the axial force and axial stress.

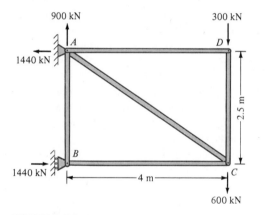

FIGURE P2.4

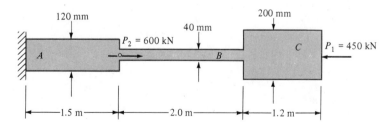

FIGURE P2.7

2.8 Refer to Fig. P2.8 and assume that two 3/4-in. bolts are used to join the three plates. Given that the axial force is 50,000 lb and the normal stress is not to exceed 18,000 psi. Compute the width of each plate to satisfy the normal stress requirements.

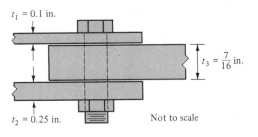

FIGURE P2.8

2.9 Assume that the bolted lap joint in Fig. 2.7 of the text is subject to a force of 225 kN, the bolt is 16 mm in diameter, the plate thickness $t = 10$ mm, and $h = 12$ mm. Compute the bearing stress between the bolt and the plates.

2.10 Four rivets are used to join three plates as shown in Fig. 2.10 of the text. The diameter of each rivet is 20 mm and the applied force P is 55 kN. Compute the average shear stress acting on each rivet.

2.11 The structural joint in Fig. 2.10 of the text is to be subjected to a force of 12,000 lb. The average shear stress for the rivets cannot exceed 10,000 psi. The normal and shear stresses for the joint material should not exceed 18,000 psi and 12,000 psi, respectively. The bearing stress cannot exceed 15,000 psi.
a. Compute the diameter of the rivets.
b. Compute the area of the plates using both the bearing stress criterion and the normal stress criterion.
c. Compute the distance h required to develop the shear stress.

2.12 The riveted joint of Fig. P2.12 is in single shear. Compute the magnitude of the force F according to the following criteria.
a. The average normal stress on a net section should not exceed 22,000 psi.
b. The average shear in the rivet should not exceed 12,000 psi. The rivets are 3/4 in. in diameter.
c. The bearing stress between the plates and the rivet should not exceed 18,000 psi.

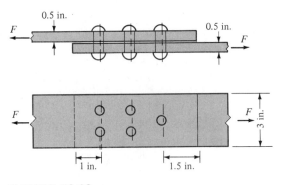

FIGURE P2.12

d. The average shear or tearing stress between the rivets and end of the plate should not exceed 12,000 psi. Which value of F governs? Why?

2.13 A short steel post is used to transmit an axial force of 550 kN through a concrete footing to the ground. The soil bearing pressure is assumed to be uniform and should not exceed 125 kPa.
a. Compute the required area of the footing.
b. Assume an allowable stress of 170 MPa for steel in compression and compute the area of the steel post.

2.14 The link of Fig. P2.14 is connected at end A to a rigid body using a pin-connection as illustrated. The other end, B, is connected to a structural member using a clevis. The axial force in the link is 135 kN.
a. Determine the diameter of the pins to be used at ends A and B. Assume that the stress is not to exceed 85 MPa in shear.
b. Assume the holes at A and B are 2 mm larger than the pins of part **a**. Assume that the bearing stress is not to exceed 100 MPa and compute the required thicknesses t_A for the link at end A and t_B at end B.

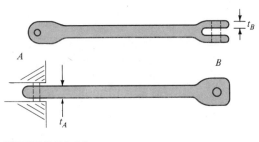

FIGURE P2.14

2.15 The frame of Fig. P2.15 is pin-connected at all joints. Neglect the weight of the frame and compute the following.

a. The diameter of the pin connectors. Assume that the pins at A and B are in single shear and the pin at D is in double shear. Assume an allowable shear stress of 70 MPa.

b. The area of member AB for allowable normal stresses of 110 MPa in compression and 140 MPa in tension.

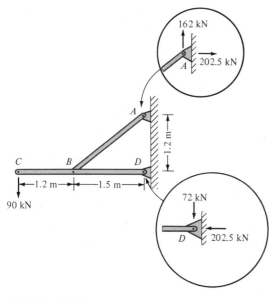

FIGURE P2.15

2.16 The frame of Fig. P2.16 is pin-connected at all joints. Neglect the weight of the members; assume that all pins are 3/4 in. in diameter. The area of member AB is 2.0 in.². Compute the following.

a. The average shear stress in the pins; assume A and C are single shear and B is double shear.

b. The normal stress in member AB.

c. The bearing stress between the pin and the support at A.

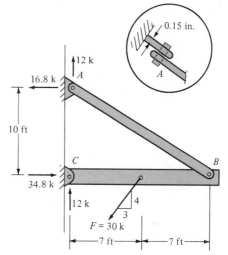

FIGURE P2.16

2.17 The frame of Fig. P2.17 is pin-connected at all joints. Neglect the weight of the members and assume all pins are 7/16 in. in diameter and in double shear. Compute the following.

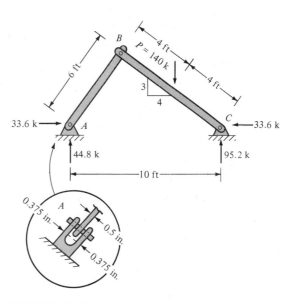

FIGURE P2.17

a. The average shear stress in the pin at *A*.

b. The bearing stress between the pin and the support at *A*.

c. The area of member *AB* for an allowable normal stress of 16,000 psi.

2.18 The structure of Fig. P2.18 is pin-connected at *A* and *B* and can be assumed to be weightless. The dimensions and loading are given with the figure. Compute the following.

a. The bearing stress at *C*.

b. The diameter of the pins at *A* and *B* for single shear and an allowable average shear stress of 12,000 psi.

c. The elongation of rod *AB*.

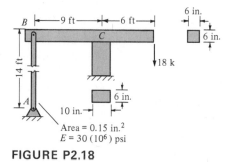

FIGURE P2.18

2.19 The short, square wooden post illustrated in Fig. P2.19 is subject to a compressive force of 40 kip. Assume a failure plane of $\theta = 67$ degrees and compute

the normal and shear stress on the failure plane. Compare these stresses to the stress acting on a plane normal to the axis of the post ($\theta = 90$ degrees).

2.20 The compression specimen of Fig. P2.20 is subjected to a load *P*. Compute the normal and shear stresses in terms of area *A* for failure planes defined by $\theta = 30, 40, 50, 60, 70$, and 90 degrees. Plot your results as stress versus θ on the axis shown. Draw some conclusions concerning the stresses on the plane defined by $\theta = 45$ degrees.

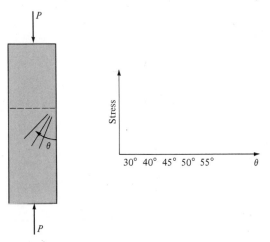

FIGURE P2.20

2.21 A length of steel cable 300 m long and 12 mm in diameter elongates 600 mm when subjected to a tensile load of 45 kN. Compute the unit strain and modulus of elasticity. Convert your answer to English units.

2.22 A block of material has an axial dimension of 12 in. and a cross section of 6 in. by 8 in. Compute the axial elongation and lateral contraction for an axial tensile load of 60,000 lb. Assume $E = 6(10^6)$ psi and $v = 0.34$. Convert your results to SI units.

2.23 An aluminum bar 50 mm by 100 mm is 0.3 m long and is subjected to a tensile load of 540 kN. The length increases by 0.46 mm and the 100 mm side decreases by 0.038 mm. Compute Poisson's ratio and Young's modulus.

2.24 A circular rod 40 mm in diameter is 0.67 m long. A compressive load of 67.5 kN is applied along the axis

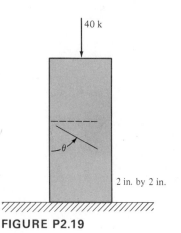

FIGURE P2.19

of the rod. Compute the change in length and the change in diameter. Assume $E = 48$ GPa and $v = 0.3$.

2.25 The small block of material shown in Fig. P2.25 is subjected to a tensile force P in the x direction. The initial volume of material is $V = \Delta x \, \Delta y \, \Delta z$. Show that the final volume of material is $V_f = (1 - 2v\varepsilon_x + \varepsilon_x) \Delta x \, \Delta y \, \Delta z$. Note that products of strains have been neglected. Calculate the change in volume as $\Delta V = (1 - 2v)(\varepsilon_x)(\Delta x \, \Delta y \, \Delta z)$ and develop an expression for the change of volume per unit volume. Demonstrate that the cross-sectional area of the block in a plane normal to the direction of the applied force is approximately $A = (1 - 2v\varepsilon_x) \Delta y \, \Delta z$.

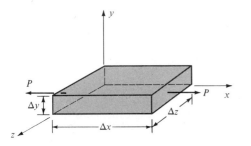

FIGURE P2.25

2.26 A circular aluminum bar 6 in. long of diameter 2 in. is subjected to an axial compressive force of 20,000 lb. Compute the following and convert your results to SI units, if $E = 10(10^6)$ psi and $v = 0.3$.
a. The average normal stress.
b. The unit strain.
c. The total elongation.
d. The change in diameter.
e. The total volume change.
f. The change in cross-sectional area.
For parts **e** and **f** refer to Problem 2.25.

2.27 Compute the change in volume per unit volume for a square steel rod 12 in. long subject to an axial tension of 20,000 lb. Assume $E = 30(10^6)$ psi, area $= 0.25$ in.^{2}, and $v = 0.33$. Does the volume increase or decrease? Note that the change in volume per unit volume is the volume change divided by the original volume. Refer to the results of Problem 2.25.

2.28 An initially straight bar is hanging vertically and is subject to its own weight. Compute the total increase

in length. Assume the area is A, modulus of elasticity is E, length is L, and unit weight of the material is γ.

2.29 Assume the rod described in Problem 2.5 is made of copper, $E = 16(10^6)$ psi. Compute and plot the axial deformation.

2.30 The axially loaded steel rod of Fig. P2.30 is subjected to the loads as illustrated. Compute and plot the axial force, normal stress, and deformation. $E = 30(10^6)$ psi.

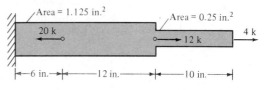

FIGURE P2.30

2.31 A circular steel tube, 26 in. long, is subjected to an axial tensile force of 25,000 lb. The outside diameter is 4 in. Compute the thickness of the tube if the total elongation is limited to 0.07 in.

2.32 The machine element of Fig. P2.7 is made of steel, with $E = 200$ GPa. Compute the deformation of the rod at each load point relative to the left end. Sketch the results as a plot of deformation versus longitudinal distance along the axis.

2.33 Assume the rod described in Problem 2.6 is made of two materials. Sections A and C are copper, $E = 110$ GPa, and section B is magnesium, $E = 45$ GPa. Compute and plot the axial deformation.

2.34 A rod, shown schematically in Fig. P2.34, is loaded with axial forces such that the axial-force dia-

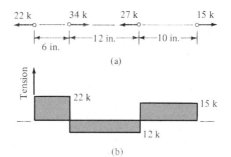

FIGURE P2.34

gram appears as in Fig. P2.34*b*. Compute the area for each section between the loads according to the following criteria: **a.** the strain in any section between the loads should not exceed 0.00067, **b.** the normal stress in compression should not exceed 18,000 psi while the tensile stress should not exceed 22,000 psi. Note that the area can change abruptly at points where the axial load changes. Assume $E = 30(10^6)$ psi and plot the corresponding axial-deformation diagram.

2.35 The rod of Fig. P2.35*a* is supported at the left end and subjected to the variable axial compression illustrated in Fig. P2.35*b*. Plot the axial-force diagram and compute and sketch the axial deformation in terms of *P*, *L*, *A*, and *E*.

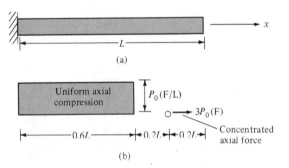

(a)

(b)

FIGURE P2.35

2.36 Work Problem 2.35 for the axial-load distribution illustrated in Fig. P2.36.

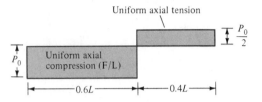

Uniform axial tension

Uniform axial compression (F/L)

P_0

$\frac{P_0}{2}$

0.6*L*   0.4*L*

FIGURE P2.36

2.37 Work Problem 2.35 using the variable axial load illustrated in Fig. P2.37.

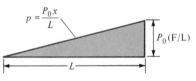

$p = \dfrac{P_0 x}{L}$

$P_0\,(F/L)$

L

Variable axial tensile load

FIGURE P2.37

2.38 Work Problem 2.35 using the sine distribution of axial compressive load illustrated in Fig. P2.38.

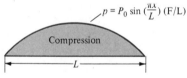

$p = P_0 \sin\left(\dfrac{\pi x}{L}\right)\,(F/L)$

Compression

L

FIGURE P2.38

CHAPTER

3

MECHANICAL PROPERTIES OF MATERIALS AND RELATED TOPICS

3.1 INTRODUCTION

An entire textbook could easily be devoted to mechanical properties of materials. This chapter is at best an introduction. Concepts of material behavior and associated terminology are important to the study of mechanics of materials. Some ideas and concepts must be accepted without proof but will become more understandable as progress is made through the remaining text.

This chapter begins with an extension of the previous chapter, outlining basic ideas and introducing terminology pertaining to material behavior. The ideas are presented using a uniaxially loaded member as the basic model. The ideas carry over to biaxial and triaxial loading conditions and consequently become more difficult to visualize. Elastic material behavior is the basis for the majority of the theoretical development of this textbook. Fortunately, most engineering materials can be considered to be elastic at low stress levels.

Two-dimensional stress states are discussed for isotropic, homogeneous materials. Their basic stress-strain relations are derived in a qualitative fashion employing the two elastic material constants that were discussed in the previous chapter.

Several related topics are introduced: fatigue, fracture, temperature stresses, and stress concentration. The chapter ends with a discussion of work and strain energy. The basic concepts of work and energy are axiomatic in nature and are sometimes difficult to accept. Physical principles that are constructed using energy methods are finding increased usefulness in engineering applications and should be mastered as rapidly as possible.

3.2 ELASTIC AND INELASTIC MATERIALS

The majority of the fundamental concepts of mechanics of materials to be developed in this textbook are based upon elastic material behavior. A steel rod, loaded axially, behaves elastically at low stress levels. The only difficulty is that the human eye cannot see this behavior. For the material to be elastic, it must meet only one criterion. After forces are applied to the material and it deforms, it must return to the original size and shape when the load is removed. Visualize a section of stiff wire. If the wire is held so that a small amount of bending is introduced, the deformation is readily apparent. After releasing the wire, it returns to its original shape. Such behavior is an example of elastic material behavior.

All materials are composed of atoms, and each material has its own particular atomic structure. When the solid body of material is at rest and no external forces are applied, the atoms are held in equilibrium with interatomic binding forces. Elastic material behavior is characterized by the atomic behavior of the material. When external forces are applied to a material to produce deformation, the internal deformation involves shifts in all of the atomic positions. The internal shifting of atoms must remain very small in order for the material to remain elastic. When the external force is removed, the atoms return to their original equilibrium position; hence, the total structure returns to its original shape.

Almost all crystalline materials—that is, materials made up of aggregates of single crystals—are elastic at reasonably low stress levels. Elastic material behavior allows for the characterization of mechanical material properties with relative ease. The relation between stress and strain has been defined in the previous chapter as Young's modulus, E, or the modulus of elasticity. Then, for uniaxial loading of a linearly elastic rod or shaft

$$\sigma = E\varepsilon \tag{3.1}$$

The material constant E is determined experimentally.

A material is said to be isotropic if the value of E in Eq. (3.1) is

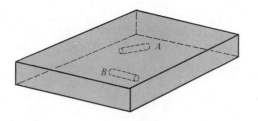

FIGURE 3.1

independent of orientation of the axis of the rod. To illustrate this idea, consider that a steel rod-shaped test specimen is to be fabricated from a piece of steel plate shown in Fig. 3.1. Imagine that the test specimen will be cut from the sheet and machined to the proper size and shape. The specimen is shown dashed in the figure at some arbitrary location, A. For an isotropic material the orientation of the specimen is immaterial; it could be as shown or oriented at 90 degrees to the location shown. The property of the test specimen—the value of the modulus of elasticity—will be the same when the specimen is tested.

Figure 3.1 can be used to illustrate the idea of homogeneity. The mechanical properties of a homogeneous material are independent of spatial location. The test specimen located at B would have identical properties to the specimen at A.

The majority of materials used in engineering design and analysis are analyzed as linear elastic, isotropic, homogeneous materials. The study of mechanics of materials in this textbook concentrates on linear material behavior.

Any material can be deformed until it is no longer elastic. The stress at which stress ceases to be linearly proportional to strain is the proportional limit. For some materials this point is well defined; for others it may be rather difficult to establish. If loading of a structure continues beyond the elastic limit, permanent deformation occurs in the structure. The idea of inelastic structural response must be introduced, and any material behavior that is not elastic is inelastic. A few special cases of inelastic material behavior will be discussed later.

Elastic material behavior may be either linear or nonlinear. Construction materials and the majority of manufacturing materials can be considered to be linear. Rubber, such as used in the tire industry, is an example of a nonlinear elastic material. Figure 3.2 illustrates the basic difference. In either case these elastic materials have the property that loading and unloading of the material follows the same path. Elastic materials are basically independent of prior load history. This statement will later be qualified since materials can suffer from fatigue loading that may cause failure at a stress well below their proportional limit.

There are two basic types of elastic materials. Ductile materials are characterized by their ability to sustain very large deformations prior

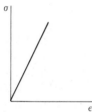

Linear

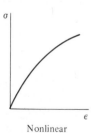

Nonlinear

FIGURE 3.2
Basic difference between elastic and inelastic behavior.

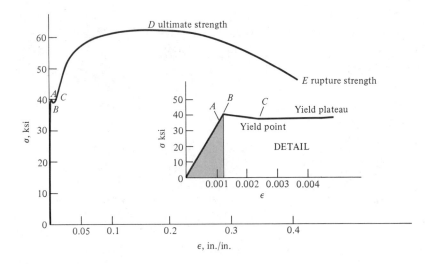

FIGURE 3.3
Typical stress-strain curve for a ductile material.

to final rupture. Steel and aluminum are common ductile materials. Ductility is a desirable feature for a material. If a ductile material is stressed well beyond the proportional limit, the situation is discernible prior to final rupture. This is not so for brittle materials. They are characterized by sudden catastrophic failure at some stress above the proportional limit. In any case, both materials behave in a linearly elastic fashion at low stress levels.

The linear portion of a stress-strain curve is oftentimes barely discernible when viewed with the complete stress-strain curve. Hence, a detail of an elastic portion of the curve is shown in Fig. 3.3.

The *proportional limit, A,* is the point where stress and strain cease to be proportional or the relation between stress and strain is no longer linear. The *elastic limit, B,* is the point where the deformation begins to produce a permanent deformation or permanent set. For the ductile material of Fig. 3.3 these points will nearly coincide. For the brittle material of Fig. 3.4 the two points may vary and be difficult to locate.

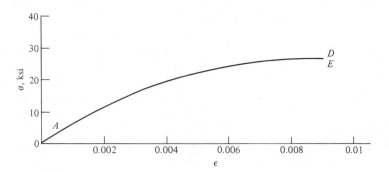

FIGURE 3.4
Typical stress-strain curve for a brittle material.

The *yield point, C*, corresponds to the stress at which deformation will continue without increase in stress. This stress level is well defined for low carbon steel. The entire horizontal section of the curve is sometimes referred to as the yield plateau or plastic plateau. Beyond the plateau, the stress again increases with strain and becomes maximum at *D* and is termed the *ultimate strength* of the material. The *rupture strength, E*, is the final point on the curve and for ductile materials is usually less than the ultimate strength.

The *modulus of resilience* of a material is a measure of the capacity of the material to absorb energy without suffering a permanent deformation. The area under the stress-strain curve defined by the proportional limit and a corresponding point on the strain axis is the modulus of resilience. The units can be thought of as energy per unit volume (lb·in./in.^{3}). The crosshatched area of Fig. 3.3 corresponds to the modulus of resilience.

The *toughness* of a material is the total area under the stress-strain diagram. Toughness is a measure of the capacity of the material to sustain permanent deformation.

Some materials do not exhibit a well-defined yield point or elastic limit. An *offset method* is often used to define a point on the stress-strain diagram corresponding to the yield point. The yield point, when established using an offset method, is usually referred to as the *yield strength*. A typical parameter is to use the strain corresponding to 0.2 percent (0.002) permanent strain; hence, a 0.2 percent offset method. The yield strength at 0.2 percent offset is obtained by locating a point on the strain axis ($\varepsilon = 0.002$) and constructing a line parallel to the initial linear portion of the stress-strain curve. The intersection of the constructed line and the stress-strain curve is taken as the yield strength.

Plasticity may be considered a special case of inelasticity. Sufficient theory and experimental results are available to support the topic and warrant its treatment as a type of material behavior. Plasticity is characterized by the loading and unloading path as a structure is repeatedly loaded and unloaded. Figure 3.5 represents an idealized stress-strain diagram for a plastic material. Visualize that a uniaxial test specimen is stressed and the stress-strain relation is linear until the proportional limit is reached. As loading continues, the curve might appear as shown from *A* to *B*. Let the load be removed and the stress will return to zero (point *C* in Fig. 3.5). The curve from *B* to *C* will be parallel, or nearly parallel, to the original loading curve. The distance *A* to *C* represents the permanent strain or permanent set that is induced into the test specimen. When loading is resumed, the curve follows *C* to *D* and continues to point *E*. Plastic behavior is characterized by the permanent strain that occurs. The loading-unloading curve, *B* to *C* to *D* in Fig. 3.5, is termed a hysteresis and indicates a slight deviation in the loading-

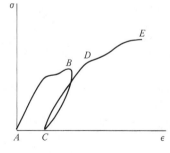

FIGURE 3.5
Idealized stress-strain diagram for a plastic material.

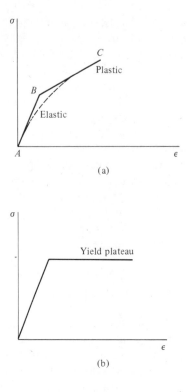

FIGURE 3.6
(a) Bilinear stress-strain
curve; (b) perfectly plastic
stress-strain curve.

unloading path. The hysteresis effect varies in different materials and can be affected by the environment of the material. For instance, it may be more pronounced at high temperatures.

The stress-strain curve for a plastic material that has a shape similar to the path *ABDE* of Fig. 3.5 can be approximated as illustrated in Fig. 3.6. The curve of Fig. 3.6*a* is termed a bilinear stress-strain relation. The linear section between *A* and *B* represents the elastic curve and the linear section between *B* and *C* represents the inelastic range or plastic range. The bilinear curve approximates the actual curve, which is shown dashed. Some materials can be termed perfectly plastic. Such a stress-strain relation is shown in Fig. 3.6*b*. The yield plateau is the second segment of a bilinear stress-strain relation.

Viscous behavior of materials has become important in recent years. Viscoelastic materials are an important category of inelastic materials. A true elastic material is characterized by an elastic spring. A purely viscous material is dependent upon the time rate at which strain occurs and is characterized by a dashpot. A viscoelastic solid exhibits both elastic and viscous reactions to loading. The material reacts to the loading directly and to the time rate of loading or unloading.

A special type of viscous behavior is called creep. The material reacts immediately to loading in an elastic way, then slowly deforms as some load level is maintained. The reverse is true. When the load is removed there is some elastic recovery. The material slowly returns toward its original shape; however, it may or may not reach its original shape since permanent set may occur. Viscous materials are sometimes affected dramatically by temperature changes. Some materials are elastic at low temperatures but become viscoelastic at higher temperatures. Needless to say, the analysis of viscoelastic structures is more complicated than that of elastic materials.

3.3 ALLOWABLE STRESS, WORKING STRESS, AND FACTOR OF SAFETY

In the design of structures, stress is the major design parameter. Allowable stress is the maximum stress to be used in the design of a structural member. The allowable stress must be low enough to represent a reasonable margin of safety and at the same time allow for efficient use of the structural material. The allowable stress is some lesser percentage of the usable strength of the material. Usable strength would correspond to a stress level just prior to structural damage. The yield point for ductile materials would usually correspond to the usable strength.

An allowable stress criterion for design and analysis is necessary for several reasons. The engineer may be unable to estimate accurately the loading to be applied to the structure. The structure may be carelessly overloaded at some future date. The material may be affected by an adverse environment, such as rusting of metal or spalling of concrete. The original material may be of inferior quality.

The working stress is the actual stress induced in the structure after it is put into service. The working stress should be less than the usable strength; however, nothing can prevent the overloading of a structure and, hence, the working stress may exceed the allowable stress.

A factor of safety is used to circumvent the possibility that a structure may become unusable after it is put into service. The factor of safety is usually the ratio of the ultimate strength of a material to the allowable stress. This would result in an ultimate-strength criterion for the design of structures.

$$\text{F.S.} = \frac{\text{Ultimate Strength for the Material}}{\text{Allowable Stress for the Material}}$$

The factor of safety may vary depending upon the loading situation. For instance, a material may have significantly different properties in tension, compression, and shear.

The factor of safety can be based upon a parameter other than ultimate strength. Consider that the factor of safety is defined as

$$\text{F.S.} = \frac{\text{Usable Strength for the Material}}{\text{Allowable Stress for the Material}}$$

In a situation where fatigue is critical, the usable strength could be the endurance limit or fatigue limit. Usable strength could be the creep limit, yield point, or some other characteristic. The design engineer must be aware of the basis for the factor of safety in order for its use to be significant.

For some materials the ultimate-strength design criterion has been refined and represents the state of the art in analysis and design. Reinforced concrete is such a material. The American Concrete Institute specifies the use of load factors for strength design. As the term implies, a load factor modifies the design loading, in contrast to the factor of safety, which modifies the usable strength. The loading for the structure is computed and then increased by multiplying by some specified number. The design strength is then the ultimate strength of the material. This approach is especially useful for reinforced concrete, which is a nonlinear, inelastic, brittle material.

3.4 HOOKE'S LAW AND BASIC EQUATIONS FOR STRESS AND STRAIN

In 1676 Robert Hooke, an English scientist, set forth the relation that bears his name today. In essence, his statement for axially loaded members was that axial force and axial deformation are proportional. His early work has been generalized and today constitutes the basis for sophisticated descriptions of material behavior. Imagine that a solid block of material could be loaded as shown in Fig. 3.7. The stress loading is shown along the x axis. The block of material can be imagined to deform in a general way; it could elongate, contract in the y and z directions, and possibly rotate about any or all coordinate axes. This total freedom of deformation due to stress along only one axis leads to the idea of a generalized Hooke's law. The normal stress along the x axis could affect the deformations or strains, both normal and shear, along all other coordinate directions.

The converse is true also; a normal or shear strain along any coordinate direction could be affected by normal and shear stresses along any and all coordinate directions. As opposed to an isotropic material, this type of material behavior is referred to as anisotropic. Admittedly,

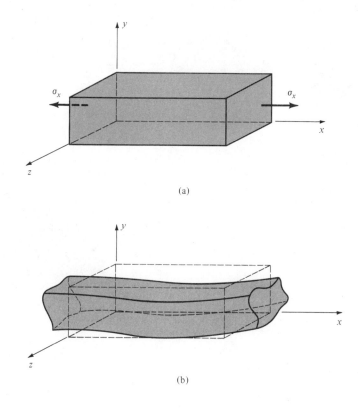

FIGURE 3.7
(a) Solid elastic block
subjected to axial stress;
(b) possible deformations
caused by axial stress.

most materials will exhibit some degree of anisotropy. Fortunately, this effect is so slight for the gross behavior of the majority of materials that an assumption of isotropy is in order and justifiable. Real materials that are used to fabricate structural components can be analyzed using the following conceptual idealization of Hooke's law. Equation (3.1) is the basis for establishing Hooke's law for isotropic material behavior. The three uniaxial loading conditions of Fig. 3.8 can each be described using the equation $\sigma = E\varepsilon$, or $\varepsilon = \sigma/E$. Recall Eq. (2.26), Poisson's ratio, which gives the relation between lateral and axial strain for a uniaxially loaded member

$$v = \frac{-\varepsilon_y}{\varepsilon_x} = \frac{-\varepsilon_z}{\varepsilon_x} \tag{3.2}$$

then $\varepsilon_y = -v\varepsilon_x$ and $\varepsilon_z = -v\varepsilon_x$ for loading along the x axis. This idea can be extended to the y and z directions as illustrated in Fig. 3.8. If the three stress conditions of Fig. 3.8 are assumed to occur simulta-

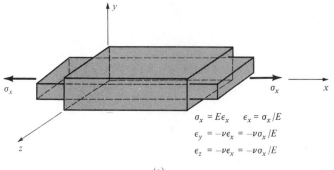

$$\sigma_x = E\epsilon_x \qquad \epsilon_x = \sigma_x/E$$
$$\epsilon_y = -\nu\epsilon_x = -\nu\sigma_x/E$$
$$\epsilon_z = -\nu\epsilon_x = -\nu\sigma_x/E$$

(a)

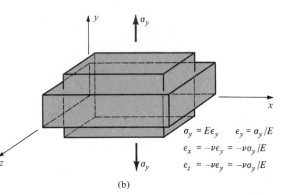

$$\sigma_y = E\epsilon_y \qquad \epsilon_y = \sigma_y/E$$
$$\epsilon_x = -\nu\epsilon_y = -\nu\sigma_y/E$$
$$\epsilon_z = -\nu\epsilon_y = -\nu\sigma_y/E$$

(b)

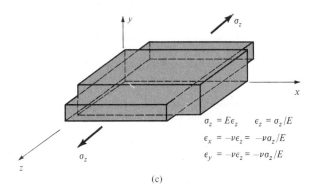

$$\sigma_z = E\epsilon_z \qquad \epsilon_z = \sigma_z/E$$
$$\epsilon_x = -\nu\epsilon_z = -\nu\sigma_z/E$$
$$\epsilon_y = -\nu\epsilon_z = -\nu\sigma_z/E$$

(c)

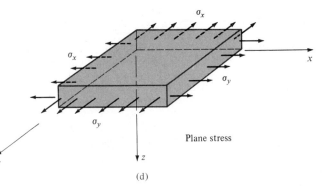

Plane stress

(d)

FIGURE 3.8

neously, the strain in the x direction is the sum of the effects of the three individual uniaxial loadings.

$$\varepsilon_x = \frac{\sigma_x}{E} - \frac{v\sigma_y}{E} - \frac{v\sigma_z}{E}$$

$$\varepsilon_x = \frac{\sigma_x - v(\sigma_y + \sigma_z)}{E} \tag{3.3}$$

Similarly

$$\varepsilon_y = \frac{\sigma_y - v(\sigma_x + \sigma_z)}{E} \tag{3.4}$$

and

$$\varepsilon_z = \frac{\sigma_z - v(\sigma_x + \sigma_y)}{E} \tag{3.5}$$

Equations (3.3)–(3.5) are general equations for an isotropic material. The next six chapters of this text are devoted to methods for computing σ for various loading conditions that occur for elementary structures. The general use of Eqs. (3.3)–(3.5) is covered in more advanced textbooks.

For loading conditions where normal stress in the third dimension can be assumed to be zero, a condition of plane stress occurs. Figure 3.8d illustrates the situation. The dimension in the z direction is small compared to the other coordinates; hence, the assumption $\sigma_z = 0$ is an excellent approximation for the actual stress condition. The equations of plane stress are obtained from Eq. (3.3).

$$\varepsilon_x = \frac{\sigma_x - v\sigma_y}{E}$$

$$\varepsilon_y = \frac{\sigma_y - v\sigma_x}{E}$$

$$\varepsilon_z = \frac{-v(\sigma_x + \sigma_y)}{E} \tag{3.6}$$

or

$$\sigma_x = \frac{E(\varepsilon_x + v\varepsilon_y)}{1 - v^2}$$

$$\sigma_y = \frac{E(\varepsilon_y + v\varepsilon_x)}{1 - v^2} \tag{3.7}$$

Note that the stress σ_z is zero for the third dimension, but ε_z is unequal to zero.

A similar condition referred to as plane strain sometimes occurs.

In this case it is justifiable to let $\varepsilon_z = 0$ and reduce Eqs. (3.3)–(3.5) to two independent equations. These equations will not be recorded since plane strain will not be discussed in later chapters.

Shear stress and shear strain are independent of normal stress and normal strain for an isotropic material. The simple definition of shear strain in Chapter 2, Fig. 2.16, can be extended to more than one dimension. The shear strains in different coordinate directions are independent of one another; hence Eq. (2.25) applies for shear stress and shear strain in any coordinate direction.

$$\tau = G\gamma \tag{3.8}$$

Shear stress is sometimes more difficult to visualize than normal stress. Further discussion of shear stress occurs in Chapter 4, a detailed discussion is in Chapter 7, and shear and normal stresses are combined in Chapters 8 and 9.

There are *two* independent engineering material constants for an elastic, isotropic, homogeneous material: E, the modulus of elasticity or Young's modulus, and v, Poisson's ratio. The shear modulus G is related to E and v as

$$G = \frac{E}{2(1 + v)} \tag{3.9}$$

This relation can be shown to be true once a more thorough understanding of the relation between normal and shear strains is attained.

One additional material constant can be derived for an elastic, isotropic material. The *bulk modulus* is given by

$$K = \frac{E}{3(1 - 2v)} \tag{3.10}$$

If a cube of material is subjected to a hydrostatic pressure, K is the ratio of that pressure to the change in volume of the cube and it is always a positive number. Equations (3.9) and (3.10) are derived in Chapter 9, Section 8.

Equations (3.9) and (3.10) give some insight into the physical limitations that must be assigned to Poisson's ratio. For any imaginable elastic isotropic material E and G must be positive. By Eq. (3.9), v cannot be less than -1. By Eq. (3.10), for K to remain positive, v cannot exceed $1/2$. Therefore,

$$-1 \leq v \leq 1/2 \tag{3.11}$$

For $v = 1/2$, $K = \infty$, or $v = 1/2$ defines an ideal material referred to as *incompressible*. An incompressible material undergoes no volume change when subjected to external forces.

3.5   STRESS CONCENTRATION

Stress concentration occurs when abrupt changes in the geometry or loading of a structural material occur. The stress is literally concentrated at a point or over a very small area. Stress concentration cannot be prevented; hence, the engineer must design against possible detrimental effects of stress concentration.

Several examples of stress concentration occur in Chapter 2—for example, the structural members of Figs. 2.5 and 2.7. The net normal stress area is shown in Fig. 2.6c and is shown again in Fig. 3.9a. The stress is assumed to be distributed uniformly over a net area given by

$$A_{net} = (w - d)t$$

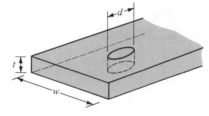

(a)

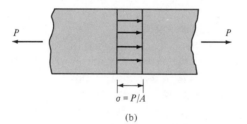

$$\sigma = P/A$$

(b)

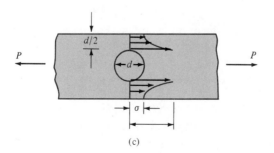

FIGURE 3.9

(a) Hole in a flat plate; (b) uniform stress distribution for a plate of constant cross section; (c) stress distribution for a plate containing a hole.

(c)

The actual stress distribution is not uniform. A flat plate containing a hole in its center has a stress distribution similar to that shown in Fig. 3.9c. The stress concentration due to the hole increases the magnitude of the uniform stress of Fig. 3.9b by a factor that can be as great as 3. The magnitude of the stress concentration can be obtained using the methods of the theory of elasticity and can be demonstrated to be correct experimentally. Note that the rivet problems of Chapter 2 are complicated even more by the presence of the rivet.

Stress concentration occurs any time there is an abrupt change in the geometry of the structure. Stresses are usually nonuniform near points where loads are applied to the structure. A more complete discussion of stress concentration is given in Chapter 15.

3.6 FRACTURE MECHANICS AND FATIGUE

Fatigue and fracture are separate topics, yet often the ultimate result of fatigue is fracture. Fracture mechanics is a current topic of research and new results are continually being added to existing knowledge.

Fracture must occur after a certain amount of deformation occurs. An exception might be a highly viscous material. Fracture occurs in different ways depending upon the material; the ultimate result is a separation of the material. Types of fracture correspond to the type of material behavior.

Brittle fracture occurs for materials that exhibit very small plastic effects. An excellent example is glass or a ceramic. Some metals tend to behave in a similar manner. Brittle fracture is best characterized by catastrophic fracture corresponding to very small deformation. *Ductile fracture* is accompanied by a high degree of plasticity throughout the structure and possible large deformations prior to final rupture. *Fatigue fracture* occurs after the structure is subjected to repeated loading. Each load is less than that which would produce fracture with a single application. Many aspects of fatigue fracture are not well understood. A ductile metal suffers fatigue fracture with very small amounts of plastic deformation. A fatigue fracture is similar to brittle fracture in that very small amounts of plastic deformation occur. *Creep fracture* is usually associated with high temperatures. Creep fracture is similar to ductile fracture since very large plastic deformation occurs prior to final rupture. The mechanism of creep rupture at the atomic level may be considerably different from ductile fracture because of the presence of higher temperatures.

Fatigue as a design criterion is not as well founded as an ultimate strength design criterion. It is not difficult to imagine fatigue fracture occurring after a structure has been subjected to a few cycles of loading that produce large plastic deformations. This phenomenon is referred to as low-cycle fatigue. Quite surprisingly, fatigue fracture can occur after thousands of repetitions of loading that produce stresses below the elastic limit. This type of fatigue failure is referred to as high-cycle fatigue. High-cycle fatigue is extremely important in this age of high technology. The mechanism of fatigue and fracture is microscopic and must be studied using very sophisticated experimental methods. This textbook is concerned with material behavior on a macroscopic scale; hence, any design criterion must be based upon experimental results. To date the only significant design criterion results from repeated laboratory tests that determine the number of cycles of repeated loading at various stress levels required for failure. While the mechanism that causes high-cycle fatigue fracture is not yet totally understood, it is possible to devise a design criterion. The result of the laboratory testing is presented as a plot referred to as an *S-N* diagram. Stress versus cycles to fatigue is usually plotted on semilog paper. An example is shown in Fig. 3.10 for an annealed steel. Usually, low-cycle fatigue is associated with failures occurring at less than 1000 cycles. The rupture stress in low-cycle fatigue corresponds roughly to the ultimate strength of the material. High-cycle fatigue occurs after 1000 cycles, and the number of cycles to failure increases as the stress decreases. A well-defined limit exists for most metals with which the structural engineer

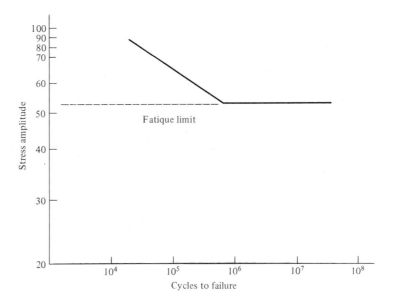

FIGURE 3.10
Typical *S-N* curve.

is concerned. The horizontal line on the diagram is the fatigue limit or *endurance limit* of the material. At stresses below this limit, failure will not occur. When an endurance limit does not exist, a *fatigue strength* is sometimes specified for a material. In the design of equipment subject to cyclic loading the factor of safety is often based upon the endurance limit or fatigue strength of the material.

3.7 TEMPERATURE

All engineering materials are affected by temperature change. Depending upon the intended use of the structure, the design engineer may or may not neglect temperature effects. Common materials expand when heated and contract when cooled. This phenomenon is well understood and is characterized by the coefficient of thermal expansion, α. For a given material, α is the strain due to a 1-degree temperature change. For a rod with length L, and temperature change ΔT, the deformation of the rod is

$$u = \alpha(\Delta T)L \tag{3.12}$$

The strain associated with the temperature change is

$$\varepsilon = \alpha(\Delta T) \tag{3.13}$$

For a rod that is unrestrained at its ends the deformation of Eq. (3.12) occurs with no corresponding axial stress. Since most structures are restrained in some manner, temperature changes causing deformation usually produce significant changes in stress. Consider the following example problem as an illustration of temperature effects.

EXAMPLE 3.1

A circular rod 2 in. in diameter and 14 in. long is made of annealed bronze and subject to a temperature decrease of 65°F.
a. Compute the change in the length of the rod.
b. Assume the ends of the rod are fixed such that the deformation of part *a* cannot occur. Compute the tensile force induced in the rod. Assume $\alpha = 9.4(10^{-6})/°F$ and $E = 15(10^6)$ psi.

Solution:
a. The change in length is computed by substituting into Eq. (3.12)

$$u = 9.4(10^{-6})(65)(14 \text{ in.}) = 8.55(10^{-3}) \text{ in.}$$

The rod decreases in length by $8.55(10^{-3})$ in.

b. If the ends of the rod were fixed such that the deformation could not occur, the rod would be attempting to contract and would cause tensile stress in the rod. Equation (2.28) can be used to obtain the axial force in the rod, as follows:

$$P = \frac{uAE}{L} = \frac{8.55(10^{-3}\ \text{in.})(\pi)(1\ \text{in.})^2(15)(10^6\ \text{psi})}{14\ \text{in.}} = 28{,}780\ \text{lb}$$

The forces in structural members caused by temperature changes can be quite large and can have detrimental effects if the design engineer is not aware of these effects. The following example illustrates a mechanism that could be used to measure a temperature change. ■

EXAMPLE 3.2

The arrangement of Fig. 3.11 could be calibrated to measure temperature change. Compute the movement of the free end of the pointer for a temperature increase of 75°F. Assume the coefficient of thermal expansion for extruded magnesium as $14.4(10^{-6})$ and for steel as $6.6(10^{-6})$. Assume that the pointer material is not affected by the temperature change.

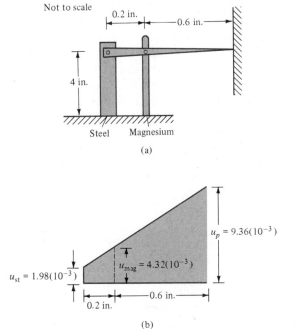

(a)

(b)

FIGURE 3.11

Solution:

The deformation of each member is computed by using Eq. (3.12).

$$u_{\text{st}} = 6.6(10^{-6})(75)(4 \text{ in.}) = 1.98(10^{-3}) \text{ in.}$$

$$u_{\text{mag}} = 14.4(10^{-6})(75)(4 \text{ in.}) = 4.32(10^{-3}) \text{ in.}$$

The movement of the pointer is obtained using similar triangles and the sketch of Fig. 3.11.

$$u_p = \frac{(4.32 - 1.98)(10^{-3})(0.8)}{0.2} = 9.36(10^{-3}) \text{ in.}$$

Hence, an upward movement of $9.36(10^{-3})$ in. indicates a temperature increase of $75°F$.   ∎

The basic idea of temperature effects have been introduced. Thermal effects in statically indeterminate structures will be discussed in Chapter 11.

3.8   WORK AND STRAIN ENERGY

Concepts of work and energy belong to a general topic that is usually referred to as energy methods. The intent of this chapter is to introduce the idea of using the balance of energy as a method for formulating a problem in mechanics. The balance of energy is a criterion of problem solving that appears in numerous engineering courses. There is basically only one energy principle. In dynamics the energy method is presented using two different viewpoints—(1) work and energy and (2) conservation of energy. It is readily apparent that these two methods are equivalent. The first law of thermodynamics is introduced early in a course of thermodynamics. The first law is the basic balance of energy equation and may appear in several different equivalent forms. In fluid mechanics or hydraulics the Bernoulli equation is introduced and usually the relation to the first law of thermodynamics is noted. Energy methods, as they are used in mechanics of materials, may be viewed as an additional application of the aforementioned energy equations.

The term *work*, as used here, refers to work done by external forces. External forces are the applied loads on a structural member. Work occurs as these loads move through a displacement.

It is straightforward to visualize external work as a force applied at a point and moving through a certain displacement to produce work. Internal work is not so simple to visualize. Internal stresses act upon areas that deform as the total member deforms. Since stress times area is force, internal work is force times deformation. The term *elastic strain energy*, or just strain energy, is used to describe internal work. Prior to developing concepts for work and strain energy it should be pointed out that these are scalar quantities. However, the quantities that are multiplied together to produce work have vector properties. It turns out that work is the scalar product of force and displacement. A brief review of the scalar product is in order.

Given a force vector **F** and a displacement vector $d\mathbf{u}$, work is defined as the scalar or dot product of **F** and $d\mathbf{u}$

$$dw = \mathbf{F} \cdot d\mathbf{u} \tag{3.14}$$

The differential form of the equation recognizes that **F** may be a function of **u**. The scalar product can be visualized as shown in Fig. 3.12. Recall that the scalar product is a mathematical operation that multiplies one vector by the component of the second vector that lies in the direction of the first vector. The process is interchangeable, as shown in Figs. 3.12*b* and *c*.

$$dw = (F \cos \theta)\, du = F(du \cos \theta) \tag{3.15}$$

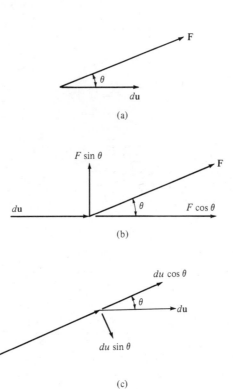

(a)

(b)

(c)

FIGURE 3.12

The strain energy associated with a bar subject to a uniaxial load is obtained first by writing the strain energy of a differential volume element and then integrating over the total volume to obtain the total strain energy. Consider the rod of Fig. 3.13 and remove a differential element as shown in Fig. 3.13b. The strain energy density is written

$$dU = \frac{1}{2}\sigma_x \varepsilon_x \, dV \tag{3.16}$$

The formulation of this equation deserves some explanation. The volume of the differential element is

$$dV = dx \, dy \, dz \tag{3.17}$$

The surface area that σ_x acts normal to is $dy\,dz$; hence, the force is $\sigma_x\,dy\,dz$. The displacement along the x axis over the length axis is $\varepsilon_x\,dx$.

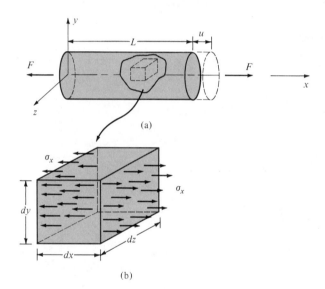

(a)

(b)

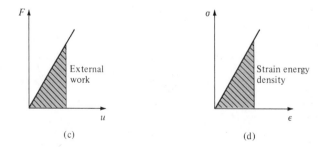

FIGURE 3.13

(c)

(d)

The constant factor of 1/2 can be explained by considering Eq. (3.14). A linear elastic rod with a uniaxial load behaves like a spring; in the elastic range the force is proportional to the displacement. Therefore, a general relation between force and displacement is

$$F = ku$$

where k is a linear spring constant. Substituting into Eq. (3.14) and integrating over the displacement gives

$$W = \int_0^u \mathbf{F} \cdot d\mathbf{u} = \int_0^u ku \, du = \frac{ku^2}{2} = \frac{Fu}{2} \tag{3.18}$$

When force is related linearly to displacement, the work done is half the force times displacement; hence, the factor of 1/2 in Eq. (3.16). Equation (3.16) can be rewritten

$$dU = \frac{1}{2}(\sigma_x \, dy \, dz)(\varepsilon_x \, dx)$$

and is 1/2 times force times displacement. The scalar product is implied since σ_x and ε_x act in the same direction. Integrating gives

$$U = \frac{1}{2}\int_v \sigma_x \varepsilon_x \, dV \tag{3.19}$$

which is the general expression for the strain energy in a uniaxially loaded rod.

A physical interpretation of work is shown in Fig. 3.13c. The cross-hatched portion of the area under the curve is numerically equal to the external work, Area = Work = $Fu/2$. Using elementary relations, $\sigma = F/A$, $E = u/L$, and $\sigma = E\varepsilon$. A stress-strain curve that relates σ and ε is shown in Fig. 3.13d. The area under the stress-strain curve is a measure of the strain energy density.

Finally, the statement for the conservation of mechanical energy for a linear elastic structural system is

$$\Delta W = \Delta U \tag{3.20}$$

The change in external work is equal to the change in strain energy. The following example problem should give some additional insight into Eq. (3.20).

EXAMPLE 3.3

Show that energy is conserved for an axially loaded rod of length L, area A, and axial load P that causes an axial deformation u.

Solution:
Refer to Eq. (2.28) that gives the linear relation between P and u, $u = PL/AE$. Equation (3.14) is written

$$W = \int_0^u P\, du = \int_0^u \frac{AEu\, du}{L} = \frac{AEu^2}{2L} = \frac{Pu}{2}$$

Equation (3.19) is evaluated as follows:

$$U = \frac{1}{2} \int_v \sigma_x \varepsilon_x\, dV = \frac{1}{2} \int_A \sigma_x\, dy\, dz \int_0^L \varepsilon_x\, dx$$

Note

$$\int_A dy\, dz = A \qquad \sigma_x = \frac{P}{A} \qquad \varepsilon_x = \frac{u}{L}$$

$$U = \frac{1}{2}(A)\frac{P}{A}\int_0^L \frac{u}{L}\, dx = \frac{Pu}{2}$$

Then $W = U$, and in Eq. (3.20) when a deformation process starts from rest, the increments of energies become the total energies, and W and U replace ΔW and ΔU, respectively.   ■

> The foregoing example illustrates, in a trivial way, that energy is conserved. Alternatively, an expression relating force and deflection can be derived.

EXAMPLE 3.4

Use conservation of energy to derive an expression relating axial force to axial deformation. Assume a constant area, constant modulus of elasticity, and length L.

Solution:
Use Eq. (3.19) to evaluate the strain energy; that is,

$$U = \frac{1}{2}\int_0^L \int_A \sigma\varepsilon\, dA\, dx = \frac{1}{2}\int_0^L \frac{\sigma^2}{E} A\, dx = \frac{1}{2}\int_0^L \frac{P^2}{AE}\, dx$$

$$U = \frac{1}{2}\frac{P^2}{AE} L$$

The external work is obtained by considering the previous argument that the axially loaded rod behaves like a spring, or $W = Pu/2$. Substituting into Eq. (3.20) gives

$$\frac{Pu}{2} = \frac{P^2 L}{2AE}$$

$$u = \frac{PL}{AE}$$ ■

This procedure becomes quite useful if A is a function of the axial coordinate. Energy methods will be studied in Chapter 10 and discussed in detail in Chapter 13.

3.9 SUMMARY

Elastic material behavior is fundamental to the analysis and design of engineering structures. The terminology is critical to the proper use and interpretation of material properties. The following summary is a capsule of the important terminology.

Ductile Material—A material that undergoes large deformation prior to failure.

Brittle Material—A material that suffers almost no plastic deformation prior to failure and is characterized by a sudden failure.

Proportional Limit or Proportional Elastic Limit—A stress level beyond which stress and strain are no longer proportional.

Elastic Limit—A stress level beyond which permanent deformation occurs.

Yield Point—A stress level at which deformation will occur without an increase in stress.

Ultimate Strength—The maximum stress that a structure will sustain.

Rupture Strength—The point on the stress-strain curve corresponding to the actual rupture of the test specimen.

Allowable Stress—The maximum stress to be used in the design of a structure.

Usable Strength—A stress level just below the point where permanent structural damage would occur.

Working Stress—The actual stress carried by the structure.

Factor of Safety—The ratio of ultimate or usable stress to allowable stress.

Modulus of Elasticity—The ratio of stress to strain in the linear elastic range.

Poisson's Ratio—The ratio of lateral strain to longitudinal strain for longitudinal loading.

Fatigue—The result of repeated loading of a structure, which may lead to failure at relatively low stress levels.

Endurance or Fatigue Limit—The limiting stress below which fatigue failure will not occur.

Fatigue Strength—A limiting stress specified for a situation when an endurance limit does not exist.

Stress Concentration—A situation caused by the loading on the structure, support condition, or structural connection or other abrupt change in the geometry of the structure that causes a significant abrupt change in the stress.

Temperature deformations are computed using the coefficient of thermal expansion, length of the structure, and temperature change. The deformation is given as

$$u = \alpha(\Delta T)L \tag{3.12}$$

The statement for conservation of mechanical energy for a structural system is

$$\Delta W = \Delta U \tag{3.20}$$

where

$$W = \int_0^u \mathbf{F} \cdot d\mathbf{u} \tag{3.18}$$

and

$$U = \frac{1}{2} \int_v \sigma_x \varepsilon_x \, dV \tag{3.19}$$

These equations are specialized for use with axially loaded members. An in-depth study of work and strain energy is reserved for a later chapter.

For Further Study

Anderson, J. C., K. D. Leaver, J. M. Alexander, and R. D. Rawlings, *Material Science*, 2nd ed., Wiley, New York, 1974.

Flinn, R. A., and P. K. Trojan, *Engineering Materials and Their Applications*, Houghton Mifflin, Boston, 1986.

3.10 PROBLEMS

3.1 A uniaxial tension test for a cylindrical bar with a diameter of 0.3 in. gives the data in Table P3.1. The gage length for the test was 2.0 in. Construct a stress-strain diagram and compute **a.** the modulus of elasticity, **b.** proportional limit, **c.** yield point, **d.** the modulus of resilience, and **e.** ultimate strength.

3.2 Data for a uniaxial tension test is given in Table P3.2. Given a diameter of 0.5 in. and gage length of 8.0 in., plot a stress-strain diagram and compute **a.** the modulus of elasticity, **b.** proportional limit, **c.** yield

TABLE P3.1

P, lb	u, in.	P, lb	u, in.
500	0.00114	3300	0.00977
1000	0.00228	3500	0.01140
1500	0.00343	3500	0.01250
2000	0.00453	3500	0.01350
2500	0.00571	3400	0.01586
3000	0.00828		

TABLE P3.2

P, lb	u, in.	P, lb	u, in.
1000	0.0014	8640	0.1480
2000	0.0028	8640	0.2560
3000	0.0043	9032	0.288
4000	0.0056	9228	0.322
5000	0.0071	9425	0.353
6000	0.0085	9520	0.416
7000	0.0099	8835	0.488
7854	0.0112		

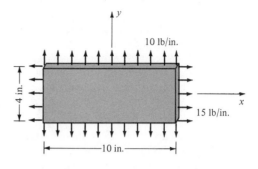

FIGURE P3.5

point, **d.** the modulus of resilience, **e.** ultimate strength, and **f.** rupture strength.

3.3 A bar 0.5 in. in diameter is tested uniaxially to obtain data as shown in Table P3.3. Assume a 2-in. gage length and plot a stress-strain curve. Compute **a.** the modulus of elasticity and **b.** yield strength at 0.2 percent offset.

TABLE P3.3

P, lb	u, in	P, lb	u, in
540	0.00037	4830	0.00441
1080	0.00074	5080	0.00515
1620	0.00111	5281	0.00585
2160	0.00147	5435	0.00905
3240	0.00211	5486	0.01785
4320	0.00294	5518	0.2735
4477	0.00368		

3.4 A thin plate is in a state of plane stress and has dimensions of 8 in. in the x direction and 4 in. in the y direction. The plate increases in length in the x direction by 0.0016 in. and decreases in the y direction by 0.00024 in. Compute σ_x and σ_y to cause these deformations. $E = 30(10^6)$ psi and $v = 0.3$.

3.5 A thin plate 0.125 in. thick is subject to the loading shown in Fig. P3.5. Compute the strain in the x and y directions. $E = 20(10^3)$ psi and $v = 0.45$.

3.6 A steel rod with cross-sectional area of 0.5 in.^{2} has an ultimate strength of 100 ksi in tension. Assume an

allowable tensile stress of 24 ksi and compute the factor of safety using an ultimate-strength criterion.

3.7 Assume a factor of safety of 5 using the ultimate-strength criterion for the rod specified in Problem 3.1 and compute the allowable axial force that can be applied to the structure.

3.8 An aluminum alloy has an ultimate tensile strength of 35 ksi and ultimate shear strength of 22 ksi. The corresponding yield strengths are 32 ksi in tension and 18 ksi in shear.
a. Compute the allowable stress for the material using an ultimate-strength criterion and a factor of safety of 2.0 for tension and 2.5 for shear.
b. Assume the yield strength as the usable strength for the material and compute the allowable stress assuming a factor of safety of 1.75 in tension and 2.25 in shear.

3.9 A certain material has a fatigue limit of 45 ksi. Assume a factor of safety of 1.75 and compute the allowable stress assuming that the fatigue limit is the usable strength.

3.10 A 26-in.-long steel rod with a cross section 1/2 in. by 1/2 in. is subject to a temperature drop of 100°F. Compute the change in length of the rod. Assume that the rod is fixed at its ends and cannot contract. What axial tensile stress will be developed in the rod? Assume $\alpha = 6.6(10^{-6})/°F$ and $E = 30(10^6)$ psi.

3.11 A circular aluminum rod is 0.75 in. in diameter and 18 in. long. A compressive load of 12 kip is applied on the rod. Compute the total deformation of the rod if the temperature drops 80°F. Assume $E = 10(10^3)$ ksi and $\alpha = 12.5(10^{-6})/°F$.

3.12 An aluminum shaft 16 in. in length is attached rigidly to a brass shaft 20 in. in length. The assembly is loaded axially with a tensile force of 1500 lb. Compute the change in temperature required to produce zero axial deformation for the composite rod. The diameter of both shafts is 0.375 in. For aluminum, $\alpha = 12.5(10^{-6})/°F$ and $E = 10(10^6)$ psi. For brass, $\alpha = 9.8(10^{-6})/°F$ and $E = 15(10^6)$ psi.

3.13 Two aluminum bars, each 1.5 in. long, are arranged as shown in Fig. P3.13. Determine the temperature increase required to cause the rods to make contact. $\alpha = 12.5(10^{-6})/°F$.

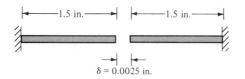

$\delta = 0.0025$ in.

FIGURE P3.13

3.14 Given the arrangement of Fig. P3.13 with a 2-in. rod of aluminum and a 1.5-in. rod of extruded magnesium and $\delta = 0.0035$ in. Compute the temperature change required to cause contact between the rods. $\alpha_{mag} = 14.4(10^{-6})/°F$.

3.15 The arrangement of Fig. P3.15 can be calibrated to measure temperature change. Assume that member

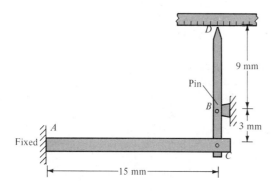

FIGURE P3.15

CBD is not affected significantly by temperature change and compute the movement of point *D* due to a 50°C decrease in temperature. $\alpha = 22.5(10^{-6})/°C$ and $E = 70$ GPa.

3.16 A homogeneous elastic rod of length *L* and of cross-sectional area *A* is acted upon by an axial force *P*. Compute the strain energy that the rod can absorb in terms of normal stress and in terms of *P*.

3.17 A rod is fabricated in two sections as shown in Fig. P3.17. The area of one section is *k* times the area of the second section. Compute the strain energy for the rod in terms of the axial force *P*, modulus of elasticity *E*, area, and length.

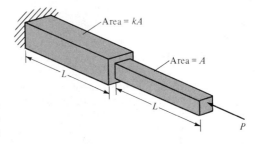

FIGURE P3.17

3.18 Two elastic rods, each of length *L*, are to absorb the same amount of energy when loaded with equal axial forces. The cross section of one is square and the other is circular with a diameter *t*. Compute the dimensions of the square rod in terms of *t*.

3.19 A steel rod, 65 mm by 100 mm, is 0.75 m long and must absorb 150 N·m as an energy load. Assume a factor of safety of 3.5 and ultimate strength of 650 MPa and compute the allowable stress.

CHAPTER 4

TORSION

4.1 INTRODUCTION

Shearing stress due to the action of a torque or twisting moment was mentioned in Chapter 2. In this chapter the topic of torsion will be discussed in detail. The action of a torque is a simple concept; the bolt of Fig. 4.1 is being twisted by the action of the wrench. The torque is given by r times F and can be represented as a vector in the y direction, say $\mathbf{M}_y$, to represent a moment vector along the y axis or $\mathbf{T}$ to indicate torsional moment.

Torsion occurs in many physical situations; for instance, the drive shaft of a vehicle, while it delivers power from one point to another, is being stressed by a torsional moment. Many times a system of pulleys might be attached to a common drive shaft and will serve to input various torques to the drive shaft.

The theoretical derivation of analytical formulas will be limited to circular cross sections in this chapter. The circle is an optimum shape to use for many design applications; hence, it is appropriate to devote a chapter to the study of the torsion of circular rods. The mathematical theory associated with noncircular cross sections is somewhat beyond the scope of this textbook.

75

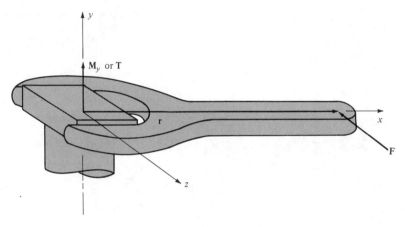

FIGURE 4.1

A torque acts on a bolt as a wrench is used to tighten the bolt.

A brief review of the vector product and its application to the analysis of torsional systems is presented in Section 4.2. In this chapter the use of vector analysis is limited to torsion. A more complete discussion of the use of vector analysis is included in Chapter 5.

The shearing stress and shearing strain equations are developed using basic principles of statics. Statements are made in Section 4.3 concerning the behavior of internal shear strain distribution—namely, its linear variation from the center of a circular rod to the outside surface. Equations that describe the angular deformation of a circular rod are discussed in Section 4.4.

The analysis of circular rods subject to variable torsional loading is discussed. The discussion is, by necessity, more mathematical in nature than the first sections of the chapter. An analogous discussion for beams subject to transverse loading is included in Chapter 5. The reader will feel more proficient with variable torque loading after having studied Chapter 5.

The chapter ends with a discussion of torsion of noncircular members. Design equations for rectangular cross sections are presented along with an illustrative example. Thin-walled members subject to torsion can be studied in an elementary manner, and the analysis is included to make the chapter more complete. Rotating shafts and power transmission are included to finish the chapter.

4.2 TORQUE

First, it is necessary to review the statics of torsional systems. The approach will be to construct a torque diagram, from which shear stress and angular-rotation diagrams will be derived. Torsion analysis will be based upon free-body diagrams that indicate the acting system of torques. The circular shaft of Fig. 4.2a has acting torques of T_1 and T_2. The sign convention is that of vector mechanics, where a couple vector oriented in the positive coordinate direction is called a positive torque vector. View the torque vector from a point on the positive axis looking toward the origin and a positive torque vector appears as a counterclockwise (ccw) action. The action of T_1 in Fig. 4.2a is clockwise (cw) and is also shown as a negative vector. A free body cut between points 1 and 2 would show an acting torque of 40 N·m, as illustrated in Fig. 4.2b. The reactive torque at the fixed end, point 3, is illustrated as the balancing torque or the torque required for equilibrium. The torque diagram is shown in Fig. 4.2c. There is no accepted sign convention for plotting the torque diagram; hence, let positive (ccw) torque appear above the baseline. The torque acting on the shaft between points 1 and 2 is a constant 40 N·m clockwise. At point 2 the torque

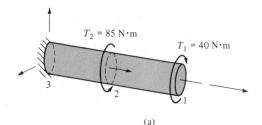

(a)

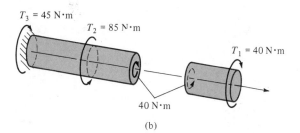

(b)

FIGURE 4.2
(a) Acting torsional system;
(b) free body indicating action and reaction; (c) a torque diagram.

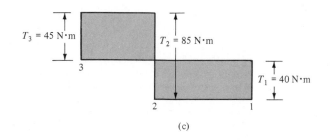

(c)

changes by an amount, 85 N·m, leaving a constant ccw torque of 45 N·m between points 2 and 3.

The vector product was introduced in statics and was used to construct equivalent force and couple systems. The real power of vector analysis is apparent when three-dimensional systems are being analyzed and its organizational power can computationally expedite the analysis of engineering problems.

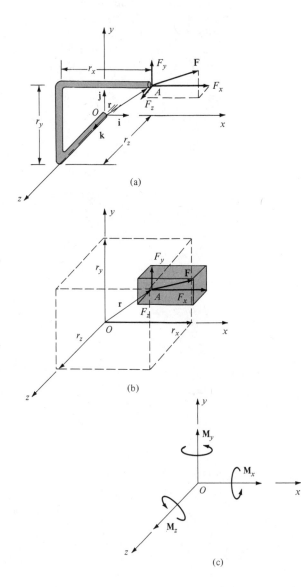

FIGURE 4.3

In this chapter only torsional moments are of interest. However, in practice the application of moment action is not always limited to one axis. A structural member subject to arbitrary loading may have moments acting about all three coordinate axes. The axial moment vector represents the torque or twisting moment, while the other moment vectors are referred to as bending moments.

A review of the vector product is in order. The crank of Fig. 4.3a has a force vector applied at A. The force vector is oriented in an arbitrary direction defined by the three components F_x, F_y, F_z. The moment about any point is the vector product of the position vector $\mathbf{r}$ connecting the point and the force vector $\mathbf{F}$. The position vector is $\mathbf{r} = r_x\mathbf{i} + r_y\mathbf{j} + r_z\mathbf{k}$. The accepted notation for cartesian unit vectors is used and is defined in Fig. 4.3a. Once the physical picture is established and coordinate points determined, the analysis turns out to be an exercise in vector analysis, as shown in Fig. 4.3b. The problem is to determine the moment of the force $\mathbf{F}$ about the point O and is computationally equivalent to

$$\mathbf{M}_O = \mathbf{r} \times \mathbf{F} = \begin{vmatrix} \mathbf{i} & \mathbf{j} & \mathbf{k} \\ r_x & r_y & r_z \\ F_x & F_y & F_z \end{vmatrix}$$

$$= (r_yF_z - r_zF_y)\mathbf{i} + (r_zF_x - r_xF_z)\mathbf{j}$$

$$+ (r_xF_y - r_yF_x)\mathbf{k} = M_x\mathbf{i} + M_y\mathbf{j} + M_z\mathbf{k} \qquad \textbf{(4.1)}$$

The result can be drawn as three components of a moment vector as illustrated in Fig. 4.3c.

The moment $\mathbf{M}_z$ acts axially along the z axis and is interpreted as a torque or twisting moment. The other moment components, $\mathbf{M}_x$ and $\mathbf{M}_y$, cause the crank to bend and may be referred to as bending moments or transverse moments. These moments will be dealt with in the next chapter. The example problems and homework problems of this chapter will deal only with the torsional moment.

EXAMPLE 4.1

In this example the use of the vector product will be demonstrated. The piping system of Fig. 4.4a is three-dimensional and has a force applied at point D. Determine the torque (axial moment) along each section of pipe.

Solution:
The pipe system can be visualized as a system of position and force vectors as shown in Fig. 4.4b. The axial moment can be obtained by connecting the axis of the structure

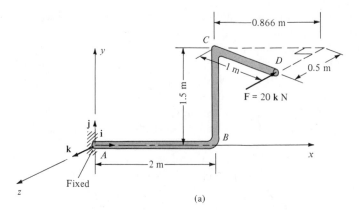

(a)

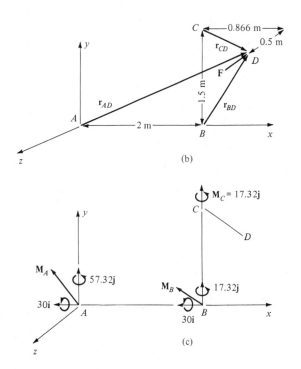

(b)

(c)

FIGURE 4.4

to the force vector with a position vector and computing a vector product. The axial moment along BC can be determined from the vector product $\mathbf{r}_{CD} \times \mathbf{F}$ or $\mathbf{r}_{BD} \times \mathbf{F}$. The calculations are as follows:

$$\mathbf{M}_C = \mathbf{r}_{CD} \times \mathbf{F} = \begin{vmatrix} \mathbf{i} & \mathbf{j} & \mathbf{k} \\ 0.866 & 0 & 0.5 \\ 0 & 0 & -20 \end{vmatrix} = 17.32\mathbf{j}$$

and

$$\mathbf{M}_B = \mathbf{r}_{BD} \times \mathbf{F} = \begin{vmatrix} \mathbf{i} & \mathbf{j} & \mathbf{k} \\ 0.866 & 1.5 & 0.5 \\ 0 & 0 & -20 \end{vmatrix} = -30\mathbf{i} + 17.23\mathbf{j}$$

The moments $\mathbf{M}_C$ and $\mathbf{M}_B$ are illustrated in Fig. 4.4c. The axial moment is the vector component along the line BC, the $\mathbf{j}$ component, or $\mathbf{M}_{BC} = 17.32\mathbf{j}$ N·m. Note that the moment $\mathbf{M}_{CD}$ along the pipe section is zero. The axial moment along the section AB is $-30\mathbf{i}$, as indicated by $\mathbf{M}_B$. The moment $\mathbf{M}_B$ can be verified by summing moments about point A.

$$\mathbf{M}_A = \mathbf{r}_{AD} \times \mathbf{F} = \begin{vmatrix} \mathbf{i} & \mathbf{j} & \mathbf{k} \\ 2.866 & 1.5 & 0.5 \\ 0 & 0 & -20 \end{vmatrix} = -30\mathbf{i} + 57.32\mathbf{j}$$

The axial moment along member AB is $-30\mathbf{i}$. ∎

4.3 SHEAR STRESS AND SHEAR STRAIN

Several assumptions are necessary before proceeding with the derivation of the shearing stress equation. The first assumption is that plane sections, before twisting occurs, will rotate but remain plane during and after twisting. To illustrate the idea, compare circular and rectangular shafts as depicted in Fig. 4.5. The noncircular cross section shows a certain amount of warping as twisting occurs. A circular cross section does not warp, which is in keeping with the actual behavior of the

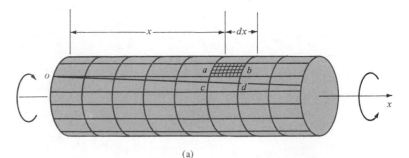

(a)

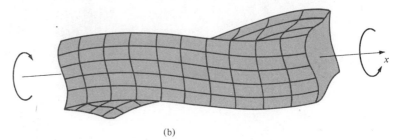

FIGURE 4.5
Torsional warping of a noncircular shaft compared to the plane rotation of a circular shaft.

(b)

shaft. Visualize a slice, dx in length, removed from the circular rod, as shown in Fig. 4.6. Define the shear strain as γ, the angle included between the lines oa and oc. The line oab, prior to the application of a torque, is a straight line on the surface of the cylinder and is parallel to the x axis. As the right end of the cylinder rotates through an angle ϕ relative to the left end, the line oab moves to the position oce. The assumption implies that point b displaces to point e in a plane that is normal to the x axis. The element, shown crosshatched in Fig. 4.6, is illustrated in detail in Fig. 4.7 in cylindrical coordinates. The notation for stress at a point in cylindrical coordinates is compared to cartesian coordinates in Fig. 4.7a. The shear strain, γ, of Fig. 4.6 is written as $\gamma_{x\theta}$ in Fig. 4.7b. The element would deform as illustrated, and a shear stress $\tau_{x\theta}$, acting as shown, would accompany the shear strain. The six-sided element of Fig. 4.7a can be visualized as deforming with six separate motions. The element could deform along each normal axis, as was discussed in the previous chapter. Each side of the element could deform angularly with respect to its adjacent side. The three possible angular deformations are illustrated in Fig. 4.7. The notation means that $\gamma_{x\theta}$ of Fig. 4.7b is the deformation of the plane normal to the x axis relative to the plane normal to the θ axis. The corresponding stress is $\tau_{x\theta}$ as illustrated in Fig. 4.7a. $\tau_{x\theta}$ is the shear stress on a plane normal to the x axis and in the direction of θ. A shear strain γ_{rx}, in the x direc-

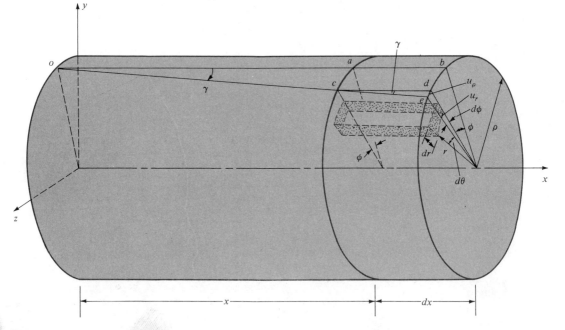

FIGURE 4.6

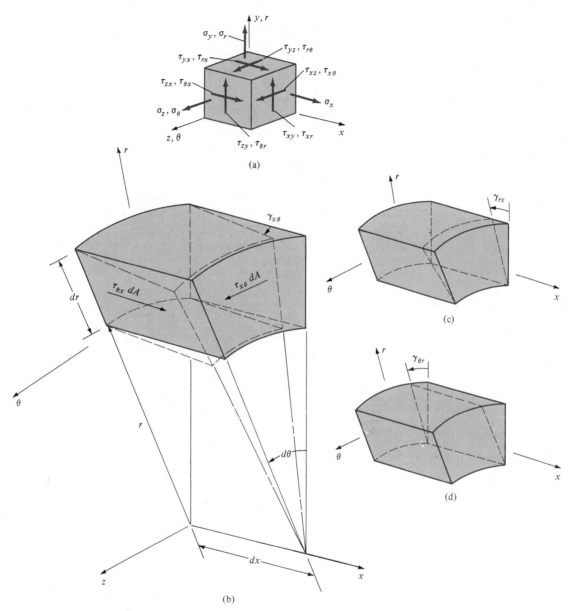

FIGURE 4.7

tion, is illustrated in Fig. 4.7c; however, the assumption does not allow for such a deformation and we conclude that $\gamma_{rx} = 0$. It follows that the corresponding shear stress τ_{rx} is also zero.

The second assumption is that the length dx of Fig. 4.6 remains constant during rotation of the element. If the element of Fig. 4.7b does not deform along the x axis, it can be concluded that the strain along the x axis is zero; that is, $\varepsilon_x = 0$.

The third assumption is that the circular element of Fig. 4.6 rotates as a rigid body, or a circular cross section before rotation is a circular cross section after rotation. The third assumption eliminates the possibility of normal strain in the r and θ direction for the element of Fig. 4.7b; hence, $\varepsilon_r = \varepsilon_\theta = 0$. The third shear strain, $\gamma_{\theta r}$, is illustrated in Fig. 4.7d; however, if the cross section remains rigid, as assumed, $\gamma_{\theta r}$ cannot exist. Therefore, $\gamma_{\theta r} = 0$ and $\tau_{\theta r} = 0$.

Since all three normal strains are zero, it is concluded, based upon the stress-strain equations of Chapter 3, that all normal stresses are zero. The state of stress is one of pure shear, as illustrated in Fig. 4.7b.

We can now establish the connection between applied torque and the internal shear stress. The force, illustrated as $\tau_{x\theta}\,dA$ in Fig. 4.7b, is shown in Fig. 4.8a as a differential force dF acting on an area dA. Note the subscripts on τ have been omitted. We employ the basic concept of Chapter 1 that the distribution of forces, dF, must balance the externally applied load. Then

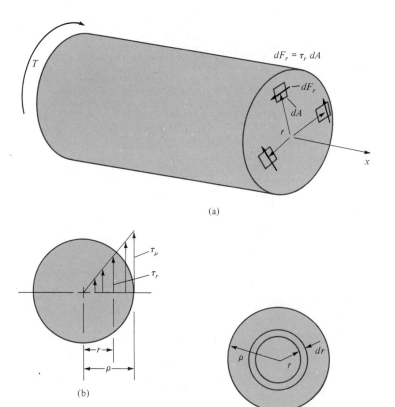

(a)

(b)

(c)

FIGURE 4.8

$$dT = r\,dF_r \tag{4.2}$$

and the sum of the differential forces times their respective position vectors would equal the torque. Integrating both sides of Eq. (4.2) gives

$$T = \int_{\text{Area}} r\,dF_r \tag{4.3}$$

Substituting $dF = \tau\,dA$ gives the torque in terms of shear stress.

$$T = \int_{\text{Area}} r\tau_r\,dA \tag{4.4}$$

where τ_r is the shear stress at r. In Chapter 2 simple shear was defined as a shear stress uniformly distributed over the area, constant with respect to the area coordinates, and integration of Eq. (2.14) proceeded independent of τ. For the case of torsional shear stress the dependence of τ on the coordinate r must be determined before integrating. It is necessary to relate shear stress to shear strain and then establish a representation for the strain distribution.

Assume the linear elastic material properties of Chapter 3 and say that stress is directly proportional to strain, $\tau = G\gamma$. The shear deformation or rotation of the cross section is given by the angle ϕ of Fig. 4.6. The shear deformation is zero at the center of the rod and varies linearly with the radial coordinate and we may establish that

$$u_r = \frac{ru_\rho}{\rho} \tag{4.5}$$

Shear strain, as it is defined in Fig. 4.7, at radius r in Fig. 4.6, is

$$\gamma_r = \frac{u_r}{dx} \tag{4.6}$$

and at radius ρ is

$$\gamma_\rho = \frac{u_\rho}{dx} \tag{4.7}$$

Substituting u_r and u_ρ into Eq. (4.5) demonstrates that shear strain also varies linearly, as follows:

$$\gamma_r = \frac{r\gamma_\rho}{\rho} \tag{4.8}$$

Substituting $\tau = G\gamma$ yields the linear relation for shear stress pictured in Fig. 4.8b; that is,

$$\tau_r = \frac{r\tau_\rho}{\rho} \tag{4.9}$$

Substitution of Eq. (4.9) into Eq. (4.4) gives

$$T = \frac{\tau_\rho}{\rho} \int_{\text{Area}} r^2 \, dA = \frac{\tau_\rho J}{\rho} \qquad (4.10)$$

Recall from statics that the polar moment of inertia is $J = \int r^2 \, dA$. The circular cross section of Fig. 4.8c can be used to evaluate the integral. The differential area is a circular strip, where $dA = 2\pi r \, dr$. Then, substituting for dA gives

$$J = 2\pi \int_0^\rho r^3 \, dr = \frac{\pi \rho^4}{2} = \frac{\pi d^4}{32}$$

where $d = 2\rho$ is the diameter of the shaft. The maximum shear stress due to a torque is given by

$$\tau = \frac{T\rho}{J} \qquad (4.11)$$

The shear stress at any location r on the cross section measured from the center, using Eq. (4.9), is

$$\tau_r = \frac{Tr}{J} \qquad (4.12)$$

The deformed element of Fig. 4.7b can be used to investigate further the behavior of shear stress. As the shaft rotates, every element tends to deform relative to its neighboring elements. The shear force on the face normal to the x axis appears to resist the angular rotation of the shaft. The shear stress on the face normal to the θ axis resists the tendency of the relative movements of elements along the longitudinal axis. It turns out that shear stresses acting on mutually perpendicular planes are equal, and the element of Fig. 4.7b can be used to prove the result. View the element as shown in Fig. 4.9 and sum moments about a radial axis r at point O.

$$\tau_{x\theta} \, dA \, dx - \tau_{\theta x} \, dA \, r \, d\theta = 0$$

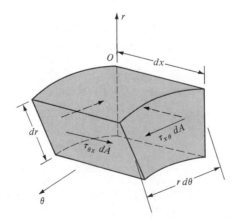

FIGURE 4.9

Substituting the respective values of dA gives

$$\tau_{x\theta} rd\,\theta dr\,dx - \tau_{\theta x}\,dx\,dr\,r\,d\theta = 0$$

or

$$\tau_{x\theta} = \tau_{\theta x} \qquad\qquad\qquad (4.13)$$

The notation for torque may appear in three different ways. In general, the accepted notation for obvious reasons is T for torque. Torque is a component of moment, as moments are studied in statics; hence, M_t is sometimes convenient, with the subscript t indicating torque. When using a vector convention, M_x, M_y, or M_z will indicate torsional moment with the subscript denoting the axis along which the vector is oriented.

EXAMPLE 4.2

A circular shaft 60 mm in diameter is subject to torques T_1 and T_2 as illustrated in Fig. 4.10. Compute the shear stress and plot the torque and shear stress diagrams.

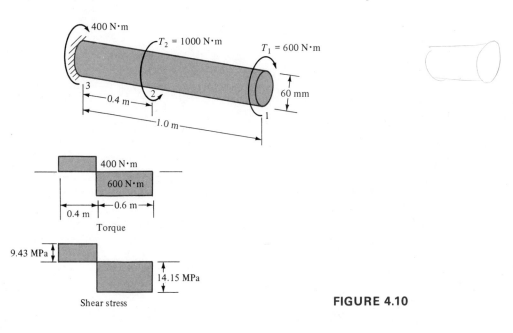

FIGURE 4.10

Solution:
Using statics, the reaction can be found as 400 N·m clockwise.

$$J = \frac{\pi(30)^4}{2} = 1.272(10^6)\ \text{mm}^4 = 1.272(10^{-6})\ \text{m}^4$$

Using Eq. (4.11),

$$\tau_{1\text{-}2} = \frac{600 \text{ N} \cdot \text{m} (0.03 \text{ m})}{1.272 (10^{-6}) \text{ m}} = 14.15 \text{ MN/m}^2 = 14.15 \text{ MPa}$$

Similarly,

$$\tau_{2\text{-}3} = \frac{(400)(0.03)}{1.272(10^{-6})} = 9.43 \text{ MPa}$$

The torque and stress diagrams are illustrated in Fig. 4.10.   ■

4.4   ANGULAR DEFORMATION

The angular deformation caused by torque applied to a circular shaft was described in the preceding section. In this section we will develop a mathematical relation between the angle of twist ϕ and the applied torque. Consider the deformation of the circular rod of Fig. 4.6 and let the shear strain at radius ρ be γ_ρ, as in Eq. (4.7). The displacement u_ρ can be described in terms of the angle γ and also in the terms of the angle $d\phi$. At the radius ρ,

$$u_\rho = \rho \, d\phi = \gamma_\rho \, dx \tag{4.14}$$

or

$$d\phi = \frac{\gamma_\rho \, dx}{\rho} \tag{4.15}$$

Substituting Eq. (4.8) gives

$$d\phi = \frac{\gamma_r \, dx}{r} \tag{4.16}$$

Again, assume linear elastic material properties and substitute $\gamma_r = \tau_r/G$ into Eq. (4.16); then

$$d\phi = \frac{\tau_r \, dx}{rG} \tag{4.17}$$

Substituting Eq. (4.12) gives the desired result:

$$d\phi = \frac{T \, dx}{GJ} \tag{4.18}$$

The angle of twist between two points on a circular shaft made of a linear elastic material is obtained by integrating Eq. (4.18).

$$\phi_2 - \phi_1 = \int_{x_1}^{x_2} \frac{T \, dx}{GJ} \tag{4.19}$$

For a uniform shaft of length L with constant torque the rotation of one end relative to the other is

$$\phi = \frac{TL}{GJ}$$

(4.20)

EXAMPLE 4.3

Construct an angle-of-twist diagram for the circular shaft of Example 4.2. Assume $G = 75 \text{ GN/m}^2$.

Solution:
Use Eq. (4.20) and start from the fixed end where $\phi = 0$. The rotation is computed as follows:

$$\phi_{3\text{-}2} = \frac{TL}{JG} = \frac{(400 \text{ N·m})(0.4 \text{ m})}{1.272(10^{-6} \text{ m}^4)(75 \text{ GN/m}^2)} = 1.67(10^{-3}) \text{ rad ccw}$$

$$\phi_{2\text{-}1} = \frac{(600 \text{ N·m})(0.6 \text{ m})}{1.272(10^{-6} \text{ m}^4)(75 \text{ GN/m}^2)} = 3.77(10^{-3}) \text{ rad cw}$$

$$\phi_{3\text{-}1} = \phi_{3\text{-}2} - \phi_{2\text{-}1} = 1.67(10^{-3}) - 3.77(10^{-3}) = 2.10(10^{-3}) \text{ rad cw}$$

These results represent the relative rotation between two points. The final rotation at the free end is the accumulation of all the relative components of rotation. The rotation can be visualized as shown in Fig. 4.11b. The line bd is originally a straight line.

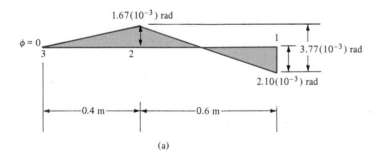

(a)

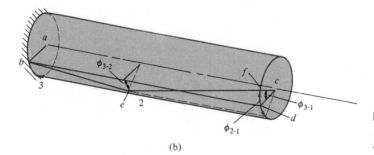

(b)

FIGURE 4.11
Rotation diagram for Example 4.3.

The plane *abdc* is rotated into the position *abefc*. Again, there is no accepted sign convention for construction Fig. 4.11*a*; therefore, counterclockwise (ccw) rotations will be plotted upward. Note that Figure 4.11*b* is highly exaggerated.   ■

4.5 DESIGN AND ANALYSIS OF CIRCULAR MEMBERS

In this section several elementary design and analysis example problems are worked out in detail. Design of circular members is based upon an allowable shear stress or a limiting value of angle of twist.

EXAMPLE 4.4

The circular shaft shown in Fig. 4.12*a* is sometimes referred to as a stepped shaft. Compute the shear stress and angle of twist caused by the application of the various torques. Plot torque, stress, and rotation diagrams. Assume $G = 80$ GPa.

Solution:
The shaft is divided into four sections and torque variation is shown in Fig. 4.12*b*. Shear stress is calculated using Eq. (4.11) for each shaft segment and is plotted in Fig. 4.12*c*.

$$\tau_{AB} = \frac{(1550 \text{ N}\cdot\text{m})(0.1125 \text{ m})}{\pi(0.1125 \text{ m})^4/2} = 0.693 \text{ MPa}$$

$$\tau_{BC} = \frac{(750)(0.1125)}{\pi(0.1125)^4/2} = 0.335 \text{ MPa}$$

$$\tau_{CD} = \frac{(750)(0.055)}{\pi(0.055)^4/2} = 2.870 \text{ MPa}$$

$$\tau_{DE} = \frac{(750)(0.040)}{\pi(0.040)^4/2} = 7.461 \text{ MPa}$$

The angle of twist is computed for each segment using Eq. (4.20) and is plotted in Fig. 4.12*d*.

$$\phi_{AB} = \frac{(1550 \text{ N}\cdot\text{m})(0.3 \text{ m})}{[80(10^9) \text{ Pa}][0.2516(10^{-3}) \text{ m}^4]} = 0.0231(10^{-3}) \text{ rad}$$

$$\phi_{BC} = \frac{(750)(0.4)}{(80)(0.2516)(10^6)} = 0.0149(10^{-3}) \text{ rad}$$

$$\phi_{CD} = \frac{(750)(0.45)}{(80)(0.1437)(10^5)} = 0.2935(10^{-3}) \text{ rad}$$

$$\phi_{DE} = \frac{(750)(0.5)}{(80)(0.4021)(10^4)} = 1.166(10^{-3}) \text{ rad}$$

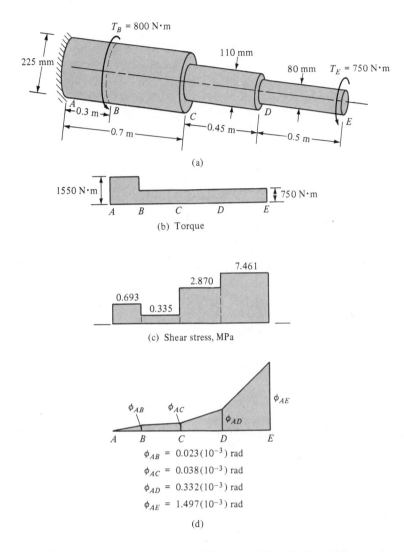

$\phi_{AB} = 0.023(10^{-3})$ rad

$\phi_{AC} = 0.038(10^{-3})$ rad

$\phi_{AD} = 0.332(10^{-3})$ rad

$\phi_{AE} = 1.497(10^{-3})$ rad

(d)

FIGURE 4.12

The rotation is zero at the left end of the shaft, which is referred to as a fixed end. The rotation at C is $\phi_{AC} = \phi_{AB} + \phi_{BC}$. Similarly, $\phi_{AD} = \phi_{AB} + \phi_{BC} + \phi_{CD}$. The rotation of the shaft is a relative quantity. For a shaft at rest there must be at least one fixed point. All rotations can be computed relative to the fixed point.   ■

EXAMPLE 4.5

A circular tube with outside diameter of 8.5 in. and length of 4 ft is subject to a torque of 300 in.·k. The shear stress should not exceed 18,000 psi, nor should the total rotation (angle of twist) exceed 0.0110 rad. Determine a minimum thickness of the tube that will satisfy the design criterion. Assume $G = 12(10^6)$ psi.

Solution:

The polar moment of the area is the parameter that contains the inside radius of the tube; hence,

$$J = \frac{\pi(d_{out}^4 - d_{in}^4)}{32}$$

$$J = (512.48 - 0.098d_{in}^4) \text{ in.}^4$$

$$18 \text{ ksi} = \frac{(300 \text{ in.} \cdot \text{k})(4.25 \text{ in.})}{(512.48 - 0.098d^4) \text{ in.}^4}$$

$$d^4 = 4506.6 \text{ in.}^4$$

$$d = 8.2$$

$$2t = d_{out} - d_{in} = 8.5 - 8.2 = 0.3 \text{ in.}$$

$$\phi = \frac{TL}{JG}$$

$$0.011 = \frac{(300)(4)(12)}{12(10^3)(512.48 - 0.098d^4)}$$

$$d^4 = 4116.2$$

$$d = 8.00$$

$$2t = d_{out} - d_{in} = 8.5 - 8.0 = 0.5 \text{ in.}$$

The criterion governing the angle of twist indicates that a thickness of 0.25 in. should be used. ∎

EXAMPLE 4.6

The solid shaft shown in Fig. 4.13 has a diameter of 75 mm and length of 1.25 m. Use vector analysis to compute the torque, then calculate the maximum torsional stress and the rotation at the free end. Assume $G = 200$ GPa.

Solution:

It is important for the coordinate system to be a right-hand system; if not, the vector sign convention or right-hand rule will not yield the correct coordinate signs. The torsional moment can be calculated by connecting the axis of the shaft to the force **F** with a position vector **r**. This vector can emanate from either point O, or point A, or any convenient point along the x axis. It is important that it is understood that the moment will be determined with respect to the point from which the position vector is directed.

To illustrate the use of the vector product the torque will be calculated using both position vectors, $\mathbf{r}_A$ and $\mathbf{r}_O$, in Fig. 4.13. It should be apparent that

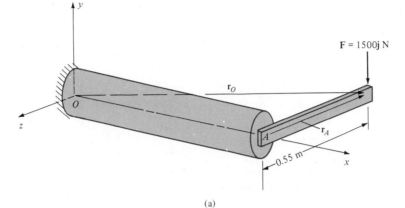

(a)

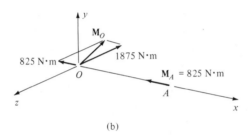

(b)

FIGURE 4.13

$$\mathbf{r}_A = -0.55\mathbf{k} \text{ m} \qquad \mathbf{r}_O = (1.25\mathbf{i} - 0.55\mathbf{k}) \text{ m}$$

$$\mathbf{F} = -1500\mathbf{j} \text{ N}$$

$$\mathbf{M}_A = \mathbf{r}_A \times \mathbf{F} = (-0.55\mathbf{k}) \times (-1500\mathbf{j}) = -825\mathbf{i} \text{ N·m}$$

The calculation of the moment is associated with the following determinant.

$$\mathbf{M}_A = \mathbf{r}_A \times \mathbf{F} = \begin{vmatrix} \mathbf{i} & \mathbf{j} & \mathbf{k} \\ 0 & 0 & -.55 \\ 0 & -1500 & 0 \end{vmatrix} = -825\mathbf{i}$$

$$\mathbf{M}_O = \mathbf{r}_O \times \mathbf{F} = (1.25\mathbf{i} - 0.55\mathbf{k}) \times (-1500\mathbf{j}) = (-1875\mathbf{k} - 825\mathbf{i}) \text{ N·m}$$

The moment vectors may be interpreted as shown in Fig. 4.13b. The axial component $(-825\mathbf{i})$ is the same either way you calculate the torsional moment. The moment M_O indicates a moment vector along the negative z axis at point O; however, only the moment multiplied by $\mathbf{i}$ lies along the axis of the shaft and is the only part of the moment that causes a twist. Then $M_t = 825$ N·m cw and the stress is

$$\tau = \frac{M_t \rho}{J} = \frac{(825 \text{ N·m})(37.5 \text{ mm})}{\pi(37.5 \text{ mm})^4/2}$$

$$\tau = 9.959(10^6) \text{ N/m}^2 = 9.959 \text{ MPa}$$

$$\phi = \frac{M_t L}{JG} = \frac{(825 \text{ N·m})(1.25 \text{ m})}{(200 \text{ GPa})\pi(37.5 \text{ mm})^4/2}$$

$$\phi = 1.659(10^{-3}) \text{ rad}$$ ∎

EXAMPLE 4.7

The structure illustrated in Fig. 4.14 is to be fabricated from sections of circular steel rod. The maximum allowable rotation that should be induced is 0.005 rad for either section. Determine the diameter of the rod to the nearest 0.10 in. and compute the corresponding shear stress. $G = 12(10^6)$ psi.

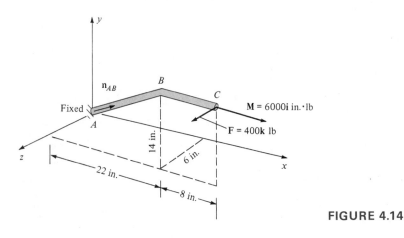

FIGURE 4.14

Solution:

$$\mathbf{M}_A = \begin{vmatrix} \mathbf{i} & \mathbf{j} & \mathbf{k} \\ 30 & 14 & 6 \\ 0 & 0 & 400 \end{vmatrix} + 6000\mathbf{i} = 11,600\mathbf{i} - 12,000\mathbf{j} \text{ in.·lb}$$

$$\mathbf{n}_{AB} = \frac{22\mathbf{i} + 14\mathbf{j} + 6\mathbf{k}}{(22^2 + 14^2 + 6^2)^{1/2}} = 0.822\mathbf{i} + 0.523\mathbf{j} + 0.224\mathbf{k}$$

$$L_{AB} = 26.76 \text{ in.}$$

$$M_{AB} = \mathbf{M}_A \cdot \mathbf{n}_{AB} = (11,600)(0.822) - (12,000)(0.523) = 3259 \text{ in.·lb}$$

$$M_{BC} = 6000 \text{ in.·lb}$$

It is assumed that member AB will twist more than member BC. The formula has T and L appearing in the numerator. The twisting moment for member BC is almost double the twisting moment for AB; however, AB is about three times as long as BC.

$$\phi_{AB} = \frac{M_{AB}L_{AB}}{JG} \quad \text{or} \quad J = \frac{M_{AB}L_{AB}}{\phi_{AB}G} = \frac{\pi d_{AB}^4}{32}$$

$$d_{AB}^4 = \frac{32M_{AB}L_{AB}}{\pi\phi_{AB}G} = \frac{(32)(3259)(26.76)}{\pi(0.005)(12)(10^6)} = 14.81$$

$$d_{AB} = 1.96 \text{ in.} \approx 2.0 \text{ in.}$$

In this example it is assumed that both sections will be of the same diameter rod. It is good design practice to check the rotation for member BC. It is possible to obtain different size rods for each section. There are other design parameters, such as economics, that must eventually be considered. In this case it has been assumed that all rod cross-sectional properties are the same.

$$J = \frac{\pi(2)^4}{32} = \frac{\pi}{2}$$

$$\phi_{BC} = \frac{(6000)(8)}{\pi(12)(10^6)/2} = 0.00254$$

which is less than the allowable.

$$\tau_{AB} = \frac{T\rho}{J} = \frac{(3259)(1)}{\pi/2} = 2075 \text{ psi}$$

$$\tau_{BC} = \frac{(6000)(1)}{\pi/2} = 3820 \text{ psi}$$ ■

4.6 CIRCULAR ROD SUBJECT TO VARIABLE TORQUE

The torsion problem is similar to the axially loaded rod problems of Chapter 2 in that applied torque, just as applied axial load, can be a function of the axial coordinate. In what follows we investigate the case when the applied torque can be described as a continuous mathematical function of the axial coordinate. Consider the circular shaft of Fig. 4.15b that is subject to the torque distribution m_t of Fig. 4.15a. Note that there are no intermediate concentrated torques. The total torque M_t shown in Fig. 4.15c is the accumulation of the applied torque loading. To demonstrate the relation between m_t and M_t consider a differential rod element as shown in Fig. 4.15d. The sum of the

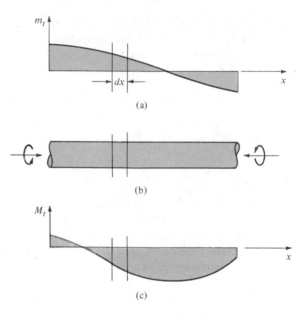

(a)

(b)

(c)

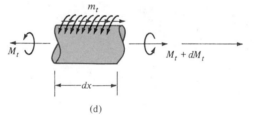

(d)

FIGURE 4.15
Rod subjected to a variable
torque.

torques applied along the element is

$$M_t + dM_t - M_t + m_t\,dx = 0$$

or

$$dM_t = -m_t\,dx \tag{4.21}$$

Then, between any two points a and b along the rod,

$$\int_{M_{t_a}}^{M_{t_b}} dM_t = -\int_{x_a}^{x_b} m_t\,dx$$

$$M_{t_b} - M_{t_a} = -\int_{x_a}^{x_b} m_t\,dx \tag{4.22}$$

The right-hand side of Eq. (4.22) is the integral expression for the negative area of the torque loading diagram of Fig. 4.15a. It is important to realize the significance of the negative integral. The torque loading of Fig. 4.15d is positive as is the increase in the total torsional moment, dM_t. In Eq. (4.21) dM_t is the torque required to balance the applied torque, $m_t\,dx$; it is also called the reactive torque. When using Eq. (4.21) or (4.22) the total torque is merely the negative of the result.

Equation (4.21) can be written

$$m_t = -\frac{dM_t}{dx}\tag{4.23}$$

In other words, the positive slope of the applied total torque diagram of Fig. 4.15c is equal in magnitude to the intensity of the applied torque at any point along the rod.

EXAMPLE 4.8

A circular shaft of length L is acted upon by a uniformly increasing positive torque distribution, as illustrated in Fig. 4.16. Determine the total torque distribution for a shaft that is fixed at the right end. Write an equation describing the variation in torque and sketch the torque diagram. Determine the angle of twist at the free end.

Solution:
Using Eq. (4.22), the reactive torque at any point $x = a$ is computed as

$$M_{t_{x=a}} = M_{t_{x=0}} - \int_0^a \frac{M_T x\,dx}{L} = \frac{-M_T a^2}{2L}\tag{a}$$

Note that $M_{t_{x=0}} = 0$.

Thus, it is established that the reactive torque at $x = a$ is $M_T a^2/2L$, as shown in Fig. 4.16d. The total applied torque at $x = a$ is shown on the right-hand free body. The reactive torque at the fixed end $x = L$ is $-M_T L/2$.

At any point x we establish, using Eq. (a), that the acting torque is

$$M_{t_x} = \frac{M_T x^2}{2L}\tag{b}$$

The torque diagram is determined by plotting M_t versus x using Eq. (b).

$$x = \frac{L}{4} \qquad\qquad x = \frac{L}{2} \qquad\qquad x = \frac{3L}{4}$$

$$M_t = \frac{M_T L}{32} \qquad M_t = \frac{M_T L}{8} \qquad M_t = \frac{9M_T L}{32}$$

The results for M_t are shown in the torque diagram of Fig. 4.16c.

The angle of twist is computed using Eq. (4.18), recognizing that the torque is a function of x.

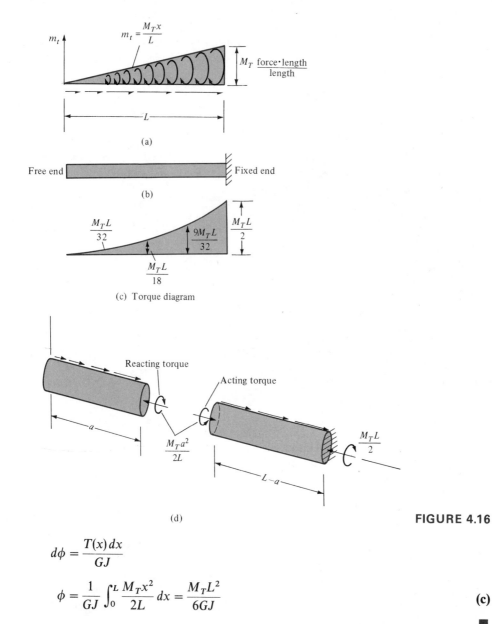

(a)

(b)

(c) Torque diagram

(d)

FIGURE 4.16

$$d\phi = \frac{T(x)\,dx}{GJ}$$

$$\phi = \frac{1}{GJ} \int_0^L \frac{M_T x^2}{2L}\,dx = \frac{M_T L^2}{6GJ}$$

(c)

∎

4.7 TORSION OF SOLID MEMBERS WITH RECTANGULAR CROSS SECTIONS

The theory of torsion for circular members has been presented in this chapter with complete derivation and illustrative applications. The theory, from a mathematical viewpoint, is exact even though it is ele-

TABLE 4.1   Constants for Torsion of a Rectangular Bar

h/b	1.0	1.2	1.5	2.0	2.5	3.0	4.0	5.0	10.0	∞
k_1	0.208	0.219	0.231	0.246	0.258	0.267	0.282	0.291	0.312	0.333
k_2	0.141	0.166	0.196	0.299	0.249	0.263	0.281	0.291	0.312	0.333

mentary in nature. A member with a noncircular cross section behaves in a much more complicated way, as illustrated in Fig. 4.5. When twisting occurs, a noncircular member will suffer some warping of the cross section. The simplifying assumption that plane sections before twisting remain plane after twisting does not apply.

The theory and solution for shear stress and angle of twist can be found in numerous textbooks dealing with the theory of elasticity.[1] The maximum shear stress is given by

$$\tau = \frac{M_t}{k_1 h b^2} \tag{4.24}$$

and the angle of twist is

$$\phi = \frac{M_t L}{k_2 h b^3 G} \tag{4.25}$$

where k_1 and k_2 are given in Table 4.1 for various ratios of h/b, G is the shear modulus, and L is the length of the member. The maximum shear stress occurs at the midpoint of the longest side, the side labeled h in Fig. 4.17. The shear stress varies nonlinearly from zero at the centroid of the cross section to the outside surface. The variation is in contrast with the linear variation for the circular cross section of Fig. 4.8. It is necessary to emphasize that Eq. (2.24) yields only the maximum shear stress. The actual stress distribution across the section is not obtainable using such a simple formula.

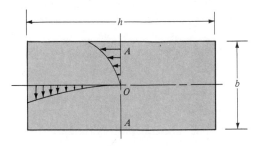

FIGURE 4.17
Shear stress variation for a shaft of solid rectangular cross section.

---

[1] S. Timoshenko and J. N. Goodier, *Theory of Elasticity*, 3rd ed., McGraw-Hill, New York, 1970, chap. 10. (First edition © 1951, chap. 11.)

EXAMPLE 4.9

A rectangular shaft 1.5 m in length is subject to a torque of 4.5 kN·m. Assume an h/b ratio of 1.5 and allowable shear stress of 120 kPa and compute the values of h and b. Compute the rotation of one end of the shaft relative to the other end.

Solution:
Table 4.1 gives $k_1 = 0.231$ for $h/b = 1.5$. Solving for h in Eq. (4.24) gives

$$h^3 = \frac{(1.5)^2 M_t}{k_1 \tau} = \frac{(2.25)(4.5 \text{ kN·m})}{(0.231)(120 \text{ kPa})}$$

$$h^3 = 0.365 \text{ m}^3$$

$$h = 0.715 \text{ m} \quad \text{and} \quad b = \frac{0.715 \text{ m}}{1.5} = 0.477 \text{ m}$$

The angle of twist is computed using Eq. (2.25) with $k_2 = 0.196$.

$$\phi = \frac{(4.5 \text{ kN·m})(1.5 \text{ m})}{(0.196)(0.715 \text{ m})(0.477 \text{ m})^3 G} = \frac{443.8(10^3)}{G}$$ ∎

4.8 TORSION OF THIN-WALLED NONCIRCULAR MEMBERS

The theoretical study of torsion has been limited to circular cross sections. Stresses in thin-walled members, illustrated in Fig. 4.18, can be determined using an elementary analysis that will yield sufficiently accurate results. The concept of shear flow in thin-walled members will aid in the development and understanding of the theory. Shear flow is defined as the internal shearing force per unit length of a thin-walled section. A thin-walled member of variable wall thickness is illustrated in Fig. 4.18a. A torque M_x is applied that produces a state of shear. The element of Fig. 4.18a is shown enlarged in the detail of Fig. 4.18b. A force is assumed to act on each face of the element. Summing forces in the x direction yields the result $S_3 = S_4$; similarly, in the z direction $S_1 = S_2$. The force on any face can be written in terms of the stress on that face. Hence,

$$S_4 = \tau_4 A_4 = \tau_4 t\, \Delta x = S_3 = \tau_3 (t + \Delta t)\, \Delta x$$

The significant result is that

$$\tau_4 t = \tau_3 (t + \Delta t) \qquad \textbf{(a)}$$

The change in thickness, Δt, occurs over an arbitrary length Δl and Eq. (a) will be valid for any length Δl or change in thickness Δt. The

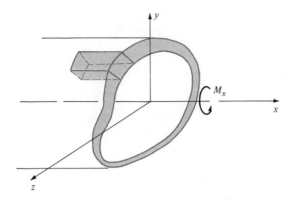

(a)

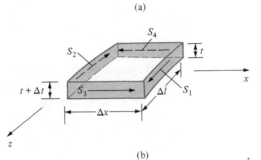

(b)

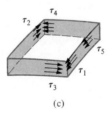

(c)

FIGURE 4.18
Thin-walled member of
variable wall thickness.

quantities in Eq. (a) are constant and represent the shear flow q that was mentioned previously.

It was demonstrated previously in Section 4.3 that shear stresses on adjacent perpendicular planes are equal in magnitude. It follows that $\tau_1 = \tau_3$. If an additional shear stress τ_5 is introduced as in Fig. 4.18c, then $\tau_4 = \tau_5$. Substituting into Eq. (a) gives

$$\tau_5 t = \tau_1(t + \Delta t) = q \qquad \textbf{(b)}$$

Equation (b) indicates that q is a constant along the entire cut section of Fig. 4.18a. The cross section of Fig. 4.18a is shown in Fig. 4.19. The

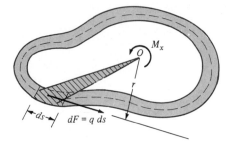

FIGURE 4.19

external moment M_x is related to the internal shear flow q by considering the differential force acting on a differential element, ds. The shear force acting on the element is $q\,ds$. The applied torque M_x is given by

$$M_x = \oint qr\,ds \qquad \text{(c)}$$

where r is the perpendicular distance from the point O to the line of action of the force dF. Equation (c) represents the integral around the center line of the thin wall. Rather than follow through with the integration of Eq. (c) for every possible cross section, a simple evaluation can be made by recognizing that q is a constant and rewriting the equation as

$$M_x = q \oint r\,ds \qquad \text{(d)}$$

where $r\,ds$ is twice the shaded area of Fig. 4.19. Imagine the differential ds to extend around the cross section and the total integral will be two times the interior area bounded by the thin wall. Let $\oint r\,ds = 2\bar{A}$ where $\bar{A}$ is the interior area. Then

$$M_x = 2\bar{A}q \qquad \text{(e)}$$

Using Eq. (b) gives

$$\tau = \frac{q}{t} = \frac{M_x}{2\bar{A}t} \qquad \textbf{(4.26)}$$

It must be emphasized that τ is the average shear stress across the thickness t. Equation (4.26) is applicable for a single closed section. Multiple closed sections represent a problem that is of a complex nature, and the reader should refer to a more advanced textbook.[2]

---

[2] See, for example, J. T. Oden, *Mechanics of Elastic Structures*, 1st ed., McGraw-Hill, New York, 1967, chap. 3.

EXAMPLE 4.10

This example should illustrate the use of Eq. (4.26) for thin-walled noncircular members and demonstrate its accuracy. Determine the wall thickness for a square, thin-walled member that will be subjected to a torque of 550 in.·lb and cannot exceed a mean width of 0.75 in. The maximum allowable shear stress is 12,000 psi.

Solution:
Assume a square, thin-walled section as shown in Fig. 4.20. The width w is shown as the dimension from the center of opposite sides. Equation (4.26) for a square cross section gives

$$12{,}000 \text{ psi} = \frac{550 \text{ in.·lb}}{2(0.75 \text{ in.})^2 t}$$

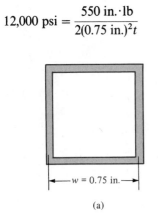

$$w = 0.75 \text{ in.}$$

(a)

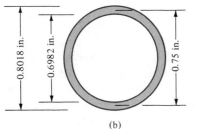

(b)

0.8018 in.   0.6982 in.   0.75 in.

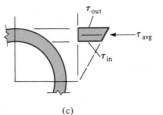

τ_{out}

τ_{avg}

τ_{in}

(c)

FIGURE 4.20

Solving for t gives

$t = 0.041$ in.

If the thin-walled member had been specified as circular with a diameter of 0.75 in., the thickness would have been calculated as

$$t = \frac{550}{2\pi(0.375)^2(12,000)} = 0.0518 \text{ in.}$$

The accuracy of the thin-walled theory as applied to the circular cross section can be determined using Eq. (4.11) and the cross section of Fig. 4.20b.

$$J = \frac{\pi[(0.8018)^4 - (0.6982)^4]}{32} = 0.0172 \text{ in.}^4$$

$$\tau = \frac{(550)(0.8018/2)}{J}$$

$\tau = 12,820$ psi at the outside surface

$$\tau = \frac{(550)(0.6982/2)}{J}$$

$\tau = 11,163$ psi at the inside surface

$\tau_{avg} = 11,992$ psi, which compares very well with the allowable value of 12,000 psi. The variation in shear stress is shown in Fig. 4.20c. ∎

4.9 ROTATING SHAFTS AND POWER TRANSMISSION

Rotating shafts, such as drive shafts for vehicles, can be designed using the elementary formulas of this chapter. The unit of power in SI notation is the watt (W), which is one joule per second (J/s) or one newton meter per second (N·m/s). Also, one horsepower (hp) is equivalent to 745.7 W. A rotating shaft is referred to as rotating with a certain frequency. In SI units frequency is given in hertz (Hz) and has units of 1/seconds. In the British gravitational system the quantity revolutions per unit of time is used.

Work will occur as torque M_t acts through a given angular rotation. There are two separate angular rotations to be considered. Consider a drive shaft for a vehicle. Power is applied at the transmission in the form of a torque. As the torque is transmitted from the transmission to the differential (rear end), a static rotation occurs that can be computed using Eq. (4.20). Power is not transmitted by this rotation. When the shaft rotates, causing the gears in the differential to move, power is being delivered from one end of the shaft to the other. The rotation of the drive shaft is a rigid body motion superposed on the deformation of the shaft. The rigid body motion is related to Newton's law as discussed in a course of dynamics. The external applied torques are equated to the angular acceleration as

$$\Sigma M_t = I\alpha \tag{4.27}$$

where I is the mass moment of inertia and α is the angular acceleration. For angular motion of the shaft let θ be the rotation, then angular velocity is

$$\omega = \frac{d\theta}{dt} \tag{4.28}$$

and angular acceleration is

$$\alpha = \frac{d\omega}{dt} = \frac{d^2\theta}{dt^2} \tag{4.29}$$

For power transmission ω is assumed to be constant; hence, the period of transient motion between initiating motion and reaching constant velocity is not considered. Constant velocity indicates zero acceleration, and Eq. (4.27) becomes

$$\Sigma M_t = 0 \tag{4.30}$$

and implies that torsional moments can be calculated using statics.

The angle change of $d\theta$ of Eq. (4.28) is related to work as

$$dW = M_t \, d\theta \tag{4.31}$$

Equation (4.31) is the differential statement of work. Power is the time rate of doing work. Differentiating Eq. (4.31) gives

$$P = \frac{dW}{dt} = \frac{M_t d\theta}{dt} \tag{4.32}$$

and by Eq. (4.28) is

$$P = M_t \omega = M_t 2\pi f \tag{4.33}$$

where f is the angular frequency. In other words, power is given by the product of torque and angular velocity.

If a shaft rotates with an angular frequency of f Hz, the angle will be $2\pi f$ rad/s. A shaft subjected to a torque of M_t would do $2\pi f M_t$ N·m of work per second. The horsepower that is supplied to a system is equal to the work per second or

$$\text{hp}(745.7)\ \text{N·m/s} = 2\pi f M_t\ \text{N·m/s} \tag{4.34}$$

$$M_t = \frac{745.7\ \text{hp}}{2\pi f} = \frac{119\ \text{hp}}{f}\ \text{N·m} \tag{4.35}$$

For a given horsepower, Eq. (4.35) yields torque in units of N·m. Frequency, f Hz, has units of 1/s.

In the British Gravitational System the quantity n revolutions per minute is commonly used and the work done by torque M_t during one revolution is $2\pi M_t$. The work done per minute is $2\pi n M_t$. One horsepower (hp) equals 33,000 ft·lb/min or 550 ft·lb/s. Then

$$\text{hp} = \frac{2\pi n M_t}{33,000} \tag{4.36}$$

for M_t in foot pounds and n in revolutions per minute. It follows that

$$M_t = \frac{63,000 \text{ hp}}{n} \text{ in·lb} \tag{4.37}$$

Equations (4.34)–(4.37) are used extensively in the design of rotating shafts, where, for instance, the shaft diameter must be determined in order to transmit a given horsepower. Consider the following example.

EXAMPLE 4.11

The power takeoff for a medium-size farm tractor will deliver 28 hp at 70 Hz. Determine the diameter of the power takeoff shaft. The shear stress should not exceed an allowable value of 45 MPa.

Solution:
Equation (4.35) yields the torque as $M_t = 119(28)/70 = 47.6$ N·m. Equation (4.11) can be used to determine the diameter of the shaft, using d as diameter.

$$\tau = \frac{M_t(d/2)}{\pi d^4/32}$$

$$d^3 = \frac{16 M_t}{\pi \tau} = \frac{(16)(47.6)}{\pi 45(10^6)} \text{ m}^3 = 5.38(10^{-6}) \text{ m}^3$$

$$d = 0.0175 \text{ m or } 17.5 \text{ mm} \qquad \blacksquare$$

EXAMPLE 4.12

The gears of Fig. 4.21 are attached to circular steel shafts as illustrated. The diameters of AB and CD are 40 mm and 30 mm, respectively. The pitch of the gear at B is 200 mm and the pitch of the gear at C is 80 mm. The torque applied at D is 800 N·m. Compute the rotation of end D relative to end A. Plot the torque diagram. Assume $G = 83$ GPa.

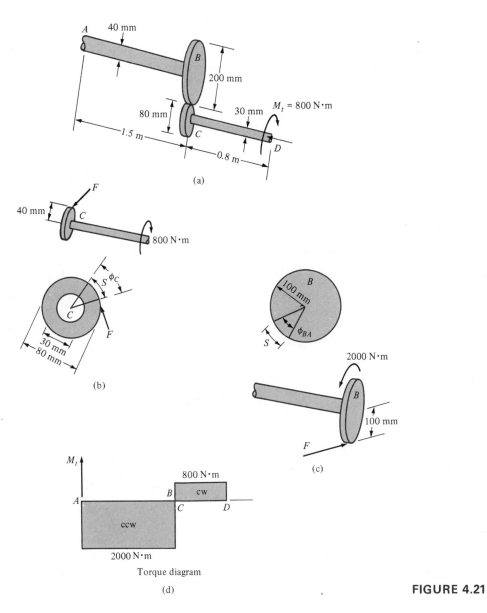

(a)

(b)

(c)

(d)

Torque diagram

FIGURE 4.21

Solution:

The free-body member of CD is sketched in Fig. 4.21b. Imagine a force F applied at the outside edge of the gear C.

$$F = \frac{800 \text{ N·m}}{0.04 \text{ m}} = 20,000 \text{ N}$$

Applying the force F to gear B gives a torque of

$$M_{t_{AB}} = (0.1 \text{ m})(20{,}000 \text{ N}) = 2000 \text{ N·m}$$

acting on AB. The total angular rotation of gear B is

$$\varphi_{BA} = \frac{ML}{JG}$$

$$\varphi_{BA} = \frac{(2000 \text{ N·m})(1.5 \text{ m})}{\pi(0.04 \text{ m})^4(83 \text{ Gpa})/32}$$

$$\varphi_{BA} = 0.144 \text{ rad}$$

The length s of Fig. 4.21c is determined as

$$s = \varphi_{BA}(100 \text{ mm}) = (0.144)(100) = 14.4 \text{ mm}$$

Assuming a point on the outside surface of gear B would move the same distance as a point on the outside surface of gear C; the rotation φ_C of Fig. 4.21b would be

$$\varphi_C = \frac{s}{40 \text{ mm}} = \frac{14.4}{40} = 0.36 \text{ rad}$$

The rotation relation between the gears is proportional to the pitch of the gears, or

$$\frac{\varphi_C}{\varphi_{BA}} = \frac{\text{pitch } B}{\text{pitch } C}$$

The rotation φ_C is applied at point C and causes the entire shaft CD to rotate an amount φ_C. The rotation of D relative to C is

$$\varphi_{DC} = \frac{(800 \text{ N·m})(0.8 \text{ m})}{\pi(0.03 \text{ m})^4(83 \text{ GPa})/32} = 0.097 \text{ rad}$$

The total rotation of D relative to A is

$$\varphi_C + \varphi_{DC} = 0.36 + 0.097 = 0.457 \text{ rad}$$

The torque diagram appears as in Fig. 4.21d. ■

EXAMPLE 4.13

The gears of Fig. 4.22 are attached to rotating shafts. A torque of 5000 in.·lb is applied at D and the angular velocity of CD is 1200 rpm. The diameters of the shafts are $AB = 1.5$ in. and $CD = 0.75$ in. The pitch of gear B is 8 in. and of gear C is 3 in. Compute the power transmitted by CD and AB. Compute the torque and angular velocity of member AB.

Solution:
The power is calculated using Eq. (4.33).

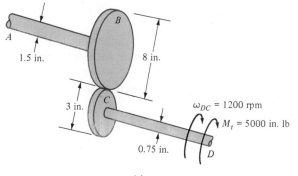

(a)

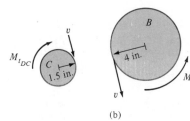

(b)

FIGURE 4.22

$$P = \left(\frac{5000 \text{ in.} \cdot \text{lb}}{12 \text{ in./ft}}\right)\left(1200 \frac{\text{rev}}{\text{min}}\right)\left(2\pi \frac{\text{rad}}{\text{rev}}\right)\left(\frac{1 \text{ min}}{60 \text{ s}}\right)\left(\frac{1 \text{ s} \cdot \text{hp}}{550 \text{ ft} \cdot \text{lb}}\right)$$

$$P = 95.2 \text{ hp}$$

The torque applied to member AB is obtained using the method of the previous example.

$$M_{t_{AB}} = M_{t_{DC}}\left(\frac{\text{Pitch } B}{\text{Pitch } C}\right) = 5000 \text{ in.} \cdot \text{lb}\left(\frac{8}{3}\right) = 13{,}333 \text{ in.} \cdot \text{lb}$$

The angular velocity is found in a similar manner. Refer to Fig. 4.22*b*. The tangential velocity is $v = r\omega$. For gear C, $v = (1.5)(1200 \text{ rpm})$, since gear B will have the same tangential velocity (note the gears are in contact; hence, their common point has the same tangential velocity),

$$\omega_{AB} = \frac{v}{r} = \frac{(1.5)(1200 \text{ rpm})}{4} = 450 \text{ rpm}$$

The power transmission for member AB can be verified as

$$P = M_{t_{AB}}\omega_{AB} = 95.2 \text{ hp}$$

This result assumes no power loss in the gear system. The power transmission remains the same; however, the torque $M_{t_{AB}}$ has increased by a factor proportional to the decrease in ω_{AB}. The reader should be able to gain some insight into design of a transmission for an automobile and the significance of the gear ratios. ■

4.10   SUMMARY

This chapter has been devoted to a thorough understanding of the torsion of circular rods or shafts. Other topics have been introduced but given less emphasis.

Equations (4.11) and (4.20) are the most important formulas for the torsion of circular rods.

$$\tau = \frac{M_t \rho}{J} \tag{4.11}$$

$$\phi = \frac{M_t L}{JG} \tag{4.20}$$

These equations for shear stress and rotation are based upon the assumption that shear strain and hence shear stress vary linearly from the center of the cross section to the outside radius. It is inherent in this assumption that the material behaves in a linearly elastic manner as discussed in Chapter 3.

The torsion of thin-walled members was discussed and the concept of shear flow introduced. Shear flow will be treated in more detail in Chapter 7. The theory of shafts of rectangular cross section subject to pure torque was recognized as being beyond the scope of this text; therefore, design formulas were merely stated. In later chapters the problems will be such that the topic will prove to be useful.

It was shown that for rotating shafts, the power is given by

$$P = M_t \omega = M_t 2\pi f \tag{4.33}$$

and that horsepower and torque are related by

$$\text{hp } (745.7) \text{ N·m/s} = 2\pi f M_t \text{ N·m/s} \tag{4.34}$$

$$M_t = \frac{119 \text{ hp}}{f} \text{ N·m} \tag{4.35}$$

$$\text{hp} = \frac{2\pi(n \text{ rpm})(M_t \text{ ft·lb})}{33,000} \tag{4.36}$$

$$M_t = \frac{63,000 \text{ hp}}{n} \text{ in.·lb} \tag{4.37}$$

The basic concepts of the vector and scalar product as they apply to torsional systems was introduced. The reader is now expected to be familiar with using vector mechanics as it was introduced in statics.

For Further Study

Hibbeler, R. C., *Engineering Mechanics*, 4th ed., Macmillan, New York, 1986.

Oden, J. T., *Mechanics of Elastic Structures*, 1st ed., McGraw-Hill, New York, 1967.

Timoshenko, S. P., and J. N. Goodier, *Theory of Elasticity*, 3rd ed., McGraw-Hill, New York, 1970. (First edition © 1951.)

4.11 PROBLEMS

4.1 Compute and plot the variation in torque for the shaft of Fig. P4.1.

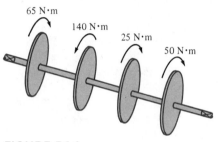

65 N·m
140 N·m
25 N·m
50 N·m

FIGURE P4.1

4.2 Use vector analysis to compute the torque that would be applied to member OA caused by the 5-kN force applied at B in Fig. P4.2.

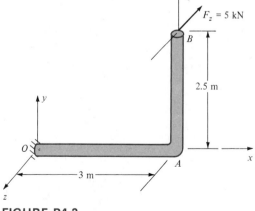

$F_z = 5$ kN
B
2.5 m
y
O
A
x
3 m
z

FIGURE P4.2

4.3 Use vector analysis to compute the torque acting on member OB due to the force **F** in Fig. P4.3.

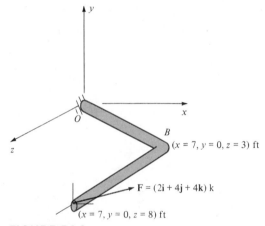

y
x
O
z
B
$(x = 7, y = 0, z = 3)$ ft
F = (2**i** + 4**j** + 4**k**) k
$(x = 7, y = 0, z = 8)$ ft

FIGURE P4.3

4.4 A solid circular bar 15 in. long with 1/2-in. diameter is subjected to a torque of 350 in.·lb. Compute the maximum shear stress.

4.5 A circular shaft with inside diameter 100 mm and outside diameter 118 mm is subjected to a torque of 1000 N·m. Compute and plot the shear stress distribution through the thickness of the pipe.

4.6 A hollow circular shaft 36 in. long is subjected to a torque of 4000 in.·lb. The outside diameter of the shaft is 3 in. Compute the wall thickness of the shaft if the shear stress is not to exceed 12,000 psi.

4.7 The circular shaft of Fig. P4.7 is fixed at the left end and is subjected to the torsional loading as illus-

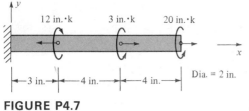

FIGURE P4.7

trated. Plot the torque diagram, and compute the corresponding shear stresses.

4.8 A circular stepped shaft is fixed at the left end and subjected to the torques illustrated in Fig. P4.8. Plot the torque diagram, compute the shear stress, and construct a shear stress diagram.

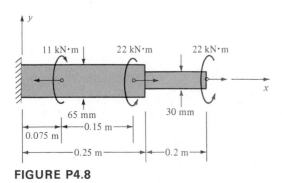

FIGURE P4.8

4.9 A solid circular shaft has a hole machined into the left end and is fixed at the right end. Torsional couples are applied as illustrated in Fig. P4.9. Compute and plot the torque and shear stress diagrams.

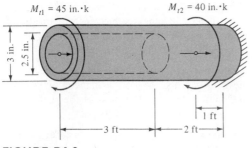

FIGURE P4.9

4.10 The circular stepped shaft illustrated in Fig. P4.10 is subjected to the torques M_t and $2M_t$ as illustrated. The shear stress should not exceed 80 MPa. Compute the maximum value of M_t.

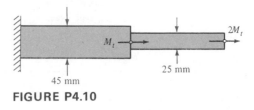

FIGURE P4.10

4.11 The circular steel column of Fig. P4.11 supports a signboard 8 ft by 8 ft that is subjected to a uniform loading of 30 lb/ft². Determine the diameter of the column if the allowable shear stress due to torsional loading is limited to 10,000 psi.

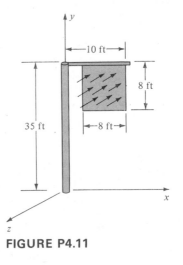

FIGURE P4.11

4.12 The structure of Fig. P4.12 is made of a solid circular rod with the following diameters: $OA = 3$ in., $AB = 2.5$ in., and $BC = 2$ in. Compute the torsional stresses in each section due to the loads F_x and F_z.

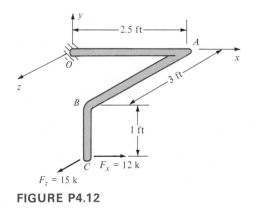

FIGURE P4.12

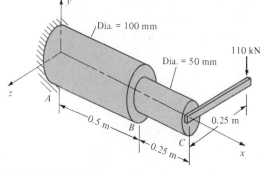

FIGURE P4.16

4.13 Assume that the circular bar of Problem 4.4 is made of steel, $G = 12(10^6)$ psi. Compute the relative rotation of one end of the bar with respect to the other.

4.14 The hollow circular shaft of Problem 4.5 is 0.4 m in length. Assume that the pipe is made of cast iron and compute the relative rotation between the ends.

4.15 Compute the shear stress and rotation of end C relative to A for the stepped shaft of Fig. P4.15. Plot stress and rotation diagrams. Assume $G = 12(10^6)$ psi.

4.17 The stepped shaft of Fig. P4.17 is loaded as shown. Plot the torque, stress, and rotation diagrams. Assume $G = 85$ GPa.

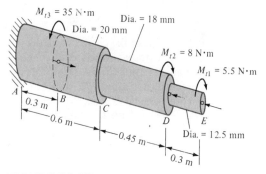

FIGURE P4.17

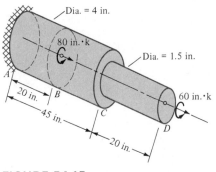

FIGURE P4.15

4.18 The total angular rotation of the shaft of Fig. P4.18 is limited to 0.020 rad. Compute the maximum value of M_t that can be applied to the shaft. Assume $G = 12(10^6)$ psi.

4.16 The stepped shaft of Fig. P4.16 is loaded as shown. Compute the applied torque, plot the shear stress, and plot the rotation between the fixed end A and the free end C. Assume $G = 26$ GPa.

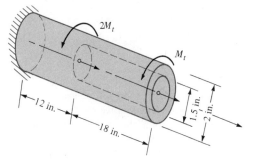

FIGURE P4.18

4.19 Plot the torque, shear stress, and rotation diagrams for the stepped shaft of Fig. P4.19. Assume the shaft is made of copper and $G = 6(10^6)$ psi.

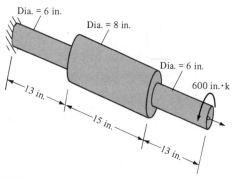

FIGURE P4.19

4.20 The outside diameter of the circular cylinder of Fig. P4.20 is to be 20 times the wall thickness. Compute the thickness and diameter using the following criteria. (a) The shear stress shall not exceed 70 MPa; (b) the total rotation shall not exceed 0.025 rad. Which answer governs the design? Why? Assume $G = 80$ GPa.

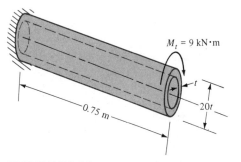

FIGURE P4.20

4.21 A circular shaft 22 in. long will be used to transmit a torque of 85 in.·k. The relative rotation is limited to 0.03 rad and maximum shear stress is limited to 11,500 psi. Compute the diameter. Which criterion governs the size of the shaft? Why? The shaft is cast brass. $G = 3.4(10^6)$ psi.

4.22 The area of a thin-walled tube can be approximated by $2\pi rt$, and the polar moment of the area can be approximated by $\pi r^3 t$, where r is the average radius. Develop the relations beginning with $r_o = r + t/2$ and $r_i = r - t/2$ as shown in Fig. P4.22. (*Hint:* For a thin-walled tube, powers of t are small compared to t and may be neglected.)

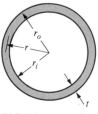

FIGURE P4.22

4.23 A copper cylinder 32 in. long is twisted through a total angle of 3.6 degrees. The diameter of the cylinder is 2.5 in. Compute the maximum shear. Assume $G = 6(10^6)$ psi.

4.24 A steel circular rod 40 mm in diameter that is 1.2 m long is subjected to a pure torsional load. The maximum allowable shear stress is 100 MPa. Compute the rotation of one end with respect to the other.

4.25 A solid circular rod 26 in. long is subjected to a torsional load. The rotation is specified to be limited to 0.5 degree per foot of axial length. Compute the diameter required to meet this criterion if the maximum shear stress is 9000 psi. Assume $G = 12(10^6)$ psi.

4.26 A hollow aluminum shaft 0.60 m long is subject to a single torsional load that produces a maximum shear stress of 40 MPa. The outside and inside diameter are 75 mm and 70 mm respectively. Compute the diameter of a solid aluminum shaft that would be stressed an equal amount. Compare the cross-sectional area of each shaft. Compute the total rotation of each shaft and compare your results. Assume $G = 26$ GPa for aluminum.

4.27 A hollow circular cylinder is subjected to the torsional loading illustrated in Fig. P4.27. The rotation per foot of axial length is specified not to exceed 0.75 degree. Compute the required thickness of the cylinder if the material is steel, with $G = 12(10^6)$ psi.

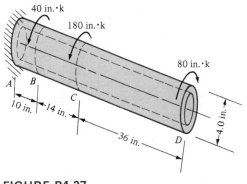

FIGURE P4.27

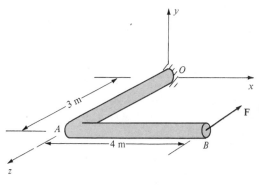

FIGURE P4.29

4.28 The circular rod of Fig. P4.28 is fabricated from two materials that are connected rigidly at point *B*. Section *AB* is copper, *G* = 40 GPa; and section *BC* is beryllium copper, *G* = 48 GPa. Compute the total rotation of *C* relative to *A*.

4.30 The structure of Fig. P4.30 is to be fabricated from sections of circular steel rod. Compute the torque in sections *OA* and *AB* due to the force **F**. Compute the shear stress in each member and the rotation of end *A* relative to *O*. The diameter of the rod is 80 mm.

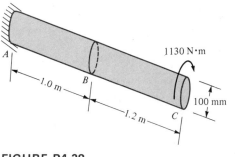

FIGURE P4.28

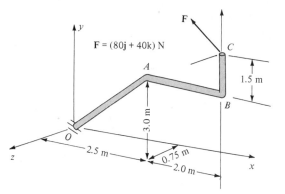

FIGURE P4.30

4.29 The structure of Fig. P4.29 is fabricated from solid circular rods, 30 mm in diameter. The force at *B* is directed from *B*, and its line of action passes through the point (x = 4.5 m, y = 0.2 m, z = 2.9 m). Compute the torsional shear stress and angular rotation for section *OA*. Compute the movement of point *B* due to the rotation of *OA*. Assume *G* is constant for the structure.

4.31 A circular shaft 0.5 m in length is acted upon by a constant torque distributed over one-half of the length as illustrated in Fig. P4.31. The beam is fixed at the right end. Compute and plot the torque distribution, shear stress, and angle of twist. *G* = 80 GPa.

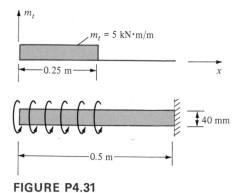

FIGURE P4.31

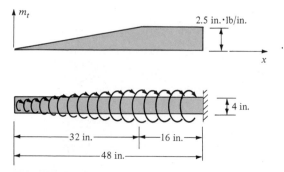

FIGURE P4.33

4.34 A circular shaft is loaded with the variable torque illustrated in Fig. P4.34. Compute and plot torque, shear stress, and rotation. $G = 80$ GPa.

4.32 A circular shaft is loaded with the variable torque illustrated in Fig. P4.32. Compute and plot torque, shear stress, and rotation. $G = 10(10^6)$ psi.

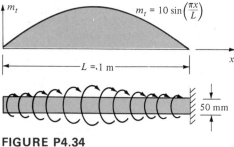

FIGURE P4.34

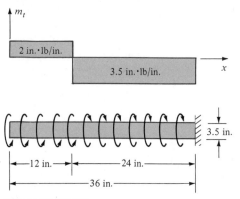

FIGURE P4.32

4.35 A solid circular shaft is tapered as illustrated in Fig. P4.35. The radius at the small end is R_1 and at the larger end is R_2. Develop an expression for the angle of twist that occurs in a length L.

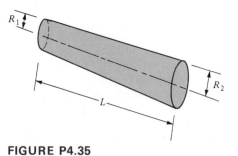

4.33 A circular shaft is loaded with the variable torque illustrated in Fig. P4.33. Compute and plot torque, shear stress, and rotation. $G = 10(10^6)$ psi.

FIGURE P4.35

4.36 A rectangular beam 12 in. wide and 18 in. deep is subjected to a torque of 25,000 ft·lb. Compute the maximum shear stress. Compute the angle of twist of one end relative to the other if the beam is 8 ft long. Assume $G = 2(10^6)$ psi.

4.37 Compute the dimensions of a rectangular beam with a width-to-depth ratio of 2 and length of 3 m. The applied torque is 2.5 kN·m and the allowable shear stress is 200 kPa. Compute the angle of twist of one end with respect to the other if $G = 14$ GPa.

4.38 A machine element is to be fabricated from steel and can be designed as a beam fixed at one end and subjected to a torque at the opposite end of 55 N·m. The thickness of the element, b, is limited to 10 mm. For an allowable shear stress of 70 MPa compute the width of the member. (*Hint:* Since h and k_1 are unknown, a trial-and-error solution is recommended.)

4.39 The structure of Fig. P4.39 is loaded as shown. The vertical member is 4 in. by 10 in. Compute the maximum shear stress and angle of twist. Assume $G = 8(10^6)$ psi.

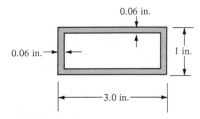

0.06 in. | 0.06 in. | 1 in. | 3.0 in.

FIGURE 4.40

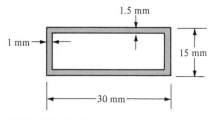

1.5 mm | 1 mm | 15 mm | 30 mm

FIGURE P4.41

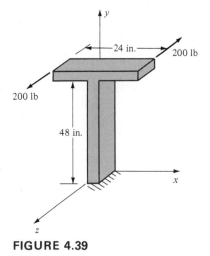

y
24 in.
200 lb
200 lb
48 in.
x
z

FIGURE 4.39

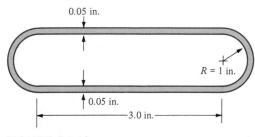

0.05 in. | 0.05 in. | R = 1 in. | 3.0 in.

FIGURE P4.42

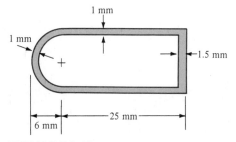

1 mm | 1 mm | 1.5 mm | 6 mm | 25 mm

FIGURE P4.43

4.40–4.43 Assume an applied torque of 50 ft·lb and compute the maximum shearing stress for the thin-walled member cross sections of Figs. P4.40–P4.43. Convert the applied torque to SI units for Figs. P4.41 and P4.43.

4.44 A solid circular shaft rotates with a frequency of 50 Hz. The power is rated at 15 hp and the allowable shearing stress is 50 MPa. Compute the diameter of the shaft.

4.45 A torque of 250 in. lb is applied to a rotating shaft. When the shaft is rotating at maximum constant angular velocity, the power output is 50 hp. Compute the angular velocity in revolutions per minute.

4.46 Compute the maximum power that can be developed in a system made up of a 15-mm-diameter shaft rotating at 30 Hz if the allowable shearing stress is 45 MPa.

4.47 The gear system of Fig. P4.47 is made of aluminum. A torque of 35 N·m is applied at end D. Compute the shear stress and rotation of end D relative to end A. Assume $G = 26$ GPa.

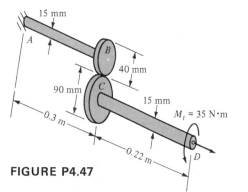

FIGURE P4.47

4.48 The gear system of Fig. P4.48 is subjected to an applied torque of 450 in.·lb. The pitch of gears B and C is 8 in. and 3 in., respectively. Compute the shear stress in each shaft and the rotation of end D relative to end A. Assume $G = 12(10^6)$ psi.

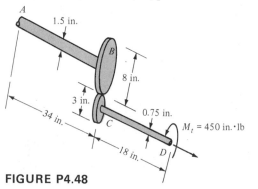

FIGURE P4.48

4.49 For the gear system of Fig. P4.49, assume no loss in the system between D and A. Compute the torque that would be applied on shaft AB and compute the diameter of shaft AB if the allowable shear stress is 10,000 psi.

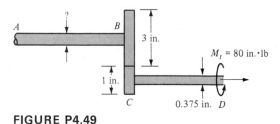

FIGURE P4.49

4.50 The power output for a gear system with gear ratio similar to that shown in Fig. P4.49 is specified to be 75,000 in.·lb/min. The maximum angular velocity that can be supplied at end D is 2500 rpm. The torque on shaft AB cannot exceed 320 in.·lb. The shear stress due to torsion cannot exceed 8000 psi. Compute the diameter of shafts CD and AB.

4.51 Shaft AB of the gear system shown in Fig. P4.51 has an angular velocity of 60 rev/s. The torque applied at end D is 100 in.·lb. Compute the power output at A.

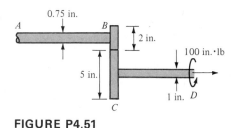

FIGURE P4.51

4.52 Compute the diameter of shaft AB of Fig. P4.51 assuming that the shear stress in AB should not exceed that of shaft CD.

CHAPTER

SHEAR AND MOMENT IN BEAMS

5.1 INTRODUCTION

The beam is one of the fundamental structural members. A beam, sometimes referred to as a flexural member, can be thought of as a structure that spans the distance between supports and is loaded primarily in a transverse direction; that is, normal to the axis of the beam. In a course of engineering mechanics, or statics, beam reactions are computed as an example of the use of equilibrium equations. The foundation for the study of beams should already exist; however, a review of statics is in order along with a discussion of the different types of support and loading conditions. The discussion will be limited to two-dimensional structures for the sake of simplicity in visualizing support conditions. The concept of shear and moment in beams will be introduced and thoroughly studied in two dimensions. A brief presentation on the use of vector analysis and shear and moment diagrams for three-dimensional structures is included at the end of the chapter.

5.2 SUPPORT AND LOADING CONDITIONS

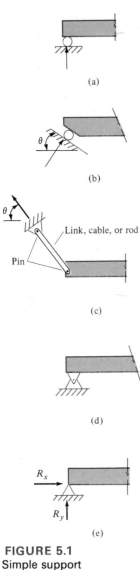

Support Conditions

Beams have been generically classified according to support conditions. There are two basic support conditions. At a point where a beam is to be supported, it is either restrained against displacement or rotation or both. Most current textbooks in engineering mechanics contain a thorough discussion concerning both two- and three-dimensional support conditions. Briefly, the important features of the various support conditions will be reviewed.

There are two primary classifications for support conditions in two dimensions: simple support and fixed support. Simple supports may be categorized further. Figure 5.1 illustrates the roller-type simple support. The reaction acts normal to the supporting surface. This support condition enables the analyst to know the direction of the support reaction. In Figs. 5.1a and b the reaction is illustrated normal to the supporting surface. In Fig. 5.1c the reaction would lie along the axis of the link. These support conditions require the evaluation of only one unknown. The roller-type support is used extensively in engineering. The roller support allows the structure to actually move in a direction normal to the reaction. This idea may seem somewhat dubious; however, the importance of this movement will eventually be explained. Figure 5.1d shows a second type of simple support, referred to as a pin or hinge. The pin will resist translation in any direction, as compared to the roller, which allows movement along the plane of the support, normal to the reaction. The pin support will have two unknown components of reaction, usually one in the vertical and one in the horizontal direction. Both the roller and the pin will allow the beam to rotate.

These simple support conditions can be viewed in actual use in some situations. The railroad bridge of Fig. 5.2a spans an interstate highway and is termed a continuous beam on five supports. The center support, between the highway lanes, is a pin-type support. As Fig. 5.2b illustrates, the support is rigidly connected to the pier and the bridge, thereby preventing any horizontal movement. The connection between the pier and the bridge is pinned or hinged, allowing for free rotation of the support. The remaining four supports are actually rockers as shown in Fig. 5.2c. The rocker is idealized as a roller in this text. The rocker allows the beam to move in the horizontal direction.

The fixed support is illustrated in Fig. 5.3 and is just as its name implies. It will resist movement in any direction in the plane and will also resist rotation of the beam. The fixed-support condition is very practical in structural engineering. It represents the situation when the beam is attached rigidly to a supporting structure. The fixed support

FIGURE 5.1
Simple support conditions.

(a)

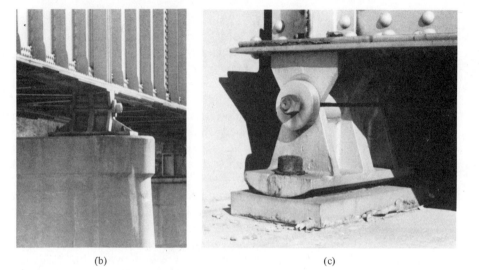

(b)   (c)

FIGURE 5.2   (a) Continuous-span bridge; (b) pin-type support; (c) rocker-type support.

may also be referred to as clamped or cantilevered, and will resist two components of displacement and any rotation of the support; hence, there are two force components and one moment component. Three equations of statics are required in order to completely define this support condition.

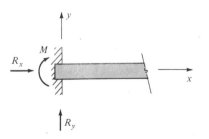

FIGURE 5.3
Fixed support.

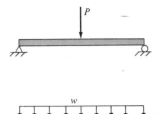

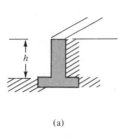

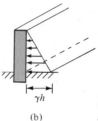

FIGURE 5.5
A type of fixed support.

(a)

(b)

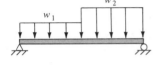

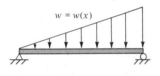

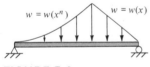

FIGURE 5.4
Loading conditions.

Loading Conditions

Loading conditions may be loosely divided into two categories—concentrated and distributed. A concentrated load is any type of loading that can be idealized as a single force acting at a point. A distributed loading is, as its name implies, some type of loading that is spread along the length of the beam and is usually approximated as a uniformly distributed load or uniformly varying load. Loading conditions are illustrated in Fig. 5.4. The magnitude of the load is equal to the area of the distributed load, and for the purpose of summing external statical moments the magnitude acts at the centroid of the distributed load.

A uniformly varying load would occur for the pressure distribution on a retaining wall holding an earth fill as illustrated in Fig. 5.5. Assume that the wall and the footing are cast in place as a single unit. The wall becomes a fixed-end beam with the pressure of the earth pushing to the left. The idealized structure is shown in Fig. 5.5b as a triangular loading acting on a fixed-end beam. The magnitude of the pressure loading at any distance h below the surface is γh, where γ is the unit weight of the soil.

Concentrated moments are sometimes applied to beams. In Fig. 5.6 the concentrated force P is applied to an extension of length a that is rigidly attached to the cantilever beam. The concentrated moment is represented schematically as shown in Fig. 5.6b.

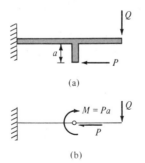

FIGURE 5.6
Example of a concentrated moment.

5.3 TYPES OF BEAMS AND BEAM REACTIONS

Beams have been classified according to their support conditions. It is well worth the effort for the reader to become familiar with the terms used, which are fairly standard. Structures may be placed in the broad categories of statically determinate or statically indeterminate. Statically determinate beams are supported in such a way that the reactions can be obtained using the equations of statics. Determinate and indeterminate beams are contrasted in Fig. 5.7. The beam of Fig. 5.7b is statically indeterminate and is usually referred to as a continuous beam. The continuous beam has five unknown external reactions, and there are only three equations of statics that can be applied to two-dimensional plane problems. Hence, the beam is said to be statically indeterminate to the second degree. The beam of Fig. 5.7a is determinate; there are three unknowns and three equations of statics. Any beam supported by a pin and a roller is called a simple beam or simply supported beam. The beam of Fig. 5.8a is a cantilever beam or fixed-free beam. Note, there are three external reactions and the cantilever beam is determinate. An overhanging beam is shown in Fig. 5.8b and is actually a form of the simple beam. The overhanging beam is statically determinate. The beam of Fig. 5.8c is referred to as a propped cantilever or a fixed-simple beam and is statically indeterminate to the first degree. The beam of Fig. 5.8d is fixed-fixed and is statically indeterminate to the third degree. Finally, the beam of Fig. 5.8e is a propped cantilever with a pin connection or hinge in the span. The pin connection physically represents a point of zero moment and may be used as a condition of equilibrium. There are four unknown reactions, three equations of statics, and the pin condition; hence, the beam of Fig. 5.8e is statically determinate.

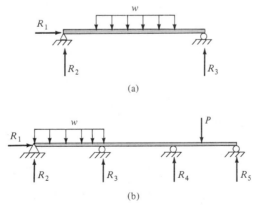

FIGURE 5.7

(a) A statically determinate beam; (b) a statically indeterminate beam.

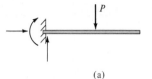

(a)

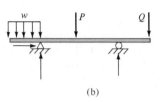

(b)

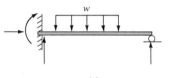

(c)

(d)

FIGURE 5.8
(a) A fixed-free beam; (b)
an overhanging beam; (c) a
propped cantilever or fixed-
simple beam; (d) a fixed-
fixed beam; (e) a propped
cantilever with internal pin.

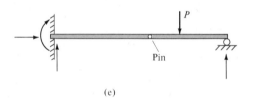

(e)

Several examples will illustrate the calculation of reactions for
statically determinate beams. These examples are planar or two-
dimensional problems similar to those discussed in statics. The equations of statics are applicable as

$$\sum F_x = 0 \qquad \sum F_y = 0 \qquad \sum M_z = 0 \tag{5.1}$$

EXAMPLE 5.1

Compute the reactions for the beam of Fig. 5.9 in terms of P, a, b, and L.

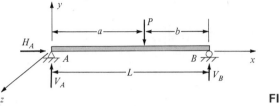

FIGURE 5.9

Solution:

The horizontal reaction at A is zero and can be verified by substituting into the first of Eqs. (5.1). Summing moments about point B gives

$$-V_A \cdot L + P \cdot b = 0$$

or

$$V_A = \frac{Pb}{L}$$

Summing forces in the y direction gives

$$V_A + V_B = P$$

Substituting V_A gives V_B as

$$V_B = P - V_A = P - \frac{Pb}{L} = \frac{P(L-b)}{L} = \frac{Pa}{L}$$

These results can be used in the future as a formula for computing the reaction for a simple beam subject to a concentrated load. ■

EXAMPLE 5.2

Compute the reactions and the force on the pin connection for the beam of Fig. 5.10.

Solution:

The section on the left, the overhanging end and up to the pin connection, will be viewed as the free body of Fig. 5.10b. The vertical force acting on the pin connection is V_C. Summing moments about point C gives the reaction V_A.

$$[(12 \text{ kN/m})(5 \text{ m})(5/2) \text{ m}] - (V_A \cdot 3 \text{ m}) = 0$$

$$V_A = 50 \text{ kN}$$

Summing forces on the free body verifies that there is no horizontal reaction at C and gives V_C as follows:

$$V_C = (12 \text{ kN/m})(5 \text{ m}) - 50 \text{ kN} = 10 \text{ kN}$$

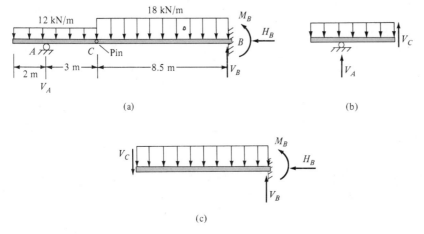

(a)

(b)

(c)

FIGURE 5.10

The remaining structure is shown in Fig. 5.10c, and the equations of statics give $H_B = 0$ and

$$V_B = V_C \text{ kN} + (18 \text{ kN/m})(8.5 \text{ m}) = 163 \text{ kN}$$

Summing moments about B gives

$$M_B + (10 \text{ kN})(8.5 \text{ m}) + (18 \text{ kN/m})(8.5 \text{ m})(8.5/2) \text{ m} = 0$$

$$M_B = -735.25 \text{ kN·m}$$

The negative sign indicates that the moment was assumed in the wrong direction on the free body and actually acts in the opposite direction. Additional discussion concerning sign conventions will follow in the next section. ■

EXAMPLE 5.3

Compute the reactions for the beam in Fig. 5.11a in terms of P, w, and L.

Solution:

$$\Sigma F_x = 0 \qquad H_A = \frac{4P}{5} \qquad \Sigma M_A = 0$$

$$\left(\frac{wL}{4}\right)\left(\frac{L}{8}\right) - \left(\frac{3P}{5}\right)\left(\frac{2L}{3}\right) + V_B L = 0$$

$$V_B = -\frac{wL}{32} + \frac{2P}{5}$$

$$\Sigma F_y = 0$$

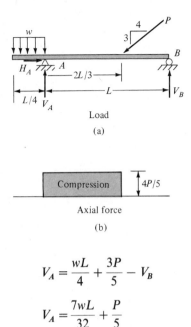

Load

(a)

Compression $4P/5$

Axial force

(b)   **FIGURE 5.11**

$$V_A = \frac{wL}{4} + \frac{3P}{5} - V_B$$

$$V_A = \frac{7wL}{32} + \frac{P}{5}$$

The horizontal and vertical reactions have been computed in terms of P, w, and L and the axial-force diagram is shown in Fig. 5.11b. ∎

EXAMPLE 5.4

Compute the reactions for the free-fixed beam of Fig. 5.12. The load is uniformly changing with the magnitude at the fixed end equal to w N/m.

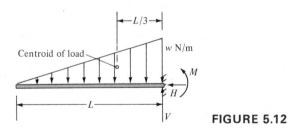

FIGURE 5.12

Solution:

$$\Sigma F_x = 0 \text{ gives } H = 0$$

$$\Sigma F_y = 0 \text{ gives } V = \text{area of the load} = \frac{wL}{2}$$

ΣM at the fixed end $= 0$ $\qquad$ $M + \left(\dfrac{wL}{2}\right)\left(\dfrac{L}{3}\right) = 0$

$M = -\dfrac{wL^2}{6}$ $\qquad$ (the negative sign indicates that the assumed moment should be reversed) $\qquad$ ■

EXAMPLE 5.5

Compute the reactions for the frame-type structure of Fig. 5.13.

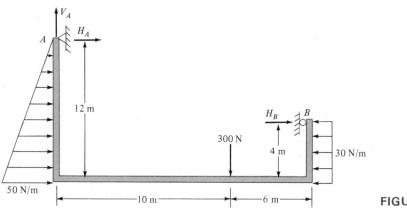

FIGURE 5.13

Solution:

V_A can be found by summing forces in the y direction.

$\qquad V_A = 300$ N

$\qquad M_A = 0$ will give H_B

$$\left[\frac{(50 \text{ N/m})(12 \text{ m})}{2}\right]\left[\frac{(2)(12 \text{ m})}{3}\right] - (300 \text{ N})(10 \text{ m})$$

$$- (30 \text{ N/m})(4 \text{ m})(12 \text{ m} - 2 \text{ m}) + H_B(12 \text{ m} - 4 \text{ m}) = 0$$

$\qquad H_B = 225$ N·m

$\qquad \Sigma F_x = 0$ will give H_A

$$H_A + \frac{(50)(12)}{2} - (30)(4) + 225 = 0$$

$\qquad H_A = -405$ $\qquad$ (the negative answer indicates H_A acts to the left) $\qquad$ ■

5.4 SIGN CONVENTION

Before proceeding with an in-depth study of shear and moment for beams, we should elaborate on the sign conventions. The external actions that cause shear and moment are represented in Chapter 1, Figs. 1.3b, d, and e, as force vectors acting adjacent to a cut section and moment vectors acting adjacent to a cut section. Since we will use vector representation for force and moment, as well as vector analysis, it is necessary to use a vector sign convention. The vector sign convention is dictated by the theory of vector analysis and we will simply apply the sign convention to mechanics of materials.

Historically, the sign convention for beam theory is based upon the physical behavior of a beam and may be termed a deformation-based sign convention. Actually, there are two deformation-type sign conventions. One is based upon the actual deformation of the beam (beam sign convention) while the other (joint sign convention) is based upon the rotation of the intersection of beams and columns. The beam sign convention is important in mechanics of materials, and the joint sign convention is used in advanced courses in structural theory.

The purpose is to show the connection between the beam sign convention and the vector sign convention. Consider the beam of Fig. 5.14a. The internal force is called the transverse shear or, commonly, just the shear. The sign convention for positive shear using the beam sign convention is shown on the free body of Fig. 5.14b. The positive face of the free body is the end that lies at the farthest point out the positive axis, end A in Fig. 5.14b. Positive shear, as illustrated in the figure, acts in the negative coordinate direction on the positive face of the free body. Admittedly, the sign convention appears inconsistent, but it is accepted by the engineering profession. *Note:* The free body is in force equilibrium and the reader must exercise care not to confuse the positive and negative ends of the free body.

Positive bending moments using the beam sign convention are illustrated in Fig. 5.14c. Positive moment causes compression at the top fibers of the beam where the top of the beam lies on the positive side of the coordinate axis. We note that the beam element is in moment equilibrium. The bending moment on both ends of the free body causes compression on their respective top edges. The bending moment that acts about the z axis is called M_z and the moment acting about the y axis is M_y.

The vector sign convention is shown in Figs. 5.15a and b. The shears, $V_y\mathbf{j}$ and $V_z\mathbf{k}$ are positive, as shown on the positive face of the free body. The moments are illustrated as vectors, $M_y\mathbf{j}$ and $M_z\mathbf{k}$, acting positive on the positive face of the free body. When comparing the sign conventions, we note that M_z acts the same for both sign conventions,

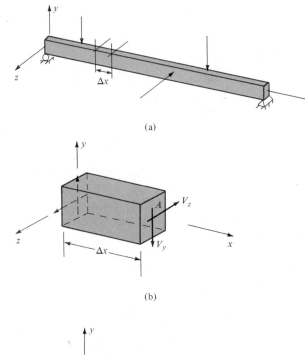

(a)

(b)

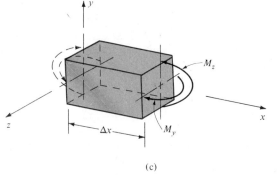

FIGURE 5.14
Positive shear and moment
for the beam sign
convention.

(c)

whereas the M_y moments appear to act in opposite directions. The vector sign convention is consistent with the coordinate system.

The beam sign convention is used primarily for two-dimensional problems that are cast in the x-y plane of Fig. 5.14. The next few sections will dwell mainly on the two-dimensional problems using the beam sign convention. The vector sign convention will be used to illustrate shear and moment equations for some elementary three-dimensional problems. We offer no rules for changing from one sign convention to the other, but suggest that it be accomplished using Figs. 5.14 and 5.15 and elementary free-body sketches. In the following discussion we demonstrate the use of both sign conventions and show the connection between the two.

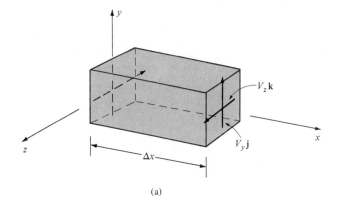

(a)

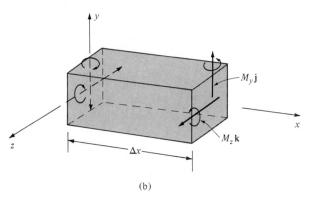

FIGURE 5.15
Positive shear and moment
for the vector sign
convention.

(b)

5.5 SHEAR AND MOMENT FOR BEAMS

Beam structures have been classified according to support conditions and loading conditions, external reactions have been computed using the basic equations of statics, and sign conventions have been discussed. The next question to be answered is: How does the reactive force vary from one support to the next? Again, the equations of statics are employed and free-body diagrams are used.

Consider the simply supported and uniformly loaded beam of Fig. 5.16a. We will write shear and moment equations that describe the behavior of internal shear and moment along the x coordinate. The coordinate origin is established at the left end of the beam and a free-body cut is made anywhere between $x = 0$ and $x = L$. The free body is shown in Fig. 5.16b and a positive shear and moment, using the beam sign convention, are assumed to act at the cut section. Force equilibrium for the free body gives the shear equation as follows:

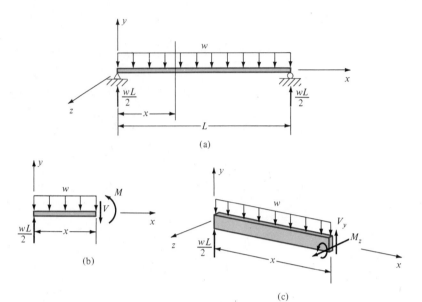

FIGURE 5.16

$$\frac{wL}{2} - wx - V = 0 \tag{a}$$

or

$$V = \frac{wL}{2} - wx \tag{b}$$

Equation (b) is the mathematical description of the internal shear between $x = 0$ and $x = L$.

The moment equation is obtained by writing the moment equilibrium equation for the free body. Moments can be summed about any point on the free body, but the most convenient point is at the cut section.

$$M + wx\left(\frac{x}{2}\right) - \frac{wLx}{2} = 0 \tag{c}$$

or

$$M = \frac{wLx}{2} - \frac{wx^2}{2} \tag{d}$$

Equation (d) will give the moment at any point x along the beam axis. We note that the shear and moment of Eqs. (b) and (d) are the internal reactive shear and moment.

Shear and moment equations using the vector sign convention are written in a similar manner, except that the shear and moment applied at the cut section must agree with the vector sign convention and are used as shown in Fig. 5.16c. Summing forces in the y direction gives the shear equation

$$V_y\mathbf{j} = wx\mathbf{j} - \frac{wL\mathbf{j}}{2} \tag{e}$$

Note that the sign of the reactive shear at the cut section is opposite to that of Eq. (b).

Summing the moments about the cut section gives

$$M_z\mathbf{k} + (-x\mathbf{i}) \times \left(\frac{wL}{2}\mathbf{j}\right) + \left(\frac{-x}{2}\mathbf{i}\right) \times (-wx\mathbf{j}) = 0 \tag{f}$$

$$M_z\mathbf{k} - \frac{wLx}{2}\mathbf{k} + \frac{wx^2}{2}\mathbf{k} = 0$$

$$M_z = \frac{wLx}{2} - \frac{wx^2}{2} \tag{g}$$

The moment equation using the vector sign convention is the same as the moment equation using the beam sign convention.

Once again, we emphasize that the action of the shear or moment must be interpreted using the convention defined in Figs. 5.14 and 5.15 and it is dependent on the sign convention that is being used.

There is a basic mathematical relationship between load and shear, and a similar relation between shear and moment. The beam of Fig. 5.17 is loaded in an arbitrary manner with a vertical load that is assumed to vary in a continuous manner. "Continuous" means that the load is described using a valid continuous mathematical function. Assume that a small element of length can be isolated as illustrated in Fig. 5.17b. The length of the element is Δx, and the positive vertical axis has been taken upward as shown. The continuous load $w(x)$ is assumed positive, and the shears and moments are assumed positive in accordance with the beam sign convention. On the left of the element both a shear V and a moment M are assumed. On the right of the element the shear has increased by an amount ΔV that can be thought of as the change in shear with respect to x along the length Δx. Similarly, the moment changes by an amount ΔM over the length Δx. Summing forces in the y direction gives

$$V - V - \Delta V + w\,\Delta x = 0 \tag{5.2}$$

or

$$w = \frac{\Delta V}{\Delta x}$$

such that in the limit, $x \rightarrow 0$,

$$dV = w\,dx \qquad\qquad (5.3)$$

Integrating over some length of beam—say, x_1 to x_2—gives

$$\int_{V_1}^{V_2} dV = \int_{x_1}^{x_2} w\,dx \qquad\qquad (5.4)$$

or

$$V_2 - V_1 = \int_{x_1}^{x_2} w\,dx = \begin{array}{l}\text{(area of the load diagram between} \\ x_1 \text{ and } x_2) \end{array} \qquad (5.5)$$

> Equation (5.5) is interpreted to mean that the area of the load diagram between any two points, x_1 and x_2, is equal to the change in shear between the same two points.

Similarly, summation of moments acting on the element of Fig. 5.17 about point A gives

$$-M + M + \Delta M - V\,\Delta x - w\,\Delta x\,\frac{\Delta x}{2} = 0 \qquad (5.6)$$

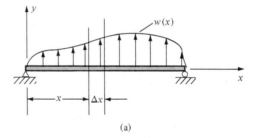

(a)

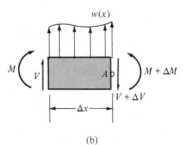

FIGURE 5.17

(b)

or

$$V = \frac{\Delta M}{\Delta x} - \frac{w\,\Delta x}{2} \tag{5.7}$$

and in the limit, as $x \to 0$,

$$dM = V\,dx \tag{5.8}$$

The term $\Delta x(\Delta x/2)$ of Eq. (5.6) is considered to be a high-order term, meaning that as a product of Δx it represents a term that is of lesser magnitude than terms containing a single Δx and it can be verified that in the limit of Eq. (5.7) the term will become zero. Integrating Eq. (5.8) over a length x_1 to x_2 gives

$$\int_{M_1}^{M_2} = M_2 - M_1 = \int_{x_1}^{x_2} V\,dx \tag{5.9}$$

> Equation (5.9) is interpreted to mean that the area of the shear diagram between any two points, x_1 and x_2, is equal to the change in moment between the same two points.

Equations (5.5) and (5.9) will be used to construct shear and moment diagrams for the beam of Fig. 5.18.

EXAMPLE 5.6

Construct shear and moment diagrams for the beam of Fig. 5.18a.

Solution:
The reactions are calculated using the equations of statics.

$$\Sigma M_A = 0 \qquad (40\text{ N})(10.5\text{ m}) = R_B\,8\text{ m}$$

$$R_B = 52.5\text{ N}$$

$$\Sigma F_y = 0$$

$$-R_A - 40\text{ N} + 52.5\text{ N} = 0$$

$$R_A = 12.5\text{ N}$$

The shear is plotted directly from the load diagram. The reaction R_A is 12.5 N down and is plotted downward on the shear diagram. Applying Eq. (5.5), the change in

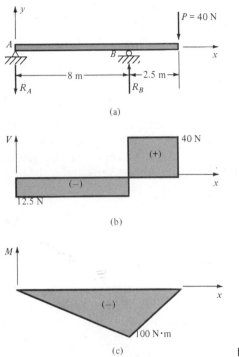

(a)

(b)

(c)

FIGURE 5.18

shear between points A and B is zero; hence, the diagrams indicate no change in shear, a horizontal line between A and B.

The change in shear at the reaction is equal to R_B and the change is 52.5 N and gives a shear of 40 N for the overhanging end. According to Eq. (5.9), the change in the moment is equal to the area of the shear diagram. Between points A and B the area of the shear diagram is -12.5 N times 8 m or -100 N·m. Starting from zero moment at A, the moment at B is -100 N·m as shown in the diagram. The change in moment between point B and the right end is equal to the area of the shear diagram, 40 N times 2.5 m, or 100 N·m. Hence, the moment at the right end is zero.

■

The question of how the moment varies between points A and B must be considered. Equations (5.3) and (5.8) are the key to understanding the relationship between load, shear, and moment. Equation (5.3) may be written

$$\frac{dV}{dx} = w \qquad\qquad (5.10)$$

and is interpreted as showing that the slope of the shear diagram is equal to the intensity of the load. The relation between load, shear, and moment is illustrated in Fig. 5.19 for three of the most common types of loading: concentrated, uniform, and uniformly varying (triangular). The concentrated-load diagram can be viewed as having $w = 0$ along the length a, a discontinuity where the load P is applied, and $w = 0$

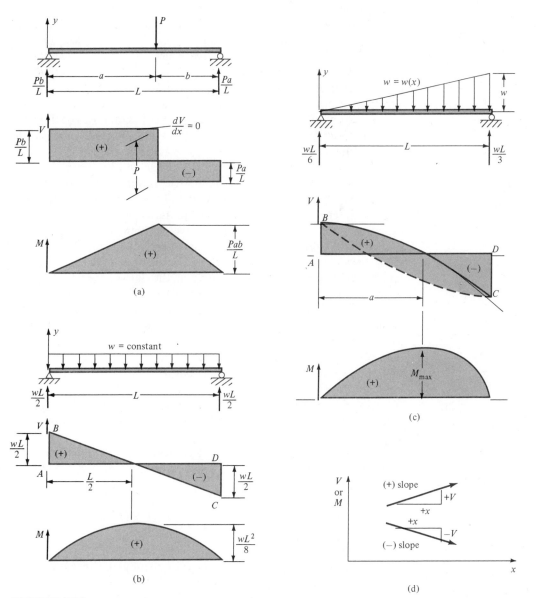

FIGURE 5.19
Comparison of (a) concentrated load; (b) uniform load; (c) uniformly varying load.

along length b. The corresponding shear diagram has $dV/dx = w = 0$ along the length a, a discontinuity of magnitude P and a slope of $dV/dx = 0$ along length b. The uniform load of Fig. 5.19b is constant and continuous along the entire length of the beam; hence, the slope of the shear diagram is constant and continuous. The diagram is plotted first by constructing the vertical ordinate $+wL/2$ and then by calculating the change in shear as the area of the load diagram, $-wL$. Since the load is negative, the change in shear is $wL/2 - wL = -wL/2$ at the right end. The right reaction is positive; hence, it is plotted upward and the final shear at the right end is zero. The sequence for plotting the diagram would be: construct a vertical line from A to B, locate C, and construct a straight line from B to C and a vertical line from C to D. The process of constructing shear and moment diagrams can be thought of as locating specific points on the diagram and connecting them with curves of the proper slope.

The uniformly varying load of Fig. 5.19c will have reactions as illustrated. The shear at the left is plotted upward. The total change in shear is $wL/2$ (the area of the load) and the shear at the right end is $-wL/3$. The reader must connect points B and C with a curve of changing slope, $dV/dx = w(x)$. The curve between B and C could be plotted by dividing the beam into increments of load and using Eq. (5.5). However, when merely sketching the diagram a curve between B and C could be either the solid curve or the dashed curve illustrated in the figure. The shape of the curve can be accurately approximated by considering Eq. (5.10). At the left end of the beam the load intensity is zero, indicating that dV/dx is zero. At the right end of the beam the load intensity is $-w$, meaning that the slope dV/dx should be negative and of value equal to w. Note, the solid curve begins with zero slope and ends with a negative slope. The dashed line would begin with a value for dV/dx and end with dV/dx equal to zero. The solid line is the obvious choice. Figure 5.19d illustrates both positive and negative slopes. The shear diagram for a uniformly varying load will be discussed in an example problem.

Moment diagrams are constructed using Eq. (5.8). Equation (5.9) is helpful in sketching the shape of the diagram. Similar to the shear equation, Eq. (5.8) can be written

$$\frac{dM}{dx} = V \tag{5.11}$$

Constructing a moment diagram using the area of the shear diagram is similar to the procedure of constructing the shear diagram using the load diagram. Referring to Fig. 5.19a, the moment at each end of the beam is zero because the moment at a simple support, *when it is located at the end of a beam*, is zero. The moment at $x = a$ is

Pab/L and varies with constant slope from $x = 0$ to $x = a$. Equation (5.11) indicates that dM/dx is constant and positive for $x = 0$ to $x = a$. According to Eq. (5.11), the slope of the moment diagram is constant and negative from $x = a$ to $x = L$. In this case the slope can be sketched and is an accurate representation of the moment variation, a straight line.

Figure 5.19*b* shows that the slope of the moment diagram between the left end and the center of the beam is uniformly changing from a positive maximum at the left end to zero at the center in exactly the same way as the intensity of the shear diagram. At the center of the beam, the shear changes sign and decreases to an extreme negative maximum value at the right end. The slope of the moment diagram behaves in the same way. The moment diagram can be sketched as a parabolic curve between $x = 0$ and $x = L$. However, the exact variation of the moment is not obtained.

The moment diagram illustrated for the beam of Fig. 5.19*c* is similar to that of Fig. 5.19*b*. The point of maximum moment does not occur at the center but at some point *a* measured from the left end. Locating the point of maximum moment will be dealt with later in this chapter. The moment diagram begins with zero at the left end and varies with positive decreasing slope until point *a* is reached. Note that the magnitude of the shear is positive and varies from $wL/6$ to zero along that part of the beam. At $x = a$ the shear becomes negative and increases negatively to the right end of the beam. The slope of the moment is similar. Again, the moment diagram has been sketched and is not an exact representation. The following rule will apply for the beam sign convention used in this text.

$$
\text{If the shear is positive}\begin{pmatrix}\text{constant}\\\text{increasing}\\\text{decreasing}\end{pmatrix}\text{the slope of the moment dia-}
$$

$$
\text{gram will be positive}\begin{pmatrix}\text{constant}\\\text{increasing}\\\text{decreasing}\end{pmatrix}
$$

$$
\text{If the shear is negative}\begin{pmatrix}\text{constant}\\\text{increasing}\\\text{decreasing}\end{pmatrix}\text{the slope of the moment dia-}
$$

$$
\text{gram will be negative}\begin{pmatrix}\text{constant}\\\text{increasing}\\\text{decreasing}\end{pmatrix}
$$

If the word *shear* is replaced with *load* and *moment* is replaced with *shear*, a rule relating *load* and *shear* will result.

The reader must master the shear and moment equation, especially moment equations. Throughout this textbook the moment equation is used as an integral part of other concepts. Thus far, moment equations have been referenced to a coordinate that has been assumed at the left end of the beam, with the x axis positive to the right. It is sometimes necessary or convenient to place the origin of the coordinate system at the right end of the beam or possibly somewhere between the extreme ends of the beam. If the origin is placed at the right end, the positive x axis would then extend to the left. The following simple rules should enable the reader to feel comfortable with any coordinate system.

1. Select the origin, any point on the beam.

2. Select a free-body diagram; cut the beam anywhere along the section where the equation is to be written. The free body should include the entire beam either to the left or right of the cut. Choose the free body that contains the coordinate origin.

3. Visualize the free body as a fixed free beam with the fixed end corresponding to the point where the cut was made. Apply a positive shear and moment at the cut section, which would correspond to the shear reaction and resisting moment at a fixed end. Sum forces and moments at the cut section. Solve for the unknown shear to obtain the shear equation. The moment equation will correspond to an equation that can be solved for the unknown moment.

To illustrate the foregoing, several examples will be presented using various locations for the coordinate origin.

EXAMPLE 5.7

---

a. Write shear and moment equations for the beam of Fig. 5.20*a* locating the origin at *A*.

b. Write shear and moment equations for the beam of Fig. 5.20*a* locating the origin at *C*.

Solution:

a. The beam is cut at section 1-1 and the left-hand free body is used since it contains the origin. The free body appears as in Fig. 5.20*b*. Imagine that the right end is fixed and V_x and M_x correspond to the reaction and resisting moment for a fixed support.

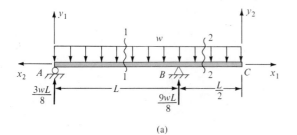

(a)

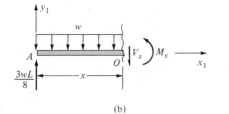

(b)

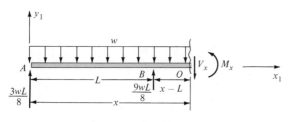

(c)

FIGURE 5.20

$$\Sigma F_y = 0$$

$$\frac{3wL}{8} - wx - V_x = 0$$

$$V_x = \frac{3wL}{8} - wx$$

$$\Sigma M_O = 0 \qquad \text{(with respect to point } O\text{)}$$

$$M_x - \frac{3wLx}{8} + \frac{wx^3}{2} = 0$$

$$M_x = \frac{3wLx}{8} - \frac{wx^2}{2}$$

Figure 5.20c is the required free body for writing the shear and moment equation between B and C. Summation of forces gives, for $L \leq x \leq 3L/2$,

$$\frac{3wL}{8} + \frac{9wL}{8} - wx - V_x = 0$$

$$V_x = \frac{3wL}{2} - wx$$

$$\Sigma M_O = 0 \qquad \text{for } L \le x \le \frac{3L}{2}$$

$$M_x - \frac{3wLx}{8} - \frac{9wL(x - L)}{8} + \frac{wx^2}{2} = 0$$

$$M_x = \frac{3wLx}{2} - \frac{9wL^2}{8} - \frac{wx^2}{2}$$

b. The origin of the coordinate system is shown at point *C* in Fig. 5.21 and labeled x_2, y_2. The positive *x* axis is assumed to the left. The free body corresponding to section 1-1 is shown in Fig. 5.21*a*. Note that M_x and V_x are applied in the positive sense using the beam sign convention.

$$\Sigma F = 0 \qquad \text{for } \frac{L}{2} \le x \le \frac{3L}{2}$$

$$V_x = wx - \frac{9wL}{8}$$

$$\Sigma M_O = 0 \qquad \text{for } \frac{L}{2} \le x \le \frac{3L}{2}$$

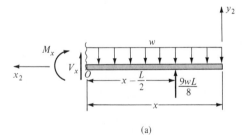

(a)

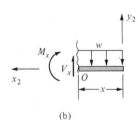

(b)   **FIGURE 5.21**

$$M_x = \frac{9wL(x - L/2)}{8} - \frac{wx^2}{2} = \frac{9wLx}{8} - \frac{9wL^2}{16} - \frac{wx^2}{2}$$

The free body of section 2-2 is shown in Fig. 5.21b.

$$\Sigma F_y = 0 \qquad 0 \le x \le \frac{L}{2}$$

$$V_x = wx$$

$$\Sigma M_O = 0 \qquad 0 \le x \le \frac{L}{2}$$

$$M_x = \frac{-wx^2}{2}$$ ∎

Shear and moment equations can be written for beam sections with the coordinates at any location. Shear and moment diagrams should be constructed beginning at the left end of the beam and progressing toward the right since the differential relations of Eqs. (5.5) and (5.9) are based upon a coordinate system oriented at the extreme left of the beam.

The concept of relating load, shear, and moment has been discussed. Shear and moment equations have been introduced as an application of the equations of statics. These topics would not be complete unless the relation between zero shear and maximum moment was included. Again consider Eq. (5.8):

$$\frac{dM}{dx} = V \qquad\qquad\qquad \textbf{(5.11)}$$

When the shear is zero, the slope on the moment diagram is zero. Consider the beam of Fig. 5.19b. The shear passes through zero at the center of the beam; similarly, the moment is maximum at the center of the beam as is the slope dM/dx zero at the center of the beam. Note that the moment is positive increasing up to the center, then changes to positive decreasing. When the moment is maximum, the slope dM/dx must be zero.

For some loadings the point of maximum moment is easily located, sometimes even by inspection; however, for a beam such as the one of Fig. 5.19c such is not the case. A sketch of the moment diagram will not give the exact location of maximum moment. However, a shear equation can be written, which when set equal to zero will yield the location of zero shear and the point of maximum moment. This will be illustrated using the beam of Fig. 5.19c.

EXAMPLE 5.8

Write shear and moment equations for the beam of Fig. 5.19c. Locate the point of zero shear and compute the maximum moment.

Solution:
Refer to Fig. 5.19c. Locate a coordinate origin at the left end and draw a free-body diagram as shown in Fig. 5.22. The magnitude of load at any location x can be found using similar triangles as $w_x/x = w/L$. The free body is shown in Fig. 5.22b. Summing forces yields the shear equation as

$$V = \frac{wL}{6} - \left(\frac{wx}{L}\right)(x)\left(\frac{1}{2}\right) \tag{a}$$

Summing moments at A gives the moment equation

$$M = \frac{wLx}{6} - \left(\frac{wx}{L}\right)(x)\left(\frac{1}{2}\right)\left(\frac{x}{3}\right) = \frac{wLx}{6} - \frac{wx^3}{6L} \tag{b}$$

To find the value of x where the shear is zero, let $V = 0$ in Eq. (a) and solve for x.

$$\frac{wL}{6} - \frac{wx^2}{2L} = 0$$

$$x = \pm\left(\frac{L^2}{3}\right)^{1/2} = \pm\frac{\sqrt{3}L}{3}$$

$$a = \frac{+\sqrt{3}L}{3} \qquad \text{(see Fig. 5.19c)} \tag{c}$$

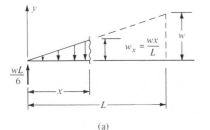

(a)

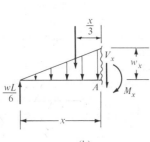

(b)

FIGURE 5.22

Choose the positive value, since the negative value of x is to the left of the origin. The magnitude of the maximum moment is found by substituting a for x in Eq. (b), as follows:

$$M_{\max} = \left(\frac{wL}{6}\right)\left(\frac{\sqrt{3}L}{3}\right) - \left(\frac{w}{6L}\right)\left(\frac{\sqrt{3}L}{3}\right)^3$$

$$M_{\max} = \frac{wL^2\sqrt{3}}{27} = 0.064wL^2 \qquad \text{at } x = 0.577L \qquad \blacksquare$$

The essentials of shear and moment equations and diagrams are complete. The following examples will illustrate the application.

EXAMPLE 5.9

Construct shear and moment diagrams, write shear and moment equations, locate the point of maximum moment, and compute the maximum moment for the beam of Fig. 5.23.

Solution:
Calculate the reactions

$$\Sigma M_B = 0$$

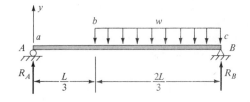

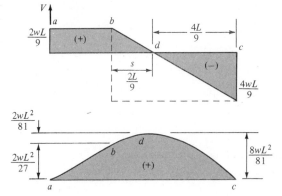

FIGURE 5.23

$$R_A L - \left(\frac{2wL}{3}\right)\left(\frac{2L}{6}\right) = 0$$

$$R_A = \frac{2wL}{9}$$

$$\Sigma F_y = 0$$

$$R_B + R_A - \frac{2wL}{3} = 0$$

$$R_B = \frac{4wL}{9}$$

Plot the shear as follows: R_A is positive (upward); hence, the shear at a is $2wL/9$. Between points a and b the load is $w = 0$; hence $dV/dx = 0$, meaning there is no change in shear. The shear changes between points b and c by an amount equal to the load ($-2wL/3$). At point c the shear is $-4wL/9$; an upward reaction R_B of $4wL/9$ plotted upward returns the shear to zero.

The moment is zero at A and B. Beginning on the left, the area of the shear diagram between a and b is positive and is the change in moment between a and b.

$$\Delta M_{a \to b} = \left(\frac{2wL}{9}\right)\left(\frac{L}{3}\right) = \frac{2wL^2}{27}$$

Since the slope is constant, plot the moment at point b and connect a and b with a straight line. The positive triangular shear area between points b and d is the next increment of change in moment. The distance s must be computed before the area can be evaluated. Formally, this should be done by writing a shear equation and finding the point of zero shear. Actually, for a triangular shear area, similar triangles can be used incorporating the large dashed triangle of Fig. 5.23 as

$$\frac{s}{2L/3} = \frac{2wL/9}{(2wL/9) + (4wL/9)}$$

$$s = \frac{2L}{9}$$

Subsequently, s will be found using the shear equation. The area between b and d is

$$\Delta M_{b \to d} = \left(\frac{2wL}{9}\right)\left(\frac{2L}{9}\right)\left(\frac{1}{2}\right) = \frac{2wL^2}{81}$$

The moment at point d is

$$M_{a \to b} + M_{b \to d} = \frac{2wL^2}{27} + \frac{2wL^2}{81} = \frac{8wL^2}{81}$$

The moment varies between b and d with continually decreasing positive slope. The slope is zero at point d. The shear area between d and c is

$$\Delta M_{d \to c} = \left(\frac{-4wL}{9} \right) \left(\frac{4L}{9} \right) \left(\frac{1}{2} \right) = \frac{-8wL^2}{81}$$

The moment diagram between d and c appears as in the figure. The increment is just enough to produce zero moment at the right end. The moment diagram must be balanced. In other words, the moment must return to zero at the right end.

The origin indicated in the figure will be used to write the shear and moment equations. Select a free body, cut anywhere between a and b as shown in Fig. 5.24a.

$$\Sigma F_y = 0 \qquad V_x = \frac{2wL}{9} \tag{a}$$

$$\Sigma M = 0 \qquad M_x = \frac{2wLx}{9} \tag{b}$$

The same origin will be used for equations between b and c. The free body is shown in Fig. 5.24b.

$$\Sigma F_y = 0 \qquad -V_x + \frac{2wL}{9} - w\left(x - \frac{L}{3} \right) = 0$$

$$V_x = \frac{5wL}{9} - wx \tag{c}$$

$$\Sigma M = 0 \qquad M_x - \frac{2wLx}{9} + w\left(x - \frac{L}{3} \right) \left(\frac{x - L/3}{2} \right)$$

$$M_x = -\frac{wL^2}{18} + \frac{5wLx}{9} - \frac{wx^2}{2} \tag{d}$$

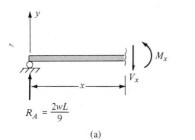

(a)

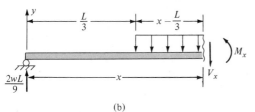

(b)   **FIGURE 5.24**

The point of zero shear is found by setting Eq. (c) equal to zero and solving for x.

$$0 = \frac{5wL}{9} - wx$$

$$x = \frac{5L}{9} \quad \text{(measured from the origin)}$$

The maximum moment would be obtained by substituting $x = 5L/9$ into Eq. (d).

∎

EXAMPLE 5.10

Construct shear and moment diagrams, write shear and moment equations, locate the point of maximum moment, and compute the maximum moment for the beam of Fig. 5.25.

Solution:

The reactions are computed as

$$R = (60 \text{ N/m})(5 \text{ m})\left(\frac{1}{2}\right) + 250 \text{ N} = 400 \text{ N}$$

$$M = \left(60 \cdot \frac{5}{2}\right)\left(1.5 + 2 + \frac{5}{3}\right) + (250)(1.5) = 1150 \text{ N·m}$$

In this problem, shear and moment equations will be written prior to constructing diagrams.

Section AB: The free body is shown in Fig. 5.25d.

$$V_x = -\left(\frac{60x}{5}\right)\left(\frac{x}{2}\right) = -6x^2 \text{ N} \tag{a}$$

$$M_x = -\left(\frac{60x}{5}\right)\left(\frac{x}{2}\right)\left(\frac{x}{3}\right) = -2x^3 \text{ N·m} \tag{b}$$

Section BC: The free body is shown in Fig. 5.25e.

$$V_x = -60\left(\frac{5}{2}\right) = -150 \text{ N} \tag{c}$$

$$M_x = -60\left(\frac{5}{2}\right)\left(x - \frac{10}{3}\right) = -150x + 500 \tag{d}$$

Section CD: The free body is shown in Fig. 5.25f.

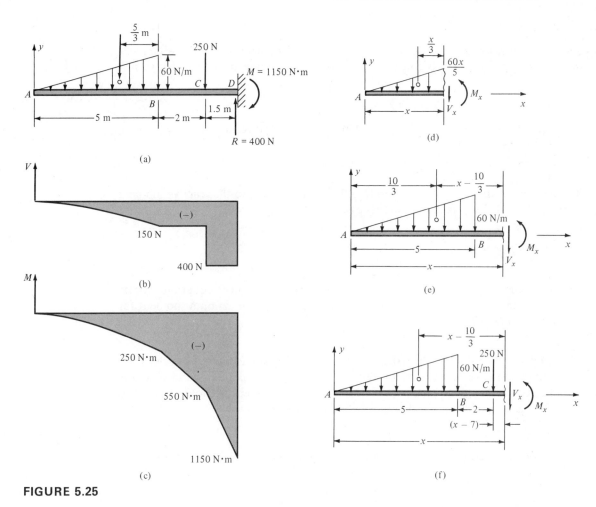

FIGURE 5.25

$$V_x = -60\left(\frac{5}{2}\right) - 250 = -400 \text{ N} \qquad \text{(e)}$$

$$M_x = -60\left(\frac{5}{2}\right)\left(x - \frac{10}{3}\right) - 250(x - 7) = -400x + 2250 \qquad \text{(f)}$$

The shear is zero at point A and changes by an amount equal to the area of the load between points A and B or $(60)(5/2) = 150$ N. The shear curve between A and B is concave downward as illustrated. The shear between B and C is constant since the load is zero. This fact is demonstrated by Eq. (c). An abrupt change occurs at point C, equal to -250 N. The shear is again constant between C and D.

The moment is zero at point A and changes by an amount equal to the shear area between A and B. The area of the parabolic portion is

$$M_{A\text{-}B} = -(5 \text{ m})(150 \text{ N})(1/3) = -250 \text{ N·m}$$

and can be verified by substituting $x = 5$ m into Eq. (b). The shape of the moment diagram may be sketched as concave downward. The change in moment between B and C is

$$M_{B\text{-}C} = (-150 \text{ N})(2 \text{ m}) = -300 \text{ N·m}$$

and gives the moment at point C as $-250 - 300 = -550$ N·m. The area between C and D is

$$M_{C\text{-}D} = (-400)(1.5) = -600 \text{ N·m}$$

The moment at the right end adds up to -1150 and is equal to the reactive moment at the right end. The maximum moment occurs at the right end. ∎

EXAMPLE 5.11

Construct shear and moment diagrams, write shear and moment equations, locate the maximum moment, and compute the magnitude of the maximum moment for the beam of Fig. 5.26.

Solution:

The reactions are computed as indicated. The shear diagram is plotted directly from the load diagram. Note that the shear is constant between points A and C and is not affected by the applied moment. The increment of moment between A and B is equal to the area of the shear diagram between A and B, or

$$M_B = (3 \text{ m})(-96 \text{ N}) = -288 \text{ N·m}$$

The applied moment changes the moment diagram abruptly by 200 N·m; therefore

$$M_B = -288 \text{ N·m} + 200 \text{ N·m} = -88 \text{ N·m}$$

The moment at C is

$$M_C = M_B - (96)(2) = -88 - 192 = -280 \text{ N·m}$$

Similarly,

$$M_D = -280 + (140)(2) = 0$$

The free bodies for shear and moment equations are shown in Figs. 5.26d, e, and f.

Section AB: See Fig. 5.26d.

$$V_x = -96 \text{ N}$$

$$M_x = -96x$$

Section BC: See Fig. 5.26e.

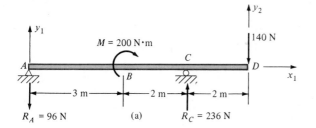

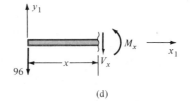

(a)

$R_A = 96$ N $\qquad R_C = 236$ N

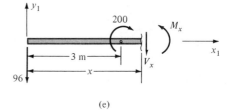

(d)

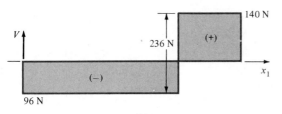

(b)

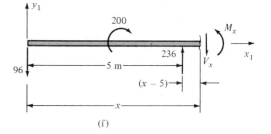

(e)

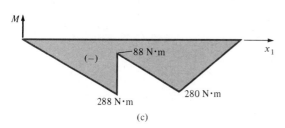

(c)

(f)

FIGURE 5.26

$$V_x = -96 \text{ N}$$

$$M_x = -96x + 200$$

Section CD: See Fig. 5.26*f*.

$$V_x = -96 + 236 = 140 \text{ N}$$

$$M_x = -96x + 200 + 236(x - 5)$$

$$M_x = 140x - 980$$

The maximum moment occurs at 3 m from the left end and is observed to be 288 N·m, as diagrammed in Fig. 5.26*c*. ∎

EXAMPLE 5.12

Construct shear and moment diagrams, write shear and moment equations, locate the point of maximum moment and compute the maximum moment for the structure of Fig. 5.27a.

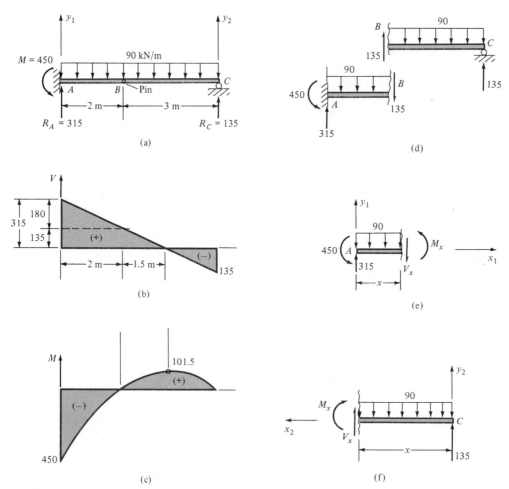

FIGURE 5.27

Solution:

The free body of Fig. 5.27d can be used to compute the reactions. The reader should be able to follow the construction of the shear and moment diagrams with no difficulty. Shear and moment equations will be written using the free-body diagrams of Figs. 5.27e and f. The beam segment between A and B is shown in Fig. 5.27e with the origin located at A.

$$V_x = 315 - 90x \tag{a}$$

$$\Sigma M = 0 \qquad M_x = -450 + 315x - \frac{90x^2}{2} \tag{b}$$

Notice that the hinge does not affect the continuity of the load. Equations (a) and (b) are valid for the entire length of the beam. In order to demonstrate the use of a different coordinate origin, refer to Fig. 5.27f. The origin is located at the right end of the beam. The shear and moment equations between C and A are

$$\Sigma F_y = 0 \qquad V_x = 90x - 135 \tag{c}$$

$$\Sigma M = 0 \qquad M_x = 135x - \frac{90x^2}{2} \tag{d}$$

The point of maximum moment can be obtained from either Eq. (a) or Eq. (c). Equation (a), for $V = 0$, yields $x = 315/90 = 3.5$ m, from the left end. Substituting into Eq. (b) gives

$$M_{max} = -450 + 315(3.5) - 90(3.5)^2 = 101.5 \text{ N·m}$$

Equation (c), for $V = 0$, yields $x = 135/90 = 1.5$ m, from the right end. Substituting into Eq. (d) gives

$$M_{max} = 135(1.5) - \frac{90(1.5)^2}{2} = 101.5 \text{ N·m}$$

and agrees with the maximum moment as computed using Eq. (b).   ■

EXAMPLE 5.13

Draw axial-force, shear, and moment diagrams for the frame of Fig. 5.28.

Solution:
Note that the support at A is a pin, and a roller is located at D. Hence, the frame is determinate. Statical summation of moments at A will give the vertical reaction at D.

$$(200)(3) + (400)(7) - (150)(12)(1/2)(12)(1/3) + V_D(10) = 0$$

$$V_D = 20 \text{ k}$$

Summation of forces in the y direction gives

$$V_A - 200 - 400 - 20 = 0$$

$$V_A = 620 \text{ k}$$

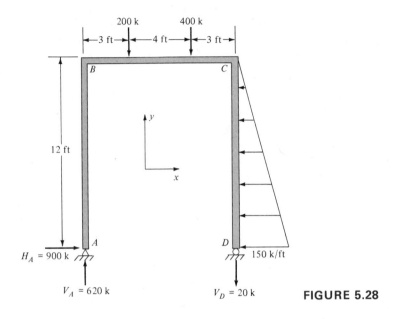

FIGURE 5.28

Finally, summing forces in the x direction gives H_A.

$$H_A = (150)(12)(1/2) = 900\,\text{k}$$

The diagrams will be constructed beginning at A and proceeding around the frame to B, C, and D. The x-y coordinates for member AB are illustrated in Fig. 5.29a; y is positive to the left and x is positive along the axis of AB.

The axial force is compressive and equal to the reaction at A, or 620 k. The diagram is plotted to the left of the member AB.

The shear is plotted to the left of the axial force diagram and is -900 k. Viewing the member AB with respect to the coordinate axis for member AB would indicate that the shear of 900 k is plotted below the axis. The load on the member AB is zero; hence there is no change in the shear between A and B.

The moment at the hinge support at A is zero. Hence the moment begins from zero and changes by an amount equal to the area of the shear diagram.

$$M_B = (-900)(12) = -10,800\,\text{ft·k}$$

The moment varies with constant slope between A and B.

The free-body diagrams of member AB, the joint at B, and member BC are shown in Fig. 5.29b. At A, the axial force and reaction, H_A, are acting on the structural member. At the top of member AB, equilibrium requires that a compressive force of 620 k act as shown, a shear of 900 k acts to the left and a negative moment of 10,800 ft·k acts at B. Note, using the coordinates for AB, that a negative moment causes compression on the right side of the column. The axial force is transferred through the joint and becomes a shear force acting on member BC. The shear of

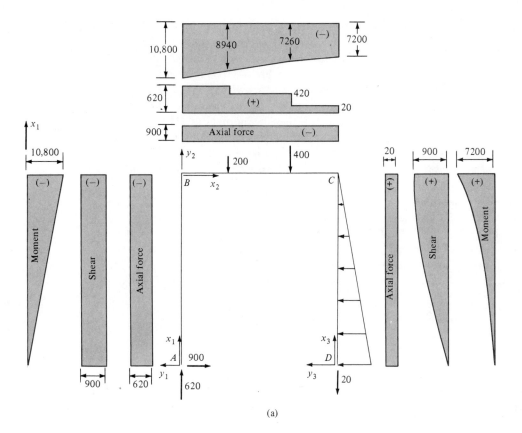

(a)

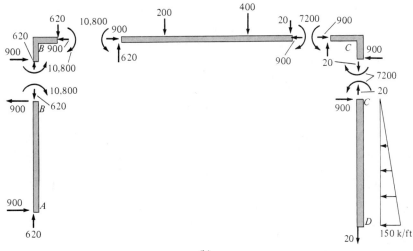

(b)

FIGURE 5.29

900 k acting at B on member AB becomes a compressive axial force acting on member BC. The moment at B, shown negative at B, is applied to the free body of the joint in an equal and opposite manner. The free body of the joint, according to the basic laws of statics, must be in equilibrium. Hence, the counterclockwise moment applied to the vertical leg of the joint must be balanced by a clockwise moment acting on the horizontal leg of the joint. The free body of member BC will then have an equal and opposite moment applied at B—in particular, the ccw moment at B of member BC. According to the beam sign convention, the moment is negative on BC using the x_2-y_2 coordinates. The diagrams for member BC will be plotted from B to C beginning with a compressive axial force of 900 k, a positive shear of 620 k, and a negative moment of 10,800 ft·k.

The axial-force diagram is shown directly above the frame in Fig. 5.29a. The shear diagram is above the axial-force diagram. A positive shear of 620 k is projected upward at B and is constant for 3 ft. The downward load of 200 k is plotted down to give a shear of 420 k. The 400-k load is plotted downward at 7 ft from B; then at C the shear of 20 is plotted downward, and the shear diagram is complete.

The moment diagram is obtained from the area of the shear diagram. The moment, 3 ft from B, is

$$-10,800 + (620)(3) = -8940 \text{ ft·k}$$

The moment at 7 ft from B is

$$-8940 + (420)(4) = -7260 \text{ ft·k}$$

The moment at point C is

$$-7260 + (20)(3) = -7200 \text{ ft·k}$$

The equilibrium of joint C is shown in Fig. 5.29b. The compressive axial force of member BC becomes a shear of 900 k acting on member DC at point C. The axial force acting on CD is 20 k tension. The moment is illustrated as it acts on BC at C. Equilibrium of joint C is shown, and the moment acting on member CD is positive at C using the x_3-y_3 coordinate system. The use of the beam sign convention for both vertical and horizontal members causes a peculiar behavior of the moment at the intersection of vertical and horizontal members. At B the negative moment on AB is negative as it acts on BC. However, the negative moment at C of member BC is positive when transferred to C of member CD. Actually, the choice of the origin at D causes the change in sign. It is customary in engineering practice to use a coordinate system similar to those at A and D when analyzing frame structures.

The coordinate origin is located at D and the diagram will be constructed from D to C. The axial-force diagram is constant as illustrated. The transverse shear is zero at D and varies according to the area of the load diagram from D to C.

$$V = \frac{(150)(12)}{2} = 900 \text{ k}$$

At C the shear changes abruptly by 900 k and returns to zero, or the shear diagram is in equilibrium. The slope of the shear diagram corresponds to the magnitude of the load. Hence, since the load is positive and approaches zero from D to C, the slope of the shear diagram behaves in a similar way.

The moment is zero at D and changes between D and C by

$$M_C = (900)(12)(2/3) = +7200 \text{ ft} \cdot \text{k}$$

The moment of 7200 ft·k at joint C balances the moment acting at the joint and the moment diagram is correct. Again, the slope of the moment diagram is determined from the magnitude of the shear diagram. ∎

5.6 STRUCTURAL MEMBERS LOADED IN BOTH TRANSVERSE DIRECTIONS

Structural members may be subject to a variety of loads. Thus far, beams have been somewhat idealized as a plane structure with the loads occurring in the plane of the beam. Actual beams are three-dimensional structures. The beam section of Fig. 5.30 is somewhat idealized. Actual beams might be rectangular, but could also have the shape of an I, a channel, or some arbitrary built-up shape. For the purpose of this discussion, the rectangular cross section will illustrate the concepts.

The beam of Fig. 5.30 is loaded vertically with a load P and horizontally with a load Q. Both loads may be termed transverse; both produce shear and bending moment. The combination of these loads produces unsymmetrical bending. Later in the text the idea of unsymmetrical bending will be dealt with in terms of the stresses produced in the beam. For the current discussion, shear and moment diagrams

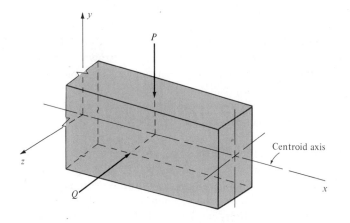

FIGURE 5.30
Beam loaded in both transverse directions.

will be constructed for this type of loading. However, two important assumptions will be made:

1. The cross section of the beam is symmetrical with respect to the axes of both cross sections; hence, the centroid lies at the center of the cross section.

2. The line of action of all forces or loading passes through the centroid of the cross section.

The discussion will be limited to the construction of shear and moment diagrams and equations. The basic method will be illustrated by the following example.

EXAMPLE 5.14

The beam of Fig. 5.31 is loaded as illustrated. Construct shear and moment diagrams and write shear and moment equations.

Solution:
Vector analysis could be used to advantage to calculate reactions and write shear and moment equations for this example problem. However, this solution will be presented using a scalar approach, and vector analysis will be discussed in the next section. The reactions will be computed first. Summing moments about a z axis at point A gives the reaction at B in the x-y plane.

$$\Sigma M_{Az} = 0 = 14V_{By} - (4)(105) \qquad \Sigma M_{By} = 0 = 14V_{Az} - (21)(6)(3)$$

$$V_{By} = 30 \text{ k} \qquad\qquad V_{Az} = 27 \text{ k}$$

$$\Sigma F_y = 0 = V_{Ay} + 30 - 105 = 0 \qquad \Sigma F_z = 0 = V_{Bz} + 27 - (21)(6)$$

$$V_{Ay} = 75 \text{ k} \qquad\qquad V_{Bz} = 99 \text{ k}$$

The shear and moment diagrams for the x-y plane are drawn below the beam in the usual manner. The diagrams for the x-z plane are drawn in a similar fashion and lie in the x-z plane. When constructing the diagrams the plot should begin at the origin and proceed along the positive axial coordinate. The sign convention for the x-z coordinates is the same as x-y coordinates. The uniform loading in the x-z plane acts in the negative z direction and the reaction at A, V_{Az}, acts in the positive z direction. The free body of Fig. 5.31b illustrates positive shear acting on the cut section. This is in accordance with the sign convention illustrated in Fig. 5.14.

Positive moment will act to cause compression at the positive edge of the beam as illustrated in Fig. 5.31b. Once the sign orientation is established, the shear diagram is constructed using Eq. (5.5), and the moment diagram is based upon Eq. (5.9).

Shear and moment equations can be written using the rules that were previously established. The free bodies of Figs. 5.31c and d illustrate positive shear and moment

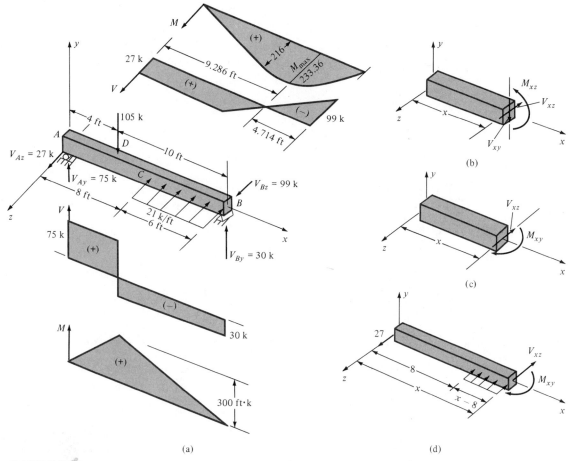

(a)   (d)

FIGURE 5.31

acting on a cut section. Note that the notation M_{xz} is shown as a positive moment at location x acting about the z axis. If M_{xz} were represented as a vector, it would be parallel to the z direction. The equations are obtained using the equations of statics.

$$\Sigma F_y = 0 \qquad V_{xy} = 75 \text{ k} \qquad (0 \le x \le 4) \tag{a}$$

$$V_{xy} = 75 - 100 = -30 \qquad (4 \le x \le 14) \tag{b}$$

Similarly,

$$\Sigma F_z = 0 \qquad V_{xz} = 27 \text{ k} \qquad (0 \le x \le 8) \tag{c}$$

$$V_{xz} = 27 - (21)(x - 8) = 195 - 21x \qquad (8 \le x \le 14) \tag{d}$$

Note the free body of Fig. 5.31d.

Figure 5.31*b* can be used as an aid to write the moment equations.

$$\Sigma M_z = 0 \qquad M_{xz} = 75x \qquad (0 \le x \le 4) \tag{e}$$

$$M_{xz} = 75x - (105)(x - 4) = 420 - 30x \qquad (4 \le x \le 14) \tag{f}$$

Visualize the moment about the *y* axis of Figs. 5.31*c* and *d*.

$$\Sigma M_y = 0 \qquad M_{xy} = 27x \qquad (0 \le x \le 8) \tag{g}$$

$$M_{xy} = 27x - \frac{(21)(x - 8)(x - 8)}{2}$$

$$= -672 + 195x - 10.5x^2 \qquad (8 \le x \le 14) \tag{h}$$

The maximum moment M_{xy} is located at the point where V_{xz} is zero. Equation (d) is solved for *x* when V_{xz} equals zero to give $x = 9.29$ ft. Substituting into Eq. (h) gives

$$M_{xy\,\text{max}} = -672 + (168)(9.29) - (10.5)(9.29)^2 = 233.36 \text{ ft} \cdot \text{k} \qquad \blacksquare$$

5.7 VECTOR ANALYSIS APPLIED TO THREE-DIMENSIONAL STRUCTURES

The plane structures of the preceding section are easily analyzed without the aid of vector analysis. However, when the loading is out-of-plane or when the structure is three-dimensional, the vector approach will usually expedite writing axial-force, shear, and moment equations. Vector shear and moment equations will be written for the simply supported beam of Fig. 5.32*a*. A uniform load is applied along the beam acting in the negative *y* direction. A concentrated force, *P*, acts in the *x* direction and is applied at the midpoint of the beam. We will first compute reactions and then write the shear and moment equations.

EXAMPLE 5.15

Use vector analysis to compute reactions and write shear and moment equations for the beam of Fig. 5.32.

Solution:
Assume positive reactions at *O* and *A* as illustrated in Fig. 5.32*b*. The vector force equation is as follows:

$$\Sigma \mathbf{F} = 0$$

$$R_{Ox}\mathbf{i} + R_{Oy}\mathbf{j} + R_{Oz}\mathbf{k} + R_{Ax}\mathbf{i} + R_{Ay}\mathbf{j} + R_{Az}\mathbf{k} + P\mathbf{i} - 2wL\mathbf{j} = 0$$

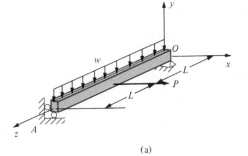

(a)

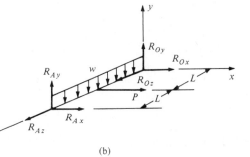

(b)

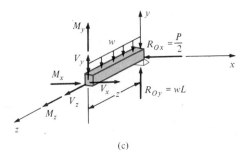

(c)

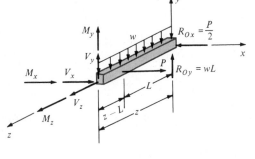

(d)

FIGURE 5.32

or

$$R_{Ox} + R_{Ax} + P = 0$$

$$R_{Oy} + R_{Ay} - 2wL = 0$$

$$R_{Oz} + R_{Az} = 0 \qquad \text{(a)}$$

Summing moments about point A gives the moment equilibrium equation.

$$\Sigma \mathbf{M}_A = 0$$

$$\begin{vmatrix} \mathbf{i} & \mathbf{j} & \mathbf{k} \\ 0 & 0 & -2L \\ R_{Ox} & R_{Oy} & R_{Oz} \end{vmatrix} + \begin{vmatrix} \mathbf{i} & \mathbf{j} & \mathbf{k} \\ 0 & 0 & -L \\ P & 0 & 0 \end{vmatrix} + \begin{vmatrix} \mathbf{i} & \mathbf{j} & \mathbf{k} \\ 0 & 0 & -L \\ 0 & -2wL & 0 \end{vmatrix} = 0$$

$$2LR_{Oy}\mathbf{i} - 2LR_{Ox}\mathbf{j} - PL\mathbf{j} - 2wL^2\mathbf{i} = 0$$

$$2LR_{Oy} - 2wL^2 = 0$$

$$2LR_{Ox} + PL = 0$$

and

$$R_{Ox} = -\frac{P}{2} \qquad R_{Oy} = wL \qquad R_{Oz} = 0$$

Combining these results with Eq. (a) give the remaining reactions.

$$R_{Ax} = -\frac{P}{2} \qquad R_{Ay} = wL \qquad R_{Az} = 0$$

The shear and moment equations for $0 \le z \le L$ are written using the free body of Fig. 5.32c. Assume positive vector reactions at the cut section.

$0 \le z \le L$

$$V_x\mathbf{i} + V_y\mathbf{j} + V_z\mathbf{k} - \frac{P\mathbf{i}}{2} + wL\mathbf{j} - wz\mathbf{j} = 0$$

$$V_x = \frac{P}{2} \qquad V_y = wz - wL \qquad V_z = 0$$

Summing moments about the cut section gives the moment equation.

$$M_x\mathbf{i} + M_y\mathbf{j} + M_z\mathbf{k} + \begin{vmatrix} \mathbf{i} & \mathbf{j} & \mathbf{k} \\ 0 & 0 & -z \\ -P/2 & wL & 0 \end{vmatrix} + \begin{vmatrix} \mathbf{i} & \mathbf{j} & \mathbf{k} \\ 0 & 0 & -z/2 \\ 0 & -wz & 0 \end{vmatrix} = 0$$

$$M_x = \frac{wz^2}{2} - wLz \qquad M_y = -\frac{Pz}{2} \qquad M_z = 0$$

Shear and moment equations for $L \le z \le 2L$ are written using the free body of Fig. 5.32d.

$L \le z \le 2L$

$$V_x\mathbf{i} + V_y\mathbf{j} + V_z\mathbf{k} + (P - P/2)\mathbf{i} + (wL - wz)\mathbf{j} = 0$$

$$V_x = -\frac{P}{2} \qquad V_y = w(z - L) \qquad V_z = 0$$

$$M_x\mathbf{i} + M_y\mathbf{j} + M_z\mathbf{k} + \begin{vmatrix} \mathbf{i} & \mathbf{j} & \mathbf{k} \\ 0 & 0 & -z \\ -P/2 & wL & 0 \end{vmatrix} + \begin{vmatrix} \mathbf{i} & \mathbf{j} & \mathbf{k} \\ 0 & 0 & -z/2 \\ 0 & -wz & 0 \end{vmatrix}$$

$$+ \begin{vmatrix} \mathbf{i} & \mathbf{j} & \mathbf{k} \\ 0 & 0 & (L-z) \\ P & 0 & 0 \end{vmatrix} = 0$$

$$M_x = \frac{wz^2}{2} - wLz \qquad M_y = \frac{Pz}{2} - PL \qquad M_z = 0 \qquad \blacksquare$$

EXAMPLE 5.16

Use vector analysis to compute reactions and write shear and moment equations for the beam structure of Fig. 5.33a.

Solution:

Assume positive reactions at O as shown in Fig. 5.33b and write a force equilibrium equation and a moment equilibrium equation with respect to point O.

$$R_{Ox}\mathbf{i} + R_{Oy}\mathbf{j} + R_{Oz}\mathbf{k} + 80\mathbf{i} - 10\mathbf{j} + 20\mathbf{k} - 50\mathbf{j} = 0$$

$$R_{Ox} = -80 \qquad R_{Oy} = 60 \qquad R_{Oz} = -20$$

$$\Sigma \mathbf{M}_O = 0 = \mathbf{M}_O + \mathbf{r}_A \times \mathbf{Q} + \mathbf{r}_B \times \mathbf{P}$$

$$M_{Ox}\mathbf{i} + M_{Oy}\mathbf{j} + M_{Oz}\mathbf{k} + (12\mathbf{i}) \times (-50\mathbf{j}) + (12\mathbf{i} + 8\mathbf{k}) \times (80\mathbf{i} - 10\mathbf{j} + 20\mathbf{k}) = 0$$

$$M_{Ox} = -80 \qquad M_{Oy} = -400 \qquad M_{Oz} = -720$$

A free body of segment OA is shown in Fig. 5.33c. Unknown forces and moments are applied at the cut section, and position vectors are drawn from the cut section to the origin with magnitude

$$(r_x, r_y, r_z) = (0, 0, 0) - (x, 0, 0)$$

and

$$\mathbf{r} = -x\mathbf{i}$$

Force and moment equilibrium for the free body give the results for shear and moment equations.

$$\Sigma \mathbf{F} = 0$$

$$V_x\mathbf{i} + V_y\mathbf{j} + V_z\mathbf{k} - 80\mathbf{i} + 60\mathbf{j} - 20\mathbf{k} = 0$$

$$V_x = 80 \qquad V_y = -60 \qquad V_z = 20$$

$$\Sigma \mathbf{M} = 0$$

$$M_x\mathbf{i} + M_y\mathbf{j} + M_z\mathbf{k} - 80\mathbf{i} - 400\mathbf{j} - 720\mathbf{k} + (-x\mathbf{i}) \times (-80\mathbf{i} + 60\mathbf{j} - 20\mathbf{k}) = 0$$

$$M_x = 80 \qquad M_y = 400 + 20x \qquad M_z = 720 + 60x$$

The free body for segment AB is shown in Fig. 5.33d. The position vectors are obtained formally by subtracting the coordinates of the cut section from the coordinates of the point of application of the applied forces.

$$(r_{Ox}, r_{Oy}, r_{Oz}) = (0, 0, 0) - (12, 0, z)$$

$$\mathbf{r}_O = -12\mathbf{i} - z\mathbf{k}$$

$$(r_{Ax}, r_{Ay}, r_{Az}) = (12, 0, 0) - (12, 0, z)$$

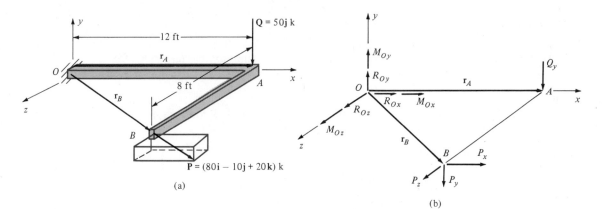

(a)

(b)

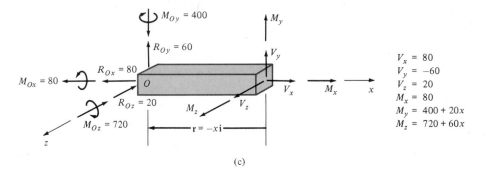

(c)

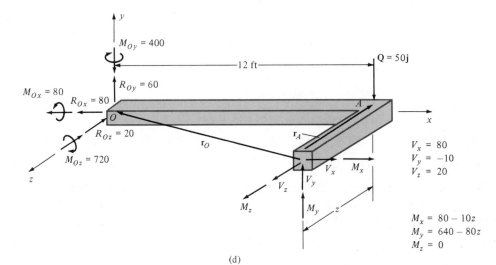

(d)

FIGURE 5.33

$\mathbf{r}_A = -z\mathbf{k}$

$\Sigma\mathbf{F} = 0$

$-80\mathbf{i} + 60\mathbf{j} - 20\mathbf{k} - 50\mathbf{j} + V_x\mathbf{i} + V_y\mathbf{j} + V_z\mathbf{k} = 0$

$V_x = 80 \qquad V_y = -10 \qquad V_z = 20$

Summing moments about the cut section gives the moment equation

$M_x\mathbf{i} + M_y\mathbf{j} + M_z\mathbf{k} - 80\mathbf{i} - 400\mathbf{j} - 720\mathbf{k} + (-12\mathbf{i} - z\mathbf{k}) \times (-80\mathbf{i} + 60\mathbf{j} - 20\mathbf{k})$
$$+ (-x\mathbf{k}) \times (-50\mathbf{j}) = 0$$

Solving for the moments gives

$M_x = -10z + 80 \qquad M_y = 640 - 80z \qquad M_z = 0$ ∎

EXAMPLE 5.17

Use vector analysis to write shear and moment equations for the beam of Fig. 5.34.

Solution:
The coordinate axis of Fig. 5.34a will be used. Consider force and moment equilibrium and compute the reactions illustrated in Fig. 5.34b.

$R_{Ox}\mathbf{i} + R_{Oy}\mathbf{j} + R_{Oz}\mathbf{k} - P\mathbf{j} = 0$

$R_{Ox} = 0 \qquad R_{Oy} = P \qquad R_{Oz} = 0$ $\qquad$ **(a)**

The position vector $\mathbf{r}$ is given as

$\mathbf{r} = -a\mathbf{i} + a\mathbf{k}$

Summing moments at point O gives

$$M_{Ox}\mathbf{i} + M_{Oy}\mathbf{j} + M_{Oz}\mathbf{k} + \begin{vmatrix} \mathbf{i} & \mathbf{j} & \mathbf{k} \\ -a & 0 & a \\ 0 & -P & 0 \end{vmatrix} = 0$$

$M_{Ox} = -Pa \qquad M_{Oy} = 0 \qquad M_{Oz} = -Pa$ $\qquad$ **(b)**

Consider the free body of Fig. 5.34c. A section is cut at an angle θ measured from the x axis. Positive reactions are assumed at the cut section, and the equation of statics gives the results for shear and moment equations. The components of the position vector are shown in Fig. 5.34d:

$V_x\mathbf{i} + V_y\mathbf{j} + V_z\mathbf{k} + P\mathbf{j} = 0$

$V_x = 0 \qquad V_y = -P \qquad V_z = 0$ $\qquad$ **(c)**

Summing moments at the cut section,

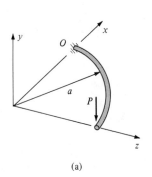

(a)

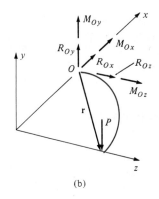

(b)

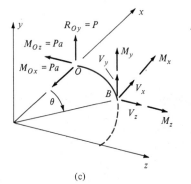

(c)

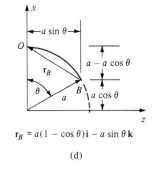

$$\mathbf{r}_B = a(1 - \cos\theta)\mathbf{i} - a\sin\theta\,\mathbf{k}$$

(d)

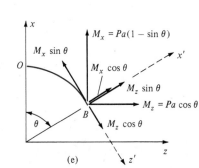

(e)

FIGURE 5.34

$$M_x \mathbf{i} + M_y \mathbf{j} + M_z \mathbf{k} - Pa\mathbf{i} - Pa\mathbf{k} + \begin{vmatrix} \mathbf{i} & \mathbf{j} & \mathbf{k} \\ a(1 - \cos \theta) & 0 & -a \sin \theta \\ 0 & P & 0 \end{vmatrix} = 0$$

$$M_x = Pa - Pa \sin \theta = Pa(1 - \sin \theta) \qquad \qquad \textbf{(d)}$$

$$M_y = 0$$

$$M_z = Pa - Pa(1 - \cos \theta) = Pa \cos \theta \qquad \qquad \textbf{(e)}$$

The vector results are shown in Fig. 5.34e. Note that M_x and M_z are parallel to the x and z axes, respectively. The results are more meaningful if the vectors are normal and tangent to the cut section such that they represent an axial torque vector and a bending moment vector adjacent to the cut section. We need merely to determine the vector components in the x'-z' coordinate system of Fig. 5.34e.

$$M'_x = M_x \cos \theta + M_z \sin \theta \qquad \qquad \textbf{(f)}$$

$$M'_z = -M_x \sin \theta + M_z \cos \theta \qquad \qquad \textbf{(g)}$$

Substitute Eqs. (d) and (e) into Eqs. (f) and (g).

$$M'_x = Pa \cos \theta - Pa \sin \theta \cos \theta + Pa \cos \theta \sin \theta$$

or

$$M'_x = Pa \cos \theta \qquad \qquad \textbf{(h)}$$

$$M'_z = -Pa \sin \theta + Pa \sin^2 \theta + Pa \cos^2 \theta$$

or

$$M'_z = Pa(1 - \sin \theta) \qquad \qquad \textbf{(i)}$$

The reader has probably realized that the result is merely a pictorial approach to formulating the scalar product to compute the components in the x'-z' coordinate system. We shall dwell upon this concept in Chapter 8. ∎

5.8 SUMMARY

The terminology and graphical representation of beam- and frame-type structures have been studied. The process of using the equations of statics to compute external reactions was reviewed.

One of the most important topics in the study of beams is the relation between load, shear, and moment. The basic relations are

$$dV = w \, dx \qquad \text{or} \qquad \frac{dV}{dx} = w \qquad \qquad \textbf{(5.3)}$$

and

$$dM = V\,dx \qquad \text{or} \qquad \frac{dM}{dx} = V \qquad\qquad \textbf{(5.8)}$$

These equations are used in constructing shear and moment diagrams. The concept of writing shear and moment equations was discussed in detail.

Shear and moment relations were applied to beams with loading in both transverse planes. Vector analysis as an engineering tool was again illustrated.

Finally, the beam and vector sign conventions, shown in Figs. 5.14 and 5.15, were introduced. It is not practical to construct shear and moment diagrams without using the beam sign convention. Other sign conventions, namely the vector sign convention, can often make the analytical work more organized and efficient.

5.9 PROBLEMS

5.1 Classify the structures of Fig. P5.1 as determinate or indeterminate. Note the degree of indeterminacy when appropriate.

5.2 Compute the reactions for the structures of Fig. P5.2.

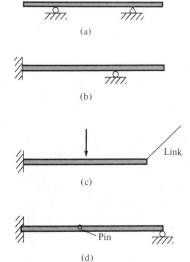

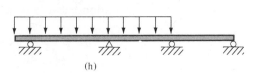

(a)

(b)

(c)

(d)

(e)

(f)

(g)

(h)

FIGURE P5.1

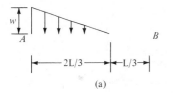

(a)

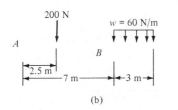

(b)

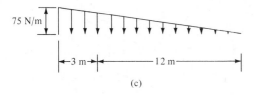

(c)

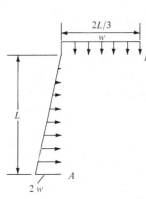

(d)

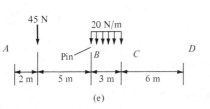

(e)

FIGURE P5.2

5.3–5.24 Write shear and moment equations for the beams of Figs. P5.3—P5.24. Construct shear and moment diagrams. Locate the point of zero shear and compute the maximum moment.

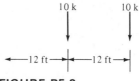

FIGURE P5.3

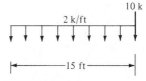

FIGURE P5.4

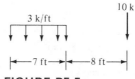

FIGURE P5.5

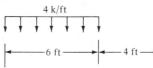

FIGURE P5.6

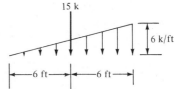

FIGURE P5.7

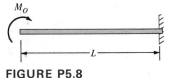

FIGURE P5.8

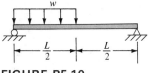

FIGURE P5.9

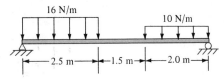

FIGURE P5.10

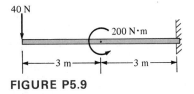

FIGURE P5.11

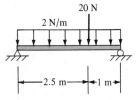

FIGURE P5.12

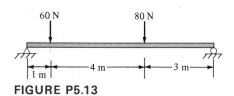

FIGURE P5.13

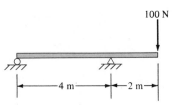

FIGURE P5.14

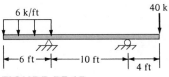

FIGURE P5.15

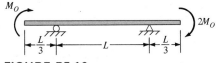

FIGURE P5.16

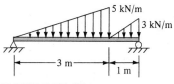

FIGURE P5.17

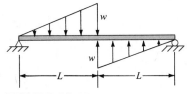

FIGURE P5.18

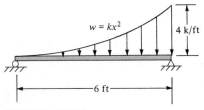

FIGURE P5.19

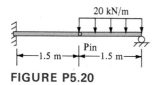

20 kN/m

Pin

|←1.5 m→|←1.5 m→|

FIGURE P5.20

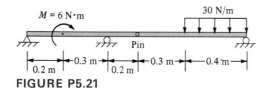

$M = 6$ N·m

30 N/m

Pin

0.2 m |←0.3 m→|←0.3 m→|←0.4 m→|
0.2 m

FIGURE P5.21

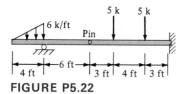

6 k/ft

5 k   5 k

Pin

| 4 ft |←6 ft→| 3 ft | 4 ft | 3 ft |

FIGURE P5.22

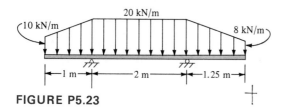

10 kN/m

20 kN/m

8 kN/m

|←1 m→|←2 m→|←1.25 m→|

FIGURE P5.23

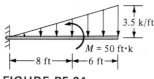

3.5 k/ft

$M = 50$ ft·k

|←8 ft→|←6 ft→|

FIGURE P5.24

5.25–5.26 Construct load and moment diagrams for the shear diagrams of Fig. P5.25 and P5.26.

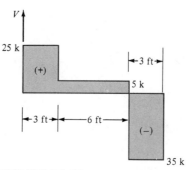

V

25 k

(+)

|←3 ft→|

5 k

|←3 ft→|←6 ft→|

(−)

35 k

FIGURE P5.25

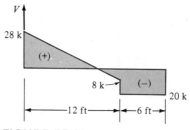

V

28 k

(+)

8 k←

(−)

20 k

|←12 ft→|←6 ft→|

FIGURE P5.26

5.27–5.28 Construct the corresponding load and shear diagrams for the moment diagrams of Figs. P5.27 and P5.28.

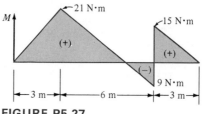

M

21 N·m

15 N·m

(+)

(+)

(−)

9 N·m

|←3 m→|←6 m→|←3 m→|

FIGURE P5.27

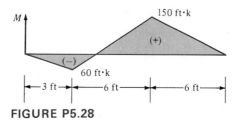

M

150 ft·k

(+)

(−)

60 ft·k

|←3 ft→|←6 ft→|←6 ft→|

FIGURE P5.28

5.29 Construct the load diagram for the shear and moment diagrams of Fig. P5.29.

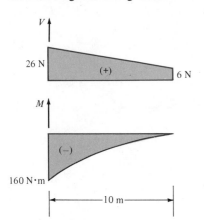

FIGURE P5.29

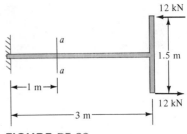

FIGURE P5.32

5.30–5.37 Write axial force, shear, and moment equations for the structures of Figs. P5.30–P5.37. Compute the axial force, shear, and moment at section *a-a*.

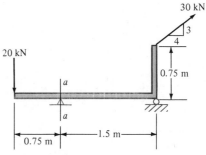

FIGURE P5.33

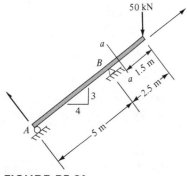

FIGURE P5.30

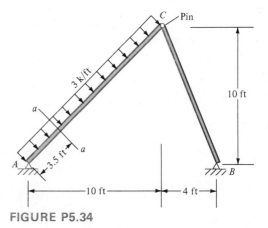

FIGURE P5.31

FIGURE P5.34

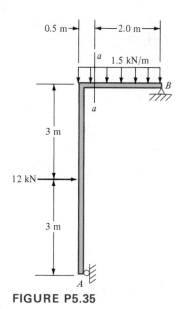

FIGURE P5.35

0.5 m

2.0 m

a 1.5 kN/m

B

a

3 m

12 kN

3 m

A

5.38–5.43 Construct axial, shear, and moment diagrams for the structures of Figs. P5.38–P5.43.

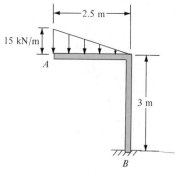

2.5 m

15 kN/m

A

3 m

B

FIGURE P5.38

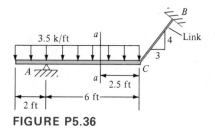

3.5 k/ft

a

B

4   Link

3

A

C

a

2.5 ft

6 ft

2 ft

FIGURE P5.36

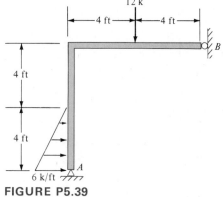

12 k

4 ft

4 ft

B

4 ft

4 ft

A

6 k/ft

FIGURE P5.39

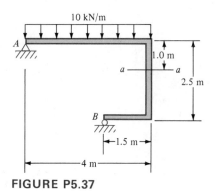

10 kN/m

A

1.0 m

a   *a*

2.5 m

B

1.5 m

4 m

FIGURE P5.37

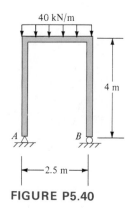

40 kN/m

4 m

A   *B*

2.5 m

FIGURE P5.40

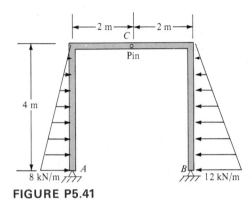

FIGURE P5.41

5.44–5.49 Construct shear and moment diagrams for the structures of Figs. P5.44–P5.49.

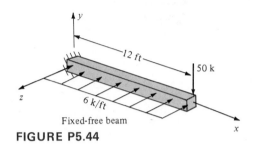

Fixed-free beam

FIGURE P5.44

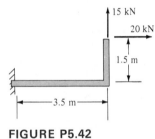

FIGURE P5.42

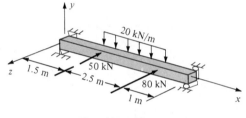

Hinged-hinged beam

FIGURE P5.45

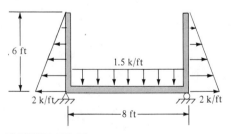

FIGURE P5.43

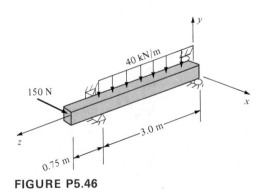

FIGURE P5.46

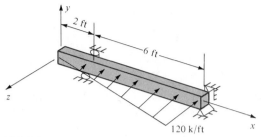

FIGURE P5.47

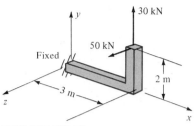

FIGURE P5.50

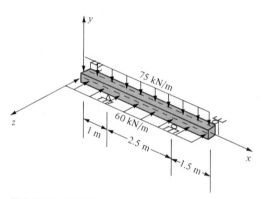

FIGURE P5.48

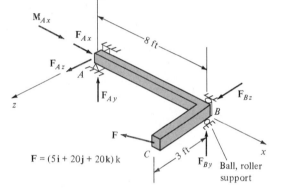

Unknown reactions: $M_{Ax}, F_{Ax}, F_{Ay}, F_{Az}, F_{By}, F_{Bz}$

FIGURE P5.51

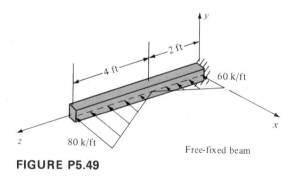

FIGURE P5.49

5.50–5.54 Use vector analysis to write shear and moment equations for the structures of Figs. P5.50–P5.54. Construct shear and moment diagrams where appropriate.

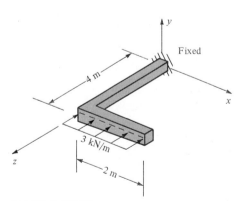

FIGURE P5.52

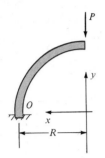

FIGURE P5.53

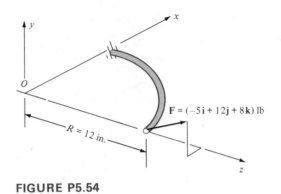

FIGURE P5.54

CHAPTER

NORMAL STRESS CAUSED BY BENDING

6.1   INTRODUCTION

In Chapter 5 the beam was studied from a statical viewpoint. Internal bending moment was characterized as a quantity that acted on a cut section to cause an equilibrium balance for the free body. The variation of the bending moment was studied using equations and diagrams. For the most part, a beam was symbolized and not given cross-sectional dimensions. In this chapter the beam will be viewed as a three-dimensional structure with cross-sectional dimensions, and the material properties will be considered. The stress produced by the bending moment acting on a cut section will be analyzed, and standard analysis and design formulas will be derived.

In the derivation of the flexural formula, the quantity that is referred to as the second moment of the area or moment of inertia arises in a natural way. The need for having previously studied this topic in both mathematics and statics will become apparent. The concept of centroid and second moment of area will be reviewed.

The flexure formula will be derived and related to the physical phenomena associated with the normal stress distribution. The engineering process of designing beams to resist bending stresses will be illustrated. The reader will be able to observe the application of moment equations and diagrams as an integral part of the design and analysis of a beam.

Again, an application of vector analysis will be demonstrated. The intent is that the reader will gain confidence in the use of vector analysis and realize that for some problems vector analysis will truly expedite the solution.

A special topic that occurs in structural design will be briefly discussed. The situation where a beam is fabricated such that it has initial curvature necessitates different considerations than the straight beam. The design formula for the stresses in a beam with initial curvature will be derived and its use illustrated.

6.2   CROSS-SECTIONAL AREA PROPERTIES FOR BEAMS

In this section the use of the first moment of the area for locating the centroid axis of a beam cross section will be reviewed. The review will extend to the second moment of the area. Both of the concepts are an integral part of the computation of normal stresses in beams due to bending.

The centroid axis of an area is that axis about which the moments of the area are balanced. Consider the rectangular area of Fig. 6.1. The centroid axis passes through the center of the area. The centroid of the area is located at the geometric center of the rectangle. In this text the location of the centroid axis will be referred to as $\bar{x}$, $\bar{y}$. The moment of the area with respect to the $\bar{x}$ axis is $(bh/2)(h/4) = bh^2/8$ above the axis and $-bh^2/8$ below the axis. Hence, the moment of the area above the $\bar{x}$ axis balances the moment below the axis. The same is true for the moment of the area about the $\bar{y}$ axis. The following observation can be made. An area that is symmetric with respect to an axis will have its centroid axis coincident with the axis of symmetry. Figure 6.2 illustrates the idea of an axis of symmetry.

The concept of the centroid of an area has been used in Chapter 5. Distributed loads on a beam were replaced with equivalent concentrated loads acting at the centroid of the loading. The idea can be extended to areas. It is worthwhile to review the basic mathematical equations that lead to locating the centroid. The area of Fig. 6.3 is arbitrary in shape, but it is assumed to be describable in terms of x-y coordinates. The total area is given by

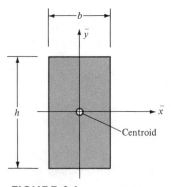

FIGURE 6.1

$$A = \int_A dA \qquad\qquad (6.1)$$

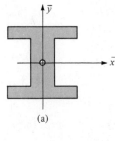

(a)

(b)

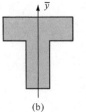

(c)

FIGURE 6.2
(a) Both $\bar{x}$ and $\bar{y}$ are axes of symmetry; (b) $\bar{y}$ is an axis of symmetry; (c) no axis of symmetry.

The first moment of the area is given by

$$\int y \, dA \qquad \text{(with respect to the } x \text{ axis)} \tag{6.2}$$

and

$$\int x \, dA \qquad \text{(with respect to the } y \text{ axis)} \tag{6.3}$$

It is helpful to view the plane area of Fig. 6.3a as illustrated in Fig. 6.3b. The differential element dA is represented as a vector $d\mathbf{A}$ and is oriented in a direction normal to the plane of the area. The magnitude of the vector is dA. The moment of the area can be expressed as

$$\int \mathbf{y} \times d\mathbf{A} \qquad \text{and} \qquad \int \mathbf{x} \times d\mathbf{A}$$

for the x and y axes, respectively.

The centroid is the point at which the total area, represented as a

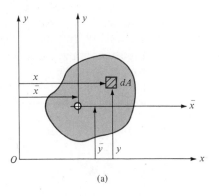

(a)

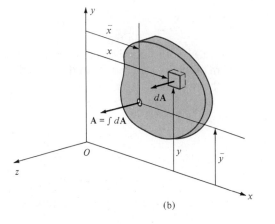

(b)

FIGURE 6.3
Representation of moment of an area as a vector quantity.

vector with magnitude A, can be concentrated in such a way that the following relation holds.

$$\bar{y} \times \int d\mathbf{A} = \int \mathbf{y} \times d\mathbf{A} \qquad (6.4)$$

or

$$\bar{y} \int dA = \int y\, dA \qquad \text{or} \qquad \bar{y} = \frac{\int y\, dA}{\int dA} \qquad (6.5)$$

Similarly,

$$\bar{x} = \frac{\int x\, dA}{\int dA} \qquad (6.6)$$

The important idea that an area can be represented as a vector has been introduced and will be used again in Chapter 9.

Equations (6.5) and (6.6) are the mathematical statement that the location of the centroid of an area is given by the integral of the moments of all differential elements of area divided by the total area. An important application of the equations is referred to as locating the centroid of composite areas. Essentially, if the area is composed of elemental geometric shapes for which the location of the centroid is already known, Eqs. (6.5) and (6.6) can be replaced by

$$\bar{y} = \frac{\displaystyle\sum_{i=1}^{N} y_i a_i}{\displaystyle\sum_{i=1}^{N} a_i} \qquad (6.7)$$

$$\bar{x} = \frac{\displaystyle\sum_{i=1}^{N} x_i a_i}{\displaystyle\sum_{i=1}^{N} a_i} \qquad (6.8)$$

where N is the number of composite areas. The term $y_i a_i$ is the area of an individual composite shape times the distance between the centroid of the area and a reference axis.

In order to illustrate the use of these equations, consider the following example.

EXAMPLE 6.1

Locate the centroid of the area of Fig. 6.4 with respect to the indicated coordinate axis.

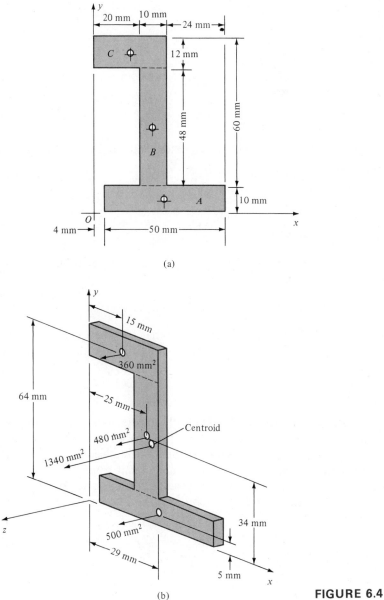

(a)

(b)

FIGURE 6.4

Solution:

The area will be divided into three composite areas as indicated by the dashed lines. The areas are represented by vectors in Fig. 6.4b. Each area vector is located at the centroid of the composite figure that it represents. The position vectors locating the

TABLE 6.1

	Area, mm^2	$\bar{x}$, mm	$\bar{y}$, mm	$A\bar{x}$	$A\bar{y}$
A	500	29	5	14,500	2,500
B	480	25	34	12,000	16,320
C	360	15	64	5,400	23,040

individual area vectors are drawn with respect to the x and y coordinates. Equations (6.7) and (6.8) can be evaluated using Table 6.1.

$$\Sigma A = 1340 \text{ mm}^2 \qquad \Sigma A\bar{x} = 31{,}900 \text{ mm}^3 \qquad \Sigma A\bar{y} = 41{,}860 \text{ mm}^3$$

$$\bar{x} = \frac{31{,}900 \text{ mm}^3}{1340 \text{ mm}^2} = 23.81 \text{ mm} \qquad \bar{y} = \frac{41{,}860 \text{ mm}^3}{1340 \text{ mm}^2} = 31.24 \text{ mm} \qquad ■$$

The second moment of the area or area moment of inertia was defined in statics as

$$I_x = \int y^2 \, dA \qquad \qquad (6.9)$$

Relating Eq. (6.9) to Fig. 6.5, it is implied that I_x is the second moment of the area with respect to an x axis passing through the centroid of the area. Similarly,

$$I_y = \int x^2 \, dA \qquad \qquad (6.10)$$

is the second moment of the area with respect to a y axis passing through the centroid of the area. Note that the second moment of the area is not represented as a vector product. The mathematical structure of the second moment of the area is such that it cannot be properly classified as a vector; hence, it cannot be the result of a vector product.

The application of Eqs. (6.9) and (6.10) can be discussed in terms of the following example.

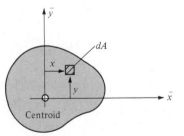

FIGURE 6.5

EXAMPLE 6.2

Compute the second moment of the area with respect to a horizontal axis passing through the centroid of the rectangle of Fig. 6.6.

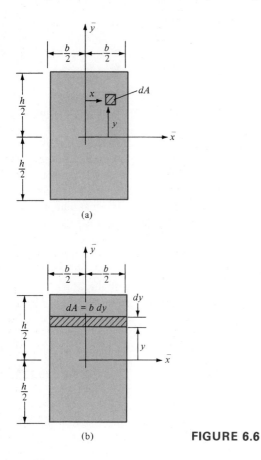

FIGURE 6.6

Solution:

In mechanics of materials the application of Eqs. (6.9) and (6.10) is such that the second moment area is almost always computed at the centroid of the geometrical figure. For the rectangle of Fig. 6.6 the centroid is at the intersection of the two axes of symmetry. In this example the second moment of area will be computed with respect to a horizontal $\bar{x}$ axis; hence, Eq. (6.9) will be used. In the strictest sense mathematically, the equation would be

$$I_x = \int_{-b/2}^{b/2} \int_{-h/2}^{h/2} y^2 \, dy \, dx$$

Since x and y have no functional relation, the integration can proceed with either variable.

$$I_x = \int_{-h/2}^{h/2} xy^2 \, dy \Big|_{-b/2}^{b/2} = \int_{-h/2}^{h/2} by^2 \, dy = \frac{by^3}{3} \Big|_{-h/2}^{h/2} = \frac{bh^3}{12} \tag{6.11}$$

Many times the computation of the area moment of inertia can be expedited by considering an elemental strip, as illustrated in Fig. 6.6b. Equation (6.9) becomes

$$I_x = \int_{-h/2}^{h/2} y^2 \, dA = \int_{-h/2}^{h/2} by^2 \, dy = \frac{bh^3}{12} \qquad ■$$

The second moment of the area for several common geometric areas is given in Appendix A.

In mechanics of materials the second moment of the area is required for the computation of flexural stresses. A beam cross section rarely requires the formal mathematical application of Eqs. (6.9) and (6.10). A beam cross-sectional area is usually made of simple geometric areas, and the area can again be referred to as a composite area. The use of the parallel-axis theorem becomes central to the computation of I_x or I_y for a composite area.

The parallel-axis theorem is used to literally transfer the second moment of area to or away from a centroid axis of a plane area. The parallel-axis theorem may be reviewed by referring to Fig. 6.7.

The second moment of area with respect to the axis a-a, according to Eq. (6.9) is

$$I_{a-a} = \int (d - y)^2 \, dA = \int (d^2 - 2dy + y^2) \, dA$$

$$= \int y^2 \, dA + d^2 \int dA - 2d \int y \, dA \tag{6.12}$$

According to the previous discussion, the term $\int y \, dA$ is zero if the moment of the area is with respect to the centroid axis. Specifying that

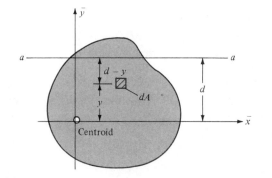

FIGURE 6.7

Eq. (6.12) is with respect to the centroid x axis dictates that the last term of Eq. (6.12) is zero, and using Eq. (6.1) gives the parallel-axis theorem as

$$I_{a\text{-}a} = I_x + Ad^2 \tag{6.13}$$

The second moment of area with respect to any arbitrary axis is equal to the second moment of area at the centroid plus the area multiplied by the transfer distance squared. The following example will illustrate the use of Eqs. (6.11) and (6.13) for a composite area.

EXAMPLE 6.3

Compute the second moment of the area with respect to the centroid axes of the composite area of Fig. 6.8.

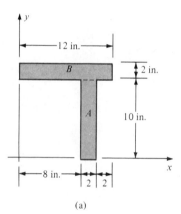

(a)

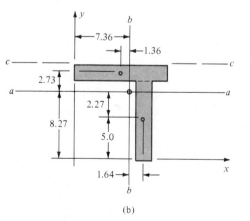

(b)

FIGURE 6.8

TABLE 6.2

	Area	$\bar{x}$	$\bar{y}$	$A\bar{x}$	$A\bar{y}$
A	20	9	5	180	100
B	24	6	11	144	264

Solution:
First the centroid must be located. Equations (6.7) and (6.8) will be used. Figure 6.8 will be divided into two areas designated A and B, and the computations are shown in Table 6.2.

$$\Sigma A = 44 \qquad \Sigma A\bar{x} = 324 \qquad \Sigma A\bar{y} = 364$$

$$\bar{x} = \frac{324}{44} = 7.36 \text{ in.} \qquad \bar{y} = \frac{364}{44} = 8.27 \text{ in.}$$

The horizontal centroid axis for the total area is referred to as *a-a* in Fig. 6.8*b*; similarly, the vertical axis is *b-b*. Equation (6.13) and the results of Example 6.2 are used as follows:

$$I_{a-a} = I_A + A_A d_A^2 + I_B + A_B d_B^2$$

$$I_{a-a} = \frac{(2)(10)^3}{12} + (2)(10)(2.27)^2 + \frac{(12)(2)^3}{12} + (2)(12)(2.73)^2$$

$$= 166.67 + 103.06 + 8.00 + 178.87 = 456.60 \text{ in.}^4$$

The second moment of area with respect to the axis *b-b* is computed in a similar manner.

$$I_{b-b} = \frac{(10)(2)^3}{12} + (10)(2)(1.64)^2 + \frac{(2)(12)^3}{12} + (2)(12)(1.36)^2$$

$$= 6.67 + 53.79 + 288.00 + 44.39 = 392.85 \text{ in.}^4 \qquad \blacksquare$$

The use of the parallel-axis theorem is limited to a transfer of the second moment of area either to the centroid or away from the centroid. For example, after computing I_{a-a} in the previous example, I_{x-x} with respect to the base of the total area could be found as

$$I_{x-x} = I_{a-a} + Ad^2 = 456.60 + (44)(8.27)^2 = 3465.89 \text{ in.}^4$$
$$\text{base}$$

If the moment of inertia I_{c-c} with respect to the top of the area is desired, the transfer must be from the centroid axis. In other words, the transfer cannot be made directly from *x-x* to *c-c*. Refer to Eq. (6.12), where the term $2h \int y\,dA$ was deleted because of the assumption that y was measured from the centroid of the area; hence, the theorem as it appears in Eq. (6.13) must always be referenced to the centroid axis.

6.3 STRESSES CAUSED BY BENDING

Thus far we have examined two of the six actions described in Fig. 1.3—namely, the axial force in Chapter 2 and axial moment in Chapter 4. In this chapter and the next we will take an in-depth look at the moment actions of Figs. 1.3*d* and *e* and the force actions of Fig. 1.3*b*.

Consider the beam segment of Fig. 6.9 and assume that the beam is initially straight, and while the cross section may be of arbitrary shape, it is constant along the axis and has both a vertical and horizontal plane of symmetry. A beam of this shape is called prismatic. Furthermore, assume that the resultant of any applied loads will pass through a plane of symmetry. These assumptions ensure that the beam will not twist during loading. If our assumptions are met, the beam is said to undergo symmetrical bending.

In preceding chapters stress has been defined as a force acting on a small element of area and we will not deviate from that definition. However, before proceeding with the derivation of the stress equation, it is in order to investigate how a beam subject to symmetrical bending actually deforms. In what follows we use a rectangular cross section for convenience in making sketches.

The beam of Fig. 6.10 is loaded to produce pure bending along the section between the concentrated loads. Pure bending occurs when the shear is zero. The deformation assumption that follows is for the case of pure bending, but we shall extend it to include the case when the shear is not zero, such as the length of beam between the supports and concentrated loads of Fig. 6.10. Consider a section of the beam that is subject to pure bending, as shown in Fig. 6.10*b*. The classical assumption for pure bending is that plane sections of the beam originally

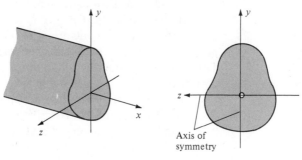

FIGURE 6.9

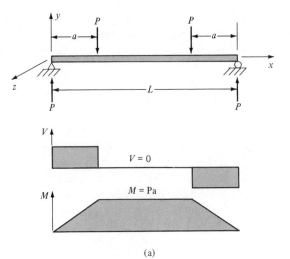

(a)

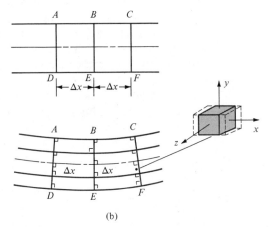

FIGURE 6.10
(a) Beam loading to produce pure bending; (b) section subjected to pure bending; (c) section subjected to transverse shear and bending; (d) deformation of the cross section.

(b)

perpendicular to the longitudinal axis of the beam remain plane and perpendicular to the longitudinal axis after deformation of the beam. The lines AD, BE, and CF of Fig. 6.10b are straight lines before and after deformation. The deformation occurs such that the top surface of the beam is in compression and the bottom surface is in tension. There is a line along the longitudinal axis of the beam that does not change in length. Actually, it represents a surface passing through the thickness of the beam. This line, which separates the compression zone from the tension zone, is called the neutral axis.

A similar beam section subject to both transverse shear and bending is shown in Fig. 6.10c, where it is illustrated that shear causes some

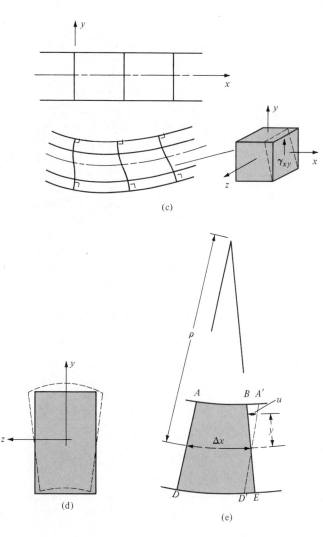

FIGURE 6.10
(continued)

distortion of the cross section. The deformation of the beam is caused primarily by bending, and the deformation caused by shear is usually neglected. However, the presence of a shear stress cannot always be neglected. The material element of Fig. 6.10c illustrates the action of the shear strain, γ_{xy}. It follows that a shear stress, τ_{xy}, also exists. It is the topic of Chapter 7.

The assumption of pure bending leads to the existence of an axial strain, ε_x, as shown for a material element in Fig. 6.10b. The flexure formula is based on the deformation that occurs when pure bending is assumed; in other words, neglect the slight warping that occurs when transverse shear is present. Before deriving the flexure formula, some comments are in order concerning transverse deformation in the y and z directions.

Refer to the cross section of Fig. 6.10d. As the beam deforms in the axial direction, the effect of Poisson's ratio would dictate an expansion of the section above the neutral axis and a contraction below the neutral axis. Recall, $\varepsilon_z = -\nu\varepsilon_x$, and the transverse strain is proportional to the axial strain. For elementary beam theory the Poisson effect is neglected and both ε_y and ε_z are taken as zero. It follows that γ_{yz} is zero for the assumption that the cross section retains its original shape. Similarly, the assumption that a plane before bending is plane after bending and remains normal to the longitudinal axis gives $\gamma_{xz} = 0$. For normal stress caused by bending we are concerned only with ε_x, the normal strain parallel to the longitudinal axis.

The section $ABDE$ of Fig. 6.10b is shown in Fig. 6.10e. The line $A'D'$ is drawn parallel to AD, and the deformation at y above the neutral axis is u. The radius of curvature for the section is denoted by ρ, as illustrated in Fig. 6.10e. The deformation u is proportional to y, the distance from the neutral axis as follows:

$$\frac{u}{y} = \frac{\Delta x}{\rho} \quad \text{or} \quad u = \frac{y\,\Delta x}{\rho} \tag{a}$$

The strain corresponding to the deformation is $\varepsilon_x = u/\Delta x$, and substituting into Eq. (a),

$$\varepsilon_x = \frac{u}{\Delta x} = \frac{y}{\rho} \tag{b}$$

The longitudinal strain is linearly proportional to the distance from the neutral axis.

Assume that the material is linearly elastic, $\sigma_x = E\varepsilon_x$, and that the modulus of elasticity is the same for both tension and compression. These assumptions lead to the conclusion that stress as well as strain, for a linear elastic material, varies linearly with the distance from the neutral axis. It follows that for symmetrical bending the neutral axis coincides with the centroid axis of the cross section.

The normal stress distribution can be idealized in a three-dimensional manner as shown in Fig. 6.11a. The total compressive force acting above the neutral axis is numerically equal to the compressive stress volume. Similarly, the tensile force acting below the neutral axis is numerically equal to the tensile stress volume. These forces can be evaluated using the elemental area of Fig. 6.11c. The stress at distance y above the neutral axis is σ_y, and since force is equal to stress times area, the differential force acting on the differential area is

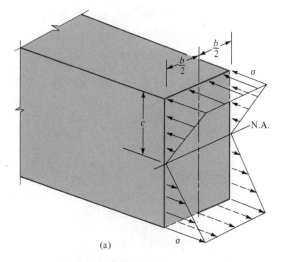

(a)

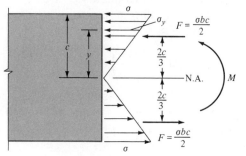

(b)

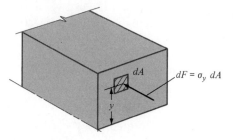

(c)

FIGURE 6.11

$$dF = \sigma_y \, dA \tag{6.14}$$

or

$$F = \int_A \sigma_y \, dA = \int_{-b/2}^{b/2} \int_0^c \sigma_y \, dy \, dx = b \int_0^c \sigma_y \, dy \tag{6.15}$$

Figure 6.11*b* can be used to write σ_y in terms of σ, y, and c. Using similar triangles,

$$\sigma_y = \frac{\sigma y}{c} \tag{6.16}$$

Substituting into Eq. (6.15) gives

$$F = b \int_0^c \frac{\sigma y \, dy}{c} = \frac{b\sigma y^2}{2c} \bigg|_0^c = \frac{\sigma bc}{2} \tag{6.17}$$

and is the volume of the stress block. The force F acts at the centroid of the stress volume as shown in Fig. 6.11*b*. The tensile force acting below the neutral axis can be evaluated in a similar manner. Summing horizontal forces on the element indicates the $\Sigma F = 0$ and satisfies that requirement of statics.

The moment illustrated in Fig. 6.11*b* is the resisting moment on a cut section as discussed in Chapter 5. The moment can be evaluated by summing moments about the neutral axis at the cut section, as follows:

$$M = F\left(\frac{2c}{3}\right) + F\left(\frac{2c}{3}\right) = \frac{2\sigma bc^2}{3} \tag{6.18}$$

Solving for σ in terms of M gives a formula for the maximum flexural stress acting on a beam with rectangular cross section

$$\sigma = \frac{3M}{2bc^2} \tag{6.19}$$

Again, Fig. 6.11*c* can be used to develop a more general approach to the calculation of flexural stresses. A differential moment can be written

$$dM = y \, dF = y\sigma_y \, dA \tag{6.20}$$

Integrating and using Eq. (6.16) will give the flexural formula

$$M = \int \frac{\sigma y^2 \, dA}{c} = \frac{\sigma}{c} \int y^2 \, dA \tag{6.21}$$

The quantity $\int y^2 \, dA$ has been previously defined as the second moment of the area with respect to the $\bar{x}$ axis, I_x. Equation (6.21) becomes

$$M = \frac{\sigma I}{c}$$

or

$$\sigma = \frac{Mc}{I} \tag{6.22}$$

Equation (6.22) gives the maximum normal stress due to bending and occurs at the maximum distance from the neutral axis. Substituting Eq. (6.16) gives the flexural stress at any location y measured from the neutral axis

$$\sigma_y = \frac{My}{I} \tag{6.23}$$

The flexural stress formula has been derived without concern for the sign of the normal stress. The coordinate y was measured upward in the positive direction; however, the internal moment M of Fig. 6.11b was also assumed positive. The resulting normal stress is observed by inspection to be compressive. In practice, many times the sense of the normal stress is determined by inspection. The following rule applies for planar beam type structures when using the flexural stress formula, Eq. (6.22).

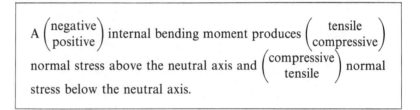

The computation of flexural stresses will be illustrated by the following examples.

EXAMPLE 6.4

Compute the maximum stress caused by bending for the beam of Fig. 6.12.

Solution:
The location of the maximum moment must be determined first. The shear and moment diagrams are constructed as shown in Fig. 6.12.

$$R_A = \frac{(20)(12)}{2} + \frac{(60)(8)}{52} = 160 \text{ k}$$

$$R_B = (20)(12) + 60 - 160 = 140 \text{ k}$$

The location of zero shear is computed using similar triangles, as illustrated in the figure. The maximum moment is computed using the area of the shear diagram, again

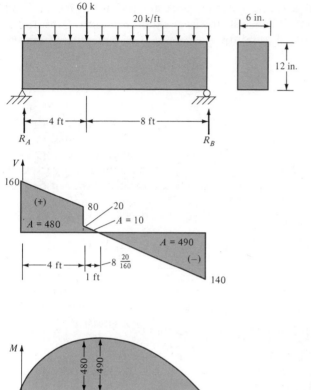

FIGURE 6.12

illustrated in Fig. 6.12, and is found to be 490 ft·k at 5 ft from the left end. The second moment of area for a rectangular beam is

$$I_x = \frac{bh^3}{12} = \frac{(6)(12)^3}{12} = 864 \text{ in.}^4$$

The distance from the neutral axis to the top or bottom of the beam, c, is 6 in. Then σ_{max} may be computed using Eq. (6.22), as follows:

$$\sigma_{max} = \frac{M_{max}c}{I} = \frac{(490 \text{ ft·k})(12 \text{ in./ft})(6 \text{ in.})}{864 \text{ in.}^4}$$

$$\sigma_{max} = 40.83 \text{ ksi} = 40,830 \text{ psi} \qquad \blacksquare$$

EXAMPLE 6.5

Compute the maximum tensile and compressive flexural stresses for the beam of Fig. 6.13.

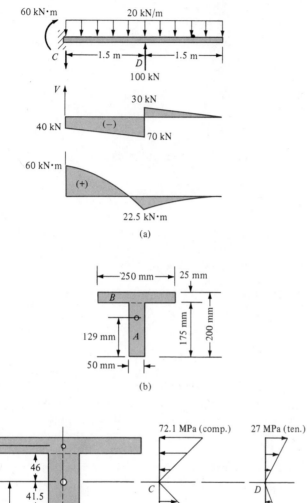

(a)

(b)

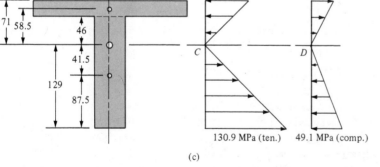

(c)

FIGURE 6.13

Solution:

Shear and moment diagrams are constructed as shown in Fig. 6.13. Maximum positive moment occurs at the fixed end and is 60 kN·m, and maximum negative moment is 22.5 kN·m at 1.5 m from the fixed end. The neutral axis is located as illustrated in Table 6.3.

TABLE 6.3

	Area	$\bar{y}$	$A\bar{y}$
A	8750	87.5	765,625
B	6250	187.5	1,171,880

$$\Sigma A = 15,000 \qquad \Sigma A\bar{y} = 1,937,505 \qquad \bar{y} = 129 \text{ mm}$$

The second moment of the area, referring to Fig. 6.13c, is computed as follows:

$$I = \frac{(50)(175)^3}{12} + (175)(50)(41.5)^2 + \frac{(250)(25)^3}{12} + (250)(25)(58.5)^2$$

$$= 5.9115(10^7) \text{ mm}^4 = 5.9115(10^{-5}) \text{ m}^4$$

The stresses at point C are calculated using Eq. (6.22) and are designated σ_t and σ_b to indicate the top and bottom fibers of the beam.

$$\sigma_t = \frac{[60(10^3) \text{ N·m}][0.071 \text{ m}]}{5.9115(10^{-5}) \text{ m}^4} = 7.21(10^7) \text{ N/m}^2 = 72.1 \text{ MPa} \qquad \text{(compression)}$$

$$\sigma_b = \frac{(60 \text{ kN·m})(0.129 \text{ m})}{5.9115(10^{-5}) \text{ m}^4} = 130.9 \text{ MPa} \qquad \text{(tension)}$$

Positive moment at point C, according to the beam sign convention, causes compression at the top of the beam and tension at the bottom. The distribution of bending stress is shown in Fig. 6.13c. Maximum negative moment occurs at point D. Even though the magnitude is less than the positive moment at C, it is a good design practice to check the stress at point D.

$$\sigma_t = \frac{(22.5 \text{ kN·m})(0.071 \text{ m})}{5.9115(10^{-5}) \text{ m}^4} = 27.0 \text{ MPa} \qquad \text{(tension)}$$

$$\sigma_b = \frac{(22.5 \text{ kN·m})(0.129 \text{ m})}{5.9115(10^{-5}) \text{ m}^4} = 49.1 \text{ MPa} \qquad \text{(compression)}$$

The stress distribution for point D is illustrated in Fig. 6.13c. In both cases the stresses at point C are the greatest. However, it is certainly possible for maximum tensile and maximum compressive stresses to occur at different locations along the beam. The situation occurs only when the distance from the neutral axis to the top fibers is unequal to the distance from the neutral axis to the bottom fibers of the beam. ∎

Example 6.5 illustrates the application of Eq. (6.22) for a T-shaped cross section. The following example is intended to give some insight into the balance of forces on the cut section and demonstrate the definition of the neutral axis.

EXAMPLE 6.6

The T-beam cross section illustrated in Fig. 6.14a is loaded such that a negative moment of 24 ft·k is acting on the section.

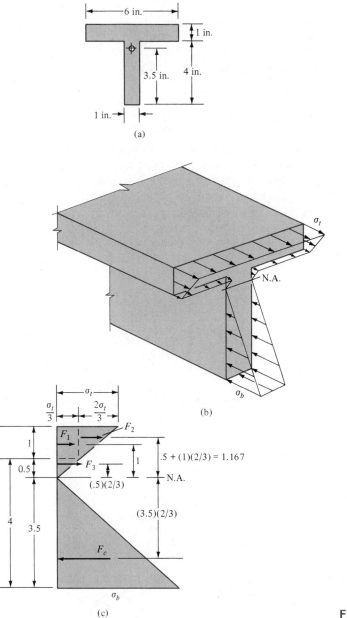

(a)

(b)

(c)

FIGURE 6.14

a. Compute the maximum tensile and compressive stresses using the volume of the stress block.

b. Compute the total tensile and compressive forces acting on the cross section.

c. Compare the result of part *a* to that obtained using the flexure formula.

Solution:

a. The stress distribution is sketched in Figs. 6.14*b* and *c*. Essentially, the volume of the stress block is equal to the force. The force acts at the centroid of the stress volume. The moment of the force with respect to the neutral axis is equal to the externally applied moment. The volume of the compressive stress block is the compressive force

$$F_c = \frac{(\sigma_b \text{ psi})(3.5 \text{ in.})(1 \text{ in.})}{2} = 1.75\sigma_b \text{ lb} \tag{a}$$

The force obtained from the tensile stress block should equal the force represented by the compressive stress volume. Recall, the sum of the forces on the section must be zero. The stress block is made up of a trapezoidal area at the flange and a triangular area at the web as shown in Fig. 6.14*c*. The trapezoidal area will be further divided into a rectangular section and a triangular section. The three forces are evaluated as follows: For the flange,

$$F_1 = \frac{\sigma_t}{3}(1)(6) = 2\sigma_t \tag{b}$$

$$F_2 = \frac{2}{3}\sigma_t \frac{(1)(6)}{2} = 2\sigma_t \tag{c}$$

For the web,

$$F_3 = \frac{\sigma_t}{3}\frac{(0.5)(1)}{2} = 0.08\sigma_t \tag{d}$$

$$F_t = F_1 + F_2 + F_3 = 4.08\sigma_t$$

Equating F_c and F_t yields a relation between σ_b and σ_t.

$$1.75\sigma_b = 4.08\sigma_t$$

$$\sigma_b = 2.33\sigma_t \tag{e}$$

The external moment must be balanced by the moment of the forces acting on the cut section. Equating these moments will give an equation in terms of maximum flexure stress. The position of each force is shown in Fig. 6.14*c*. Summing moments about the neutral axis and substituting Eqs. (a)–(d) yields

$$(1.75\sigma_b)(3.5)\left(\frac{2}{3}\right) + (2\sigma_t)(1) + (2\sigma_t)(1.167) + (0.08\sigma_t)(0.5)\left(\frac{2}{3}\right) = (24)(12)$$

where all units are in kips and inches.

$$4.08\sigma_b + 4.36\sigma_t = 288$$

Substituting Eq. (e) will give an answer for the flexural stress.

$$(4.08)(2.33\sigma_t) + (4.36\sigma_t) = 288$$

$$\sigma_t = 20.75 \text{ ksi} \tag{f}$$

$$\sigma_b = 48.36 \text{ ksi} \tag{g}$$

b. The force given by Eq. (a) is compressive and is obtained by substituting Eq. (g):

$$F_c = (1.75)(48.36) = 84.63 \text{ k}$$

The total tensile force is obtained by substituting Eq. (f) in the sum of Eqs. (a)–(c).

$$F_t = (4.08)(20.75) = 84.66 \text{ k}$$

The slight discrepancy between F_c and F_t can be attributed to the accumulated round-off error. For all practical purposes the section is in statical equilibrium.

c. The second moment of the area for the cross section can be computed referring to Fig. 6.14a.

$$I = \frac{(1)(4)^3}{12} + (1)(4)(1.5)^2 + \frac{(6)(1)^3}{12} + (6)(1)(1)^2$$

$$= 20.83 \text{ in.}^4$$

The flexure formula gives

$$\sigma_t = \frac{(24)(12)(1.5)}{20.83} = 20.74 \text{ ksi}$$

$$\sigma_b = \frac{(24)(12)(3.5)}{20.83} = 48.38 \text{ ksi} \qquad \blacksquare$$

EXAMPLE 6.7

The free-fixed beam of Fig. 6.15 is loaded with a moment and a concentrated force at the free end. The beam is rectangular in cross section with a width dimension of 6 in. The allowable stress is 6 ksi for both tension and compression.

a. Compute the depth of the member based upon the maximum moment.

b. Size the member using variable depth based upon the moment at sections *a-a* through *e-e*.

c. Compare the volume of material required using design *b* as opposed to design *a*.

Solution:

a. The moment diagram is sketched in Fig. 6.15b. The maximum moment occurs at the fixed end and is 72 ft·k. The depth of the beam, based upon the maximum moment, is obtained using Eq. (6.22).

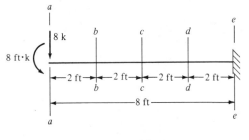

(a)

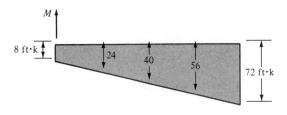

(b)

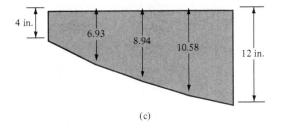

(c)

FIGURE 6.15

$$I = \frac{bh^3}{12} = \frac{Mc}{\sigma} = \frac{Mh}{2\sigma}$$

Solving for h gives

$$h = \left(\frac{6M}{b\sigma}\right)^{1/2}$$

$$h = \left[\frac{(6)(72 \text{ ft}\cdot\text{k})(12 \text{ in./ft})}{(6 \text{ in.})(6 \text{ ksi})}\right]^{1/2} = 12 \text{ in.}$$

The entire beam should be 12 in. in depth.

b. The moment at the various sections is shown in Fig. 6.15b. Use the previous as a model.

$$h = \left[\frac{(6)(12 \text{ in./ft})}{(6 \text{ in.})(6 \text{ ksi})} M \right]^{1/2} = (2M)^{1/2} \text{ in.}$$

$h_{a\text{-}a} = [(2)(8)]^{1/2} = 4 \text{ in.}$

$h_{b\text{-}b} = [(2)(24)]^{1/2} = 6.93 \text{ in.}$

$h_{c\text{-}c} = [(2)(40)]^{1/2} = 8.94 \text{ in.}$

$h_{d\text{-}d} = [(2)(56)]^{1/2} = 10.58 \text{ in.}$

$h_{e\text{-}e} = 12 \text{ in.}$

The results are illustrated in Fig. 6.15c. A straight-line variation is assumed between each section.

c. The volume of material can be computed by summing the four trapezoidal areas of Fig. 6.15c and multiplying by the width of the beam.

$$\text{Area} = \left[\frac{(4 + 6.93)}{2} + \frac{(6.93 + 8.94)}{2} + \frac{(8.94 + 10.58)}{2} + \frac{(10.58 + 12)}{2} \right] (24 \text{ in.}^2)$$

$$= 826.8 \text{ in.}^2$$

$$\text{Volume} = (826.8 \text{ in.}^2)(6 \text{ in.}) = 4960.8 \text{ in.}^3$$

The volume of the beam of part **a** is

$$(8 \text{ ft})(12 \text{ in./ft})(12 \text{ in.})(6 \text{ in.}) = 6912 \text{ in.}^3$$

The material savings is significant. In this design example the second moment of the area has been allowed to vary according to the bending moment requirements. Such an approach is common in the design of structures to resist bending loads. ■

EXAMPLE 6.8

A floor system is supported by 2- by 10-in. members as illustrated in Fig. 6.16. The floor loading is 40 lb/ft^2 and the beams span 14 ft. Determine the spacing for 2- by 8-in. members if the maximum flexural stress is to remain the same as that for 2- by 10-in. members at 16-in. centers.

Solution:

The uniform loading on one 2- by 10-in. member can be computed by considering a 16-in. strip as shown in Fig. 6.16a.

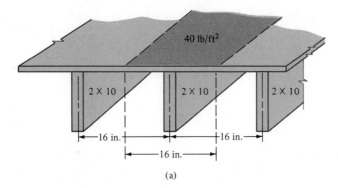

(a)

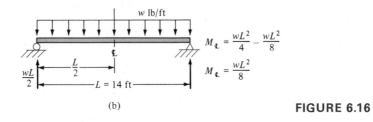

(b)

FIGURE 6.16

$$40 \text{ lb/ft}^2 \cdot \frac{16 \text{ in.}}{12 \text{ in./ft}} = 53.33 \text{ lb/ft}$$

The maximum flexure stress for a 2- by 10-in. member loaded with 53.33 lb/ft and spanning 14 ft is obtained from Fig. 6.16b.

$$M_{\mathfrak{t}} = \frac{wL^2}{8} = \frac{(53.33)(14)^2}{8} = 1306.7 \text{ ft·lb}$$

$$\sigma_{max} = \frac{(1306.7 \text{ ft·lb})(12 \text{ in./ft})(5 \text{ in.})}{(2)(10^3)/12 \text{ in.}^4} = 470.4 \text{ psi}$$

The spacing s, in inches, for 2- by 8-in. members is to be determined. The load on one 2- by 8-in. member, in pounds per foot would be

$$40 \text{ lb/ft}^2 \cdot \frac{s \text{ in.}}{12 \text{ in./ft}} = 3.33s \text{ lb/ft}$$

The maximum moment is

$$M_{\mathfrak{t}} = (3.33s \text{ lb/ft})(14 \text{ ft})^2/8 = 51.58s \text{ lb·ft}$$

The flexural stress formula can be written

$$M = \frac{\sigma I}{c}$$

$$s = \frac{(470 \text{ psi})}{(81.58)(12)} \frac{(2)(8)^3}{12} \frac{1}{4} = 10.25 \text{ in.}$$

The center-to-center spacing for 2- by 8-inch beams would be 10.25 in. ∎

The preceding examples illustrate the use of moment diagrams as a very practical method for determining moment distribution along the axis of a beam. The stress is directly dependent upon the magnitude of the moment. The following discussion is intended to point out an error that sometimes occurs when evaluating the moment.

The equations of statical equilibrium are used extensively in mechanics of materials. They are, of course, the same equations that were introduced in engineering mechanics. There is one basic difference. In statics the equations were used to determine external forces and moments—that is, external to the total structural system. In mechanics of materials the application is to determine internal effects, stress distribution along the axis of the member. Hence, the exact variation in moment along the member axis is required. By way of illustration, consider the beam of Fig. 6.17a. The beam is of length L with uniform load w. The reactions are $wL/2$ at each support. The beam of Fig. 6.17b is statically equivalent to the beam of Fig. 6.17a, where the uniform load has been replaced by a concentrated force equal in magnitude to the total uniform load, wL. The concentrated force has been placed at the corresponding centroid of the uniform load. The two systems have the same reactions and are externally equivalent.

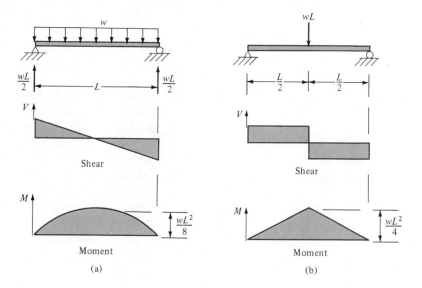

FIGURE 6.17

(a)

(b)

Compare the shear and moment diagrams; they are significantly different. The maximum moment for Fig. 6.17a is $wL^2/8$; however, for Fig. 6.17b the maximum moment is twice that amount, $wL^2/4$. The shear and moment equations are totally different for the two beams. Hence, the reader should be aware of the following observation. For the purpose of computing flexural stresses, replacing a force system with its statical equivalent can cause gross errors.

6.4   EFFICIENT STRUCTURAL CROSS SECTIONS

It has been shown that the second moment of the area plays an important role in the calculation of flexure stress. The larger the second moment of the area, the smaller the stress. It would behoove the structural designer to use a structural cross-sectional shape that would yield the greatest second moment of area using the least amount of material. Judging from the previous examples, the depth of a beam is the significant dimension. Also, the parallel-axis theorem indicates that as an area is moved away from the centroid axis, the transfer term in the parallel-axis theorem increases according to the square of the distance. Then, an efficient structural shape would have the least area near the centroid. The area should be largest at the greatest distance away from the centroid. Structural shapes, referred to as rolled sections, take advantage of this.

The properties of selected structural shapes have been reproduced from the American Institute of Steel Construction Manual and appear as Appendix C. Representative data is given for the four most popular cross sections.

The American Standard I-beam is designated with the letter S. The American wide-flange beam is called a W shape. The S and W shapes are compared in Fig. 6.18. The wide-flange shape has greater flange area; hence, for a given depth will have a larger second moment of area than the S shape. However, the greater flange area usually means that the W structural member will weigh more than an S structural member, assuming they are the same depth. Steel structural members are usually sold according to weight. The design engineer must always be aware of economics of design and strive to conserve weight.

The American Standard channel is abbreviated using the letter C. A channel cross section is shown in Fig. 6.18. Steel angle sections are manufactured in two basic shapes: equal leg and unequal leg angles. The angle cross section is sometimes designed with the symbol L.

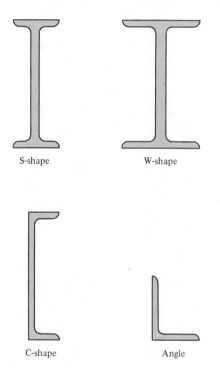

S-shape

W-shape

C-shape

Angle

FIGURE 6.18

Rolled structural sections are specified in terms of the depth, a shape symbol, and the weight per foot. A W14x53 would specify a 14-in.-deep wide flange weighing 53 lb/ft. Alternatively, the old designation 14W53 is sometimes, but rarely, used. This beam can be found in Appendix C by looking in the left-hand column. All 14-in. members are grouped together and there is only one weighing 53 lb/ft. All properties for analysis and design are given in Appendix C.

A flexure formula, Eq. (6.22), is dependent upon two cross-sectional parameters, I and c. The quantities I, S, and r are given in the Appendix for both centroid axes. I is the second moment of the area; S is the section modulus, I/c. It is convenient to write Eq. (6.22) as

$$\sigma = \frac{M}{S} \tag{6.24}$$

Equation (6.24) gives the extreme fiber stress. The radius of gyration $r = (I/A)^{1/2}$ is a column design parameter and will be used in a later chapter.

The design of structural members using American Standard beams can best be illustrated with examples.

EXAMPLE 6.9

A W10x39 beam spans 8 ft and carries a uniform load of 8 k/ft.
a. Compute the maximum flexure stress.
b. Include the weight of the beam and compute the maximum flexure stress.
c. If the uniform load increases to 11.0 k/ft, select a 10-in. W-member that will not be stressed more than the beam of part *a.*

Solution:
a. The maximum moment occurs at the center and is computed as

$$M = \frac{wL^2}{8} = \frac{8(8)^2}{8} = 64 \text{ ft·k}$$

$$S = 42.1 \text{ in.}^3$$

Note: There are two values of S in Appendix C. Refer to the figure in the Appendix; under the heading AXIS X-X the properties are given referenced to the *x-x* axis. In the problem the beam would be used such that the depth dimension would be 10 in. The section modulus for the AXIS Y-Y is significantly smaller: $S = 11.3 \text{ in.}^3$. Equation (6.24) will yield the solution for stress

$$\sigma = \frac{M}{S} = \frac{(64)(12)(10^3)}{42.1} = 18,200 \text{ psi}$$

b. The weight of the member is 39 lb/ft, or 0.039 k/ft. Repeating the calculations of part *a* gives

$$M = \frac{8.039(8)^2}{8} = 64.31 \text{ ft·k}$$

$$\sigma = \frac{(64.31)(12)(10^3)}{42.1} = 18,290 \text{ psi}$$

The increase in stress due to the weight of the member is practically negligible.
c. The moment for 11 k/ft, neglecting the weight of the member, would be

$$M = \frac{(11)(8)^2}{8} = 88 \text{ ft·k}$$

$$S = \frac{M}{\sigma} = \frac{(88)(12)(10^3)}{18,200} = 58.02 \text{ in.}^3$$

Checking the Appendix under the heading AXIS X-X, the S value of 66.7 in.^{3} will satisfy the requirement. The next highest value above the required 58.02 in.^{3} is selected.

The section modulus corresponds to a W10x60. It is good design practice to check the final solution.

$$\sigma = \frac{M}{S} = \frac{(88)(12)(10^3)}{66.7} = 15,830 \text{ psi}$$

A more complete set of beam tables might allow the selection of a lighter member that would be closer to the results of part *a*. ■

EXAMPLE 6.10

The cross section of Fig. 6.19 is often referred to as a built-up section. The beam is built up from two channels, C12x20.7, with 1- by 12-in. steel plates attached rigidly to the channels. Compute the second moment of the area with respect to the *x-x* axis.

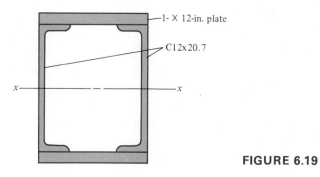

1- X 12-in. plate

C12x20.7

x ————————— x

FIGURE 6.19

Solution:
The properties of the channels are obtained from Appendix C. The second moment of the area with respect to the *x-x* axis is 129 in.⁴. The parallel-axis theorem will be used to transfer the second moment area of the cover plates to the *x-x* axis.

$$I_{x\text{-}x} = (2)(129) + 2\left[\frac{(12)(1)^3}{12} + (1)(12)(6.5)^2\right]$$

$$I_{x\text{-}x} = 258 \text{ in.}^4 + 1014 \text{ in.}^4 = 1272 \text{ in.}^4$$

The area of the two channels is 12.18 in.² as obtained from Appendix C. The area of the steel plates is 24 in.². The use of the steel plates has increased the area of the beam by a factor of 3, while the second moment of the area has increased by a factor of about 5. ■

EXAMPLE 6.11

A simply supported steel beam 12 ft long is loaded with concentrated loads as illustrated in Fig. 6.20a. The only available rolled section is a S12x31.8. Compute the thickness and length of a steel plate the same width as the beam that could be attached to the top and bottom of the beam to increase the second moment of the area so that the flexure stress will not exceed 18 ksi tension or compression.

Solution:
The moment diagram is shown in Fig. 6.20a. The maximum moment is 88 ft·k. The section modulus for a S12x31.8 is 36.4 in.^{3}. The flexure formula gives

$$\sigma = \frac{(88)(12)}{36.4} = 29.00 \text{ ksi}$$

The beam is overstressed by 11.0 ksi. Cover plates t in thickness and 5 in. wide will increase the second area moment as follows:

$$I = 218 \text{ in.}^4 + 2\left[\frac{(5)(t)^3}{12} + (5)(t)\left(6 + \frac{t}{2}\right)^2\right] \tag{a}$$

The flexure formula will give the required value

$$I = \frac{Mc}{\sigma} = \frac{(88)(12)(6 + t)}{18} = 352 + 58.67t \tag{b}$$

Solving Eqs. (a) and (b) involves finding the roots of a cubic equation. A quicker, but less sophisticated, method of solution would be to assume t and check the results or use a trial-and-error method of solution. Assume $t = 5/16$ in. Equation (b) gives the I required.

$$I = 370.33 \text{ in.}^4$$

Equation (a) gives the I available; that is,

$$I = 218 \text{ in.}^4 + 2\left[\frac{(5)(0.3125)^3}{12} + (0.3125)(5)(6.156)^2\right]$$

$$= 336.43 \text{ in.}^4$$

A plate 5/16 in. thick will not satisfy the stress criterion. Steel plate is available in thickness increments of 1/16 in. Next, a thickness of 1/2 in. will be assumed. Assume $t = 1/2$ in.

$$I \text{ (required)} = 381.33 \text{ in.}^4 \qquad I \text{ (available)} = 413.31 \text{ in.}^4$$

A plate 1/2 in. thick will satisfy the stress criterion. The actual flexure stress is

$$\sigma = \frac{(88)(12)(6 + 0.5)}{413.31} = 16.61 \text{ ksi}$$

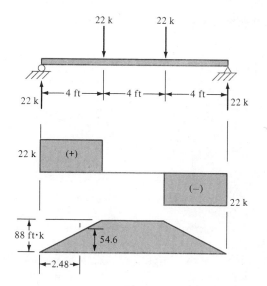

(a)

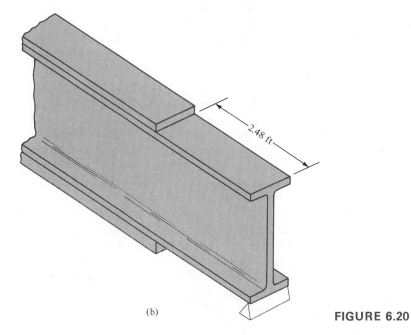

(b)

FIGURE 6.20

The cover plate is required in the center portion of the beam. The beam without cover plates will provide for a moment of

$$M = \sigma S = \frac{(18)(36.4)}{12} = 54.6 \text{ ft} \cdot \text{k}$$

The moment equation with origin at the left end of the beam is $M_x = 22x$. Solving for x gives

$$x = \frac{54.6}{22} = 2.48 \text{ ft}$$

The cover plate should be $12 - 2(2.48) = 7.04$ ft long and 1/2 in. thick, as illustrated in Fig. 6.20b. ■

6.5   BEAMS OF TWO OR MORE MATERIALS

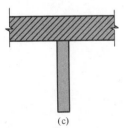

FIGURE 6.21
Beams of two materials.

Beams of two or more materials are historically referred to as composite beams. The term *composite beam* has a broader meaning than the derivation that will follow. A beam made of two or more separate materials where the materials occupy a definite position on the cross section of the beam is a specific type of composite beam. A beam made of a fiber reinforced material, such as metal fibers cast in an epoxy base, is also a specific type of composite beam. The fibers represent a percentage of the cross-sectional area and usually are not assigned a specific location on the cross section of the beam. This section will not be concerned with the latter type of composite. The general theory of composite beams is not beyond the grasp of the reader at this stage of the development of mechanics of materials. We choose to omit the general topic and refer the reader to the reference at the end of the chapter. The analysis of a beam of two or more materials will be discussed in the remainder of this section.

Consider the beam cross sections shown in Fig. 6.21. A bimetallic beam is a beam made of two metals that are bonded together as illustrated in Fig. 6.21a. A sandwich beam usually consists of a beam with a core material sandwiched between layers of relatively stiff material. The material with the greater modulus of elasticity is the stiffer material. The core might be wood and the outer layers metal. The core could be a material as soft as styrofoam and the outer material wood, but some care must be given to the selection of the material. Thus far in this text we have only considered elastic materials. A material such as styrofoam may be inelastic, and the methods of Chapter 14 would need to be incorporated in the analysis. The beam cross section of Fig. 6.21c might occur as a floor support rigidly connected to a floor. A section of the floor is assumed to act with the support to form a T-beam of two materials.

Consider a beam of two materials with the cross section of Fig. 6.22a. The beam is assumed to be subjected to pure bending and, hence, the discussion of Section 6.3 that is depicted in Fig. 6.10 remains valid. Essentially, plane sections normal to the axis of the beam prior to

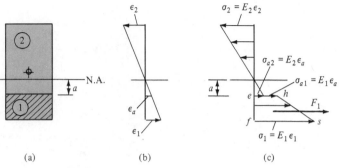

FIGURE 6.22
(a) Cross section; (b) strain distribution; (c) stress distribution.

bending of the beam remain plane after bending. The result is that the strain varies linearly over the cross section and is zero at the neutral axis. Note that the neutral axis and the centroid axis will not usually coincide for beams of two materials. Subsequently, we will discuss the question of how to locate the neutral axis. The linear strain distribution is shown in Fig. 6.22b. Let the modulus of elasticity for materials 1 and 2 be E_1 and E_2, respectively. Assume $E_2 < E_1$ and compute the stresses corresponding to the strains of Fig. 6.22b as

$$\sigma_1 = E_1 \varepsilon_1 \qquad \textbf{(a)}$$

$$\sigma_2 = E_2 \varepsilon_2 \qquad \textbf{(b)}$$

The stresses are shown in Fig. 6.22c. The abrupt change in E causes an abrupt change in the stress. Equation (b) of Section 6.3 states that the longitudinal strain is linearly proportional to the distance from the neutral axis, as illustrated in Fig. 6.22b. Stress is assumed to be linearly proportional to strain. The stress at the intersection of the two materials, the distance a in Fig. 6.22c, is given by

$$\sigma_{a1} = E_1 \varepsilon_a = E_1 \left(\frac{a}{\rho} \right) \qquad \textbf{(c)}$$

or

$$\sigma_{a2} = E_2 \varepsilon_a = E_2 \left(\frac{a}{\rho} \right) \qquad \textbf{(d)}$$

where ρ was defined as the radius of curvature. Equations (c) and (d) define the abrupt change in stress at the intersection of the two materials.

The actual stress is shown to be distributed over the cross section in Fig. 6.23a. Recall, as shown in Fig. 6.11 and Example 6.6 that the stress volume above the neutral axis was equal to the stress volume below the neutral axis so that the sum of the forces would balance. The force corresponding to the trapezoidal area, *efgh*, of Fig. 6.22c is

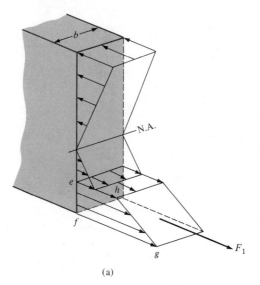

(a)

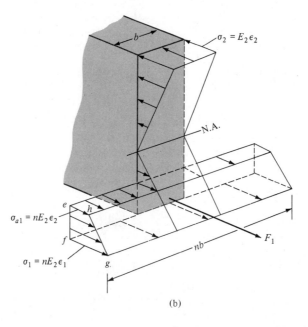

(b)

FIGURE 6.23
(a) Actual stress distribution;
(b) stress distribution for
transformed area.

shown as a trapezoidal volume in Fig. 6.23a. Imagine that we replace the volume defined by *efgh* of Fig. 6.23a with the volume of *efgh* of Fig. 6.23b and impose the restriction that the volume remain constant. The volume remaining constant implies that the axial force represented by the volume remains constant. Defining the ratio E_1/E_2 as n,

$$n = \frac{E_1}{E_2} \tag{e}$$

or

$$E_1 = nE_2 \tag{f}$$

Substitute Eq. (f) into Eqs. (a) and (c):

$$\sigma_1 = nE_2\varepsilon_1 \tag{g}$$

$$\sigma_{a1} = nE_2\varepsilon_a \tag{h}$$

as illustrated in Fig. 6.23b.

The process of going from Fig. 6.23a to Fig. 6.23b is called computing a *transformed cross-sectional area*. Material 1 is replaced with an equivalent amount of material 2 in order that material 2 can be used to compute the location of the neutral axis and the second moment of the area. The flexural formula is used to compute the stresses.

$$\sigma_2 = \frac{My}{I} \quad \text{(material 2)} \tag{i}$$

$$\sigma_1 = \frac{nMy}{I} \quad \text{(material 1)} \tag{j}$$

EXAMPLE 6.12

A timber beam is 4 in. by 8 in. as shown in Fig. 6.24. A steel plate is securely fastened to the top surface. Assume $E_w = 1.4(10^6)$ psi for wood and $E_s = 30(10^6)$ for steel and an applied positive bending moment of $65(10^3)$ in.·lb. Compute the stress at the top and bottom of the total section and the stress in each material at the intersection of the two materials.

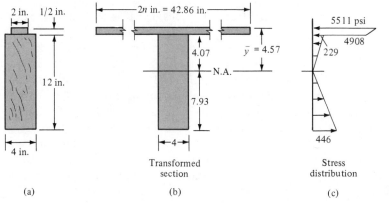

<table>
<tr><td>(a)</td><td>(b)
Transformed
section</td><td>(c)
Stress
distribution</td><td>**FIGURE 6.24**</td></tr>
</table>

Solution:

The ratio $n = E_s/E_w = 30(10^6)/1.4(10^6) = 21.43$, and it is assumed that the steel area will be transformed into an equivalent area of wood, as shown in Fig. 6.24b. Compute the location of the neutral axis as

$$\bar{y} = \frac{(42.86)(0.5)(0.25) + (4)(12)(6.5)}{(42.86)(0.5) + (4)(12)} = 4.57 \text{ in.}$$

The second moment of the area with respect to the neutral axis is

$$I = \frac{(4)(7.93)^3}{3} + \frac{(4)(4.07)^3}{3} + \frac{(42.86)(0.5)^3}{12} + (42.86)(0.5)(4.57 - 0.25)^2$$

$$= 1155.18 \text{ in.}^4$$

The stress at the top of the section is computed using Eq. (j); that is,

$$\sigma_t = \frac{(21.43)(65,000)(4.57)}{(1155.18)} = 5511 \text{ psi}$$

Similarly, use Eq. (i) to compute the stress at the bottom of the section.

$$\sigma_b = \frac{(65,000)(7.93)}{(1155.18)} = 446 \text{ psi}$$

At the intersection of the steel and wood the stresses are

$$\sigma_w = \frac{(65,000)(4.07)}{(1155.18)} = 229 \text{ psi}$$

$$\sigma_s = (229)(21.43) = 4908 \text{ psi}$$

The stress distribution is shown in Fig. 6.24c. ∎

6.6 BEAMS WITH INITIAL CURVATURE

A beam with initial curvature is any beam that is cast, fabricated, or manufactured such that it is curved before loading is applied. It is assumed that the beam lies in a plane and its cross section has an axis of symmetry. An example of such a beam is an ordinary C-clamp that might be found in a home workshop. The assumptions for flexure of straight beams are still valid. Plane sections before bending remain plane during and after bending, the material is linearly elastic, and the modulus of elasticity is the same in both tension and compression.

An equation similar to the flexure formula for straight beams will be derived for curved beams. Consider the beam section of Fig. 6.25a. For simplicity assume the cross section is rectangular as illustrated in Fig. 6.25b.

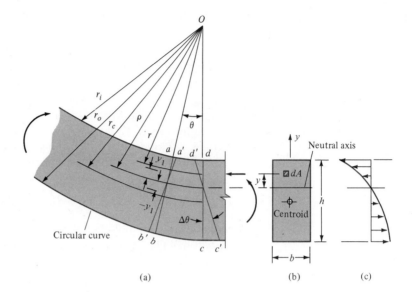

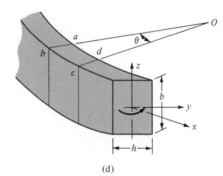

FIGURE 6.25
Beam with initial curvature.

(d)

The point O represents the center of curvature; r_o and r_i are the outside and inside radii, respectively. The distance r_c locates the centroid, ρ is the distance to the neutral axis, and r locates a point at a distance y from the neutral axis. The beam element $abcd$ is defined by the angle θ. The element deforms to the position $a'b'c'd'$ due to the application of the bending moment. Recall that the neutral axis is the line on the element that does not change in length during deformation. The neutral axis and centroid axis do not coincide for curved beams as they do for straight beams. The deformation of beam fibers can be determined along the line cd by defining a small angle, $\Delta\theta$. The deformation normal to the curved axis of the beam varies linearly from the neutral axis to the outermost fibers of the beam. The deformation is linear but the *strain is not*. If two points are located equidistant above and below the neutral axis, denoted by y_1 in Fig. 6.25a, the deformation would be $y_1 \Delta\theta$. However, the strain would be the deformation divided by the original length of the element.

The strain on the positive side of the neutral axis is $y_1 \Delta\theta/(\rho - y_1)\theta$. Note that the positive y axis is directed toward the center of the curvature. The strain on the negative side of the neutral axis is $y_1 \Delta\theta/(\rho + y_1)\theta$. The strain would vary from $(\rho - r_i)\Delta\theta/r_i\theta$ at the inside radius to $(r_o - \rho)\Delta\theta/r_o\theta$ at the outside radius. Figure 6.25c illustrates the manner in which the strain might vary.

The strain at some location y measured from the neutral axis is

$$\varepsilon_y = \frac{y\,\Delta\theta}{r\theta} = \frac{(\rho - r)\,\Delta\theta}{r\theta} \tag{a}$$

where r locates the differential element dA with respect to the center of curvature. The normal force acting on a longitudinal fiber element at dA is

$$dF = \sigma\,dA \tag{b}$$

but

$$\sigma = E\varepsilon_y = \frac{E(\rho - r)\,\Delta\theta}{r\theta} \tag{6.25}$$

Substituting Eq. (6.25) into Eq. (b) and integrating gives

$$F = \int_A \frac{E(\rho - r)\,\Delta\theta}{r\theta}\,dA \tag{6.26}$$

where the integration is over the cross-sectional area.

Statical equilibrium, $\Sigma F = 0$, requires that compressive forces, shown on the positive side of the neutral axis, must balance the tensile forces below the neutral axis. Therefore, Eq. (6.26) can be set equal to zero. Note that E, θ, $\Delta\theta$, and ρ are constants unequal to zero. It follows that

$$\frac{E\,\Delta\theta}{\theta}\int_A \frac{(\rho - r)\,dA}{r} = 0 \tag{c}$$

or

$$\rho\int_A \frac{dA}{r} - \int_A dA = 0 \tag{d}$$

The second term can be integrated and ρ, the location of the neutral axis, can be evaluated as

$$\rho = \frac{A}{\int_A dA/r} \tag{6.27}$$

As in the derivation of the flexure formula for straight beams, the next step is the summation of the moment of the forces acting on the cross section. The moment is about the z axis as shown in Fig. 6.25d. The force is taken as the stress acting on the differential area dA, and the position vector is referenced to the center of curvature.

$$dM = \sigma(\rho - r)\,dA$$

Substituting Eq. (6.25) and integrating gives

$$M = \int_A \frac{E(\rho - r)^2\,\Delta\theta}{r\theta}\,dA = \frac{E\,\Delta\theta}{\theta}\int_A \frac{(\rho - r)^2}{r}\,dA \qquad \textbf{(6.28)}$$

An expression for the constant term in Eq. (6.28) can be obtained from Eq. (6.25) as

$$\frac{E\,\Delta\theta}{\theta} = \frac{\sigma r}{\rho - r} \qquad \textbf{(e)}$$

Substituting Eq. (e) into Eq. (6.28) and simplifying leads to an expression for M in terms of σ.

$$M = \frac{\sigma r}{\rho - r}\int_A \frac{(\rho^2 - 2\rho r + r^2)}{\cdot r}\,dA \qquad \textbf{(f)}$$

$$M = \frac{\sigma r}{\rho - r}\left\{\left[\int_A \frac{\rho^2}{r}\,dA - \int_A \rho\,dA\right] - \int_A \rho\,dA + \int_A r\,dA\right\} \qquad \textbf{(g)}$$

The two terms inside the brackets are the same as Eq. (d) since ρ is a constant. The third term $\int_A \rho\,dA$ is simply ρA. The last integral can be represented as $r_c A$, the moment of the area with respect to the centroid. Then

$$M = \frac{\sigma r}{\rho - r}(r_c A - \rho A) \qquad \textbf{(h)}$$

The flexural stress for a curved beam at a location r measured from the center of curvature is

$$\sigma = \frac{M(\rho - r)}{rA(r_c - \rho)} \qquad \textbf{(6.29)}$$

where ρ is given by Eq. (6.27).

EXAMPLE 6.13

Compute the flexural stresses in a curved beam with cross-sectional dimensions 6 in. deep and 4 in. wide. The distance from the centroid of the beam cross section to the center of curvature is 6.0 in. The moment applied to the beam is 2 ft·k. Compare your results to an analysis using the straight-beam flexure formula.

Solution:

A cross section of the beam is illustrated in Figs. 6.26b and c. Referring to Fig. 6.25d, the dimensions are $h = 4$ in. and $b = 6$ in. To use Eq. (6.29), the value of ρ must be determined from Eq. (6.27). The integral of Eq. (6.27) is to be evaluated for the cross section of the curved beam. The elemental area of Fig. 6.26c can be used since b is a constant.

$$\rho = \frac{A}{\displaystyle\int_A \frac{dA}{r}} = \frac{bh}{\displaystyle\int_{r_i}^{r_o} \frac{b\,dr}{r}} = \frac{h}{\ln r \Big|_{r_i}^{r_o}} = \frac{h}{\ln r_o - \ln r_i}$$

$$= \frac{h}{\ln(r_o/r_i)} = \frac{4 \text{ in.}}{\ln(8/4)} = \frac{4 \text{ in.}}{0.693} = 5.77 \text{ in.}$$

Equation (6.29) gives, for the inside radius, $r_i = 4$ in.

$$\sigma_i = \frac{(2 \text{ ft·k})(12 \text{ in./ft})(5.77 - 4.0) \text{ in.}}{(4 \text{ in.})(24 \text{ in.}^2)(6.0 - 5.77) \text{ in.}} = 1.92 \text{ ksi} \qquad \text{(compression)} \qquad \textbf{(a)}$$

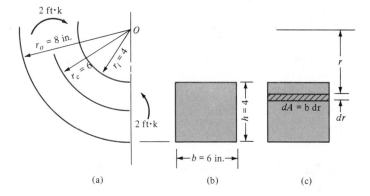

(a)   (b)   (c)

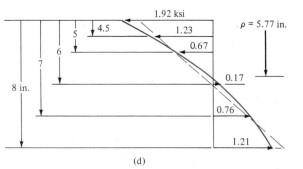

(d)

FIGURE 6.26

The stress at the outside radius for $r_o = 8$ is

$$\sigma_o = \frac{(2)(12)(5.77 - 8.0)}{(8)(24)(0.23)} = 1.21 \text{ ksi} \qquad \text{(tension)} \qquad \textbf{(b)}$$

The action of the stress is obtained from the figure. A sign convention has not been established. The applied moment acts to put compression on the inside of the curved beam, and the positive y axis is assumed to be directed toward the center of curvature. If r_c is imagined to approach infinity, whereby the curved beam becomes straight, then the applied moment would be considered to be positive. A positive moment causes compression in the top fibers of the beam. The negative sign for σ_o is an indicator that the stress at the outside of the beam is opposite that at the inside. Figure 6.26d shows a plot of the stresses at several locations through the section. The straight-beam formula yields

$$\sigma_{max} = \frac{Mc}{I} = \frac{(2)(12)(2)(12)}{(6)(4^3)} = 1.5 \text{ ksi} \qquad \textbf{(c)}$$

The dashed line of Fig. 6.26d represents the results of the straight-beam formula.

∎

The difference between the two flexure theories can be significant. In the next example, following a discussion of curved-beam correction factors, a different design situation will be illustrated. Table 6.4, Correction Factors for Straight-Beam Formulas, presents an abridged table of correction factors.

The correction factors were obtained by computing the stresses at the extreme fibers using a curved-beam formula, then dividing by the corresponding straight-beam results to obtain the K factor. The stress is computed using

$$\sigma = \frac{KMc}{I} \qquad \textbf{(6.30)}$$

All of the assumptions for the derivation of the previous beam formulas apply for Eq. (6.30). Note, K factors are given only for the inside and outside extreme fiber stresses.

The K factors for the previous example problem would be obtained by computing $r_c/(r_c - r_i)$ and selecting the values of K from the table. The correction factor is multiplied by the results of Eq. (c). These results for Example 6.13 are

$$\frac{r_c}{r_c - r_i} = \frac{6 \text{ in.}}{(6 - 4) \text{ in.}} = 3$$

For the rectangular cross section of Table 6.4, $K_i = 1.30$ and $K_o = 0.81$.

TABLE 6.4 Correction Factors for Straight-Beam Formula*

Values of K for Different Sections and Different Radii of Curvature

Section	$\dfrac{r_c}{r_c - r_i}$	Factor K K_i Inside Fiber	K_o Outside Fiber
K the same for circle and ellipse and independent of dimensions			
	1.2	3.41	0.54
	1.4	2.40	0.60
	1.6	1.96	0.65
	1.8	1.75	0.68
	2.0	1.62	0.71
	3.0	1.33	0.79
	4.0	1.23	0.84
	6.0	1.14	0.89
	8.0	1.10	0.91
	10.0	1.08	0.93
K independent of section dimensions			
	1.2	2.89	0.57
	1.4	2.13	0.63
	1.6	1.79	0.67
	1.8	1.63	0.70
	2.0	1.52	0.73
	3.0	1.30	0.81
	4.0	1.20	0.85
	6.0	1.12	0.90
	8.0	1.09	0.92
	10.0	1.07	0.94
	1.2	3.26	0.44
	1.4	2.39	0.50
	1.6	1.99	0.54
	1.8	1.78	0.57
	2.0	1.66	0.60
	3.0	1.37	0.70
	4.0	1.27	0.75
	6.0	1.16	0.82
	8.0	1.12	0.86
	10.0	1.09	0.88

* Reproduced with permission, *Advanced Mechanics of Materials*, F. B. Seely and J. O. Smith, John Wiley & Sons, Inc., 2nd ed., New York, 1952, p. 149. Originally published as "A Simple Method of Determining Stresses in Curved Beams," by B. J. Wilson and J. F. Quereau, Circular 16, Engineering Experiment Station, University of Illinois, Urbana, Ill., 1928.

Hence

$$\sigma_i = 1.30(1.5 \text{ ksi}) = 1.95 \text{ ksi}$$

$$\sigma_o = 0.81(1.5 \text{ ksi}) = 1.21 \text{ ksi}$$

and compares favorably with the results of Example 6.13.

EXAMPLE 6.14

A circular ring is made from a cylindrical beam segment with diameter of 12 mm and has a bending moment of 6.925 N·m applied so that the outside fibers are in compression. Compute the smallest radius of curvature for the ring if the flexure stress is not to exceed 80 MPa.

Solution:
Equation (6.30) is to be used. The value of I, referring to Appendix A, is

$$I = \frac{\pi r^4}{4} = \frac{\pi(6 \text{ mm})^4}{4} = 1018 \text{ mm}^4$$

Substituting into Eq. (6.30) and being careful to convert units properly, gives

$$80(10^6) \text{ N/m}^2 = \frac{K_i(6.925 \text{ N·m})(6)(10^{-3}) \text{ m}}{(1.018)(10^{-9}) \text{ m}}$$

$$K_i = 1.96 \qquad \text{(see Table 6.4)}$$

The K_i correction factor gives the largest stress; hence K_i must govern the analysis. The value of K_i corresponds to $r_c/(r_c - r_i)$ of 1.6.

$$(r_c - r_i) = 6 \text{ mm} \qquad \text{(the radius of the beam cross section)}$$

Then

$$r_c = (1.6)(6 \text{ mm}) = 9.6 \text{ mm}$$

The stress carried by the outside fibers is, using $K_o = 0.65$,

$$\sigma = \frac{(0.65)(6.925)(6)(10^{-3})}{(1.018)(10^{-9})} = 26.53 \text{ MPa} \qquad \blacksquare$$

6.7 SUMMARY

The theory associated with normal stress in beams subject to transverse loading has been presented. Two basic methods of analysis were developed: an equilibrium method and the flexural stress formula. The equilibrium method illustrates the physical effect of internal forces in beams.

The flexure formulas

$$\sigma = \frac{Mc}{I} \qquad \textbf{(6.22)}$$

and

$$\sigma_y = \frac{My}{I} \qquad \textbf{(6.23)}$$

are the basis of stress analysis for beams, both in the classroom and in applied engineering work. The use of the flexure formula dictates the need for a practical application of the theory associated with centroids and second moments of area.

The industry that supplies the structural materials for steel and aluminum construction supplies structural shapes that are intended to be most efficient for the least investment. The flexure formula illustrates the efficiency of these structural shapes.

It is important to mention the sign convention once again. The beam sign convention—positive moment causes compression in the top fibers of a beam—is used throughout the profession for the design of beams. Traditionally, beam theory is presented in a two-dimensional fashion. The beam sign convention was developed for three-dimensional loading in Chapter 5. Mention of it will occur in Chapter 7, and considerable applications are illustrated in Chapter 8. It is important for the design engineer to become accustomed to analysis situations when transverse loading occurs in both planes, and an interpretation of the sign convention is necessary for the third dimension.

Finally, the point has been made that a system of forces and moments acting on a structure cannot be replaced with some equivalent system for the purpose of stress analysis. To replace a uniform load with an equivalent external loading system can lead to considerable error when studying stress distributions.

For Further Study

Jones, R. M., *Mechanics of Composite Materials*, McGraw-Hill, New York, 1975.

Seely, F. B., and J. O. Smith, *Advanced Mechanics of Materials*, 2nd ed., John Wiley, New York, 1952.

Whitney, J. M., I. M. Daniel, and R. B. Pipes, *Experimental Mechanics of Fiber Reinforced Composite Materials*, Society for Experimental Stress Analysis, Brookfield Center, Conn., Prentice-Hall, Englewood Cliffs, N.J., 1982.

Wilson, B. J., and J. F. Quereau, "A Simple Method of Determining Stresses in Curved Beams," Circular 16, Engineering Experiment Station, University of Illinois, Urbana, 1928.

6.8 PROBLEMS

6.1–6.2 Compute the location of the centroid with respect to the base of the geometric areas of Figs. P6.1 and P6.2.

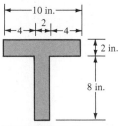

FIGURE P6.1

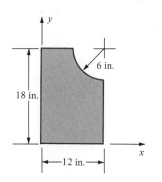

FIGURE P6.4

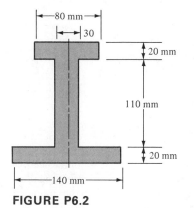

FIGURE P6.2

6.3–6.5 Locate the centroid with respect to both the x and y axes for the geometric areas illustrated in Figs. P6.3–P6.5.

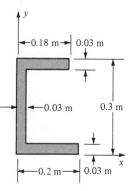

FIGURE P6.5

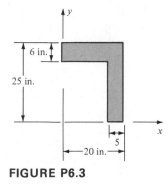

FIGURE P6.3

6.6–6.10 Compute the second moment of the area with respect to the centroid axes for the geometric areas of Figs. P6.1–P6.5.

6.11 A cantilever beam 9 ft long is loaded uniformly with 600 lb/ft. The cross section is rectangular, 6 by 9 in. Compare the maximum flexure stress, assuming **a**. the beam is 6 in. in height and **b**. the beam is 9 in. in height.

6.12 Compute the maximum flexure stress, both tension and compression, for the beam of Fig. P6.12.

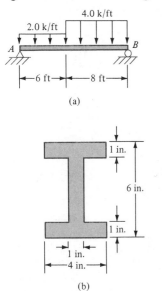

FIGURE P6.12

6.13 Compute the maximum fiber stress 4 ft from the right end for the beam of Fig. P6.13.

FIGURE P6.13

6.14 Locate the point of maximum moment and compute the maximum fiber stress for the beam of Fig. P6.14. Compare your results with the stress at midspan.

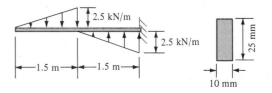

FIGURE P6.14

6.15 A cantilever beam loaded as shown in Fig. P6.15 is 20 mm wide. Compute the required depth if the flexural stress is not to exceed 35 MPa.

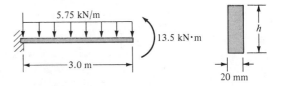

FIGURE P6.15

6.16 Compute the maximum fiber stress for the beam of Fig. P6.16. The beam is made by rigidly laminating three 50- by 200-mm members. Choose depth of the laminated beam to give the most efficient flexural member.

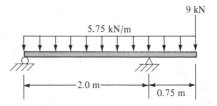

FIGURE P6.16

6.17 A circular pipe 6 ft long, 8 in. in outside diameter, and 1/2 in. thick is securely fastened in an upright position. A concentrated force of 6 k is applied transversely at the free end. Assume the pipe to be fixed free and compute the maximum flexure stress.

6.18 A beam 3 ft long is simply supported and subject to a concentrated force of 1000 lb at midspan. The cross section of the beam is triangular in shape, as illustrated in Fig. P6.18. Compute the maximum tensile and compressive flexural stresses.

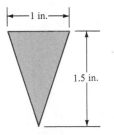

FIGURE P6.18

6.19 A beam with the cross section illustrated in Fig. P6.19 supports a maximum negative moment of 76 kN·m. Compute the maximum tensile and compressive flexural stresses.

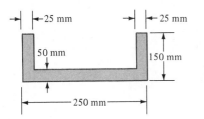

FIGURE P6.19

6.20 The simply supported beam of Fig. P6.20 is loaded with two concentrated forces as illustrated. The cross

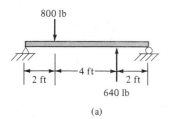

(a)

FIGURE P6.20a

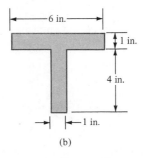

(b)

FIGURE P6.20b

section of the beam is shown in Fig. P6.20*b*. Compute the maximum positive and negative flexure stresses. Indicate their location on the beam.

6.21 The beam of Fig. P6.21 is specified to be 100 mm in width and rectangular in cross section. Compare the required depth of the member at the supports to the required depth at the center. Assume an allowable flexure stress of 40 MPa.

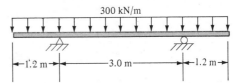

FIGURE P6.21

6.22 The beam of Fig. P6.22 is loaded as illustrated. The beam is to be rectangular and 50 mm in width. Compute the depth of the beam if the allowable compressive stress cannot exceed 12 MPa and the allowable tensile stress cannot exceed 8 MPa.

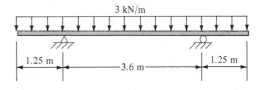

FIGURE P6.22

6.23 A square beam 6 by 6 in. is subject to a given bending moment. Compute the diameter of a beam with circular cross section that would be stressed an equal amount when subjected to the same bending moment.

6.24 A simply supported beam 15 ft long is made of four 2- by 8-in. members. Assume a uniform loading of 5 k/ft and compare the maximum flexural stress for the four cross-sectional configurations shown in Fig. P6.24. Assume that the members are rigidly connected to each other.

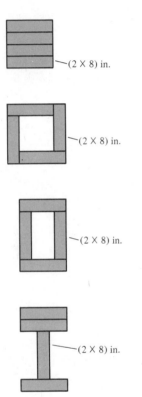

(2 × 8) in.

(2 × 8) in.

(2 × 8) in.

(2 × 8) in.

Actual dimensions

FIGURE P6.24

6.25 A beam 2.4 m long is made of circular pipe with outside diameter of 100 mm and wall thickness of 6 mm. The beam is simply supported and carries a single concentrated load of 2.5 kN at its midpoint. Compute the maximum compressive flexural stress acting on the cross section of the beam.

6.26 A beam is built up using five 2- by 4-in. members as shown in Fig. P6.26. The maximum normal stress due to bending is 2000 psi for both tension and compression.

Compute the normal force acting on each 2- by 4-in. member.

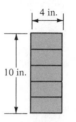

4 in.

10 in.

FIGURE P6.26

6.27 A beam with the cross section illustrated in Fig. P6.27 is loaded such that the bending moment acting on the section is 70 kN·m positive.
a. Compute the total compressive force acting on the section.
b. Compute the flexural stress using the equilibrium method.
c. Compute the flexural stress using the flexure formula.

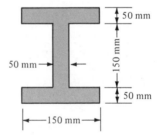

50 mm

50 mm

150 mm

50 mm

150 mm

FIGURE P6.27

6.28 A beam with the cross section of Fig. P6.28 is subjected to a bending moment of 50 ft·k. Compute the normal force acting on the shaded area.

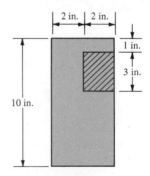

2 in. | 2 in.

1 in.

3 in.

10 in.

FIGURE P6.28

6.29 The uniformly loaded beam of Fig. P6.29 is rectangular in cross section, with a width of 6 in. Plot the maximum tensile flexure stress at 2-ft intervals along the beam.

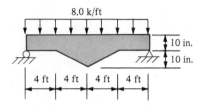

FIGURE P6.29

6.30 The beam of Fig. P6.30 is loaded as shown. The allowable flexure stress is 10 ksi. The beam is rectangular in cross section and the width is 5 in.
a. Compute the depth of the beam using the maximum moment.
b. Compute the depth at sections *a-a* through *e-e* and assume a variable depth beam with linear variation in depth between sections.
c. Compare the volume of material required for designs *a* and *b*.

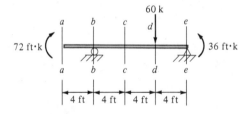

FIGURE P6.30

6.31 The beam of Fig. P6.31 has a variable depth given by the equation $d = 15 - x^2/1600$. Note that the *d* axis is

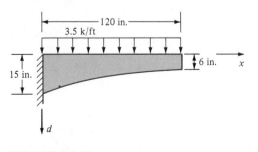

FIGURE P6.31

positive downward and *x* should be in inches. Compute the flexure stress at intervals of 30 in. along the beam. Assume a rectangular cross section 5 in. in width.

6.32 A beam is fabricated using two C8x18.75 American Standard channels with cover plates as illustrated in Fig. P6.32. Compute the second moment of the area of the section.

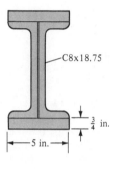

FIGURE P6.32

6.33 A composite beam is fabricated using four equal leg angles, L6x6x3/4, attached to a steel plate as illustrated in Fig. P6.33. Compute the second moment of the area of the section with respect to the neutral axis.

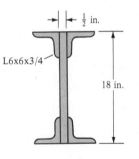

FIGURE P6.33

6.34 A box beam is to be made of four angles and four steel plates as illustrated in Fig. P6.34. The angles are L4x4x3/4 and the steel plates are 1/2 in. thick. Compute the second moment of the area with respect to the neutral axis.

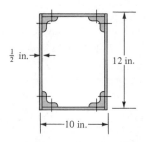

FIGURE P6.34

6.35 A simply supported W14x74 wide-flange beam spans 14 ft and carries a single concentrated force of 100 k located 6 ft from the left end. Compute the maximum flexure stress.

6.36 A fixed-free beam 6 ft long is expected to carry a concentrated force of 16.5 k at the free end. Select the most economical American Standard S-shape beam if the flexure stress is not to exceed 20,000 psi for either tension or compression.

6.37 A simply supported beam spans 11 ft and carries a uniform load of 3.5 k/ft. Select a wide-flange beam with a depth of 8 in. that will have an allowable flexure stress of less than 15,000 psi.

6.38 Two wide-flange beams are rigidly connected as shown in Fig. P6.38. The composite beam will span 11 ft. Compute the maximum uniform load that the beam will carry if the normal stress is not to exceed 20,000 psi. Assume a simply supported beam.

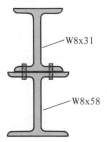

FIGURE P6.38

6.39 A simply supported beam 10 ft long is subjected to a uniform load of 25 k/ft. Select a wide-flange member if the normal stress is not to exceed 20 ksi and the depth of the member is specified to be 21 in. Can a more economical section be used if the depth requirement is ignored?

6.40 A beam is subjected to a bending moment of 150 ft·k. The cross section is built up using two C8x18.75 channel sections and an S-shape section 12 in. in depth. Select the most economical S-shape member for an allowable flexure stress of 18,000 psi.

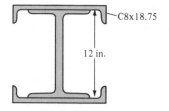

FIGURE P6.40

6.41 A wide-flange member, W14x34, is subjected to a bending moment of 90.0 ft·k. Compute the thickness of cover plates 6.5 in. in width that could be attached to the top and bottom of the member such that the maximum normal stress would not exceed 15,000 psi.

6.42 The S18x70 beam of Fig. P6.42 is loaded as illustrated. Compute the thickness of 6-in.-wide cover plates required to produce an allowable normal stress for both tension and compression of 18,000 psi.

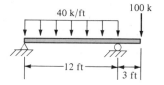

FIGURE P6.42

6.43 A fixed free beam as shown in Fig. P6.43 carries a uniform load of 2 k/ft. Compute the length and thickness of a cover plate 12 in. wide that will be attached to the top and bottom of the member to yield an allowable stress of 15,000 psi.

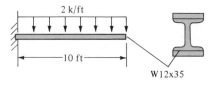

FIGURE P6.43

6.44 A floor system is supported by 2- by 12-in. members spaced 24 in. center to center. The floor loading is 40 lb/ft² and the beams are simply supported and span 18 ft. Compute the maximum flexural stress for the beams.

6.45 A floor measuring 12 by 18 ft is to be supported using 2- by 10-in. beams. Compute the spacing of the beams for the 12-ft length and compare to the required spacing for the 18-ft length. Assume a floor loading of 80 lb/ft² and an allowable flexure stress of 2200 psi for the beams.

6.46 A floor system carries a uniform load of 300 lb/ft². The supporting beams are simply supported and span 12 ft. Assume a spacing of 24 in. and select American Standard S-beams that would support the floor and remain within an allowable flexure stress of 16,000 psi.

6.47 A floor system supports a uniform load of 125 lb/ft² and is supported using 2- by 8-in. members. The beams are supported as shown in Fig. P6.47. Compute the spacing of the beams if the allowable compressive flexural stress is 2200 psi and the allowable tensile flexure stress is 1500 psi.

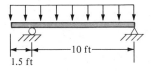

FIGURE P6.47

6.48 A wooden beam, 2 by 10 in., is used to span 12 ft and carries a load of 500 lb/ft. Assume that the stress concentration caused by the cutouts can be ignored.
a. Compute the maximum flexure stress.
b. Assume that a 5-in.-diameter hole is drilled at the center of the beam as illustrated in Fig. P6.48*b*. Compute and plot the flexural stress through the depth of the beam at the center of the span.
c. Assume that two semicircular holes are cut into the section as illustrated in Fig. P6.48*c*. Compute and plot the flexural stress through the depth of the beam at the center of the beam.
d. For parts **c** and **d** the same amount of material was removed from the section. Discuss the reasons for the significant difference in the resulting stresses.

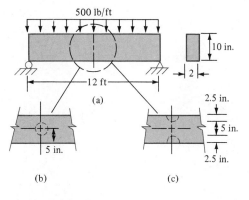

FIGURE P6.48

6.49 Repeat Example 6.12 using a transformed area for the wood section.

6.50 A beam with the cross section of Fig. P6.50 is subjected to a positive bending moment of 55 in.·k. Compute the maximum normal stress in each material. $E_{al} = 10(10^3)$ ksi; $E_w = 1.8(10^3)$ ksi.

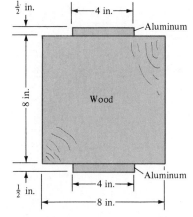

FIGURE P6.50

6.51 A 50-mm by 200-mm wood beam (actual dimensions) is reinforced with brass strips as shown in Fig. P6.51. Assume the modulus of elasticity as $E_w = 8$ GPa and $E_b = 100$ GPa. Compute the maximum normal stress in each material for an applied positive bending moment of 3.5 N·m.

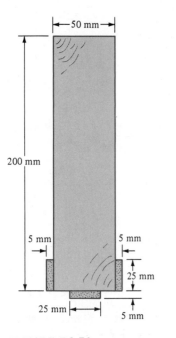

FIGURE P6.51

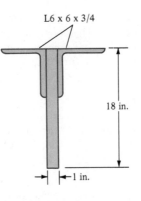

FIGURE P6.53

ing moment of 60 ft·k and compute the maximum compressive stress in each material. $E_{al} = 10(10^6)$ psi; $E_s = 30(10^6)$ psi. (*Hint:* Transform the aluminum area.)

6.54 A structural model box beam is fabricated using four wood members as shown in Fig. P6.54. There are three materials, with moduli of elasticity $E_1 = 200$ ksi, $E_2 = 900$ ksi, and $E_3 = 1400$ ksi. All members are 14 in. long with a cross section of 1.0 in. by 0.1 in. (actual di-

6.52 A wood beam is laminated using three different species of lumber as shown in Fig. P6.52. Assume $E_1 = 1000$ ksi, $E_2 = 700$ ksi, and $E_3 = 1200$ ksi and an applied negative bending moment of 5 in.·k. Compute the maximum stress in each material.

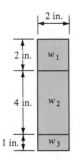

FIGURE P6.52

6.53 A T-beam is fabricated by attaching two steel equal leg angles, L6x6x3/4, to an aluminum plate as shown in Fig. P6.53. Assume an applied positive bend-

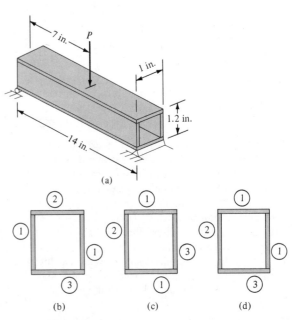

FIGURE P6.54

mensions). There are two members made of material 1 and one each of the other two materials. All materials have a working stress of 1200 psi in tension and compression. Compute the load P that can be applied to the beam for each of the cross-sectional configurations shown in Figs. P6.54b–d. (Neglect the possibility that configurations c and d may tend to warp because of the uneven stiffness of the transformed section; i.e., the transformed section is unsymmetrical. Assume that P is applied through the shear center as discussed in Section 7.6.)

6.55 A curved flexural member has cross-sectional dimensions of 100 mm by 35 mm. The distance from the centroid of the cross section to the center of curvature is 200 mm. Compute the flexural stress for an applied moment of 3500 N·m.

6.56 Compute the extreme fiber stresses for the beam of Problem 6.55 using the correction factors for the straight-beam formula (Table 6.4).

6.57 Derive an expression for ρ, the location of the neutral axis relative to the center of curvature, for a circular cross section of radius c using Eq. (6.27). (*Ans.:* $\rho = (1/2)[r_c + (r_c^2 - c^2)^{1/2}]$)

6.58 A circular ring is made from a cylindrical beam segment with diameter of 0.5 in. and has a bending moment of 12 ft·lb applied such that the inside fibers are in tension. Compute the extreme fiber flexure stresses for a radius of curvature of 1.25 in. Use the results of Problem P6.57 and Eq. (6.29). Compare to similar results obtained using the correction factors.

6.59 A circular beam segment with radius of curvature of 32 mm is square in cross section, 18 by 18 mm, and subject to a bending moment of 9.0 N·m, causing compression on the inside fibers. Compute the extreme fiber stresses using correction factors.

6.60 Use the correction factors to compute the maximum bending moment that can be applied to the curved beam segment of Fig. P6.60. Assume an allowable flexure stress of 80 MPa.

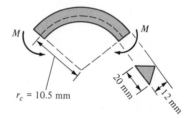

$r_c = 10.5$ mm

20 mm

12 mm

FIGURE P6.60

CHAPTER

SHEAR STRESS CAUSED BY TRANSVERSE LOADING

7.1 INTRODUCTION

In the previous chapter the relationship between bending moment and the resulting flexural stress was studied. When a beam is loaded transversely—that is, when the load is applied normal to the beam axis, producing bending moment and corresponding normal stress— internal shear occurs and causes shear stresses. The shear forces were discussed in Chapter 1, Fig. 1.3b, and are the remaining two actions that are to be discussed. Usually, a transverse shear force will accompany a bending moment. The transverse shear force will produce a shear stress whose behavior is similar to the average shear stress introduced in Chapter 2 and the torsional shear stress of Chapter 4. Even though the action is similar, the origin and method of computation of shear stress due to transverse loading is significantly different from the previously studied shear stresses. This entire chapter will be devoted to the analysis and design of beams subject to transverse shear.

It will be demonstrated once again that a transverse shear stress occurs in conjunction with a shear stress that is parallel to the longitudinal axis, and the two are equal in magnitude on any mutually perpendicular planes. In Section 7.2 it is shown that shear force, shear flow, and shear stress can be computed using equilibrium methods, which is an elementary method of analysis intended to demonstrate the dependence of shear stress computations on the bending moment. Transverse shear occurs only when the bending moment is variable along the axis of the beam.

A theoretical derivation of the shear flow formula and shear stress formula will demonstrate the relation between longitudinal shear stress and transverse shear stress. Several examples will illustrate the use of the formulas.

The design of structures using standard structural sections is still another goal of this chapter. The process of design always entails defining limiting parameters and performing an analysis to determine the unknown parameters. This chapter, like previous chapters, is concerned more with analysis than actual design.

The chapter ends with a discussion of shear center. The computations pertaining to shear center are not pertinent to the remainder of the textbook; however, the examples can be used to strengthen the understanding of shear force, shear flow, and shear stress.

7.2   HORIZONTAL SHEAR IN A BEAM SUBJECT TO BENDING

In Chapter 5 an elementary discussion was given concerning how shear forces must be present along vertical planes in a beam. Just as vertical shear forces must be present to distribute beam reactions from one support to another support, horizontal shear forces are acting along the axis of a beam. To visualize these forces refer to Fig. 6.11 and Eq. (6.17). For a single section cut through the member, the horizontal forces are balanced. The total tensile force is equal to the total compressive force. However, if a free-body section of finite dimension is taken horizontal from the beam section of Fig. 6.11b, an apparent imbalance of horizontal forces can be shown, thus verifying that internal horizontal shear forces do exist.

Consider the simply supported beam of Fig. 7.1. A single concentrated force is applied at 6 ft from the left end. The shear and moment diagrams are constructed and indicate a constant distribution of vertical shear and a uniformly varying moment along the beam. A free-body section 2 ft long will be studied and arbitrarily removed 2 ft from the left end as illustrated in Fig. 7.1b. The free body of Fig. 7.1b shows the positive vertical shear and positive bending moments acting on the free body. It can be verified that the equations of statics are satisfied. The moments acting on the free body can be converted to their equivalent normal-force representations as illustrated in Figs. 7.1c and d. The normal force is equal to the volume of the stress block and acts at the centroid of the stress block. The maximum stress at point A using the flexure formula is

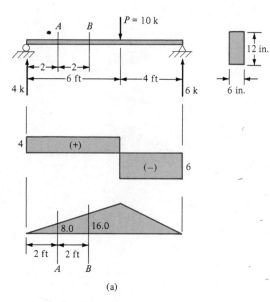

(a)

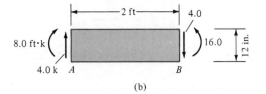

(b)

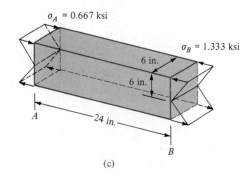

(c)

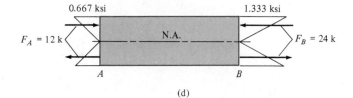

FIGURE 7.1

(d)

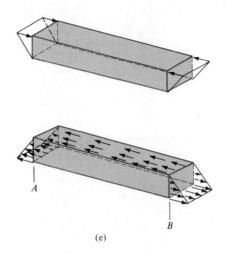

(e)

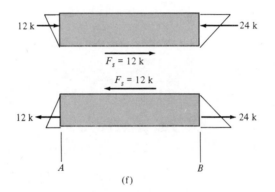

$F_s = 12$ k

$F_s = 12$ k

(f)

FIGURE 7.1
(continued)

$$\sigma_A = \frac{(8.0 \text{ ft} \cdot \text{k})(12 \text{ in./ft})(6 \text{ in.})}{(6 \text{ in.})(12 \text{ in.})^3/12} = 0.667 \text{ ksi} \qquad \textbf{(a)}$$

At section B,

$$\sigma_B = \frac{(16.0)(12)(6)}{(6)(12)^3/12} = 1.333 \text{ ksi} \qquad \textbf{(b)}$$

The forces at sections A and B, using the volume of the stress block, are

$$F_A = \frac{(0.667 \text{ ksi})(6 \text{ in.})(6 \text{ in.})}{2} = 12 \text{ k} \qquad \textbf{(c)}$$

$$F_B = \frac{(1.333)(6)(6)}{2} = 24 \text{ k}$$

These forces are illustrated in Fig. 7.1d acting at the centroids of the stress blocks. Again, the equilibrium of forces and moments can be verified.

Now, imagine that a free-body section is removed from the free body of Figs. 7.1c and d by cutting the section along the neutral axis. The new free bodies appear as Figs. 7.1e and f. Summing forces on the bottom section indicates an imbalance of horizontal forces. The force required for equilibrium, for the lower body, would be a horizontal force of 12 k acting to the left that is referred to as F_s in the figure. An equal and opposite force acts on the upper free body. The force is illustrated as an equivalent stress distribution in Fig. 7.1e. The force F_s represents the horizontal shear force that must act on an internal section of the beam. The force F_s varies throughout the depth of the beam and can be verified to be variable if the beam section is sliced horizontally at another location. The balancing shear force would not be equal to 12 k.

The physical concept of the shear force acting at the interface between imaginary layers of a beam can be demonstrated using the beams of Fig. 7.2. Assume each beam is laminated using three members as illustrated in the figure. The beams of Fig. 7.2a are merely placed one on top of the other. However, those of Fig. 7.2b are securely glued or bolted together. A vertical force is applied, hence deforming the structures. As the beams of Fig. 7.2a deflect downward, they slide relative to one another. Shear forces, other than friction, are not developed

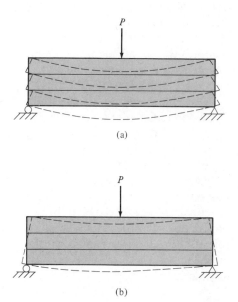

(a)

(b)

FIGURE 7.2

(a) Beam made of unconnected layers; (b) beam made of rigidly connected layers.

at the interface between component members. The beam of Fig. 7.2*b* deforms as a single member. Shear forces are developed at the interface between laminations, and the component members cannot slide relative to each other.

A solid beam would want to slide apart as shown in Fig. 7.2*a*; however, the internal material structure causes a behavior similar to that shown in Fig. 7.2*b*. A question to be answered in this chapter is: What magnitude of shear, caused by the external load *P*, is required to cause a solid structural member to tend to slide apart along horizontal planes?

In this section there are two practical applications for the theory that is to be developed. One occurs when two or more structural members are to be connected to form one rigid member. The required number of nails, bolts, rivets, or connectors is dependent upon the shear force that will act at the interface between the component members. For the study of connectors a quantity referred to as the shear flow will be computed. The second application involves the analysis and design of a member to withstand a given shear stress. All materials have limitations on the amount of shear stress they can withstand; hence, it is necessary to be able to compute with confidence the shear stress at internal locations in a structural member.

The computation of internal horizontal shear force is based upon the theory of bending. In the previous chapters the analysis of stress was based on assumptions concerning the strain distribution. Strain was assumed to be a function of a cross-sectional coordinate in every case: axial stress, torsional stress, and normal stress due to bending.

Shear stress that is associated with bending stress will be studied using the theory that has been developed for normal stress due to bending. Hence, all of the assumptions and limitations set forth in Chapter 6 for the derivation of the flexure formula apply equally to the theory to be developed in this chapter.

Before proceeding with the study of shear stress we will review our discussion of Chapter 6 that described the behavior of a beam subject to bending. Figure 6.10*c* illustrates the warping that occurs as the beam deforms. Recall, we established that all strains were zero or could be neglected except the normal strain ε_x and a shear strain γ_{xy}. The behavior of shear strain and shear stress was discussed in Section 4.3, where it was established that shear stresses on mutually perpendicular planes are equal. As we proceed with the next example and subsequent derivation of equations describing shear stress, it will be apparent that the role of horizontal shear and vertical shear are interchanged. The reader must remain aware of the deformation behavior that was established in Chapter 6.

An example problem illustrating the distribution of horizontal shear through the depth of the beam is in order prior to developing the shear stress formula. Example 7.1 will incorporate a design concept that illustrates the practical importance of computing the horizontal shear force or shear flow.

EXAMPLE 7.1

A beam loaded and supported as in Fig. 7.1 is to be laminated using 12 individual 1- by 6-in. members. The members are to be rigidly connected resulting in a homogeneous beam 6 in. wide and 12 in. high.

a. Compute and plot the horizontal shear force at the interface between laminations using the 2-ft section of Fig. 7.1*b*. Convert the force into shear flow, then into shear stress.

b. Assume that the laminations are to be connected using connectors arranged in one row along the axis of the member. Assume that each connector withstands a horizontal shear force of 1200 lb when loaded in single shear. (See Fig. 2.5 to review the connector subjected to a single-shear-type load.)

Solution:

a. The free bodies of Figs. 7.1*b* and d can be used to visualize the solution process. A slice 1-in. thick is removed from the bottom of the free body of Fig. 7.1*d* and is shown in Fig. 7.3*a*. The stress block is trapezoidal, and the forces can be evaluated using the volume of a right trapezoid or the block can be divided into rectangular and triangular shapes. The magnitude of the normal stress 5 in. below the neutral

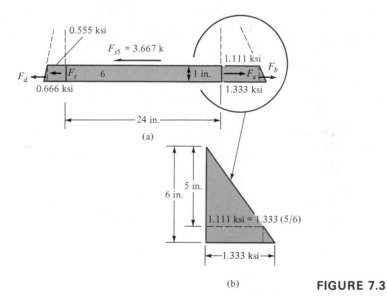

(b)   **FIGURE 7.3**

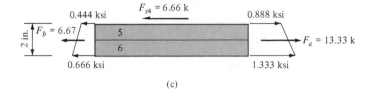

(c)

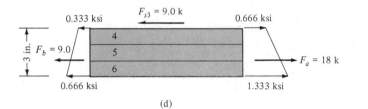

(d)

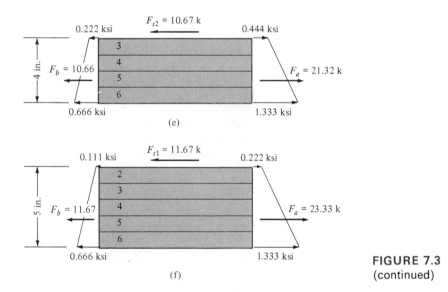

(e)

0.111 ksi $\quad F_{s1} = 11.67$ k $\quad$ 0.222 ksi

$F_b = 11.67$ $\quad$ 2 / 3 / 4 / 5 / 6 $\quad F_a = 23.33$ k

0.666 ksi $\quad$ 1.333 ksi

(f)

FIGURE 7.3
(continued)

axis is computed using similar triangles, as illustrated in Fig. 7.3*b*. The forces F_a and F_b are equal to the corresponding volumes of the stress blocks.

$$F_a = (1.111 \text{ ksi})(1 \text{ in.})(6 \text{ in.}) = 6.666 \text{ k}$$

$$F_b = \frac{(1.333 - 1.111) \text{ ksi } (1 \text{ in.})(6 \text{ in.})}{2} = 0.666 \text{ k}$$

Similarly,

$$F_c = (0.555)(1)(6) = 3.333 \text{ k}$$

$$F_d = \frac{(0.666 - 0.555)(1)(6)}{2} = 0.333 \text{ k}$$

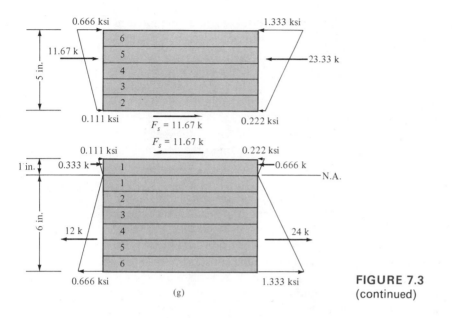

FIGURE 7.3
(continued)

(g)

Summing the forces acting on the free body yields the magnitude of F_{s5}, the shear force at the interface 5 in. below the neutral axis as

$$F_{s5} - 6.666 \text{ k} - 0.666 \text{ k} + 3.333 \text{ k} + 0.333 \text{ k} = 0$$

$$F_{s5} = 3.667 \text{ k}$$

The shear force at the interface of the next laminate is computed in a similar way. Refer to Fig. 7.3c. The normal stress distributions will be used to evaluate the shear forces using the trapezoidal volume.

$$F_a = \frac{(1.333 + 0.888)}{2} \text{ ksi } (2 \text{ in.})(6 \text{ in.}) = 13.33 \text{ k}$$

$$F_b = \frac{(0.666 + 0.444)}{2} (2)(6) = 6.67 \text{ k}$$

$$F_{s4} = 13.33 - 6.67 = 6.66 \text{ k}$$

Figure 7.3d can be used to compute the shear force 3 in. below the neutral axis.

$$F_{s3} = \frac{(1.33 + 0.666)}{2} (3)(6) - \frac{(0.666 + 0.333)}{2} (3)(6) = 9 \text{ k}$$

Figures 7.3e and f correspond to the shear force 2 in. and 1 in. below the neutral axis, respectively. These values are

$$F_{s2} = 10.67 \text{ k} \qquad F_{s1} = 11.67 \text{ k}$$

The horizontal shear at the neutral axis is $F_s = 12$ k, as shown in Fig. 7.1f.

The shear forces at laminate interfaces above the neutral axis are computed in a similar manner. It can be verified that the shear distribution for this structural member is symmetrical with respect to the neutral axis. Consider Fig. 7.3g. The lower free body can be used to illustrate that the shear force 1 in. above the neutral axis is the same as the shear stress 1 in. below the neutral axis. The top free body of Fig. 7.3g verifies that either free body can be used to arrive at the same result. The final results are plotted in Fig. 7.4a in kips. The shear force variation is not linear as contrasted to the normal force variation, which is linear. The horizontal shear force is zero at the top and bottom fibers of the beam and is maximum at the neutral axis. Normal force due to bending is the opposite: zero at the neutral axis and maximum at the outermost beam fibers.

Shear flow can be thought of as horizontal shear force per unit length of the member. The results plotted in Fig. 7.4a are for a 24-in. length of beam. Hence, shear flow would be computed as the results of Fig. 7.4a divided by 24 in. The shear flow distribution is shown in Fig. 7.4b in pounds per inch. The shear flow in units of force per length of beam is shown in Fig. 7.5a and is assumed to be constant from a to b, across the width of the beam. The shear stress in units of kips per square inch is obtained by dividing the shear flow by the width of the beam in inches and is shown in Fig. 7.4c. Figure 7.5b shows the horizontal shear stress as horizontal shear force per square inch. The process used to arrive at the results shown in Fig. 7.4 has been somewhat tedious and lengthy. Eventually a shear stress formula will be derived that will expedite the computation of shear flow and shear stress.

b. The individual laminations will be connected assuming that all layers are connected with one set of connectors. The shear flow as plotted in Fig. 7.4b is greatest at the neutral axis and is 500 lb/in. A single connector will withstand a force of 1200 lb. The spacing S for these connectors is shown in Fig. 7.6 and can be computed using $q = F/S$, or

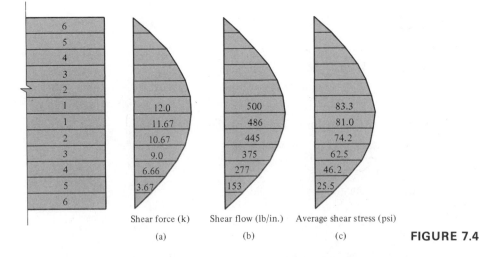

Shear force (k)   Shear flow (lb/in.)   Average shear stress (psi)

(a)   (b)   (c)   **FIGURE 7.4**

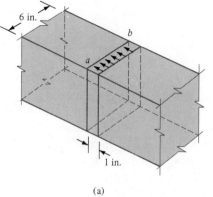

6 in.

b

a

1 in.

(a)

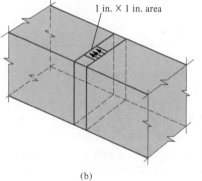

1 in. × 1 in. area

FIGURE 7.5
(a) Shear flow; (b) shear
stress.

(b)

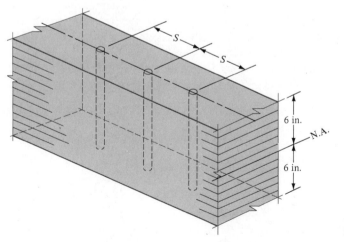

S

S

6 in.

N.A.

6 in.

FIGURE 7.6
A section of the beam of
Example 7.1.

$$\text{Shear flow (lb/in.)} = \frac{\text{Force supported by connectors, } F}{\text{Spacing of connectors, } S}$$

then $S = (1200 \text{ lb})/(500 \text{ lb/in.}) = 2.4$ in. Then the connection for the laminated beam would require one connector spaced every 2.4 in. ∎

The purpose of the foregoing example is to determine the shear stress behavior for a beam subject to bending. In addition, the problem illustrates the use of the shear flow concept for computing connector spacing. The method used for individually connecting the 12 laminates may not be the most practical or most efficient. For instance, if the laminates were a material such as wood, the force per connector might be controlled by allowable bearing stresses that could be sustained by the wood.

7.3   THE SHEAR STRESS FORMULA

In the preceding discussion and example several concepts were introduced. Important to the derivation of the shear stress formula is the variation in bending moment along the beam axis. The fact that the bending moment was unequal at the ends of the free-body section leads to the evaluation of the balancing shear force. Imagine that the beam of Fig. 7.7 is loaded in an arbitrary manner, producing a vertical shear and a variable moment. A small element, Δx in length, will be removed and viewed as a free body. Assume a positive moment M that acts on the left-hand side of the free body and increases to $M + \Delta M$ acting on the right-hand side of the free body. The normal stress due to bending on the left of the section is σ while that on the right is $\sigma + \Delta\sigma$. Only the stresses acting at the top fibers need be considered. Slice the free body of Fig. 7.7b along section $a\text{-}a$ and view the free body as shown in Figs. 7.7c and d. The section $a\text{-}a$ defines an area $abcd$, as shown in Fig. 7.7c.

Define a differential area dA located y above the neutral axis as shown in Figs. 7.7c and e. The force acting on the area dA is

$$dF = \sigma \, dA$$

or

$$F = \int_{A_{abcd}} \sigma \, dA \qquad \text{(on the left)} \tag{7.1}$$

and

$$d(F + \Delta F) = \sigma \, dA + \Delta\sigma \, dA$$

or

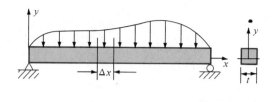

(a)

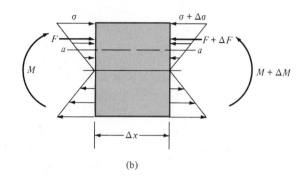

(b)

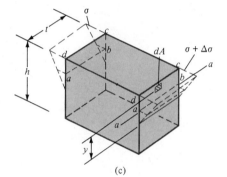

(c)

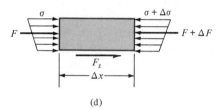

(d)

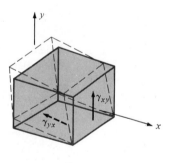

(e)

(f)

FIGURE 7.7

$$F + \Delta F = \int_{A_{abcd}} (\sigma + \Delta\sigma)\, dA \qquad \text{(on the right)} \tag{7.2}$$

Summing forces on the free body of Fig. 7.7c or d shows that the horizontal shear force F_s is ΔF; or formally summing forces gives

$$F + F_s - F - \Delta F = 0$$

or

$$F_s = \Delta F \tag{7.3}$$

In deriving equations to describe physical phenomena, the ultimate aim is to represent the unknown in terms of physical quantities that are easily describable. Note that ΔF of Eq. (7.3) can be written in terms of stress $\Delta\sigma$ using Eqs. (7.1) and (7.2). Stress can be evaluated in terms of moment as discussed in Chapter 6, and moment relates to shear as presented in Chapter 5. With this in mind, the derivation proceeds as follows: Equations (7.1) and (7.2) can be combined to show that

$$\Delta F = \int_{A_{abcd}} \Delta\sigma \, dA \tag{7.4}$$

The flexure formula of Chapter 6, $\sigma = My/I$, is employed to write $\Delta\sigma$ in terms of ΔM. Refer to Fig. 7.7b. The stress on the left of the free body is $\sigma = My/I$ and, that on the right is

$$\sigma + \Delta\sigma = \frac{(M + \Delta M)y}{I}$$

Evaluate the increase in stress, $\Delta\sigma$, that is dependent only upon ΔM, or

$$\Delta\sigma = \frac{\Delta M y}{I} \tag{7.5}$$

Substituting into Eq. (7.4) gives

$$\Delta F = \frac{\Delta M}{I} \int_{A_{abcd}} y \, dA \tag{7.6}$$

The incremental change in the force is distributed over the beam length Δx. Shear flow has been previously computed as the shear force divided by the length of beam along which the shear force acts. Hence, dividing both sides of Eq. (7.6) by Δx,

$$\frac{\Delta F}{\Delta x} = \frac{\Delta M}{I \Delta x} \int_{A_{abcd}} y \, dA \tag{7.7}$$

The shear flow q is obtained by evaluating Eq. (7.7) as Δx approaches zero, or

$$q = \lim_{\Delta x \to 0} \frac{\Delta F}{\Delta x} = \lim_{\Delta x \to 0} \frac{\Delta M}{I \Delta x} \int_{A_{abcd}} y \, dA \tag{7.8}$$

then

$$q = \frac{dF}{dx} = \frac{dM}{dx} \frac{1}{I} \int_{A_{abcd}} y \, dA \tag{7.9}$$

Equation (5.11) states that $V = dM/dx$ and, hence, the shear flow is formally related to the vertical shear. The integral of Eq. (7.9) is moment of the area as represented in Fig. 7.7e. Traditionally, the integral is replaced with Q; that is,

$$Q = \int_A y \, dA \tag{7.10}$$

Q is the first moment with respect to the neutral axis of the area between the location on the beam cross section where shear flow or shear stress is to be computed and the top or bottom of the beam cross section.

The shear flow formula is

$$q = \frac{VQ}{I} \tag{7.11}$$

The horizontal shear stress is the shear flow divided by the thickness t of the beam at the location on the cross section where the shear stress is to be computed (section a-a in Fig. 7.7). The shear stress formula is

$$\tau = \frac{VQ}{It} \tag{7.12}$$

Once again we emphasize the behavior of shear stress. A material element is shown in Fig. 7.7f, where the dashed lines indicate an idealized deformed shape. The shear stress τ of Eq. (7.12) corresponds to the shear strain γ_{xy}. Similarly, the shear strain γ_{yx} corresponds to a shear stress τ_{yx}, but as we have shown in Chapter 4, $\tau_{xy} = \tau_{yx}$ and we have replaced both with the notation τ.

The following examples will illustrate applications for Eqs. (7.10)–(7.12).

EXAMPLE 7.2

Compute the shear flow and shear stress at a point 3 in. below the neutral axis and 3 ft from the left end using the laminated beam of Example 7.1.

Solution:
The cross section of the beam is shown in Fig. 7.8. The quantity Q of Eq. (7.10) must be evaluated prior to computing q and τ. The area defined by the section a-a and denoted by A_{abcd} is shaded in Fig. 7.8. Formally, the differential area is $dA = t \, dy$ and Q is

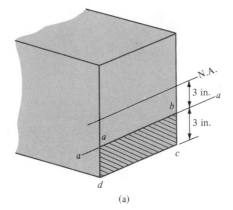

(a)

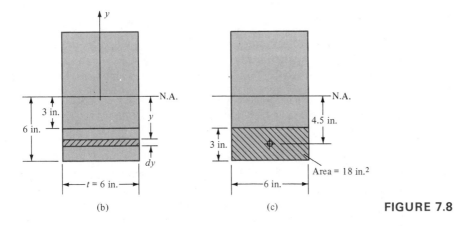

(b)   (c)   **FIGURE 7.8**

$$Q = \int_{-3}^{-6} yt\,dy = \frac{ty^2}{2}\Big|_{-3}^{-6} = \frac{6 \text{ in.}}{2}(36 - 9)\text{ in.}^2 = 81 \text{ in.}^3$$

Q can be computed by recognizing that $\int y\,dA = \bar{y}A$, or the shaded area multiplied by the distance from the neutral axis to the centroid of the shaded area. Then, by Fig. 7.8c,

$$Q = (4.5 \text{ in.})(18 \text{ in.}^2) = 81 \text{ in.}^3$$

The shear flow 3 in. below the neutral axis is

$$q = \frac{VQ}{I} = \frac{(4 \text{ k})(81 \text{ in.}^3)}{(6 \text{ in.})(12 \text{ in.})^3/12} = 0.375 \text{ k/in.} = 375 \text{ lb/in.}$$

(*Note:* The value of $V = 4$ k was obtained from the shear diagram of Fig. 7.1a.) The shear stress, according to Eq. (7.12), is

$$\tau = \frac{VQ}{It} = \frac{q}{t} = \frac{375 \text{ lb/in.}}{6 \text{ in.}} = 62.5 \text{ psi}$$

Both of these results agree with Example 7.1. ∎

EXAMPLE 7.3

A beam with an I-shaped cross section is to be fabricated using three wooden planks as shown in Fig. 7.9. The maximum vertical shear force acting on the beam is 11 kN. Common nails are to be used as connectors. Compute the spacing S for one row of nails assuming that each nail is capable of withstanding a shear force of 3.5 kN.

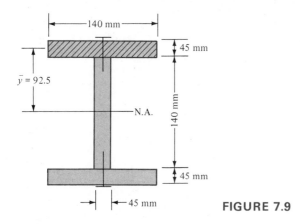

FIGURE 7.9

Solution:
The connector spacing is dependent on the shear flow at the section where the members are to be connected. Equation (7.11) will be used to compute the shear flow. The second moment of the area with respect to the centroid axis is computed as

$$I = \frac{(45 \text{ mm})}{12} (140 \text{ mm})^3 + 2\left[\frac{140}{12}(45)^3 + (140)(45)(92.5)^2\right] \text{mm}^4$$

$$= 1.202(10^8) \text{ mm}^4 = 1.202(10^{-4}) \text{ m}^4$$

The statical moment of the shaded area of Fig. 7.9 with respect to the centroid axis is

$$Q = (140)(45)(92.5) \text{ mm}^3 = 5.828(10^5) \text{ mm}^3 = 5.828(10^{-4}) \text{ m}^3$$

The shear flow is

$$q = \frac{VQ}{I} = \frac{(11000 \text{ N})(5.828)(10^{-4}) \text{ m}^3}{1.202(10^{-4}) \text{ m}^3} = 5.333(10^4) \text{ N/m}$$

The spacing is obtained by dividing the load capacity of one nail by the shear flow.

$$S = \frac{3.5(10^3) \text{ N}}{5.333(10^4) \text{ N/m}} = 65.6(10^{-3}) \text{ m}$$

and can be specified as 65.6 mm.

EXAMPLE 7.4

Oak beams of actual size 2 by 8 in. are used to span 10 ft and support a floor system that will be required to carry a load of 200 lb/ft². Assuming that the beams are simply supported and placed on 16-in. centers, compute the maximum shear stress in the beam. Neglect the dead weight of the beams and floor system.

Solution:

The system described in the problem statement is pictured in Fig. 7.10a. Each beam supports a floor section 16 in. wide and 10 ft long. The loading on a strip 1 ft in width is 200 lb/ft. The loading on a section 16 in. wide is

$$W = 200 \text{ lb/ft} \left(\frac{16 \text{ in.}}{12 \text{ in.}} \right) = 266.7 \text{ lb/ft}$$

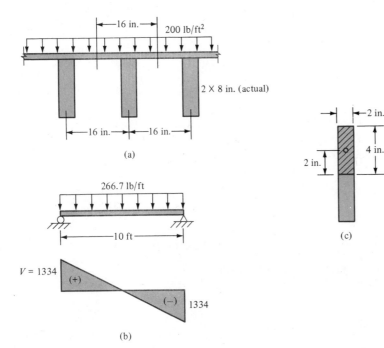

(a)

(b)

(c)

FIGURE 7.10

The shear diagram of Fig. 7.10*b* indicates that the maximum shear is 1334 lb and occurs at the supports. The value of *V* to be used in Eq. (7.12) is 1334 lb. The maximum shear stress occurs at the neutral axis of the beam. The quantity *Q* is computed using the shaded area of Fig. 7.10*c*.

$$Q = Ay = (2 \text{ in.})(4 \text{ in.})(2 \text{ in.}) = 16 \text{ in.}^3$$

$$I = \frac{(8 \text{ in.})^3(2 \text{ in.})}{12} = 85.33 \text{ in.}^4$$

$$\tau = \frac{VQ}{It} = \frac{(1334)(16)}{(85.33)(2)} = 125 \text{ psi}$$

The value of *t* is 2 in. Note that the *t* is constant in this problem. However, *t* is the thickness of the section at the location where τ is to be computed. For some cross sections *t* may not be constant. ∎

EXAMPLE 7.5

A special-purpose section is to be fabricated with the cross section shown in Figure 7.11. The maximum vertical shear acting on the beam is 3.5 kN.

a. Compute the spacing of the nails required to connect the members at locations *A-A* and *B-B* in Fig. 7.11*a*. Assume one row of nails and an allowable shear force per nail of 1500 N.

b. Compute the maximum shear stress for the section.

Solution:

a. This example problem differs from Example 7.3 in that the shear flow acts on a vertical surface at section *A-A*. However, the solution follows the same pattern. The

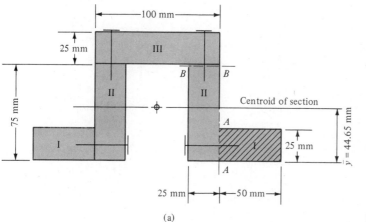

(a)

FIGURE 7.11

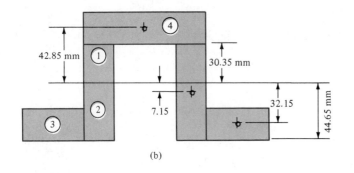

(b)

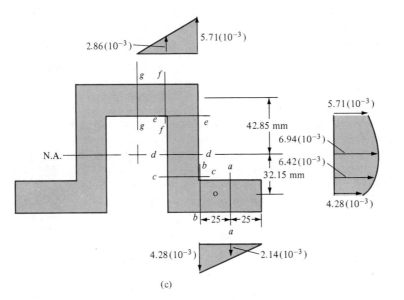

(c)

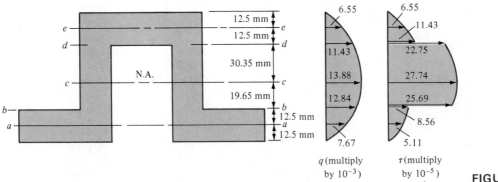

(d)

FIGURE 7.11
(continued)

TABLE 7.1

	Area, mm^2	$\bar{y}$, mm	$A\bar{y}$, mm^3
I	2500	12.5	$31.25(10^3)$
II	3750	37.5	$140.63(10^3)$
III	2500	87.5	$218.75(10^3)$

centroid must be located and the second moment of the area computed. The Q is evaluated and all quantities are substituted into the shear flow equation. The location of the centroid is computed using Table 7.1 as follows:

$$\Sigma A = 8.75(10^3) \qquad \Sigma A\bar{y} = 390.63(10^3)$$

$$\bar{y} = \frac{390.63(10^3)}{8.75(10^3)} = 44.65 \text{ mm}$$

The second moment of the area is computed referring to Figs. 7.11a and b.

$$I_1 = \frac{(2)(25 \text{ mm})(30.35 \text{ mm})^3}{3} = 46.59(10^4)$$

$$I_2 = \frac{(2)(25 \text{ mm})(44.65 \text{ mm})^3}{3} = 148.36(10^4)$$

$$I_3 = \frac{(2)(50 \text{ mm})(25 \text{ mm})^3}{12} + (2)(50 \text{ mm})(25 \text{ mm})(32.15 \text{ mm})^2$$

$$= 271.43(10^4)$$

$$I_4 = \frac{(100 \text{ mm})(25 \text{ mm})^3}{12} + (100 \text{ mm})(25 \text{ mm})(42.85 \text{ mm})^2$$

$$= 472.05(10^4)$$

$$I = I_1 + I_2 + I_3 + I_4$$

$$I = 938.43(10^4) \text{ mm}^4$$

The shear flow at section A-A involves only the shaded area I. The statical moment of the area, Q, is computed using the shaded area I. (*Note:* Only one area is used, as contrasted to the computation of $\bar{y}$ and I when both of the I areas were used.)

$$Q = (50 \text{ mm})(25 \text{ mm})(32.15 \text{ mm}) = 40.19(10^3) \text{ mm}^3$$

$$q = \frac{VQ}{I} = \frac{(3.5(10^3) \text{ N } 40.19(10^3) \text{ mm}}{938.43(10^4)} = 15 \text{ N/mm}$$

$$S = \frac{\text{Force per nail}}{q} = \frac{1500 \text{ N}}{15 \text{ N/mm}} = 100 \text{ mm}$$

The shear flow at section *B-B* is obtained using areas I and II of Fig. 7.11*a*.

$$Q = \sum_{i=1}^{2} A_i \bar{y}_i = (25)(50)(32.15) + (25)(75)(7.15) = 53.59(10^3) \text{ mm}^3$$

The shear flow at section *B-B* is

$$q = \frac{VQ}{I} = \frac{3.5(10^3)(53.59)(10^3)}{938.43(10^4)} = 20 \text{ N/mm}$$

$$S = \frac{1500 \text{ N}}{20 \text{ N/mm}} = 75 \text{ mm}$$

b. The maximum shear stress occurs at the neutral axis. The quantity Q must be computed using the total cross-sectional area, either above or below the neutral axis. Using the area below the neutral axis.

$$Q = (2)(25)(50)(32.15) + (2)(25)(44.65)(22.33) = 130.23(10^3) \text{ mm}^3$$

The thickness t to be used when computing shear stress is the total thickness at the location on the cross section where τ is to be computed; hence $t = (2)(25) = 50$ mm. Note that Q was computed using the total area below the neutral axis.

$$\tau = \frac{VQ}{It} = \frac{3.5(10^3)(130.23)(10^3)}{938.43(10^4)(50)} = 0.97 \text{ N/mm}^2 \qquad ■$$

The computation of shear flow and shear stress is dependent upon the orientation of the beam cross section with regard to the direction of loading and the location on the cross section where the analyst requires the value of shear flow or shear stress. Consider again the cross section of Fig. 7.11 and the following example.

EXAMPLE 7.6

A beam with the cross section of Fig. 7.11 is loaded such that a unit shear, 1 N, acts on the section.
a. Plot the shear flow along the cross section using the sections shown in Fig. 7.11*c*.
b. Plot the shear flow and shear stress along the cross section using the sections shown in Fig. 7.11*d*.

Solution:
The solution to the problem is somewhat long and tedious. All computations will not be shown. The main objective of this example is to demonstrate the method of computing shear flow and shear stress and to illustrate the variation of these quantities over the cross section. Once this illustrative example is mastered, the reader will have mastered the computation of shear flow and shear stress.
a. The second moment of the area was computed in the previous example as $I =$

$938.43(10^4)$ mm^4. The shear flow at section *a-a* of Fig. 7.11*c* is dependent upon Q, which is computed using the area between section *a-a* and the end of the flange. Visualize that the beam section to the left of section *a-a* is trying to move relative to the beam section on the right of *a-a*.

$$Q_{a-a} = (25 \text{ mm})(25 \text{ mm})(32.15 \text{ mm}) = 20.10(10^3) \text{ mm}^3$$

$$q_{a-a} = \frac{(1 \text{ N})(20.10)(10^3) \text{ mm}^3}{938.43(10^4) \text{ mm}^4} = 2.14(10^{-3}) \text{ N/mm}$$

$$Q_{b-b} \text{ (see Example 7.5)} = 40.19(10^3) \text{ mm}^3$$

$$q_{b-b} = \frac{40.19(10^3)}{938.43(10^4)} = 4.28(10^{-3}) \text{ N/mm}$$

$$q_{c-c} = \frac{(25)(75)(32.15)}{938.43(10^4)} = 6.42(10^{-3}) \text{ N/mm}$$

$$q_{d-d} \atop (N.A.) = \frac{(25)(50)(32.15) + (25)(44.65)(22.33)}{938.43(10^4)} = 6.94(10^{-3}) \text{ N/mm}$$

$$q_{e-e} = q_{d-d} - \frac{(25)(30.35)(15.18)}{938.43(10^4)} = 5.71(10^{-3}) \text{ N/mm}$$

$$q_{f-f} = q_{e-e} - \frac{(25)(25)(42.85)}{938.43(10^4)} = 2.86(10^{-3}) \text{ N/mm}$$

$$q_{g-g} = q_{f-f} - \frac{(25)(25)(42.85)}{938.43(10^4)} = 0$$

The shear flow can be treated as a cumulative quantity as the computation proceeds along the section. The moment of the area is computed with respect to the neutral axis (N.A.); hence, after passing section *d-d* the moment of the area changes sign and q begins to decrease. The shear flow is zero at the axis of symmetry.

b. In this part of the problem the sections are all parallel to the neutral axis. The computation will begin at section *a-a* of Fig. 7.11*d* and proceed upward, over the cross section.

$$q_{a-a} = \frac{2(1 \text{ N})(12.5 \text{ mm})(75 \text{ mm})(38.40 \text{ mm})}{938.43(10^4) \text{ mm}^4} = 7.67(10^{-3}) \text{ N/mm}$$

$$\tau_{a-a} = 7.67(10^{-3})/150 = 5.11(10^{-5}) \text{ N/mm}^2$$

$$q_{b-b} = q_{a-a} + \frac{(2)(12.5)(75)(25.90)}{938.43(10^4)} = 12.84(10^{-3}) \text{ N/mm}$$

$$\tau_{b-b} = \frac{12.84(10^{-3})}{150} = 8.56(10^{-5}) \text{ N/mm}^2$$

and

$$\tau_{b\text{-}b} = \frac{12.84(10^{-3})}{50} = 25.69(10^{-5}) \text{ N/mm}^2$$

At section *b-b* there are two choices for the thickness: the flange thickness and the web thickness. Hence, there is an abrupt change in shear stress at section *b-b*.

$$q_{c\text{-}c}_{\text{(N.A.)}} = \frac{Q}{I} = \frac{130.23(10^3)}{938.43(10^4)} = 13.88(10^{-3}) \text{ N/m}$$

Q was computed in Example 7.5*b*.

$$\tau_{c\text{-}c} = \frac{13.88(10^{-3})}{50} = 27.74(10^{-5}) \text{ N/mm}^2$$

$$q_{d\text{-}d} = q_{c\text{-}c} - \frac{(2)(30.35)(25)(15.18)}{938.43(10^4)} = 11.43(10^{-3})$$

$$\tau_{d\text{-}d} = \frac{11.48(10^{-3})}{50} = 22.85(10^{-5})$$

and

$$\tau_{d\text{-}d} = \frac{11.43(10^{-3})}{100} = 11.43(10^{-5})$$

$$q_{e\text{-}e} = q_{d\text{-}d} - \frac{(12.5)(100)(36.6)}{938.43(10^4)} = 6.55(10^{-3})$$

This value of shear flow has been computed using the statical moment of the area between section *e-e* and the base of the section. For comparison the area above section *e-e* will be used.

$$q_{e\text{-}e} = \frac{(12.5)(100)(49.10)}{938.43(10^4)} = 6.55(10^{-3})$$

It has been verified numerically that either area has the same statical moment with respect to the neutral axis. Finally,

$$\tau_{e\text{-}e} = \frac{6.55(10^{-3})}{100} = 6.55(10^{-5}) \text{ N/mm}^2$$

The results are plotted in Fig. 7.11*d*. The shear flow varies according to the selection of a particular section. Shear flow should be visualized as a force per unit length causing the cross section to tend to slide apart at a particular section. ∎

EXAMPLE 7.7

The beam of Fig. 7.12*a* is loaded transversely in both the *x-y* plane and the *x-z* plane. Assume that the line of action of the loads passes through the centroid of the cross section and compute the maximum shear stress caused by each load.

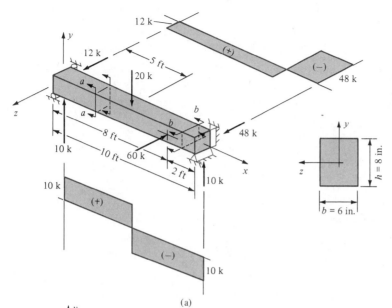

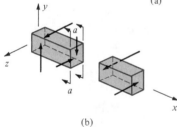

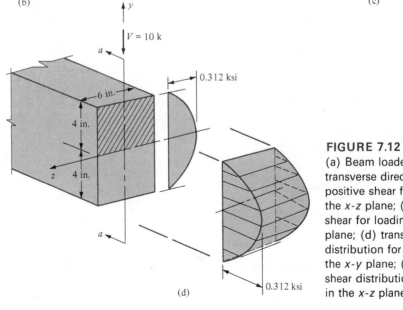

(a)

(b)

(c)

(d)

FIGURE 7.12
(a) Beam loaded in both transverse directions; (b) positive shear for loading in the x-z plane; (c) positive shear for loading in the x-y plane; (d) transverse shear distribution for loading in the x-y plane; (e) transverse shear distribution for loading in the x-z plane.

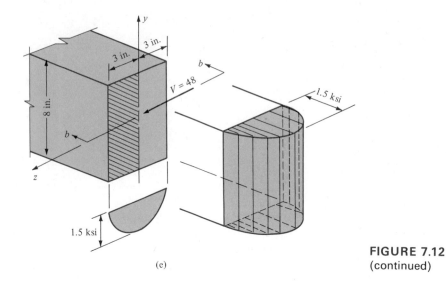

(e)

FIGURE 7.12
(continued)

Solution:

In this three-dimensional problem a transverse shear acts parallel to the y axis caused by the 20-k force and a transverse shear acts parallel to the z axis caused by the 60-k force. Shear diagrams are constructed in Fig. 7.12a using the methods of Chapter 5. Positive shear is illustrated on the free body of Fig. 7.12b. The shear sign convention for the force lying in the x-z plane is the same as that for the x-y plane (Fig. 7.12c) except that it must be viewed in the x-z plane. A free body of the beam taken at section a-a is shown in Fig. 7.12d. The direction of the transverse shear parallel to the y axis is normal to the z centroid axis as in the derivation of the shear stress formula; hence, the second moment of the area is computed with respect to the z axis. Also, Q is computed with respect to the z centroid axis.

$$\tau = \frac{(10 \text{ k})(4 \text{ in.})(6 \text{ in.})(2 \text{ in.})}{(6 \text{ in.})(8 \text{ in.})^3(6 \text{ in.})/12} = 0.312 \text{ ksi}$$

The distribution of the shear stress over the cross section is illustrated in Fig. 7.12d.

The transverse shear force parallel to the z axis is shown in Fig. 7.12e acting normal to the y centroid axis at section b-b. The second moment of the area and Q must be computed with respect to the y axis.

$$\tau = \frac{(48 \text{ k})(8 \text{ in.})(3 \text{ in.})(1.5 \text{ in.})}{(8 \text{ in.})(6 \text{ in.})^3(8 \text{ in.})/12} = 1.5 \text{ ksi}$$

The resulting shear stress distribution is shown in Fig. 7.12e.   ∎

7.4 EFFICIENT STRUCTURAL CROSS SECTIONS

In Chapter 6 it was demonstrated that rolled structural shapes were both efficient and economical for design of beams subject to bending.

It turns out that they are also ideally suited for use in the design of structures to resist shear, which will be illustrated using the following example.

EXAMPLE 7.8

An S12x31.8 beam is supported and loaded as shown in Fig. 7.13a. Compare the shear stress in the web to the shear stress in the flange.

Solution:

The maximum shear is equal to the reaction and is 100 k. The dimensions and properties of the section are obtained from Appendix C. For the purpose of this example

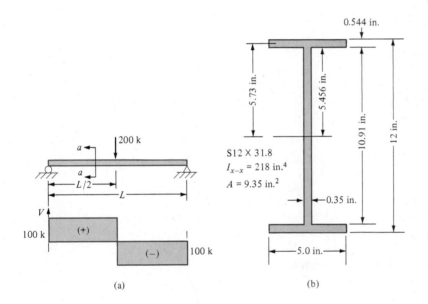

S12 × 31.8
$I_{x-x} = 218$ in.^{4}
$A = 9.35$ in.^{2}

(a)

(b)

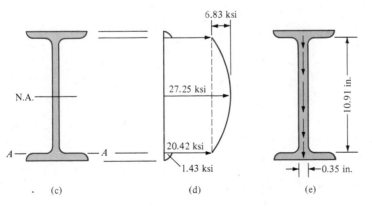

(c)

(d)

(e)

FIGURE 7.13

some approximations are necessary. The fillets and rounds will be neglected, and the dimensions will be used as shown in Fig. 7.13b. The shear stress will be computed at the neutral axis and at section A-A as shown in Fig. 7.13c. At section A-A,

$$Q_{A\text{-}A} = (5)(0.544)(5.73) = 15.58 \text{ in.}^3$$

At the neutral axis

$$Q_{\text{N.A.}} = Q_{A\text{-}A} + (5.456)(0.35)(5.456/2) = 20.79 \text{ in.}^3$$

$$\tau_{\text{N.A.}} = \frac{(100)(20.79)}{(218)(0.35)} = 27.25 \text{ ksi}$$

$$\tau_{A\text{-}A} = \frac{(100)(15.58)}{(218)(0.35)} = 20.42 \text{ ksi}$$

and

$$\tau_{A\text{-}A} = \frac{(100)(15.58)}{(218)(5)} = 1.43 \text{ ksi}$$

The variation in shear stress is illustrated in Fig. 7.13d. It can be seen that the web of the S-member carries practically all of the shear stress. The vertical shear force is shown in Fig. 7.13e. The force is shown acting downward and would be the direction of positive shear acting on section a-a of Fig. 7.13a. The shear force actually carried by the web can be evaluated using the stress distribution of Fig. 7.13d as it acts on the web area of Fig. 7.13e.

$$V_{\text{web}} = (20.42 \text{ ksi})(0.35 \text{ in.})(10.91 \text{ in.}) + (2/3)(6.83)(0.35)(10.91)$$
$$= 95.36 \text{ k}$$

To compute V_{web} the area of the stress distribution was divided into a rectangular area and a parabolic area. The result indicates that the web carries about 95 percent of the 100-k shear, which leads to an approximation that is used in engineering with little loss of accuracy. For a W- or S-shape member, the shear area is assumed to be the area of the web, and an average shear can be computed as

$$\tau_{\text{avg}} = \frac{V}{A_{\text{web}}} \tag{7.13}$$

Applying this equation to the data of Example 7.8 gives

$$\tau_{\text{avg}} = \frac{100 \text{ k}}{(10.91 \text{ in.})(0.35 \text{ in.})} = 26.18 \text{ ksi} \qquad \blacksquare$$

Note that the web area was not allowed to extend into the flanges. An additional example is in order illustrating an application of shear flow in rolled sections.

EXAMPLE 7.9

Two wide-flange members are connected as shown in Figure 7.14. The vertical shear is 25 k and 1/2-in. bolts are to be used as connectors. The bolts are subject to single shear and have an allowable shear stress of 18 ksi. Determine the spacing of the bolts along the longitudinal axis of the member.

Solution:

The properties and dimensions of the wide-flange members can be obtained from Appendix C. The first step is to locate the neutral axis for the composite member. The computations are shown in Table 7.2.

$$A = 16.96 \qquad A\bar{y} = 164.91 \qquad \bar{y} = 9.72 \text{ in.}$$

The second moment of the area with respect to the neutral axis is obtained using the data from Appendix C and the parallel-axis theorem.

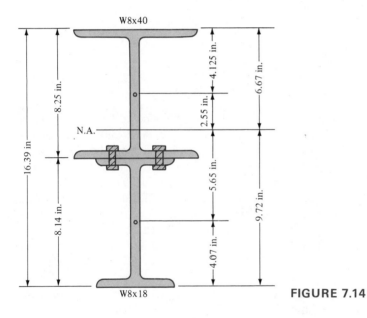

FIGURE 7.14

TABLE 7.2

Member	Area, in.^{2}	$\bar{y}$ with Respect to Base	$A\bar{y}$, in.^{3}	I, $in.^4$
W8x40	11.70	12.265	143.50	146.0
W8x18	5.26	4.070	21.41	61.9

$$I_{\text{N.A.}} = 146 + (11.7)(2.55)^2 + 61.9 + (5.26)(5.26)^2 = 451.89 \text{ in.}^4$$

The shear flow is to be computed at the section where the two wide-flange members are connected. The quantity Q must be computed for the area between that section and an outermost surface of the composite beam. The lower section is chosen, the W8x18 member.

$$Q = (5.26 \text{ in.}^2)(5.65 \text{ in.}) = 29.72 \text{ in.}^3$$

$$q = \frac{VQ}{I} = \frac{(25.0)(29.72)}{(451.89)} = 1.64 \text{ k/in.}$$

The shear force carried by each bolt is obtained using the methods of Chapter 2. The bolts are in single shear, and the shear force is given by

$$V_{\text{bolt}} = \tau_{\text{allow}} A_{\text{bolt}} = (18 \text{ ksi})(\pi)(0.5 \text{ in.})^2/4 = 3.53 \text{ k}$$

However, there are two rows of bolts; hence, the allowable shear force is $(2)(3.53 \text{ k}) = 7.06$ k.

$$S = \frac{7.06 \text{ k}}{1.64 \text{ k/in.}} = 4.31 \text{ in.} \qquad\blacksquare$$

7.5   THE BEHAVIOR OF SHEAR STRESS AND SHEAR FLOW

The shear stress equation is most accurate when used to compute the shear stress distribution for narrow rectangular cross sections such as the web of a wide-flange or standard rolled section. The formula disagrees with physical boundary conditions for some applications. The situation in question is illustrated in Fig. 7.15a. The surface located at A along the bottom edge of the beam, physically, must have zero shear stress. A surface is always free of stress, either normal or shear, unless a force is applied directly on the surface. The shear stress formula yields a calculated value of zero shear stress at A as shown in Fig. 7.13d and illustrated again in Fig. 7.15b. However, point B lies on a free surface where the shear stress must physically be zero. The shear stress formula and the diagram of Fig. 7.15b would appear to yield a value for shear stress along the interior surface of the flange, point B in the figure. The formula is not intended to give valid results along the flange when the shear loading is normal to the flange. The results shown in Fig. 7.15b should be interpreted to be valid for the section of web that extends into the flange area.

Consider the symmetrical cross section of Fig. 7.16a. The shear load is in the vertical direction and, as has been demonstrated, the web of the section is predominant in resisting the shear. Hence, stresses in the flange will be small and the actual magnitude is of little consequence

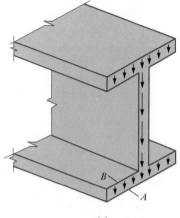

(a)

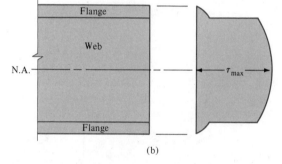

(b)

FIGURE 7.15
(a) Wide-flange member subjected to shear stress. Points *A* and *B* are external surfaces and should have zero shear stress; (b) computed shear stress distribution for a wide-flange member.

when designing using a shear stress criterion. If the shear load was horizontal, as in Fig. 7.16*b*, the flanges would carry the shear force. The neutral axis would be along the vertical centroid axis and bending would occur about the vertical axis. In general, for a thin-walled member, the thin sections whose longitudinal axes are parallel to the shear load are predominant in resisting the shear.

Some engineering applications require an understanding of how the shear flow varies along the sections of a thin-wall member. Assume a symmetrical cross section such as the one shown in Fig. 7.17*b*. The beam is oriented as shown in Fig. 7.17*a*. The load *P* of Fig. 7.17*a* acts on the cross section, as shown in Fig. 7.17*b*. The shear flow at section *a-a* is obviously zero. At any other section, such as *b-b*, along the top flange between *a-a* and the web, the shear flow increases as the area increases.

For the section illustrated, the V, I, t, and $\bar{y}$ are constant; hence, the shear flow increases as the distance or area along the flange. This

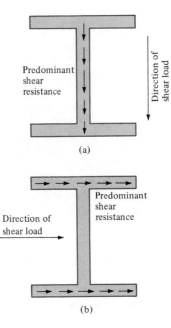

(a)

(b)

FIGURE 7.16

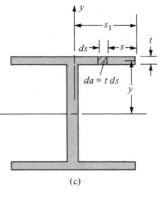

(a)

(c)

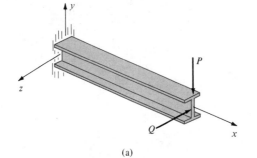

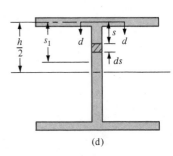

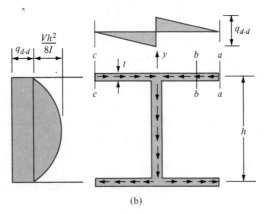

FIGURE 7.17

(b)

(d)

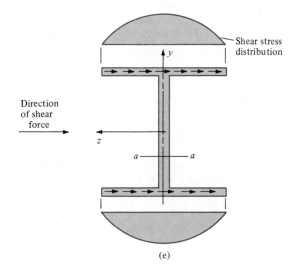

Shear stress distribution

Direction of shear force

(e)

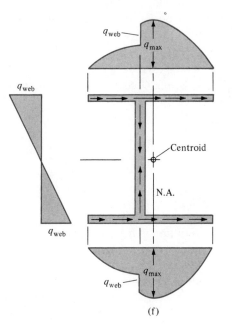

(f)

FIGURE 7.17
(continued)

is termed a linear variation and is illustrated in Fig. 7.17b. The variation beginning from the left end of the flange, section c-c, is also linear as shown in the figure.

The foregoing statements may be shown formally by referring to Fig. 7.17c. The shear flow is given by VQ/I, and Q for the differential area is

$$Q = \int_0^{s_1} \bar{y}\,da = \int_0^{s_1} \bar{y}t\,ds = \bar{y}ts\Big|_0^{s_1} \qquad \textbf{(a)}$$

Hence, Q varies as s varies, and the variation may be viewed as described previously.

At the web the shear flows in the flanges meet and can be imagined to turn a corner and flow down the web. The variation in shear flow along the web is parabolic, $\bar{y}$ is no longer a constant, and the area also varies as the y coordinate varies. The situation is shown in Fig. 7.17d. The change in shear flow along the web can be written as the shear flow at section d-d plus the change along the web, or

$$q_{\text{web}} = q_{d\text{-}d} + \frac{VQ}{I}\bigg|_{\text{web}} \qquad \textbf{(b)}$$

where

$$Q = \int \bar{y}\,da = \int_0^{s_1} \left(\frac{h}{2} - s\right)t\,ds \qquad \textbf{(c)}$$

The coordinate s may be thought of as a local coordinate measured from section d-d toward the neutral axis. Equation (c) gives

$$Q = \left[\frac{hs}{2} - \frac{s^2}{2}\right]_0^{s_1} \qquad \textbf{(d)}$$

For $s_1 = h/2$, Q corresponds to the neutral axis and Q is a maximum. Equation (b) becomes

$$q_{\text{web}} = q_{d\text{-}d} + \frac{V}{2I}\left[hs - s^2\right]_0^{s_1} \qquad \textbf{(e)}$$

and, for $s_1 = h/2$,

$$q_{\text{web}} = q_{d\text{-}d} + \frac{V}{2I}\left[\frac{h^2}{4}\right] \qquad \textbf{(f)}$$

The variation in q along the web is shown in Fig. 7.17b and is parabolic in nature.

A result similar to Eq. (f) will be illustrated using a different approach in Problem 7.29 at the end of the chapter.

A general statement concerning the behavior of shear flow is in order.

> In general the shear flow in a thin-wall member varies linearly along sections normal to the applied shear force and varies parabolically along sections parallel to the applied shear force.

To further illustrate the behavior of shear stress and shear flow, consider the applied force Q of Fig. 7.17a. The bending would be about the y axis; hence, for the symmetrical section, a vertical axis through the centroid is the neutral axis, as shown in Fig. 7.17e. The computation for shear flow along the flanges is straightforward and will result in a parabolic distribution of shear flow, as shown in Fig. 7.17e. Note that the flanges are parallel to the direction of shear force. A section *a-a* located anywhere in the web section will show that the shear flow is zero along the web. The computation for $Q = a\bar{z}$ would be the area below section *a-a* multiplied by $\bar{z}$. But $\bar{z}$ is zero and the shear flow along the web is zero. If the section were unsymmetrical, as shown in Fig. 7.17f, the shear flow would not be zero but would vary in a linear fashion along the web as illustrated in Example 7.12.

Assume that the beam of Fig. 7.17a is fabricated using two channel sections as shown in Fig. 7.18. The problem is to design the connectors. Again, the shear flow concept is to be used, but the problem must be viewed differently than the previous discussion. Consider the following problem.

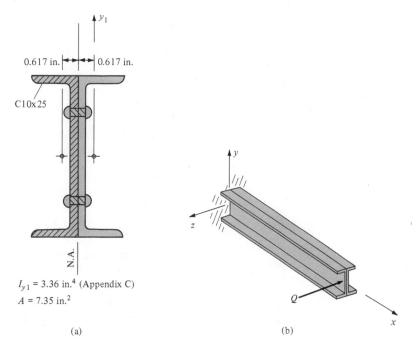

$I_{y1} = 3.36$ in.^{4} (Appendix C)

$A = 7.35$ in.^{2}

FIGURE 7.18   (a)   (b)

EXAMPLE 7.10

Two C10x25 members are connected using two rows of 1-in.-diameter bolts as shown in Fig. 7.18. The loading is illustrated in Fig. 7.18b. Assume the maximum shear is 16 k and compute the spacing of the bolts if the shear stress in each bolt is not to exceed 15 ksi.

Solution:

The shear force to be carried by each bolt is obtained using the methods of Chapter 2.

$$V_b = \tau A = \frac{(15 \text{ ksi})(\pi)(1 \text{ in.})^2}{4} = 11.78 \text{ k}$$

For two bolts,

$$V_b = 23.56 \text{ k}$$

Since bending occurs about the y axis, the neutral axis is vertical and the second moment of the area is computed with respect to the vertical centroid axis. Properties for the channel are obtained from Appendix C.

$$I_{\text{N.A.}} = 2[3.36 \text{ in.}^4 + (7.35 \text{ in.}^2)(0.617 \text{ in.})^2] = 12.32 \text{ in.}^4$$

Q is computed for the shaded area with respect to the neutral axis. The shear flow is to be computed along the neutral axis as opposed to normal to the neutral axis as in the previous discussion.

$$Q = a\bar{z} = (7.35 \text{ in.}^2)(0.617 \text{ in.}) = 4.53 \text{ in.}^3$$

$$q = \frac{VQ}{I} = \frac{(16 \text{ k})(4.53 \text{ in.}^3)}{12.32 \text{ in.}^4} = 5.88 \text{ k/in.}$$

$$S = \frac{23.56 \text{ k}}{5.88 \text{ k/in.}} = 4.00 \text{ in.}$$

The next section in which the concept of shear center is discussed will require an understanding of the basics of shear flow in thin-walled members.

7.6 SHEAR CENTER

Throughout Chapter 6 and thus far in Chapter 7 the assumption has been made that the beam cross section was symmetrical with respect to the line of action of the transverse loading. If the loading was vertical, which was the case for most problems, the cross section contained a vertical line of symmetry. For a horizontal load the cross section contained a horizontal axis of symmetry. All of these beam problems have in common that the line of action of the loading passes through the axis of symmetry, which has been coincident with the centroid axis. In this section consideration is given to the case when the cross section is not symmetrical with respect to the plane of the loads.

To compare a symmetrical section to an unsymmetrical section consider the beam of Fig. 7.19. The cross section is symmetrical, and it will be assumed that the load passes through the centroid. The shear flow distribution is shown in Fig. 7.19*b*. Summing forces on the cut section would show that the shear flow acting along the web would exactly balance the externally applied shear. The shear flows in the flanges would themselves be balanced. Hence, the force equations of statics would be satisfied. Summing moments of the distributed shear force at the centroid, point *A*, or any other point on the cross section would verify that the moment equation of statics is satisfied. The beam is indeed subject to a state of symmetrical bending.

In order to illustrate the effect of an unsymmetrical cross section consider the cross section of Fig. 7.20. Assume the channel section to be loaded such that the line of action of the vertical load passes through the web. Again the shear flow in the web can be summed and will exactly balance the applied shear *V*. The shear flows in the flanges are equal and opposite in direction. The force equations of statics are satisfied. Summing moments about point *A*, or any other point, will give an unbalanced moment on the cross section.

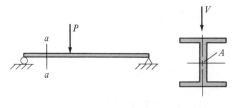

(a)

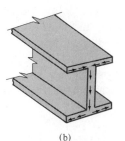

(b)

FIGURE 7.19
(a) Symmetrical cross section; (b) shear flow acting on section *a-a*.

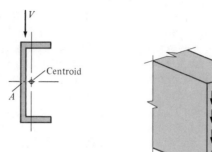

FIGURE 7.20
Shear flow acting on an
unsymmetrical cross section.

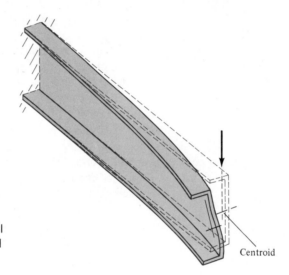

FIGURE 7.21
Warping of an unsymmetrical
cross section with shear load
applied at the centroid.

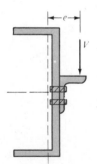

FIGURE 7.22
Shear load applied at the
shear center.

An unbalanced moment cannot occur physically; the equations of statics must be satisfied. The beam will correct the situation by twisting so that the internal shear forces can redistribute and satisfy the equations of statics. The actual situation is illustrated in Fig. 7.21. Twisting of the beam is an undesirable effect. The situation is corrected by placing the external shear at a location such that the moment of internal shear is balanced by the moment of the external shear. The point in the plane of the cross section where the external shear is applied is referred to as the shear center. The correction may be accomplished as illustrated in Fig. 7.22. An angle has been attached to the channel, and the external shear is applied to the angle. The location of the external shear is computed as outlined in Example 7.11.

EXAMPLE 7.11

The channel section of Fig. 7.23 will carry an external shear of 1000 lb. Locate the shear center with respect to the center of the web of the channel.

Solution:
The shear flow must be computed; then the shear force in each section can be calculated. Then the summation of moments about point A will give the value of e,

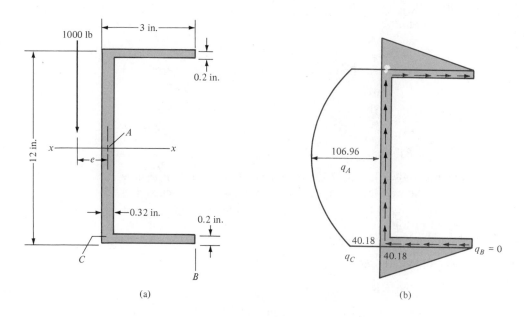

(a)

(b)

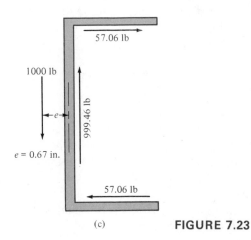

(c)

FIGURE 7.23

the location of the shear center. The shear loading is normal to the x axis and causes bending about the x axis. Hence, the second moment of the area should be computed with respect to the x axis.

$$I_x = \frac{(0.32)(11.6)^3}{12} + 2\left[\frac{(3.0)(0.2)^3}{12} + (3.0)(0.2)(5.9)^2\right]$$

$$= 83.40 \text{ in.}^4$$

The shear flow q is zero at point B. At point C,

$$q_C = \frac{VQ}{I} = \frac{(1000 \text{ lb})[(2.84)(0.2)(5.9)] \text{ in.}^3}{83.40 \text{ in.}^4}$$

$$= 40.18 \text{ lb/in.}$$

The shear flow at A is equal to the shear flow at C plus the increase between C and A; that is,

$$q_A = q_C + (1000)\frac{[(5.9)(0.32)(5.9/2)]}{83.40}$$

$$= 40.18 + 66.78 = 106.96 \text{ lb/in.}$$

The variation in shear flow is shown in Fig. 7.23b. Along the flanges the shear flow variation is linear and along the web it is parabolic. This variation was discussed in the preceding section. The shear force in each segment of the thin-walled member is equal to the area illustrated in Fig. 7.23b. For the web the shear will be computed using a rectangular area and a parabolic area.

$$V_{\text{web}} = (40.18 \text{ lb/in.})(11.8 \text{ in.}) + (66.78 \text{ lb/in.})(11.8 \text{ in.})(2/3)$$

$$= 474.12 + 525.34 = 999.46 \text{ lb}$$

The result is slightly less than 1000 but close enough to indicate that the summation of forces in the vertical direction is satisfied. The shear force in the flange is computed using a triangular shear flow variation.

$$V_{\text{flange}} = (40.18 \text{ lb/in.})(2.84 \text{ in.})(1/2) = 57.06 \text{ lb}$$

The shear forces are illustrated in Fig. 7.23c. Summing moments about the center of the web gives

$$1000e = (57.05 \text{ lb/in.})(5.9 \text{ in.})(2)$$

$$e = 0.67 \text{ in.}$$

The applied shear force should be placed 0.67 in. to the left of the centerline of the web.

In the computation of shears, the length of web was assumed to be from the center of the bottom flange to the center of the top flange. Similarly, the length of the flange was assumed to extend to the center of the web. ∎

EXAMPLE 7.12

The unsymmetrical, thin-walled section of Fig. 7.24 is subject to shear loading from either transverse direction. Locate the shear center with respect to the centroid axis of the section. Assume unit shear forces, plot the shear flow, and check for force equilibrium on the section.

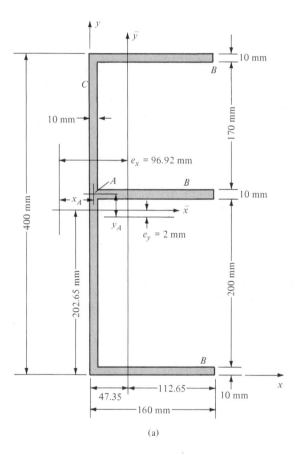

(a)

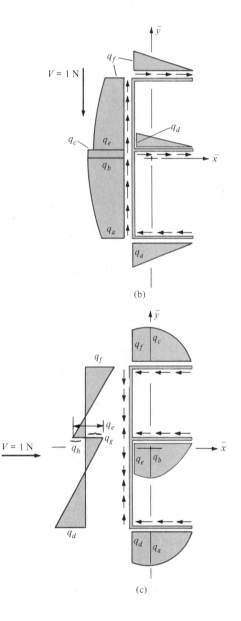

(b)

(c)

FIGURE 7.24

Solution:
First, locate the centroid with respect to the x-y axis. Table 7.3 illustrates the computations.

$$\Sigma A = 8.5(10^3) \qquad \Sigma xA = 4.025(10^5) \qquad \Sigma yA = 1.7225(10^6)$$

$$\bar{x} = \frac{4.025(10^5)}{8.5(10^3)} = 47.35 \text{ mm}$$

$$\bar{y} = \frac{1.7275(10^6)}{8.5(10^3)} = 202.65 \text{ mm}$$

The second moment of the area is computed as follows:

$$I_{\bar{x}} = \frac{(3)(150)(10)^3}{12} + (1500)(197.65)^2 + (1500)(12.35)^2$$

$$+ (1500)(192.35)^2 + \frac{(10)(202.65)^3}{3} + \frac{(10)(197.35)^3}{3}$$

$$= 1.677(10^8) \text{ mm}^4$$

$$I_{\bar{y}} = \frac{(3)(10)(112.67)^3}{3} + \frac{(3)(10)(37.35)^3}{3} + \frac{(400)(10)^3}{12} + (4000)(42.35)^2$$

$$= 2.202(10^7) \text{ mm}^4$$

A unit shear load will be assumed in the vertical y direction as shown in Fig. 7.24b. Bending would be about the $\bar{x}$ axis, Q is computed with respect to the $\bar{x}$ axis, and the second moment of the area is $I_{\bar{x}}$. For areas B, V and $I_{\bar{x}}$ are constant but Q varies.

$$q_a = \frac{VQ}{I_{\bar{x}}} = \frac{(1 \text{ N})(155 \text{ mm})(10 \text{ mm})(197.65 \text{ mm})}{1.677(10^8) \text{ mm}^4}$$

$$= 1.829(10^{-3}) \text{ N/mm}$$

Note that the area dimension is assumed to extend to the centerline of the vertical web member.

TABLE 7.3

	Area, mm^2	x, mm	y, mm	xA, mm^3	yA, mm^3
B	1500	85	5	1.275(10^5)	7.500(10^3)
B	1500	85	215	1.275(10^5)	3.225(10^5)
B	1500	85	395	1.275(10^5)	5.925(10^5)
C	4000	5	200	2.000(10^4)	8.000(10^5)

$$q_b = q_a + \frac{(1)(197.65)(10)(197.65/2)}{1.677(10^8)}$$

$$= 1.829(10^{-3}) + 1.165(10^{-3}) = 2.994(10^{-3}) \, \text{N/mm}$$

The shear flow is maximum at the centroid axis (neutral axis) and then decreases above the neutral axis. The small section of web between the neutral axis and the center flange will be considered next.

$$q_c = q_b - \frac{(1)(12.35)(10)(12.35/2)}{1.677(10^8)}$$

$$= 2.994(10^{-3}) - 4.547(10^{-6}) = 2.989(10^{-3}) \, \text{N/mm}$$

The shear flow at q_c will decrease by the amount of shear that flows out of the center flange. Hence, compute the shear flow in the center flange.

$$q_d = \frac{(1)(150)(10)(12.35)}{1.677(10^8)} = 1.104(10^{-4}) \, \text{N/mm}$$

$$q_e = q_c - q_d = 2.989(10^{-3}) - 1.104(10^{-4}) = 2.879(10^{-3}) \, \text{N/mm}$$

$$q_f = q_e - \frac{(1)(180)(10)(102.35)}{1.677(10^8)}$$

$$= 2.879(10^{-3}) - 1.098(10^{-3}) = 1.781(10^{-3}) \, \text{N/mm}$$

The shear flow q_f should be balanced by the shear flow in the top flange, or

$$q_f = \frac{(1)(155)(10)(192.35)}{1.677(10^8)} = 1.778(10^{-3}) \, \text{N/mm}$$

The difference is negligible; hence, assume that the computations are correct thus far. The shear flows in the flanges should be equal; a statical check will verify that they are. The shear flows in the web should add up to balance the 1-N shear.

Keeping in mind that the shear flows in the vertical direction vary as a quadratic, the shear flows add as

$$V_{\text{web}} = 197.65q_a + (q_b - q_a)(197.65)(2/3) + (q_b + q_c)(12.35)(1/2)$$
$$+ (q_e - q_f)(180)(2/3) + 180q_f$$

The small area between q_b and q_c has been approximated as a trapezoid.

$$V_{\text{web}} = (197.65)(1.829)(10^{-3}) + (1.165)(10^{-3})(197.65)(2/3)$$
$$+ (5.983)(10^{-3})(12.35)(1/2) + (1.098)(10^{-3})(180)(2/3)$$
$$+ 180(1.778)(10^{-3}) = 1.004 \, \text{N}$$

agrees well enough with the applied shear of 1 N to assume that the analysis is correct.

The shear center can be located by summing moments about any point on the cross section. A convenient point is the intersection of the web and center flange.

$$\Sigma M_A = 0$$

$$(1.829)(10^{-3})(155)(210)(1/2) + (1.778)(10^{-3})(155)(180)(1/2) = (1)(x_A)$$

x_A is the location of the shear center with respect to point A.

$$x_A = 54.57 \text{ mm}$$

$$e_x = 54.57 + 42.35 = 96.92 \text{ mm}$$

to the left of the centroid.

A unit shear will be assumed to act horizontally as shown in Fig. 7.24c. The shear flows in each flange will act as illustrated to oppose the direction of the applied shear. For the loading, bending would occur about the $\bar{y}$ axis; hence, $I_{\bar{y}}$ is used in the calculation. The shear flow is zero at the end of the flanges and increases as a parabola to the $\bar{y}$ axis and then decreases.

For the flanges, to the right of the $\bar{y}$ axis,

$$q_a = q_b = q_c = \frac{(1)(112.65)(10)(112.65/2)}{2.202(10^7)}$$

$$= 2.881(10^{-3}) \text{ N/mm}$$

From the $\bar{y}$ axis to the center of the web the shear flow decreases, for q_d

$$q_d = q_a - \frac{(1)(42.35)(10)(42.35/2)}{2.202(10^7)}$$

$$q_d = q_e = q_f = 2.881(10^{-3}) - 4.072(10^{-4}) = 2.474(10^{-3}) \text{ N/mm}$$

At q_d the shear flow tends to flow around the corner and decreases; note that for the Q computation the centroid locations of the areas are negative with respect to the $\bar{y}$ axis.

$$q_g = q_d - \frac{(1)(210)(10)(42.35)}{2.202(10^7)}$$

$$= 2.474(10^{-3}) - 4.038(10^{-3}) = -1.564(10^{-3}) \text{ N/mm}$$

$$q_h = q_f - \frac{(1)(180)(10)(42.35)}{2.202(10^7)}$$

$$= 2.474(10^{-3}) - 3.462(10^{-3}) = -0.988(10^{-3}) \text{ N/mm}$$

At the intersection of the web and center flange the negative and positive shear flows should be in equilibrium.

$$q_e + q_g + q_h = 2.474(10^{-3}) - 1.564(10^{-3}) - 0.988(10^{-3})$$
$$= -0.078(10^{-3})$$

which is about 3 percent of q_e or accurate enough for our use of the centerline of the section to establish the length dimensions.

Along the web the shear flows should be balanced, which is to say that the areas of the triangular shear flow distribution along the web, shown in Fig. 7.24c, should be in equilibrium.

The horizontal shear flows should sum to equal 1 N.

$$3[q_a(112.65)(2/3) + q_d(42.35) + (q_a - q_d)(42.35)(2/3)]$$
$$= 3[2.881(10^{-3})(112.65)(2/3) + 2.474(10^{-3})(47.35) + 4.072(10^{-4})(42.35)(2/3)]$$
$$= 3(0.345) = 1.035$$

which is in error by about 3.5 percent. Assume this error to be tolerable considering the many computations. The quantity in the brackets is the force in each flange; hence, summing moments about A gives

$$(0.345 \text{ N})(210 \text{ mm}) - (0.345 \text{ N})(180 \text{ mm}) - (1 \text{ N})(y_A) = 0$$

$$y_A = 10.35 \text{ mm} \qquad \text{(as shown in Fig. 7.24a)}$$

$$e_y = 12.35 - 10.35 = 2.0 \text{ mm}$$

on the positive side (above) of the centroid axis. Then $e_x = 96.92$ mm and $e_y = 2.0$ mm, as shown in Fig. 7.24. ■

7.7 SUMMARY

This chapter, dealing with transverse shear stress produced by bending-type loads, completes the development and application of the basic theory of stress analysis in mechanics of materials.

The equilibrium method illustrated in Section 7.2 served the purpose of introducing the physical phenomena of shear flow and horizontal shear stress. The equilibrium method yields the average shear stress and is dependent on the choice of free-body section. The computations are somewhat tedious; hence, it is recommended that Eqs. (7.11) and (7.12) be used to compute shear flow and shear stress.

The significant concepts in this chapter center about the internal effects caused by transverse shear loading. Shear flow, given as

$$q = \frac{VQ}{I} \qquad\qquad (7.11)$$

is important for determining the spacing of connectors for composite beams. The evaluation of shear flow is important as an intermediate step in locating the shear center.

The quantity Q is given as

$$Q = \int y\, dA \tag{7.10}$$

and is the first moment of the area between the outside surface of the beam cross section and the point where shear stress is to be computed.

The transverse shear stress is given by

$$\tau = \frac{VQ}{It} \tag{7.12}$$

The significant result that shear stresses on mutually perpendicular planes are equal in magnitude has been dealt with again.

Still another word of caution is in order. Equations (7.11) and (7.12) are approximate. However, they are very accurate for thin, deep structural members and should be used accordingly. The general theory of shear stress distribution can be considered to be an advanced topic reserved for a more in-depth treatment than this text allows.

For Further Study

Boresi, A. P., and O. M. Sidebottom, *Advanced Mechanics of Materials*, 4th ed., John Wiley, New York, 1985.

Cowper, G. R., "The Shear Coefficient in Timoshenko's Beam Theory," *Journal of Applied Mechanics*, June 1966, pp. 335–340.

Kuhn, P., *Stresses in Aircraft and Shell Structures*, McGraw-Hill, New York, 1956.

7.8 PROBLEMS

7.1 The beam of Fig. P7.1 is loaded and supported as shown. Compute the shear force and shear flow acting 100 mm above the neutral axis for the beam section between *a-a* and *b-b*. Use the equilibrium method.

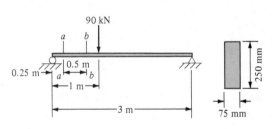

7.2 Assume that the beam of Fig. P7.1 has the cross-sectional shape shown in Fig. P7.2. Compute the shear flow at section *A-A* using the equilibrium method.

FIGURE P7.1

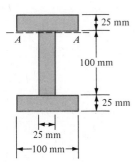

FIGURE P7.2

7.3 The beam of Fig. P7.3 is loaded as illustrated and has a solid circular cross section. Compute the shear force and shear flow acting at the neutral axis for the 1-ft section between *a-a* and *b-b*.

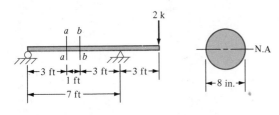

FIGURE P7.3

7.4 Assume that the beam of Fig. P7.3 has a T-shape cross section as shown in Fig. P7.4. The flange section will be securely fastened to the web using connectors that will withstand a shear force of 400 lb each. Compute the spacing for a single row of these connectors. Use the beam of Fig. P7.3 and the section *a-a* to *b-b*.

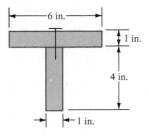

FIGURE P7.4

7.5 A section of a box beam is shown in Figs. P7.5*a* and *b*. Compute the spacing of the connectors, as illustrated, if each connector carries an allowable shear force of 600 lb.

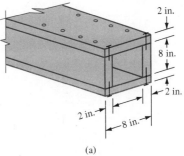

(a)

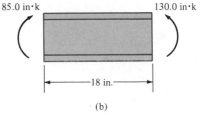

(b)

FIGURE P7.5

7.6–7.10 Locate the neutral axis and show that the statical moment of the shaded area is the same as the statical moment of the unshaded area, with respect to the neutral axis, for the beam cross sections of Figs. P7.6–P7.10.

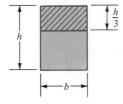

FIGURE P7.6

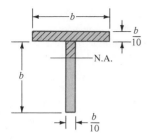

FIGURE P7.7

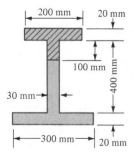

FIGURE P7.8

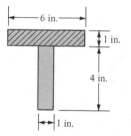

FIGURE P7.9

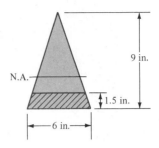

FIGURE P7.10

7.11 A beam is to be fabricated using three wooden planks as shown in Fig. P7.11. The maximum transverse shear has been computed to be 10 kN. Determine the spacing of connectors if each connector has an allowable shear load of 400 N.

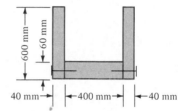

FIGURE P7.11

7.12 Assume the beam cross section of Fig. P7.11 with connectors specified to be spaced at 125 mm. The maximum transverse shear acting on the section is 10 kN. Compute the shear force carried by each connector.

7.13 The cross section of Fig. P7.13 is subjected to a transverse shear of 2 k.
a. Assume that the 1- by 2-in. blocks are nailed in place and each nail will withstand a shear force of 350 lb. Compute the nail spacing.

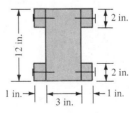

FIGURE P7.13

b. Assume that the 1- by 2-in. blocks are glued in place and compute the shear stress on the glue joint.

7.14 A composite beam is fabricated as shown in Fig. P7.14. The shear on the section is 4000 lb. Compute the spacing of the bolts if each bolt will carry an allowable shear of 1000 lb.

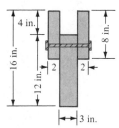

FIGURE P7.14

7.15 The cross section of Fig. P7.14 is fastened together with bolts spaced at 4 in. If the allowable shear force per bolt is 1600 lb, compute the maximum transverse shear that can be applied to the section.

7.16 A laminated beam with cross section shown in Fig. P7.16 carries a maximum transverse shear of 12 kN. Compute the minimum value of the dimension h in order to limit the shear stress on the glue joint between laminates to 350 kPa.

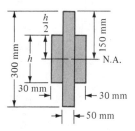

FIGURE P7.16

7.17 An I-shape structural member is to be built up using three sections as shown in Fig. P7.17. The maximum vertical shear is 8 kN. Compute the spacing of shear connectors if the maximum allowable shear per connector is 800 N.

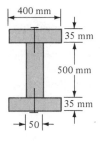

FIGURE P7.17

7.18 The beam of Fig. P7.18 is simply supported and carries a uniformly varying load. The cross section is square and built up from two members as illustrated in the figure. The two sections are securely connected using two rows of bolts. Compute the spacing of the bolts at 2-ft intervals along the length of the member. Assume that each bolt will withstand an allowable shear of 2500 lb.

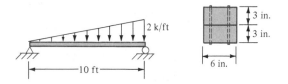

FIGURE P7.18

7.19 A beam with the cross section shown in Fig. P7.19 is subjected to a transverse shear force of 2000 N. Compute the shear flow at sections a-a, b-b, and c-c.

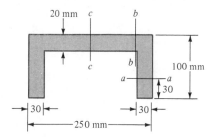

FIGURE P7.19

7.20 Plot the shear flow for the symmetrical cross section shown in Fig. P7.20. Assume a shear loading of 1 N acting through the axis of symmetry. Show that the shear flow in the vertical sections balance the applied shear.

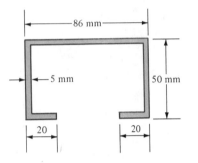

FIGURE P7.20

7.21 Compute the maximum shear stress for the beam of Fig. P7.21.

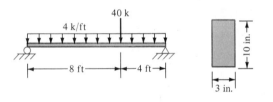

FIGURE P7.21

7.22 Compute the maximum shear stress for the beam of Fig. P7.22. The beam is a standard-weight 3-in. steel pipe. See Appendix C for properties of steel pipe.

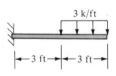

FIGURE P7.22

7.23 Compute the maximum shear stress for the beam of Fig. P7.23.

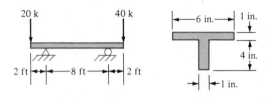

FIGURE P7.23

7.24 A 2- by 10-in. beam is subjected to a transverse shear of V lb. A 3-in.-diameter hole is drilled through the beam at the neutral axis, as illustrated in Fig. P7.24. Compare the maximum shear stress for the section with the hole to a solid section.

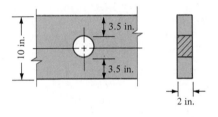

FIGURE P7.24

7.25 A floor system similar to the one of Example 7.4 is illustrated in Fig. P7.25. The uniform load on the floor is 120 lb/ft². The supporting beams are 12 ft long and simply supported. Compute the spacing S of the supporting beams if the maximum shear stress in each beam is not to exceed 300 psi.

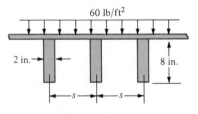

FIGURE P7.25

7.26 A floor system is simply supported and spans 10 ft using 2- by 12-in. members spaced 24 in. center to center. The floor loading is 60 lb/ft². Compute the maximum shear stress for each beam.

7.27 A simply supported floor system spans 3.5 m and is subjected to a uniform load of 3 kPa. The floor system is supported with rectangular beams, 50 by 250 mm, spaced 0.4 m center to center. Compute the spacing of 50- by 200-mm beams if the shear stress is not to exceed the maximum shear developed in the deeper members.

7.28 A floor system is supported as illustrated in Fig. P7.28. The supporting members are 2 by 6 in., and the floor loading is 450 lb/ft². Compute the spacing of the beams if the allowable shear stress is not to exceed 400 psi.

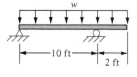

FIGURE P7.28

7.29 Use the beam cross section of Fig. P7.29 to derive an expression for the variation in shear stress over the cross section as

$$\tau = \frac{V}{2I}\left[\left(\frac{h}{2}\right)^2 - h_1^2\right]$$

Compare this result to Eq. (f) of Section 7.5.

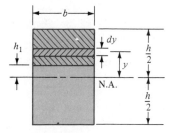

FIGURE P7.29

7.30 A simply supported beam 8 ft long is subjected to uniform loading as illustrated in Fig. P7.30. Compute the maximum transverse shear stress and note its location on the cross section.

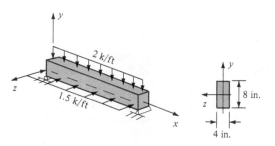

FIGURE P7.30

7.31 The beam of Fig. P7.31 is loaded in both transverse directions and requires that transverse shear stress be evaluated as it acts both vertically and horizontally. Compute the value of h if the allowable transverse shear stress is 1200 psi for both directions.

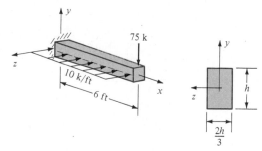

FIGURE P7.31

7.32 A simply supported beam 9 ft long is loaded as shown in Fig. P7.32. The forces pass through the centroid. Select a wide-flange member of minimum weight that will satisfy the transverse shear criterion for both transverse directions. Assume an allowable shear stress of 18,000 psi.

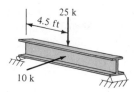

FIGURE P7.32

7.33 A W14x74 wide-flange member is subjected to a vertical shear of 9 k. Compute the maximum shear stress in the member.

7.34 Two structural channels, C12x20.7, are connected as illustrated in Fig. P7.34. Compute the maximum shear stress caused by a vertical shear of 40 k

FIGURE P7.34

7.35 Two identical wide-flange members, W8x40, are to be securely fastened as illustrated in Fig. P7.35. A maximum vertical shear of 30 k will be applied to the section. Compute the spacing of 1/2-in. bolts if the allowable shear stress per bolt is not to exceed 18,000 psi.

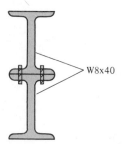

FIGURE P7.35

7.36 Two wide-flange members are to be connected as illustrated in Fig. P7.36. The composite beam is to sup-port a maximum shear of 60 k. Compute the spacing of 3/4-in. connectors if the allowable shear stress per connector is 15,000 psi.

FIGURE P7.36

7.37 A composite beam is illustrated in Fig. P7.37. Compute the shear flow at section *a-a* for a vertical shear loading of 12 k.

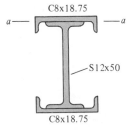

FIGURE P7.37

7.38 A rectangular box beam is illustrated in Figs. P7.38*a* and *b*. The angles are L4x3x1/2, with the longer leg parallel to the *y* axis. Cover plates are 1/2-in. thick. The beam is loaded as shown in Fig. P7.38*b*. Compute the shear flow at sections *a-a* and *b-b*.

7.39 The composite beam section of Fig. P7.39 has connectors 15 mm in diameter evenly spaced in two rows every 125 mm. The allowable shear stress per connector is 85 MPa. Compute the maximum shear flow that the connectors will carry, and then determine the maximum vertical shear that can be applied to the section.

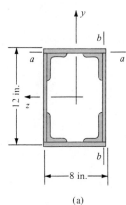

(a)

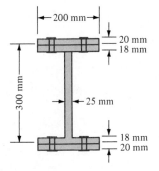

FIGURE P7.38

FIGURE P7.39

7.40–7.43 Compute the location of the shear center with respect to the centroid axis for Problems 7.40–7.43. Assume that a unit shear force acts on the section. Plot the distribution of the shear flow.

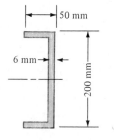

FIGURE P7.40

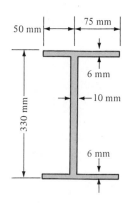

FIGURE P7.41

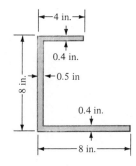

FIGURE P7.42

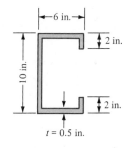

FIGURE P7.43

CHAPTER 8

COMBINED STRESSES

8.1   INTRODUCTION

In the previous chapters the six basic actions of statics were studied. Each action was considered separately, and basic formulas were derived for computing individual stresses caused by a particular force system. In this chapter a systematic method for combining the effects of two or more of these individual stresses will be discussed. The formulas are presented below for ready reference.

Normal stress due to an axial force was discussed in Chapter 2 and is given by

$$\sigma = \frac{P}{A} \tag{8.1}$$

It is important to recall that in order for this equation to be used, the force must be applied through the centroid of the cross section of the member. If the force is not applied through the centroid it must be transferred to the centroid. This is a process that was dealt with in engineering mechanics-statics and involves creating an equivalent force couple system. Subsequently, this concept will be reviewed.

287

Normal stress, referred to as flexural stress, due to transverse loading on a beam-type structure is given by the bending-stress formula

$$\sigma = \frac{My}{I} \tag{8.2}$$

Flexural stresses occur as longitudinal stress in a beam structure.

A shear stress usually accompanies the flexural stress. Transverse shear stress caused by transverse loads is given by

$$\tau = \frac{VQ}{It} \tag{8.3}$$

This equation may very well be the most difficult for students to remember and use properly.

Finally, shear stress due to an applied torque or axial moment is

$$\tau = \frac{M_t\rho}{J} \tag{8.4}$$

Equation (8.4) is limited in applicability. It is valid only for a shaft of circular cross section.

The primary purpose of this chapter is to illustrate the combined effect of these stresses. The method of analysis is twofold. The force and moment system acting at a particular point on the structure must be identified; then the stresses produced by the force and moment system must be computed. Results will be presented using a stress element.

Thin-wall pressure vessel theory will be discussed. The state of stress produced by internal pressure acting on a pressure vessel will be combined with the previous equations in order to illustrate combined stress effects.

8.2   STRESS AT A POINT

Stress at a point is terminology that means exactly what it says. Refer to Fig. 8.2 of Example 8.1. Stress at a point C would imply that the stresses at point C are to be computed. The point must be drawn large enough for it to be visualized. Therefore the concept of a stress block or material element is necessary. The stress block is the point enlarged for the practical purpose of drawing it and is referred to as a stress element since its actual size is elemental. Point C of Fig. 8.2 is shown enlarged in Fig. 8.1a. Physically, the element can be visualized as a small square near the outside surface of the beam located at point C.

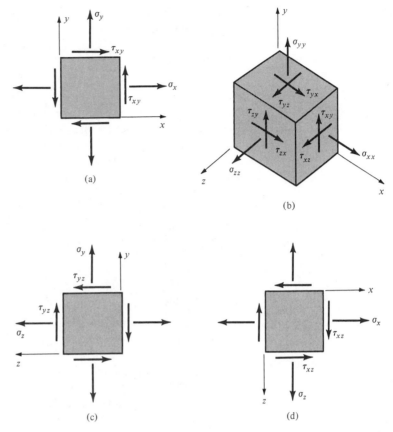

(a)

(b)

(c)

(d)

FIGURE 8.1
(a) Element viewed along the z axis; (b) positive stresses acting on an element; (c) element viewed along the x axis; (d) element viewed along the y axis.

The stresses are identified as acting on the edge of the elemental square and is the standard concept of stress at a point in two dimensions.

Actually, the elemental square is an elemental cube located at point C. The cube is shown in Fig. 8.1b. The subscript notation identifies the direction of the stress at the face, or area, of the cube on which it acts. For instance, the stress σ_{xx} is the normal stress acting on the face perpendicular to the x axis. The first subscript identifies the face, or area, of the cube. The second subscript identifies the direction of the stress. A shear stress on the face of the cube normal to the y axis and acting in the x direction would be τ_{yx}, as shown in Fig. 8.1b. The complete three-dimensional description of stress at a point is illustrated in the figure.

The discussion in the previous chapters concerning shear stresses on mutually perpendicular planes would imply that $\tau_{xy} = \tau_{yx}$, $\tau_{yz} = \tau_{zy}$, and $\tau_{xz} = \tau_{zx}$. The subscript notation lends itself toward identifying this equivalence. The nine components shown in Fig. 8.1b are conveniently presented using the following format.

$$\boldsymbol{\sigma} = \begin{pmatrix} \sigma_{xx} & \tau_{xy} & \tau_{xz} \\ \tau_{yx} & \sigma_{yy} & \tau_{yz} \\ \tau_{zx} & \tau_{zy} & \sigma_{zz} \end{pmatrix} \tag{8.5}$$

The use of the boldface $\boldsymbol{\sigma}$ will represent the stress at a point. Equation (8.5) represents the stress, $\boldsymbol{\sigma}$, as a nine-component quantity and is referred to as a stress tensor, which has certain mathematical properties that are useful in advanced studies.

Stress at a point will be viewed two dimensionally in this chapter. Figure 8.1a is actually the cube as it is viewed along the z axis; therefore, even though the stresses appear to be applied along the edge of the element, they are actually applied to a surface that is perpendicular to the plane of the page. The stress components of Fig. 8.1a will be written as

$$\boldsymbol{\sigma} = \begin{pmatrix} \sigma_{xx} & \tau_{xy} \\ \tau_{yx} & \sigma_{yy} \end{pmatrix} \quad \text{or} \quad \begin{pmatrix} \sigma_{x} & \tau_{xy} \\ \tau_{xy} & \sigma_{y} \end{pmatrix} \tag{8.6}$$

The second subscript has been dropped since the notation σ identifies a normal stress and the subscript identifies the surface upon which it acts. The shear stresses are equal in magnitude, $\tau_{xy} = \tau_{yx}$, and no confusion should occur by replacing τ_{yx} with τ_{xy}.

In some computations the element will be viewed along the x or y axis. In these instances stress at a point will be written

$$\boldsymbol{\sigma} = \begin{pmatrix} \sigma_{y} & \tau_{yz} \\ \tau_{yz} & \sigma_{z} \end{pmatrix}$$

corresponding to Fig. 8.1c and

$$\boldsymbol{\sigma} = \begin{pmatrix} \sigma_{x} & \tau_{xz} \\ \tau_{xz} & \sigma_{z} \end{pmatrix} \tag{8.7}$$

corresponding to Fig. 8.1d.

The directions of the stresses are positive in Fig. 8.1 for the purpose of representing stress at a point. The actual directions of the stresses for a structure such as the beam of Fig. 8.2 are determined using the beam sign convention of Chapter 5.

The stresses at a point are caused by moments and forces acting at the section. The analysis of stress for a given loading condition applied to a structure can be broken down into two basic steps. A free body of the structure with the point in question lying on the cut section must be visualized and the proper internal forces and moments computed. Second, the stresses must be computed using the methods of the previous chapters. The six actions, three forces and three moments, are computed using the methods of statics. In mechanics of materials these actions are always interpreted as an axial force, two shears perpendicular to the axial force, a torque or axial moment, and two bending moments. The solution of a problem containing combined stresses will best illustrate the method.

EXAMPLE 8.1

The beam of Fig. 8.2 (next page) is loaded with a single, 300-kN concentrated force. Compute the stresses at points A, B, and C.

Solution:
The beam is analyzed in Fig. 8.2b showing axial-force, shear, and moment diagrams. The sign convention for the vertical section of the member is the same as in Chapter 5. When the vertical member is viewed from the right edge of the page toward the left, compression in the top of the member is positive bending moment.

A free-body section cut 0.9 m from the left end is illustrated in Fig. 8.2c. The axial force is 300 kN, and the bending moment, applied about the z axis, is 135 kN·m positive. The axial stress is computed using Eq. (8.1).

$$\sigma = \frac{300 \text{ kN}}{(0.1)(0.25) \text{ m}^2} = 12(10^3) \text{ kN/m}^2 = 12 \text{ MPa} \qquad \text{(compression)} \qquad \textbf{(a)}$$

The axial stress is compressive and constant at A, B, and C as illustrated in Fig. 8.2c.
The bending stress is given by Eq. (8.2).

$$\sigma = \frac{(135 \text{ kN·m})y}{(0.1)(0.25)^3/12} = 1.037(10^6)y \text{ kN/m}^2 = 1037y \text{ MPa} \qquad \textbf{(b)}$$

At point A, using Eq. (b),

$$\sigma = (1037)(0.125) \text{ MPa} = 129.6 \text{ MPa} \qquad \text{(tension)} \qquad \textbf{(c)}$$

The state of stress at point A is shown in Fig. 8.2d. The two normal stresses at A are added algebraically to give $\sigma_A = 117.6$ MPa (tension). At point B, using Eq. (b) with $y = 0.125$ m gives $\sigma = 129.6$ MPa (compression). The stresses at B are shown in Fig. 8.2e. At point C, $y = 0$, and Eq. (b) gives zero bending stress at point C. The result is shown in Fig. 8.2f.

Formally, the results may be written using the notation for stress at a point given by Eq. (8.6). That is,

$$\sigma_A = \begin{pmatrix} 117.6 & 0 \\ 0 & 0 \end{pmatrix} \qquad \sigma_B = \begin{pmatrix} -141.6 & 0 \\ 0 & 0 \end{pmatrix} \qquad \sigma_C = \begin{pmatrix} -12 & 0 \\ 0 & 0 \end{pmatrix}$$

The results, as presented in Fig. 8.2c, lead to an important concept, the superposition of stresses. A short commentary on this concept is in order. ■

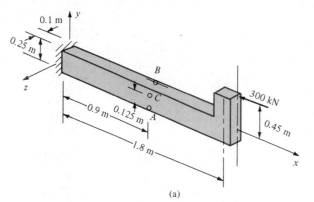

(a)

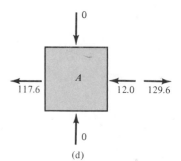

(d)

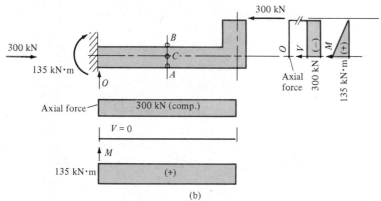

(b)

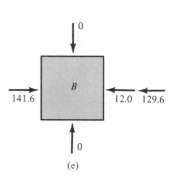

(e)

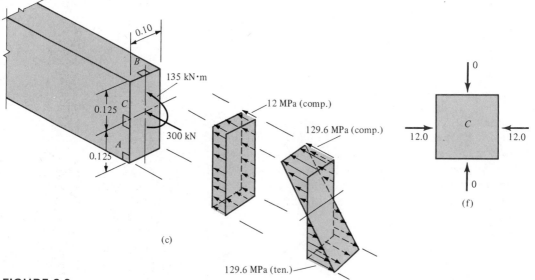

(c)

(f)

FIGURE 8.2

8.3   THE SUPERPOSITION OF STRESSES

Superposition of stresses is a concept that was used in the preceding example problem. Superposition means algebraically adding the stresses at a point. The stress distributions illustrated in Fig. 8.2c could be added or superposed to produce a total stress distribution. The idea of superposition is valid as long as the stress remains 'inear with respect to the parameters that are used to compute the stress. For example, the flexure formula, $\sigma = My/I$, can be analyzed for this linearity. For a given point on a structure, point A of Example 8.1 shown in Fig. 8.2, y and I are always constant. If the concentrated force is doubled, the moment increases by a factor of two and σ doubles. Hence, the stress is linear in its relation to the moment, while all other quantities remain constant. For the stress Eqs. (8.1)–(8.4), a linear relation holds between the stress and the load that produces the stress for the purpose of computing stress at a point.

The linearity described for stress is also true for strains and deformation. Superposition of stresses is valid as long as the application is for small deformations of elastic structures. The assumption of linearity and smallness leads to what might be thought of as an engineering theory, as opposed to an actual or exact theory, of mechanics of materials.

To illustrate a situation when superposition is not used, consider an axially loaded member. The stress is given as $\sigma = P/A$. Assume the material is steel, $E = 30(10^6)$ psi and Poisson's ratio $\nu = 0.33$. Assume the member to be square, 1 by 1 in., and 10 in. in length. For an axial force of 30,000 lb tension, Eq. (8.1) gives

$$\sigma = \frac{P}{A} = \frac{30{,}000}{(1)(1)} = 30{,}000 \text{ psi}$$

Assume that the load is to be applied in 10,000-lb increments and intermediate computations will be made to correct for the change in area and length. Table 8.1 is convenient for organizing the computations, which are based upon equations of Chapter 2. Notations at the bottom of each column in the table explain the operation required for that step. The stress at the end of the third increment is 30,013 psi compared to 30,000 psi using the axial-stress equation and one 30,000-lb increment. There are two conclusions to be drawn from the results: (1) If the change in area is considered as well as the change in load, the result is not linear; in other words, in the strictest sense the stresses should not be directly superposed. (2) The effect of neglecting the change of area as deformation occurs is not significant. There are limitations with this engineering approach for computing axial stresses. All calculations have been for stresses that would be below the yield point of the material. The material has remained elastic. If the material is stressed

TABLE 8.1

1	2	3	4	5	6, Eq. (2.28)	7, Eq. (2.22)	8, Eq. (2.27)	9	10	11	12
Increment, lb	Total Load, lb	Area, in.²	Stress, P/A, psi	Length, in.	Deformation $u = PL/AE$, in.	Longitudinal Strain, $\varepsilon_{long} = u/L$	$\varepsilon_{lat} = \nu\varepsilon_{long}$	$u_{lat} = (\varepsilon_{lat})$ (width), in.	Width, 1 in. $- u_{lat}$	Area, in.²	Length, 10 in. $+ u$ in.
1 10,000	10,000	1.0	10,000	10.0	$3.333(10^{-3})$	$3.333(10^{-4})$	$1.1099\,(10^{-4})$	$1.1099(10^{-4})$	0.9998	0.9997	10.0033
2 10,000	20,000	0.9997	20,004	10.0033	$6.671(10^{-3})$	$6.668(10^{-4})$	$2.2207(10^{-4})$	$2.2202(10^{-4})$	0.9997	0.9995	10.0067
3 10,000	30,000	0.9995	30,013	10.0067	$10.011(10^{-3})$	$10.004(10^{-4})$	$3.3315(10^{-4})$	$3.3315(10^{-4})$	0.9996	0.9993	10.0101
		Col. 11	$\dfrac{\text{Col. 2}}{\text{Col. 3}}$	10 in. + (Col. 6 of previous increment)	$\dfrac{(\text{Col. 2})(\text{Col. 5})}{(\text{Col. 3})(E)}$	$\dfrac{\text{Col. 6}}{\text{Col. 5}}$	$(\nu)(\text{Col. 7})$	(Col. 8)(Col. 10) of previous increment	1 in. $-$ Col. 9	(Col. 10)²	10 in. + Col. 6

294

beyond the yield point, the computation in Column 6 of Table 8.1 would be different, as discussed in Chapter 14. Young's modulus of $30(10^6)$ psi would decrease significantly.

8.4 GENERAL ANALYSIS FOR COMBINED STRESSES

In this discussion a general approach for stress analysis will be developed. Stress analysis can be viewed in a systematic way. If the reader follows the steps and confidently applies the methodology, the most complicated-looking problem can be resolved.

The first step is to determine the forces and moments at a section. This operation was referred to as finding an equivalent force system in statics. To illustrate and review, Example 8.1 will be repeated using a vector analysis approach.

EXAMPLE 8.2

Compute the forces and moments acting at a section 0.9 m from the fixed end of the beam of Fig. 8.2a.

Solution:

The section in question contains the points A, B, and C. The problem is to visualize the beam in a way that vector analysis can be applied. The beam is illustrated again in Fig. 8.3. Point O is at the centroid of the cross section. The beam is redrawn as a coordinate system containing the point O and the force $\mathbf{F} = 300$ kN $\mathbf{i}$ and a position vector $\mathbf{r}$. The equivalent system at point O is the force moved to point O and the moment of the force with respect to point O. The moment is computed with respect to the centroid of the cross section. Note, all of the stress equations, given by Eqs. (8.1)–(8.4), were derived with respect to the centroid of a cross section. The equivalent system is

$$\mathbf{F}_O = -300 \text{ kN } \mathbf{i}$$

$$\mathbf{M}_O = \mathbf{r} \times \mathbf{F} = (0.9\mathbf{i} + 0.45\mathbf{j}) \times (-300\mathbf{i}) = (0.45)(300)\mathbf{k} = 135 \text{ kN·m}\mathbf{k}$$

The vector system is shown in Fig. 8.3c. The shear and moment of Fig. 8.3c are, of course, the same as those in Fig. 8.2c. The vector approach gives a systematic method for obtaining the force system. Completion of the problem follows the technique of Example 8.1. ∎

The use of an equivalent-force system merely avoids the computations that are required for the reactions at the fixed end of the structure. The free body of Fig. 8.3d will illustrate the equivalence between acting- and reacting-force systems. The free body on the left is in equilib-

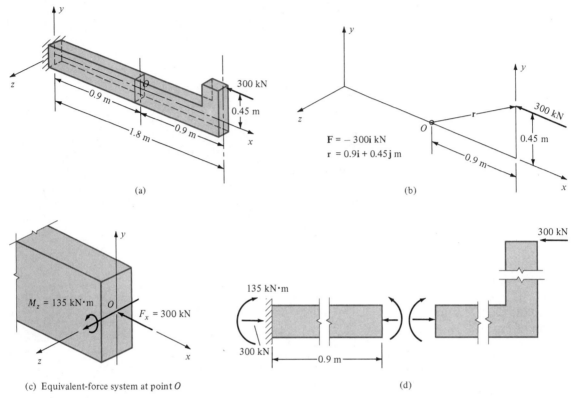

(a)

$$F = -300i \text{ kN}$$
$$r = 0.9i + 0.45j \text{ m}$$

(b)

$M_z = 135$ kN·m

$F_x = 300$ kN

(c) Equivalent-force system at point O

135 kN·m

300 kN

0.9 m

(d)

FIGURE 8.3

rium. The force and couple on the cut section represent the internal reactions required to balance the fixed-end reaction. The equivalent force-couple system that was computed using Figs. 8.3b and c is exactly the same as the internal reactions on the left-hand free body of Fig. 8.3d. The equivalent force-couple system, as it will be used, avoids the extra computations for calculating the reactions at the fixed end.

In general, six actions can occur at a given section of a structure, three forces and three moments. Obviously, not all six must occur. In the previous example only two actions were present. The general case is illustrated in Fig. 8.4. The three forces cause an axial normal stress and two shear stresses. The force and the corresponding stress are identified in Fig. 8.4a. The stress distributions are shown in Fig. 8.4b. For instance, τ_{xz} in Fig. 8.4a is caused by a force F_z acting in the z direction; Q_y is the moment of an area with respect to the y axis; I_y is the moment of inertia with respect to the y axis; and h is the depth dimension of the beam. A similar analysis is shown for the three moments.

The moment M_z corresponds to the bending moment for the two-dimensional problems of Chapter 5. The stress distribution for σ_x,

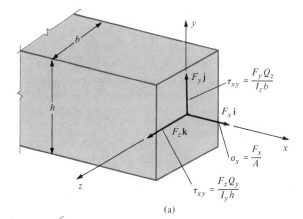

$$\tau_{xy} = \frac{F_y Q_z}{I_z b}$$

$$\sigma_x = \frac{F_x}{A}$$

$$\tau_{xy} = \frac{F_z Q_y}{I_y h}$$

(a)

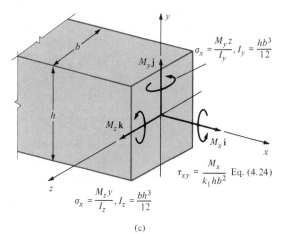

$$\sigma_x = \frac{M_y z}{I_y}, I_y = \frac{hb^3}{12}$$

$$\tau_{xy} = \frac{M_x}{k_1 h b^2} \quad \text{Eq. (4.24)}$$

$$\sigma_x = \frac{M_z y}{I_z}, I_z = \frac{bh^3}{12}$$

(c)

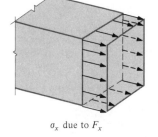

σ_x due to F_x

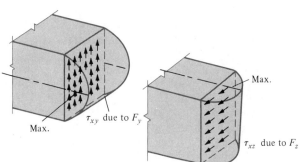

Max.

τ_{xy} due to F_y

Max.

τ_{xz} due to F_z

(b)

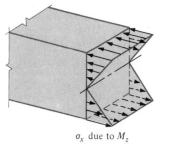

σ_x due to M_z

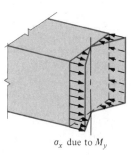

σ_x due to M_y

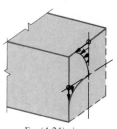

Eq. (4.24) gives τ_{xy}
due to M_x for
noncircular members

(d)

FIGURE 8.4 Actions and corresponding stress distributions.

caused by M_z, is shown in Fig. 8.4d. Note that the bending is about the z axis and the second moment of the area is with respect to the z axis. The moment M_y is interpreted in a similar manner in Fig. 8.4d. A stress diagram has not been sketched for the torque vector M_x since a general formula for torsion of a noncircular member is not available from Chapter 4.

The next example illustrates all six vector actions that are shown in Fig. 8.4. The reader should master the example before going further.

EXAMPLE 8.3

The pipe structure of Fig. 8.5a is loaded at the free end as illustrated. The dimensions are constant along the entire length.

a. Compute the state of stress at points a and b on a section at the fixed end.

b. Compute the state of stress at points c and d on a section at A. Show the results on a stress block and record the stress using the notation for stress at a point.

Solution:

a. Recall that the stress equations are referenced to the centroid of the cross section. Hence, equivalent-force systems at sections O and A should be with respect to the centroid at those sections. As a first step, redraw the structure of Fig. 8.5a as a coordinate system containing force vectors and position vectors.

The force system acting at point O is given by transferring the force vector to point O and computing the corresponding couple.

$$\mathbf{F}_O = -10\mathbf{i} + 5\mathbf{j} + 4\mathbf{k}$$

$$F_{Ox} = -10 \text{ lb} \qquad F_{Oy} = 5 \text{ lb} \qquad F_{Oz} = 4 \text{ lb}$$

$$\mathbf{M}_O = \begin{vmatrix} \mathbf{i} & \mathbf{j} & \mathbf{k} \\ 4 & 3 & 2 \\ -10 & 5 & 4 \end{vmatrix} = 12\mathbf{i} - 20\mathbf{j} + 20\mathbf{k} + 30\mathbf{k} - 16\mathbf{j} - 10\mathbf{i}$$

$$\mathbf{M}_O = 2\mathbf{i} - 36\mathbf{j} + 50\mathbf{k}$$

$$M_{Ox} = 2 \text{ ft·lb} \qquad M_{Oy} = -36 \text{ ft·lb} \qquad M_{Oz} = 50 \text{ ft·lb}$$

These forces and moments are applied as shown in Fig. 8.5c. Note that F_{Ox} and M_{Oy} are negative and hence are applied along the negative-coordinate direction. The right-hand rule is used to establish the action caused by the moment vectors.

Element a is shown in Fig. 8.5d. Only the four actions that actually cause stresses at point a are illustrated in Fig. 8.5d. These four individual stresses are shown as they act on the element. The stress distribution can be visualized through comparison with Fig. 8.4. The stresses are computed as follows:

$$I_y = I_z = \frac{\pi(3)^4}{4} = 63.62 \text{ in.}^4$$

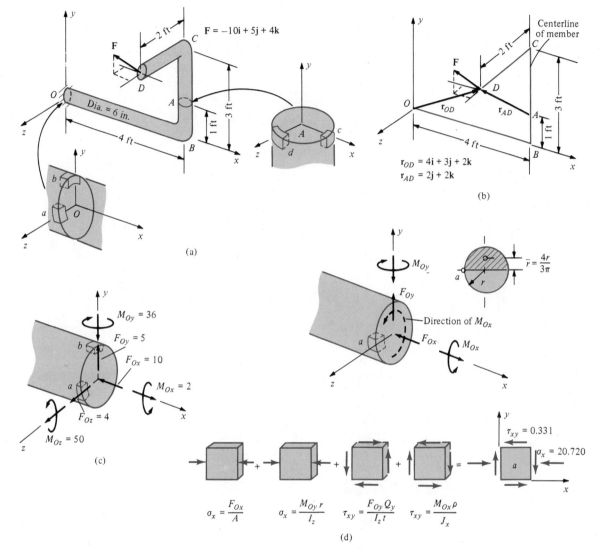

FIGURE 8.5

$$J_x = \frac{\pi(3)^4}{2} = 127.24 \text{ in.}^4$$

$$\text{Area} = \pi(3)^2 = 28.27 \text{ in.}^2$$

$$\sigma_x = \frac{F_{Ox}}{\text{Area}} = \frac{10}{28.27} = 0.349 \text{ psi} \qquad \text{(compression)}$$

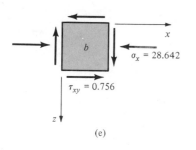

(e)

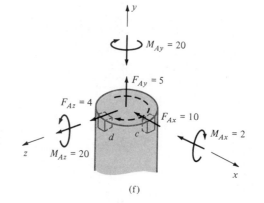

(f)

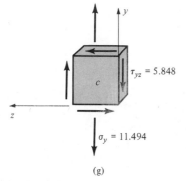

(g)

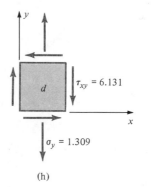

(h)

FIGURE 8.5 (continued)

$$\sigma_x = \frac{M_{Oy}z}{I_y} = \frac{(36)(12 \text{ in./ft})(3)}{63.62} = 20.371 \text{ psi} \qquad \text{(compression)}$$

$$\tau_{xy} = \frac{F_{Oy}Q_z}{I_z(\text{dia.})} = \frac{(5)[(\pi)(3)^2/2][(4)(3)/(3)\pi]}{(63.62)(6)} = 0.236 \text{ psi}$$

This value of $\tau_{xy} = 0.236$ psi is acting upward in the same direction as F_{Oy}.

$$\tau_{xy} = \frac{M_{Ox}\rho}{J_x} = \frac{(2)(12 \text{ in./ft})(3)}{127.24} = 0.567 \text{ psi}$$

This value of $\tau_{xy} = 0.567$ psi is acting downward on the element in the same manner as the torque shown acting on the face of the cut section.

The stresses at point a are formally written as

$$\sigma_a = \begin{pmatrix} -20.720 & -0.331 \\ -0.331 & 0 \end{pmatrix} \qquad \text{in the } x\text{-}y \text{ plane} \tag{a}$$

Note that a negative shear corresponds to the sign convention established in Fig. 8.1a.

The element at b has normal stresses caused by F_{Ox} and M_{Oz} and shear stresses caused by F_{Oz} and M_{Ox}. The element at b is shown in Fig. 8.5e as if it were viewed from above looking down the y axis. The four contributions to the stresses are computed individually and superposed.

$$\sigma_x = \frac{10}{28.27} = 0.349 \text{ psi} \quad \text{(compression)}$$

$$\sigma_x = \frac{(50)(12)(3)}{63.62} = 28.293 \text{ psi} \quad \text{(compression)}$$

$$\tau_{xz} = \frac{(4)[(\pi)(3)^2/2][(4)(3)/(3)(\pi)]}{(63.62)(6)} = 0.189 \text{ psi}$$

$\tau_{xz} = 0.189$ psi acts from negative z toward positive z in the same direction as F_{Oz}.

$$\tau_{xz} = \frac{(2)(12)(3)}{127.24} = 0.567 \text{ psi}$$

$\tau_{xz} = 0.567$ psi acts in the positive z direction at element b, as illustrated by the action of the torque shown dashed in Fig. 8.5d.

The stresses at point b are

$$\sigma_b = \begin{pmatrix} -28.642 & 0.756 \\ 0.756 & 0 \end{pmatrix} \quad \text{in the } x\text{-}z \text{ plane} \tag{b}$$

Again the directions of the stresses acting on element b, in Fig. 8.5e, are determined by the actions of the forces and moments. The signs used in Eq. (b) are determined by comparing Fig. 8.5e to the sign convention of Fig. 8.1d.

b. The forces and moments acting at section A are computed using Fig. 8.5b. That is,

$$\mathbf{F}_A = -10\mathbf{i} + 5\mathbf{j} + 4\mathbf{k}$$

$$\mathbf{M}_A = \begin{vmatrix} \mathbf{i} & \mathbf{j} & \mathbf{k} \\ 0 & 2 & 2 \\ -10 & 5 & 4 \end{vmatrix} = -2\mathbf{i} - 20\mathbf{j} + 20\mathbf{k}$$

The vector system is illustrated in Fig. 8.5f. The y component of the force vector acts axially at section A to cause tension. The x and z force components are shears. The axial moment vector is a torque, whereas M_{Ax} and M_{Az} cause bending.

For point c the stresses are computed using the formulas. Note that the quantity $Q = 18$ in.^{3} is constant for all cross sections.

$$\sigma_y = \frac{5}{28.27} = 0.117 \text{ psi} \quad \text{(tension)}$$

$$\sigma_y = \frac{(20)(12)(3)}{63.62} = 11.317 \text{ psi} \quad \text{(tension)}$$

$$\tau_{yz} = \frac{(4)(18)}{(63.62)(6)} = 0.189 \text{ psi} \quad \text{(acting in the positive } z \text{ direction)}$$

$$\tau_{yz} = \frac{(20)(12)(3)}{127.24} = 5.659 \text{ psi} \qquad \text{(acting in the positive } z \text{ direction)}$$

Combining the stresses yields

$$\sigma_c = \begin{pmatrix} 11.494 & 5.846 \\ 5.848 & 0 \end{pmatrix} \qquad \text{(in the } y\text{-}z \text{ plane)}$$

The stress at point c is illustrated in Fig. 8.5g.
Finally the point d is analyzed as follows:

$$\sigma_y = 0.177 \text{ psi} \qquad \text{(tension)}$$

$$\sigma_y = (2)(12)(3)/63.62 = 1.132 \qquad \text{(tension)}$$

$$\tau_{xy} = \frac{(10)(18)}{(63.62)(6)}$$

$$= 0.472 \text{ psi} \qquad \text{(acting in the negative } x \text{ direction)}$$

$$\tau_{xy} = 5.659 \text{ psi} \qquad \text{(acting in the negative } x \text{ direction)}$$

Combining the stresses yields

$$\sigma_d = \begin{pmatrix} 0 & -6.131 \\ -6.131 & 1.309 \end{pmatrix} \qquad \text{(in the } x\text{-}y \text{ plane)}$$

The results are shown in Fig. 8.5h. ∎

> This illustrative example is as complicated as any that will be encountered in this text. It is very important to master the concept employed. Using sketches of force and moment systems acting on free bodies is the most direct method of determining which actions cause stress at a given point. This example should be thoroughly understood before continuing with this chapter.

EXAMPLE 8.4

A concrete dam that is trapezoidal in cross section as illustrated in Fig. 8.6 is used to contain water. Compute the maximum stresses at the base of the dam at points A, B, and C.

Solution:
The dam illustrated in Fig. 8.6a is a long structure with constant cross section and constant loading. In practice such a structure is simplified by removing a unit strip as shown in Fig. 8.6b. The strip appears as a beam-type structure fixed at the base and free at the top. The weight of the dam acts axially along the x axis, and the force of the water acts as a lateral load causing bending at the base and transverse shear at the base.

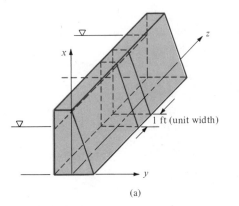

(a)

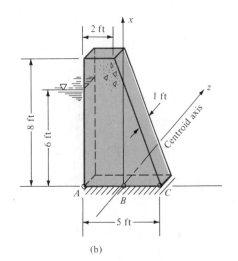

(b)

(c)

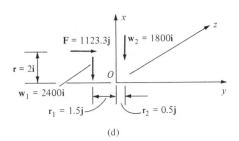

(d)

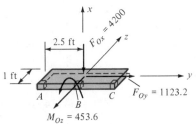

(e)

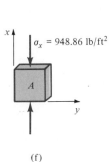

$\sigma_x = 948.86 \ \text{lb/ft}^2$

A

(f)

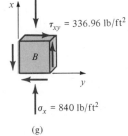

$\tau_{xy} = 336.96 \ \text{lb/ft}^2$

B

$\sigma_x = 840 \ \text{lb/ft}^2$

(g)

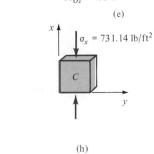

$\sigma_x = 731.14 \ \text{lb/ft}^2$

C

(h)

FIGURE 8.6

The weight of the dam is considered to act through the *centroid of the distributed weight* of the dam. The simplest method of analysis would be to consider a rectangular and a triangular section. In order to compute stresses at the base, consider the forces and moments with respect to the *centroid of the cross section* at the base. The force of the water is determined by considering the distributed weight of the water.

The pressure caused by the water at a given depth below the surface is computed as the unit weight of water times the distance below the surface of the water. At the base of the dam the pressure is $(62.4 \text{ lb/ft}^3)(6 \text{ ft})$ or 374.40 lb/ft^2, as shown in Fig. 8.6c. An equivalent force can replace the distributed pressure configuration and can be assumed to act at the *centroid of the pressure distribution*.

The force of the water is equal to the volume of the pressure distribution and acts as shown in Fig. 8.6c.

$$F = \frac{(62.4 \text{ lb/ft}^3)(6 \text{ ft})(6 \text{ ft})(1 \text{ ft})}{2} = 1123.2 \text{ lb}$$

The unit weight of concrete is 150 lb/ft^3, and the total weight of the dam section is equal to the volume of the section acting as shown in Fig. 8.7c.

$$W_1 = (150 \text{ lb/ft}^3)(8 \text{ ft})(2 \text{ ft})(1 \text{ ft}) = 2400 \text{ lb}$$

$$W_2 = \frac{(150 \text{ lb/ft}^3)(8 \text{ ft})(3 \text{ ft})(1 \text{ ft})}{2} = 1800 \text{ lb}$$

In keeping with the vector analysis, the force system is shown in Fig. 8.6d. Move the forces to the centroid of the dam section where the stresses are to be computed and sum moments about the centroid. The equivalent force is

$$\mathbf{F}_O = -4200\mathbf{i} + 1123.2\mathbf{j} \text{ lb}$$

The moment with respect to the centroid is

$$\mathbf{M}_O = \mathbf{r}_1 \times \mathbf{w}_1 + \mathbf{r}_2 \times \mathbf{w}_2 + \mathbf{r} \times \mathbf{F}$$

$$\mathbf{M}_O = (-1.5\mathbf{j} \times -2400\mathbf{i}) + (0.5\mathbf{j} \times -1800\mathbf{i}) + (2\mathbf{i} \times 1123.2\mathbf{j})$$

$$\mathbf{M}_O = -3600\mathbf{k} + 900\mathbf{k} + 2246.4\mathbf{k} = -453.6\mathbf{k} \text{ ft·lb}$$

The equivalent system is shown in Fig. 8.6e. Normal stresses due to both F_{Ox} and M_{Oz} occur at points A and C. F_{Oy} causes shear stress at B.

$$\sigma_x = \frac{F_{Ox}}{\text{Area}} = \frac{4200}{(5)(1)}$$

$$= 840 \text{ lb/ft}^2 \qquad \text{(compression at } A, B, \text{ and } C\text{)}$$

$$\sigma_x = \frac{(453.6)(2.5)}{(1)(5)^3/12}$$

$$= 108.86 \text{ lb/ft}^2 \qquad \text{(compression at } A, \text{ tension at } C\text{)}$$

$$\tau_{xy} = \frac{(1123.2 \text{ lb})(1 \text{ ft})(2.5 \text{ ft})(2.5 \text{ ft}/2)}{[(1 \text{ ft})(5 \text{ ft})^3/12][1 \text{ ft}]}$$

$$= 336.96 \text{ lb/ft}^2 \qquad \text{(acting in the positive } y \text{ direction)}$$

$$\sigma_A = \begin{pmatrix} -948.86 & 0 \\ 0 & 0 \end{pmatrix}$$

$$\sigma_B = \begin{pmatrix} -840.0 & 336.96 \\ 336.96 & 0 \end{pmatrix}$$

$$\sigma_C = \begin{pmatrix} -731.14 & 0 \\ 0 & 0 \end{pmatrix}$$

The elements are shown in Figs. 8.6*f*, *g*, and *h*. ∎

EXAMPLE 8.5

A short column is shown in Fig. 8.7. A short column can be analyzed for stresses using the theory that has been developed thus far. Longer columns are discussed in Chapter 12. For the short column of Fig. 8.7 compute the stress at each corner and locate the neutral axis.

Solution:

Because of the loading there are no shear stresses for this problem. An important feature of the analysis is the illustration of the location of the neutral axis. Three linear stress distributions are superposed to yield a neutral axis that is skewed on the cross section.

View the load as shown in Fig. 8.7*b* acting in a coordinate system and transfer the force to the centroid and compute the bending moments to yield the system of Fig. 8.7*c*.

$$\mathbf{F} = F_z\mathbf{k} = -900\mathbf{k}$$

$$\mathbf{M}_O = (0.25\mathbf{i} + 0.075\mathbf{j}) \times (-900\mathbf{k}) = (-67.5\mathbf{i} + 255\mathbf{j}) \text{ kN·m}$$

The stress distributions are obtained by referring to Fig. 8.4. Note, the axis of Fig. 8.4 corresponds to the *z* axis of Fig. 8.7.

$$\sigma_{zI} = \frac{M_x y}{I_x} = \frac{(67.5)(0.2)}{(0.5)(0.4)^3/12} = 5.06 \text{ MPa}$$

$$\sigma_{zII} = \frac{M_y x}{I_y} = \frac{(255)(0.25)}{(0.4)(0.5)^3/12} = 13.50 \text{ MPa}$$

$$\sigma_{zIII} = \frac{F_z}{A} = \frac{900}{(0.4)(0.5)} = 4.50 \text{ MPa}$$

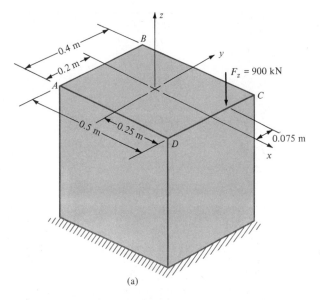

$F_z = 900$ kN

0.4 m
0.2 m
0.5 m
0.25 m
0.075 m

(a)

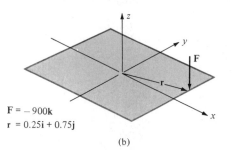

$\mathbf{F} = -900\mathbf{k}$
$\mathbf{r} = 0.25\mathbf{i} + 0.75\mathbf{j}$

(b)

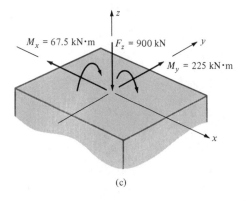

$M_x = 67.5$ kN·m
$F_z = 900$ kN
$M_y = 225$ kN·m

(c)

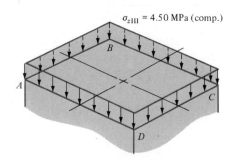

$\sigma_{zIII} = 4.50$ MPa (comp.)

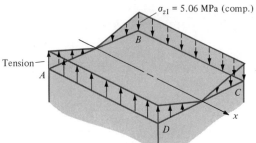

$\sigma_{zI} = 5.06$ MPa (comp.)

Tension

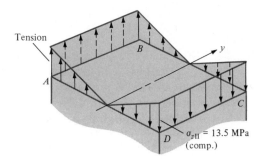

Tension

$\sigma_{zII} = 13.5$ MPa (comp.)

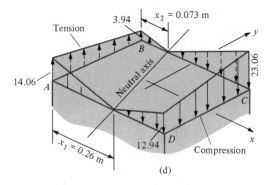

Tension
3.94
$x_2 = 0.073$ m
23.06
14.06
Neutral axis
$x_1 = 0.26$ m
12.94
Compression

(d)

FIGURE 8.7

The three stress distributions and the superposed final stress distribution are illustrated in Fig. 8.7d. The final stress distribution is characterized by an equation of the form

$$\sigma = \frac{F_z}{A} \pm \frac{M_x y}{I_x} \pm \frac{M_y x}{I_y} \tag{a}$$

To use Eq. (a), interpretation must be given to the sign convention. The beam sign convention has some shortcomings since the usual interpretation is that positive moment causes compression in the top fibers of the beam and establishing the top fiber of the beam is usually done in an arbitrary manner.

Let the positive coordinate axis be directed toward the top fibers of the beam. Using the beam sign convention, M_x, a negative vector moment, causes compression on the positive side of the y axis; hence, it would be a positive moment using the beam sign convention. M_y causes compression on the positive side of the x axis and is a positive moment. Positive moment using the beam sign convention produces compressive stress; hence, if we arbitrarily assume a (+) sign to be associated with compression, Eq. (a) should be written

$$\sigma = \frac{F_z}{A} + \frac{M_x}{I_x}(\pm y) + \frac{M_y}{I_y}(\pm x) \tag{b}$$

For point A, $x = -0.25$ and $y = -0.2$. Equation (b) gives

$$\sigma_A = \frac{F_z}{A}(\text{comp.}) + \frac{M_x}{I_x}(-0.2)(\text{ten.}) + \frac{M_y}{I_y}(-0.25)(\text{ten.}) \tag{c}$$

Equation (c) agrees with the stress distribution sketches of Fig. 8.7d. It is far easier to establish tension or compression using sketches and inspection than attempting to relate the beam sign convention to more than one dimension.

The final stress distribution of Fig. 8.7d was obtained by inspection.

$\sigma_A = 4.5\ (\text{comp.}) + 5.06\ (\text{ten.}) + 13.5\ (\text{ten.}) = 14.06\ (\text{ten.})$

$\sigma_B = 4.5\ (\text{comp.}) + 5.06\ (\text{comp.}) + 13.5\ (\text{ten.}) = 3.94\ (\text{ten.})$

$\sigma_C = 4.5\ (\text{comp.}) + 5.06\ (\text{comp.}) + 13.5\ (\text{comp.}) = 23.06\ (\text{comp.})$

$\sigma_D = 4.5\ (\text{comp.}) + 5.06\ (\text{ten.}) + 13.5\ (\text{comp.}) = 12.94\ (\text{comp.})$

The location of the neutral axis can be obtained using similar triangles. For side AD and BC,

$$\frac{x_1}{14.06} = \frac{0.5}{(14.06 + 12.94)} \qquad x_1 = 0.260\ \text{m}$$

$$\frac{x_2}{3.94} = \frac{0.5}{(3.94 + 23.06)} \qquad x_2 = 0.073\ \text{m}$$

The neutral axis is skewed with respect to the centroid axis. This problem could also be solved by establishing a set of principal axes coinciding with the neutral axis and resolving the moment vectors into components along the principal axes. This basic approach is referred to as unsymmetrical bending and is employed primarily for unsymmetrical cross sections. The concept of unsymmetrical bending and transformation of the second moments of area are discussed in the next chapter. A major concept illustrated by this problem is that the resultant bending occurs about a skewed neutral axis.   ∎

> The scalar or dot product was introduced in statics and used primarily for obtaining the projection of a vector in the direction of a second vector. A practical use of the scalar product will be demonstrated in the following example.

EXAMPLE 8.6

A fixed-free beam fabricated from standard-weight 4-in. pipe (nominal diameter) is illustrated in Fig. 8.8. A force of 2500 lb is applied to the free end by means of the loading system shown in the figure. Compute the stress at points a and b of section A. Point b is on a vertical y axis and point a is on the horizontal z' axis. The free-body section at A is cut normal to the axial direction of the rod. Neglect the weight of the member. Sketch the results on elements at a and b.

Solution:
The forces and moments acting at the centroid of section A must be computed. The position vector is obtained using Fig. 8.8b. The vector diagram gives the equation

$$\mathbf{r}_{OB} = \mathbf{r}_{OA} + \mathbf{r}_{AB} \tag{a}$$

Solving for $\mathbf{r}_{AB}$ gives

$$\mathbf{r}_{AB} = \mathbf{r}_{OB} - \mathbf{r}_{OA} = (4\mathbf{i} + 1.5\mathbf{j} + 2\mathbf{k}) - (2\mathbf{i} + \mathbf{k}) = 2\mathbf{i} + 1.5\mathbf{j} + \mathbf{k} \tag{b}$$

The force is directed along the cable BC. According to Fig. 8.8b, the vector $\mathbf{r}_{BC}$ is evaluated as

$$\mathbf{r}_{BC} = \mathbf{r}_{OC} - \mathbf{r}_{OB} = (6\mathbf{i} + 6\mathbf{j}) - (4\mathbf{i} + 1.5\mathbf{j} + 2\mathbf{k}) = 2\mathbf{i} + 4.5\mathbf{j} - 2\mathbf{k} \tag{c}$$

The unit vector $\mathbf{n}_{BC}$ is illustrated in Fig. 8.8c. Recall from statics that the unit vector components are the cosines of the angles between $\mathbf{r}_{BC}$ and the x, y, and z coordinates. The magnitude of r_{BC} is

$$|r_{BC}| = [(2)^2 + (4.5)^2 + (-2)^2]^{1/2} = 5.315 \text{ ft}$$

$$\mathbf{n}_{BC} = \frac{\mathbf{r}_{BC}}{|r_{BC}|} = \frac{2\mathbf{i} + 4.5\mathbf{j} - 2\mathbf{k}}{5.315} = 0.376\mathbf{i} + 0.847\mathbf{j} - 0.376\mathbf{k} \tag{d}$$

The components of the 2500-lb force are obtained by multiplying the force times the unit vector $\mathbf{n}_{BC}$.

$$\mathbf{F} = (2500 \text{ lb})(0.376\mathbf{i} + 0.847\mathbf{j} - 0.376\mathbf{k}) = 940\mathbf{i} + 2117.5\mathbf{j} - 940\mathbf{k} \tag{e}$$

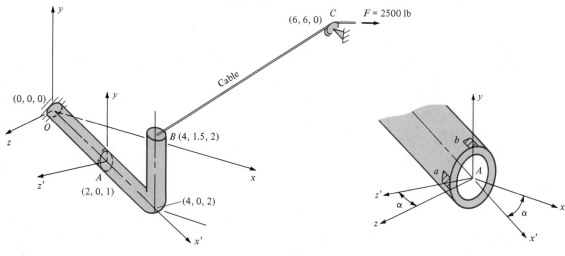

(a)

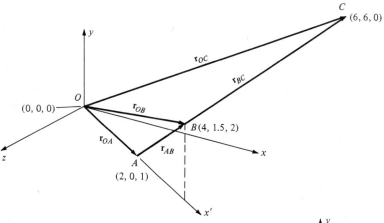

(b)

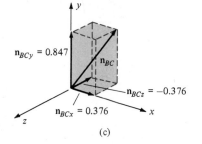

(c)

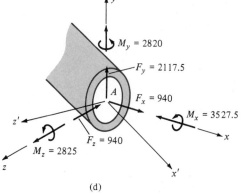

(d)

FIGURE 8.8

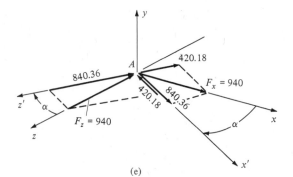

(e)

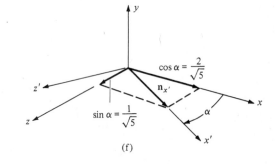

(f)

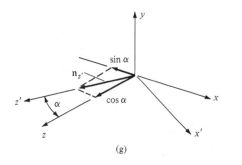

(g)

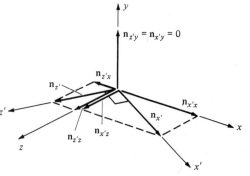

(h)

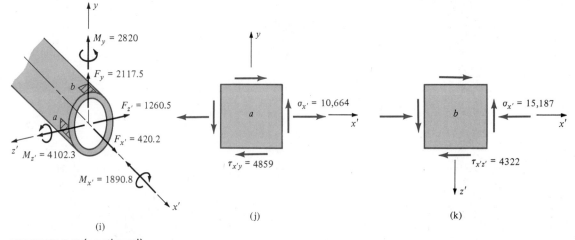

(i)

(j)

(k)

FIGURE 8.8 (continued)

The moment at section A is computed using Eqs. (b) and (e).

$$\mathbf{M}_A = \begin{vmatrix} \mathbf{i} & \mathbf{j} & \mathbf{k} \\ 2 & 1.5 & 1 \\ 940 & 2117.5 & -940 \end{vmatrix} = -3527.5\mathbf{i} + 2820\mathbf{j} + 2825\mathbf{k} \tag{f}$$

Also,

$$\mathbf{F}_A = 940\mathbf{i} + 2117.5\mathbf{j} - 940\mathbf{k} \tag{g}$$

The computational power of vector analysis is illustrated by Example 8.6. The vector components of Eqs. (f) and (g) act in the x-y-z system. However, as shown in Figs. 8.8a and d, the x' is the axis normal to section A while y and z' are adjacent to the cut section. To compute stresses, the force system in the x-y-z coordinates must be transferred into a force system acting in the x'-y-z' coordinate system. Both F_x and F_z must be combined to yield $F_{x'}$ and $F_{z'}$. F_y is already oriented in the proper direction. In the schematic drawing the vector that looks the longest is not necessarily the resultant. The resultant is equal to the sum of its components. The force vector along x' is the sum of the components of F_x and F_z that lie along x'. Again, the scalar product is employed to compute the x' component of the force. A unit vector along x' is computed using Figs. 8.8a and b.

$$\mathbf{n}_{x'} = \frac{\mathbf{r}_{OA}}{|\mathbf{r}_{OA}|} = \frac{2\mathbf{i} + \mathbf{k}}{[(2)^2 + (1)^2]^{1/2}} = 0.984\mathbf{i} + 0.447\mathbf{k} \tag{h}$$

$$F_{x'} = \mathbf{F} \cdot \mathbf{n}_{x'} = (940\mathbf{i} + 2117.5\mathbf{j} - 940\mathbf{k}) \cdot (0.894\mathbf{i} + 0.447\mathbf{k})$$

$$F_{x'} = 840.36 - 420.18 = 420.2 \text{ lb} \qquad \text{(along the } x' \text{ axis)}$$

The scalar product has been used to find the projection of F_x and F_z along the x' axis. The results of Eq. (h) are illustrated in Fig. 8.8f. The components of the unit vector along x' correspond to the length of the components as shown along x and z. The angle α can be computed, $\tan \alpha = 1/2$, or $\alpha = 26.56$ degrees. Then $\cos(25.56°) = 0.894$, which is the same as the ratio of $2/\sqrt{5}$. Similarly, $\sin(26.56°)$ is 0.447.

The unit vector that defines the z' direction can be obtained from Fig. 8.8g. The angle between z and z' is the same angle, α, of Fig. 8.8f. The component of $n_{z'}$ in the z direction is

$$n_{z'} \cos \alpha = (1) \cos 26.56° = 0.894$$

In the x direction the component is

$$n_{z'} \sin \alpha = (1) \sin 26.56° = 0.447$$

and is in the negative x direction. Then,

$$\mathbf{n}_{z'} = -0.447\mathbf{i} + 0.894\mathbf{k} \tag{i}$$

In this problem the geometry of the unit vectors is easily obtainable from Fig. 8.8g. A more formal approach can be used and will be demonstrated. However, the problems in this text will be cast in such a way that geometry can be obtained primarily from sketches.

Consider Fig. 8.8h. The unit vectors $\mathbf{n}_{x'}$ and $\mathbf{n}_{z'}$ are illustrated along the x' and z' axis. Their components are shown in the x-y-z system. Formally,

$$\mathbf{n}_{x'} = n_{x'x}\mathbf{i} + n_{x'y}\mathbf{j} + n_{x'z}\mathbf{k} \tag{j}$$

as shown in the figure. Since $n_{x'}$ lies in the x'-z' plane, $n_{x'y} = 0$. Also

$$\mathbf{n}_{z'} = n_{z'x}\mathbf{i} + n_{z'z}\mathbf{k} \tag{k}$$

The unit vector $\mathbf{n}_{x'}$ was computed as in Eq. (h). Since $\mathbf{n}_{x'}$ and $\mathbf{n}_{y'}$ are perpendicular, their scalar product is zero. Recall that a scalar product is the product of two vectors times the cosine of the angle between them. Using Eqs. (h) and (k) gives

$$(\mathbf{n}_{x'}) \cdot (\mathbf{n}_{z'}) = (0.894\mathbf{i} + 0.447\mathbf{k}) \cdot (n_{z'x}\mathbf{i} + n_{z'z}\mathbf{k}) = 0.894n_{z'x} + 0.447n_{z'z} = 0$$

or

$$n_{z'x} = -0.5n_{z'z} \tag{l}$$

Any vector is equal to the square root of the sum of its squared components. Since $n_{z'}$ equals one in magnitude,

$$(n_{z'x})^2 + (n_{z'y})^2 + (n_{z'z})^2 = 1 \tag{m}$$

where $n_{z'y} = 0$.

Substituting Eq. (l) into Eq. (m) gives

$$n_{z'z} = \pm 0.894$$

Noting the figure, we choose the positive value. From Eq. (l),

$$n_{z'x} = -0.5n_{z'z} = -0.447$$

By Eq. (k),

$$\mathbf{n}_{z'} = -0.447\mathbf{i} + 0.894\mathbf{k} \tag{n}$$

the same as Eq. (i).

Continuing the problem, find the force along the z' direction or take the scalar product between Eqs. (g) and (n).

$$F_{z'} = (940\mathbf{i} + 2117.5\mathbf{j} - 940\mathbf{k}) \cdot (-0.447\mathbf{i} + 0.894\mathbf{k})$$

$$F_{z'} = -1260.5 \text{ lb} \quad \text{(along the } z \text{ axis)}$$

also, $F_y = 2117.5$ lb since the y force component is the same in either coordinate system.

The moment vector at A as given by Eq. (f) must also be resolved into components in the x'-y-z' system. The scalar product between $\mathbf{M}_A$ and $\mathbf{n}_x$ yields the moment along x'.

$$M_{x'} = (-3527.5\mathbf{i} + 2820\mathbf{j} + 2825\mathbf{k}) \cdot (0.894\mathbf{i} + 0.447\mathbf{k})$$

$$M_{x'} = -1890.8 \text{ ft} \cdot \text{lb}$$

Similarly,

$$M_{z'} = (-3527.5\mathbf{i} + 2820\mathbf{j} + 2825\mathbf{k}) \cdot (-0.447\mathbf{i} + 0.894\mathbf{k})$$

$$M_{z'} = 4102.3 \text{ ft} \cdot \text{lb}$$

and

$$M_y = 2820 \text{ ft} \cdot \text{lb}$$

The resultant force and moment system is shown in Fig. 8.8*i*. The computation for stress proceeds as in previous examples.

The properties of a 4-in. standard-weight pipe are obtained from Appendix C.

Area $= 3.17 \text{ in.}^2$

$I_y = I_{z'} = 7.23 \text{ in.}^4$

$J_{x'} = I_y + I_{z'} = 14.46 \text{ in.}^4$

Outside diameter $= 4.5$ in.

$t = 0.237$ in.

For element *a*, referring to Fig. 8.8*i*,

$$\sigma_{x'} = \frac{420.2 \text{ lb}}{3.17 \text{ in.}^2} = 133 \text{ psi} \qquad \text{(tension)}$$

$$\sigma_{x'} = \frac{(2820 \text{ ft} \cdot \text{lb})(12 \text{ in./ft})(2.25 \text{ in.})}{7.23 \text{ in.}^4} = 10{,}531 \text{ psi} \qquad \text{(tension)}$$

The moment of the area to compute $Q_{z'} = Q_y$ is $(A/2)(\bar{r})$, where $\bar{r}$ is obtained from Appendix A. For the shear force F_y,

$$Q_{z'} = Q_y = \frac{(3.17 \text{ in.}^2)(2)(2.25 - 0.118)}{2\pi} = 2.15 \text{ in.}^3$$

$$\tau_{x'y} = \frac{(2117.5 \text{ lb})(2.15 \text{ in.}^3)}{(7.23 \text{ in.}^4)(2)(0.237 \text{ in.})} = 1328 \text{ psi} \qquad \text{(in the positive } y \text{ direction)}$$

For the axial moment $M_{x'}$,

$$\tau_{x'y} = \frac{(1890.8 \text{ ft} \cdot \text{lb})(12 \text{ in./ft})(2.25 \text{ in.})}{14.46 \text{ in.}^4}$$

$$= 3531 \text{ psi} \qquad \text{(in the positive } y \text{ direction)}$$

The stress at point *a* is illustrated acting on an element in Fig. 8.8*j*; also,

$$\sigma_a = \begin{pmatrix} 10{,}664 & 4859 \\ 4859 & 0 \end{pmatrix}$$

The stress state at point b is computed in a similar fashion.

$$\sigma_{x'} = 133 \text{ psi} \qquad \text{(tension)}$$

$$\sigma_{x'} = \frac{(4102.3)(12)(2.25)}{7.23} = 15,320 \text{ psi} \qquad \text{(compression)}$$

$$\tau_{x'z'} = \frac{(1260.5)(2.15)}{7.23)(2)(0.237)} = 791 \text{ psi} \qquad \text{(in the negative } z' \text{ direction)}$$

$$\tau_{x'z'} = 3531 \text{ psi} \qquad \text{(in the negative } z' \text{ direction)}$$

The results are shown in Fig. 8.8k.

$$\sigma_b = \begin{pmatrix} -15,187 & -4322 \\ -4322 & 0 \end{pmatrix}$$

Computationally the foregoing example is somewhat tedious. It is representative of a real engineering problem. Currently, piping analysis for power plants, ships, and so forth are some of the most challenging analysis problems with which engineers are confronted.

Force vectors can be visualized as they are transformed into components and oriented in different directions. Moment, without the use of vector analysis, can be perplexing when attempting to visualize their action about a rotated coordinate system. The analysis presented here is limited to a rotation about the y axis; hence, a fully three-dimensional use of vector analysis has not been demonstrated. The extension to a general rotation is not difficult, but the magnitude of computations increases significantly.

The next example will demonstrate another situation where vector analysis is an excellent computational tool. Also, a problem in stress analysis can be resolved in a somewhat simple manner.

In previous examples involving shearing stresses for circular members the location of elements has been judiciously chosen to lie on a coordinate axis. Such a problem is illustrated in Fig. 8.5c. The loading produces six actions that cause normal stresses and shear stresses at various locations on the cross section. At element a of Fig. 8.5c the shear stresses act vertically, whereas at element b all shear stresses act horizontally. Consider the circular cross section of Fig. 8.9. Assume that six actions occur. An element oriented at a 30-degree angle above the z axis would be subject to a horizontal and vertical shear caused by the shear forces and a shear stress caused by the torque that acts normal to the 30-degree line. This state of shear stress cannot be analyzed *directly* using the methods of mechanics of materials illustrated in the previous examples. The state of stress illustrated in Fig. 8.9 can

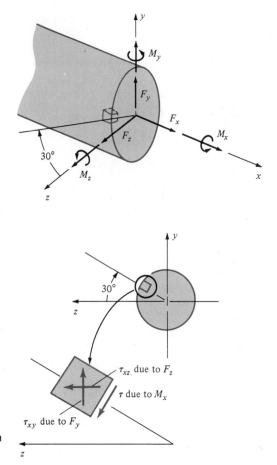

FIGURE 8.9
Stress at an arbitrary point on
a circular cross section.

be analyzed using the methods of mechanics of materials if the shear
forces are transformed to act adjacent to the sides of the element, as
will be illustrated in the next example.

EXAMPLE 8.7

The solid circular rod shown in Fig. 8.10 is loaded as illustrated. The pipe is 75 mm
in diameter and made of bronze. Compute the state of stress on an element at A
located as shown in the figure.

Solution:

The dimensions and loading conditions are shown in Fig. 8.10. The actions at section
B are computed as

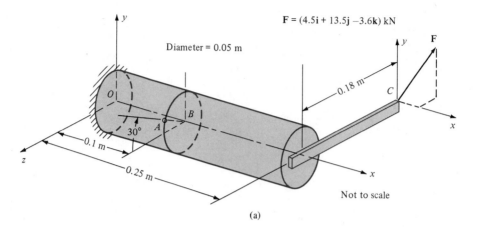

$\mathbf{F} = (4.5\mathbf{i} + 13.5\mathbf{j} - 3.6\mathbf{k})$ kN

Diameter = 0.05 m

0.18 m

0.1 m

0.25 m

Not to scale

(a)

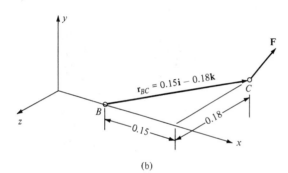

$\mathbf{r}_{BC} = 0.15\mathbf{i} - 0.18\mathbf{k}$

0.15

0.18

(b)

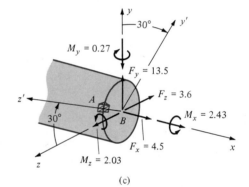

$M_y = 0.27$

$F_y = 13.5$

$F_z = 3.6$

$M_x = 2.43$

$F_x = 4.5$

$M_z = 2.03$

30°

(c)

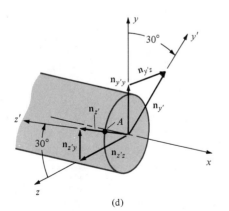

30°

$\mathbf{n}_{y'z}$

$\mathbf{n}_{y'y}$

$\mathbf{n}_{y'}$

$\mathbf{n}_{z'}$

$\mathbf{n}_{z'y}$

$\mathbf{n}_{z'z}$

30°

(d)

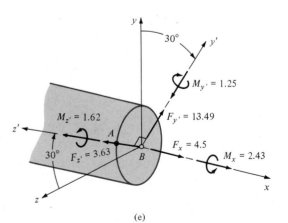

30°

$M_{y'} = 1.25$

$F_{y'} = 13.49$

$M_{z'} = 1.62$

$F_x = 4.5$

$F_{z'} = 3.63$

$M_x = 2.43$

30°

(e)

FIGURE 8.10

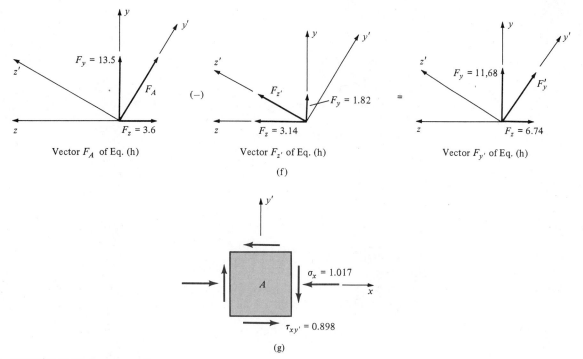

Vector F_A of Eq. (h)   Vector $F_{z'}$ of Eq. (h)   Vector $F_{y'}$ of Eq. (h)

(f)

(g)

FIGURE 8.10 (continued)

$$\mathbf{F}_B = (4.5\mathbf{i} + 13.5\mathbf{j} - 3.6\mathbf{k}) \text{ kN} \tag{a}$$

$$\mathbf{M}_B = \begin{vmatrix} \mathbf{i} & \mathbf{j} & \mathbf{k} \\ 0.15 & 0 & -0.18 \\ 4.5 & 13.5 & -3.6 \end{vmatrix} = (2.43\mathbf{i} - 0.27\mathbf{j} + 2.03\mathbf{k}) \text{ kN·m} \tag{b}$$

The actions given by Eqs. (a) and (b) are shown in Fig. 8.10c.

The force and moment system will be rotated to a new coordinate system defined by x-y'-z'. In other words, the actions will be transformed so that all vectors act either parallel or normal to the element at A. Once the transformation has been accomplished, the standard equations can be applied as in previous problems. The unit vectors are illustrated in Fig. 8.10d and are computed by inspection as follows:

$$\mathbf{n}_{z'} = n_{z'y}\mathbf{j} + n_{z'z}\mathbf{k} = n_{z'} \sin 30° \, \mathbf{j} + n_{z'} \cos 30° \, \mathbf{k}$$

$$\mathbf{n}_{z'} = 0.5\mathbf{j} + 0.866\mathbf{k} \tag{c}$$

$$\mathbf{n}_{y'} = n_{y'y}\mathbf{j} - n_{y'z}\mathbf{k} = n_{y'} \cos 30° \, \mathbf{j} - n_{y'} \sin 30° \, \mathbf{k}$$

$$\mathbf{n}_{y'} = 0.866\mathbf{j} - 0.5\mathbf{k} \tag{d}$$

The shear force along z' is the scalar product of Eqs. (a) and (c).

$$F_{z'} = (4.5\mathbf{i} + 13.5\mathbf{j} - 3.6\mathbf{k}) \cdot (0.5\mathbf{j} + 0.866\mathbf{k}) = 3.63 \text{ kN} \qquad \text{(e)}$$

Similarly,

$$F_{y'} = (4.5\mathbf{i} + 13.5\mathbf{j} - 3.6\mathbf{k}) \cdot (0.866\mathbf{j} - 0.5\mathbf{k}) = 13.49 \text{ kN} \qquad \text{(f)}$$

The transformed shear forces are illustrated in Fig. 8.10e.

It is worthwhile to digress from the computations and discuss still another method for obtaining $F_{y'}$ after $F_{z'}$ has been computed. The results of Eq. (e) can be written in x-y-z components by multiplying the scalar $F_{z'}$ by the unit vector $\mathbf{n}_{z'}$.

$$F_{z'}\mathbf{n}_{z'} = (3.63)(0.5\mathbf{j} + 0.866\mathbf{k}) = 1.82\mathbf{j} + 3.14\mathbf{k} \qquad \text{(g)}$$

Given that $\mathbf{F}_{z'}$ is normal to $\mathbf{F}_{y'}$, then

$$\mathbf{F}_{y'} = \mathbf{F}_A - \mathbf{F}_{z'} = (4.5\mathbf{i} + 13.5\mathbf{j} - 3.6\mathbf{k}) - (1.82\mathbf{j} + 3.14\mathbf{k})$$

$$\mathbf{F}_{y'} = 4.5\mathbf{i} + 11.68\mathbf{j} - 6.74\mathbf{k} \qquad \text{(h)}$$

The y and z components are shown in Fig. 8.10f. The magnitude of $F_{y'}$ is $[(11.68)^2 + (-6.74)^2]^{1/2} = 13.49$, which agrees with Eq. (f). The angle may be verified to be 30 degrees. This vector concept can be used to advantage in the next chapter.

Return to the primary problem and compute the moments $M_{z'}$ and $M_{y'}$.

$$M_{z'} = (2.43\mathbf{i} - 0.27\mathbf{j} + 2.03\mathbf{k}) \cdot (0.5\mathbf{j} + 0.866\mathbf{k}) = 1.62 \text{ kN} \cdot \text{m} \qquad \text{(i)}$$

$$M_{y'} = (2.43\mathbf{i} - 0.27\mathbf{j} + 2.03\mathbf{k}) \cdot (0.866\mathbf{j} - 0.5\mathbf{k}) = -1.25 \text{ kN} \cdot \text{m} \qquad \text{(j)}$$

The moments are illustrated in Fig. 8.10e.

The stresses on element A are computed in the usual manner.

$$\sigma_x = \frac{4.5 \text{ kN}}{(\pi)(0.025 \text{ m})^2} = 2.29(10^3) \text{ kN/m}^2 \qquad \text{(tension)}$$

$$\sigma_x = \frac{M_{y'}z'}{I_{y'}} = \frac{(1.25)(0.025)}{\pi(0.025)^4/4} = 10.18(10^6) \text{ kN/m}^2 \qquad \text{(compression)}$$

$$\tau_{xy'} = \frac{F_{y'}Q_{z'}}{I_{y'}t} = \frac{(13.49)[(\pi)(0.025)^2/2][(4)(0.025)/3\pi]}{[(\pi)(0.025)^4/4][0.05]}$$

$$\tau_{xy'} = 9.16(10^3) \text{ kN/m}^2 \qquad \text{(acting in the positive } y' \text{ direction)}$$

$$\tau_{xy'} = \frac{M_x\rho}{J_x} = \frac{(2.43)(0.025)}{\pi(0.025)^4/2}$$

$$= 0.099(10^6) \text{ kN/m}^2 \qquad \text{(acting in the negative } y' \text{ direction)}$$

The stress at point A is

$$\sigma_A = \begin{pmatrix} -1.017 & -0.898 \\ -0.898 & 0 \end{pmatrix} \text{GPa}$$

and is illustrated in Fig. 8.10g.   ■

The practical application of the scalar product has been illustrated as an aid in the analysis of stress at a point. An additional mathematical tool should be illustrated. Matrix theory is a mathematical subject that is worthy of study for its own sake. An elementary concept, borrowed from matrix theory, can be used to advantage to organize some of the computations of the previous example problems.

The scalar products of Eqs. (e) and (f) of Example 8.7 can be written as a transformation matrix multiplied by the corresponding force vector, written as a column matrix. There is no computational advantage for Example 8.7; however, the computations already performed in the example will aid in describing how the method is used.

A part of the example was to transform the force vector in the x-y-z system

$$\mathbf{F} = 4.5\mathbf{i} + 13.5\mathbf{j} - 3.6\mathbf{k} \tag{a}$$

to an equivalent force vector in the x-y'-z' system

$$F_x = 4.5 \qquad F_{y'} = 13.49 \qquad F_{z'} = 3.63 \qquad [\text{Eqs. (e) and (f)}]$$

The mathematical operation can be referred to as a matrix multiplication of the force vector by a transformation matrix. A detailed discussion of the transformation is given as an appendix to this chapter. The multiplication is illustrated by the following equation. Elementary matrix operations are defined and illustrated in Appendix D.

$$\begin{Bmatrix} F_x \\ F_{y'} \\ F_{z'} \end{Bmatrix} = \begin{vmatrix} 1 & 0 & 0 \\ 0 & 0.866 & -0.5 \\ 0 & .5 & 0.866 \end{vmatrix} \begin{Bmatrix} F_x \\ F_y \\ F_z \end{Bmatrix}$$

$$= \begin{vmatrix} 1 & 0 & 0 \\ 0 & 0.866 & -0.5 \\ 0 & 0.5 & 0.866 \end{vmatrix} \begin{Bmatrix} 4.5 \\ 13.5 \\ -3.6 \end{Bmatrix}$$

Performing the indicated matrix operations gives $F_x = 4.5$, remaining unchanged.

$$F_{y'} = 0 + (0.866)(13.5) + (-0.5)(-3.6) = 13.49$$

$$F_{z'} = 0 + (0.5)(13.5) + (0.866)(-3.6) = 3.63$$

These are exactly the same computations as the scalar products of Eqs. (f) and (e) of Example 8.7. Matrix multiplication is a standard mathematical operation. The terms are placed in the transformation matrix in a proper order such that the transformed forces are the result of the multiplication.

The examples thus far have dealt with a number of different situations and have employed the fixed-free beam structure. The methods of analysis will now be applied to problems that require the computation of reactions. A major part of a problem in analysis can be the computations required for finding reactions. The necessary theory was developed in statics. The next two examples will deal with problems where equilibrium methods of statics are used to compute reactions prior to computing stresses.

EXAMPLE 8.8

Compute the reactions and the maximum normal and shear stress at section *a-a* for the structure of Fig. 8.11. There is a pin connection at point *C*. Neglect the weight of the structure.

Solution:
The reactions cannot be computed without isolating at least one segment of the structure as a free body. By inspection, and experience in problem solving, choose member *CB* for a free body as shown in Fig. 8.11*b*. The reaction at *B* and the shear at *C* are both 24.0 k upward. Place the force at the pin as a force acting at *C* on the free body of member *AC* as shown in Fig. 8.11*c*.

The reactions at *A* are obtained in the usual manner. Imagine a coordinate system with origin at *A*. The unknown reactions are assumed positive using a vector sign convention.

$$H_A = 0 \tag{a}$$

$$V_A = 24.0 \text{ k} \tag{b}$$

$$M_A \mathbf{k} + (10\mathbf{i} + 5\mathbf{j}) \times (-24.0\mathbf{j}) = 0$$

$$M_A = 240.0 \text{ ft·k} \quad \text{(acting as shown in Fig. 8.11c)} \tag{c}$$

The actions at section *a-a* can be obtained from a free body of section *AD* or section *DC*. Using section *AD* as shown in Fig. 8.11*d* will result in the following analysis. Evaluate a unit vector $\mathbf{n}_{AD}$ or $\mathbf{n}_{x'}$ along section *AD* as shown in Fig. 8.11*d*.

$$\mathbf{n}_{x'} = \frac{2\mathbf{i} + \mathbf{j}}{(5)^{1/2}} = 0.894\mathbf{i} + 0.447\mathbf{j} \tag{d}$$

The normal force in member *AC* is

$$N_{x'} = (24.0\mathbf{j}) \cdot (0.894\mathbf{i} + 0.447\mathbf{j}) = 10.73 \text{ k} \tag{e}$$

also the unit vector along *y'* is, by inspection,

$$\mathbf{n}_{y'} = \frac{-\mathbf{i} + 2\mathbf{j}}{(5)^{1/2}} = -0.447\mathbf{i} + 0.894\mathbf{j} \tag{f}$$

$$V_{y'} = (24.0\mathbf{j}) \cdot (-0.447\mathbf{i} + 0.894\mathbf{j}) = 21.46 \text{ k} \tag{g}$$

Applying the equations of statics to the free body of Fig. 8.11*e* gives the acting forces and moment at section *a-a*.

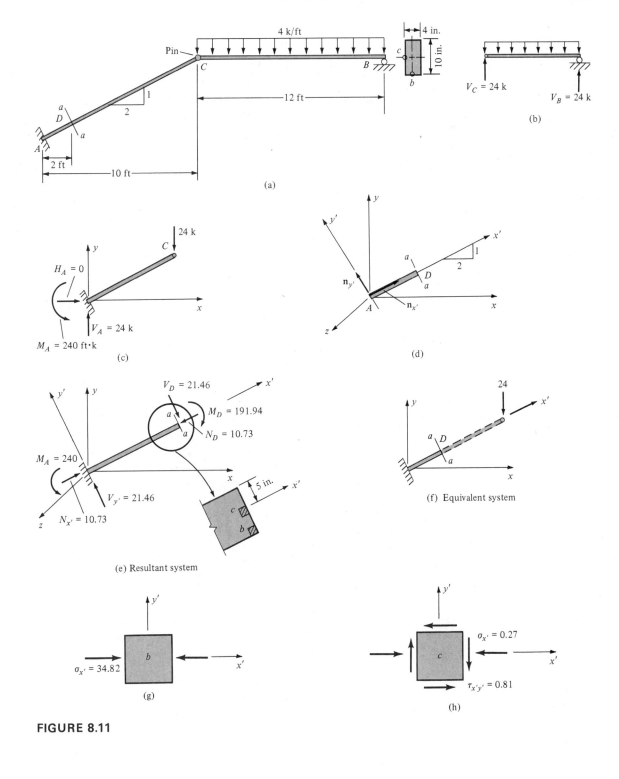

FIGURE 8.11

$$V_D = 21.46 \text{ k} \qquad \text{(as shown in Fig. 8.11}e) \qquad \textbf{(h)}$$

$$N_D = 10.73 \text{ k} \qquad \text{(compression)} \qquad \textbf{(i)}$$

$$M_D = -191.94 \text{ ft·k} \qquad \text{(as shown in Fig. 8.11}e) \qquad \textbf{(j)}$$

These actions were obtained by putting the free body of Fig. 8.11e in equilibrium. The same result would have been obtained by considering the force of Fig. 8.11f as an equivalent system acting at point D.

The cross section of the member is shown in Fig. 8.11a. Maximum normal stress occurs at point b, where both the bending moment and the normal force cause compression.

$$\sigma_b = \frac{(10.73)}{(4)(10)} + \frac{(191.94)(12)(5)}{(4)(10)^3/12} = 34.82 \text{ ksi} \qquad \text{(compression)} \qquad \textbf{(k)}$$

$$\sigma_c = \frac{10.73}{(4)(10)} = 0.27 \text{ ksi} \qquad \text{(compression)}$$

$$\tau_c = \frac{(21.46)(4)(5)(5/2)}{(4)(10)^3/12(4)}$$

$$= 0.81 \text{ ksi} \qquad \text{(acting in the direction of } V_D \text{ of Fig. 8.11}e) \qquad \textbf{(l)}$$

$$\sigma_b = \begin{pmatrix} -34.82 & 0 \\ 0 & 0 \end{pmatrix} \text{ksi} \qquad \sigma_c = \begin{pmatrix} -0.27 & -0.81 \\ -0.81 & 0 \end{pmatrix} \text{ksi} \qquad \textbf{(m)}$$

The stresses are illustrated in Figs. 8.11g and h. ∎

The computation of reactions for three-dimensional structures is usually lengthy. For this reason, the examples have been chosen such that the computational effort to find reactions has been kept to a minimum. The problems with which the practicing engineer must deal are not so considerate. It will suffice here to point out that the analysis techniques that have been discussed are completely general and applicable to any structure, no matter how complicated the loading or support condition.

One last example will illustrate the organizational power of vector methods. An analysis for a structure that lies in a two-dimensional plane but loaded in the third dimension will be illustrated.

EXAMPLE 8.9

The circular arch of Fig. 8.12 is fixed at end A and pinned in the center. The support at B is restrained in the y and z directions but free to translate in the x direction and rotate in all directions. A distributed horizontal load acts on one section of the

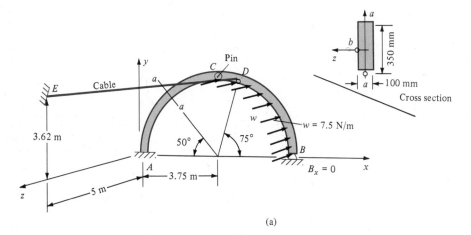

(a)

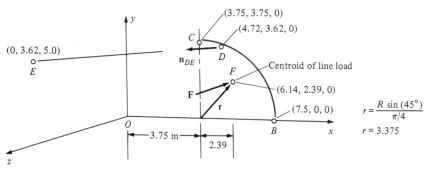

(b) Coordinates (x, y, z)

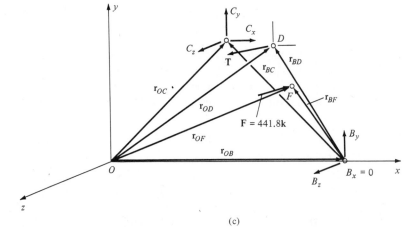

(c)

FIGURE 8.12

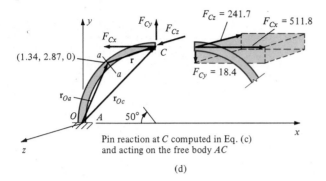

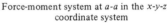

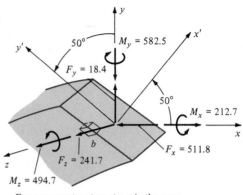

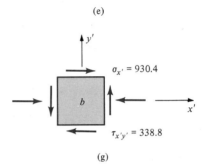

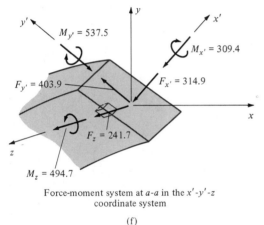

Pin reaction at C computed in Eq. (c) and acting on the free body AC

(d)

Force-moment system at a-a in the x-y-z coordinate system

(e)

Force-moment system at a-a in the x'-y'-z coordinate system

(f)

(g)

FIGURE 8.12 (continued)

arch. For stability in the z direction, a cable is connected at D. The arch is fabricated such that the dimension along the radius of the arch is 350 mm and its thickness is 100 mm. Compute the stresses at point b of section a-a. The radius of the arch is large compared to the cross-sectional dimensions. The curved beam formula of Chapter 6 does not apply.

Solution:
A study of Fig. 8.12a will indicate that there are six unknown reactions at A and two at B and an unknown cable tension at D. An equivalent force system can be computed at section a-a once the pin reactions at B or C are calculated. A free body of section BC, shown in Fig. 8.12b, will serve to compute the reactions. The methods of vector statics will be used. The resultant of the uniform load is labeled F and placed at the centroid of load shown in Fig. 8.12b. The uniform load is being treated as a

uniform load on a line segment, and the formula of Appendix A has been used as illustrated in Fig. 8.12b.

$$F = \frac{(75 \text{ N/m})(\pi)(3.75 \text{ m})}{2} = -441.8 \text{ N}$$

Coordinate locations of forces and reactions are shown in Fig. 8.12b. The direction of the cable tension is known; hence, T can be written in terms of a resultant vector. A unit vector acting from D to E along the cable can be computed using a position vector from D to E as follows:

$$\mathbf{r}_{DE} = \mathbf{r}_{OE} - \mathbf{r}_{OD} = (3.62\mathbf{j} + 5\mathbf{k}) - (4.72\mathbf{i} + 3.62\mathbf{j})$$
$$= -4.72\mathbf{i} + 5\mathbf{k}$$

$$\mathbf{n}_{DE} = \frac{-4.72\mathbf{i} + 5\mathbf{k}}{6.88} = -0.686\mathbf{i} + 0.727\mathbf{k} \tag{a}$$

$$\mathbf{T} = T\mathbf{n}_{DE} = -0.686\mathbf{i}T + 0.727\mathbf{k}T \tag{b}$$

Using Fig. 8.12c,

$$\Sigma \mathbf{F} = 0$$

$$(C_x - 0.686T)\mathbf{i} = 0 \tag{c}$$

$$(C_y + B_y)\mathbf{j} = 0 \tag{d}$$

$$(C_z + B_z + 0.727T - 441.8)\mathbf{k} = 0 \tag{e}$$

Since reactions are desired at C, sum moments about point B using the position vectors of Fig. 8.12c.

$$\mathbf{r}_{BF} = \mathbf{r}_{OF} - \mathbf{r}_{OB} = (6.14\mathbf{i} + 2.39\mathbf{j}) - (7.5\mathbf{i}) = -1.36\mathbf{i} + 2.39\mathbf{j} \tag{f}$$

$$\mathbf{r}_{BD} = \mathbf{r}_{OD} - \mathbf{r}_{OB} = (4.72\mathbf{i} + 3.62\mathbf{j}) - (7.5\mathbf{i}) = -2.78\mathbf{i} + 3.62\mathbf{j} \tag{g}$$

$$\mathbf{r}_{BC} = \mathbf{r}_{OC} - \mathbf{r}_{OB} = (3.75\mathbf{i} + 3.75\mathbf{j}) - (7.5\mathbf{i}) = -3.75\mathbf{i} + 3.75\mathbf{j} \tag{h}$$

$$\Sigma M_B = 0$$

$$\begin{vmatrix} \mathbf{i} & \mathbf{j} & \mathbf{k} \\ -1.36 & 2.39 & 0 \\ 0 & 0 & -441.8 \end{vmatrix} + \begin{vmatrix} \mathbf{i} & \mathbf{j} & \mathbf{k} \\ -2.78 & 3.62 & 0 \\ -6.86 & 0 & 0.727 \end{vmatrix}T + \begin{vmatrix} \mathbf{i} & \mathbf{j} & \mathbf{k} \\ -3.75 & 3.75 & 0 \\ C_x & C_y & C_z \end{vmatrix} = 0$$

Performing the indicated computations gives the three moment equations

$$2.63T + 3.75C_z = 1055.9 \quad (x \text{ direction}) \tag{i}$$

$$2.02T + 3.75C_z = 600.8 \quad (y \text{ direction}) \tag{j}$$

$$2.48T - 3.75C_x - 3.75C_y = 0 \quad (z \text{ direction}) \tag{k}$$

Solving Eqs. (i) and (j) gives T and C_z.

$$T = 746.1 \text{ N} \qquad C_z = -241.7 \text{ N}$$

Equation (c) gives

$$C_x = 511.8 \text{ N}$$

Equation (k) gives

$$C_y = -18.4 \text{ N}$$

Similarly, B_y and B_z can be calculated. The forces acting on free body BC are

$$\mathbf{F} = -441.8\mathbf{k}$$

$$\mathbf{F}_C = 511.8\mathbf{i} - 18.4\mathbf{j} - 241.7\mathbf{k} \tag{l}$$

$$\mathbf{T} = -511.8\mathbf{i} + 542.4\mathbf{k} \tag{m}$$

$$\mathbf{F}_B = 18.4\mathbf{j} + 141.1\mathbf{k} \tag{n}$$

Construct the free body of the structure between A and point C as shown in Fig. 8.12d. The force $\mathbf{F}_C$ acts as shown in Fig. 8.12d. All that remains is to compute an equivalent force system at section a-a. The coordinate location of section a-a is found by inspection. The position vector between a-a and C is

$$\mathbf{r} = \mathbf{r}_{OC} - \mathbf{r}_{Oa} = (3.75\mathbf{i} + 3.75\mathbf{j}) - (1.34\mathbf{i} + 2.87\mathbf{j})$$

$$= 2.41\mathbf{i} + 0.88\mathbf{j}$$

$$\mathbf{M}_{a\text{-}a} = \begin{vmatrix} \mathbf{i} & \mathbf{j} & \mathbf{k} \\ 2.41 & 0.88 & 0 \\ -511.8 & 18.4 & 241.7 \end{vmatrix} = 212.7\mathbf{i} - 582.5\mathbf{j} + 494.7\mathbf{k} \tag{o}$$

$$\mathbf{F}_{a\text{-}a} = -511.8\mathbf{i} + 18.4\mathbf{j} + 241.7\mathbf{k} \tag{p}$$

The force system of Eqs. (o) and (p) acts in the x-y-z coordinate system as illustrated in Fig. 8.12e. Prior to using the stress formulas the force system of Fig. 8.12e must be transformed (rotated) such that the vectors act normal and adjacent to section a-a. The vectors $\mathbf{F}_z$ and $\mathbf{M}_z$ are already acting adjacent to section a-a. The transformation given by Eq. (A8.9) in the appendix of this chapter will be used with $\theta = 50$ degrees.

$$\begin{Bmatrix} F_{x'} \\ F_{y'} \end{Bmatrix} = \begin{bmatrix} -511.8 \cos(50°) + 18.4 \sin(50°) \\ 511.8 \sin(50°) + 18.4 \cos(50°) \end{bmatrix} = \begin{Bmatrix} -314.9 \\ 403.9 \end{Bmatrix} \text{ N} \tag{q}$$

$$\begin{Bmatrix} M_{x'} \\ M_{y'} \end{Bmatrix} = \begin{bmatrix} (212.7)(0.643) - (582.5)(0.766) \\ -(212.7)(0.766) - (582.5)(0.643) \end{bmatrix} = \begin{Bmatrix} -309.4 \\ -537.5 \end{Bmatrix} \text{ N·m} \tag{r}$$

The results of Eqs. (q) and (r) are shown in Fig. 8.12f. Figure 8.12f may be compared to Fig. 8.4, and the stress produced by the loading can be computed as in previous examples. At point b,

$$\sigma_{x'} = \frac{314.9 \text{ N}}{(0.1)(0.35) \text{ m}^2} = 9.0 \text{ kPa} \qquad \text{(compression)} \qquad \text{(s)}$$

$$\sigma_{x'} = \frac{(537.5 \text{ N} \cdot \text{m})(0.05)}{(0.35)(0.10)^3/12 \text{ m}^4} = 921.4 \text{ kPa} \qquad \text{(compression)} \qquad \text{(t)}$$

$$\tau_{x'y'} = \frac{(403.9 \text{ N})(0.10 \text{ m})(0.175 \text{ m})(0.175/2 \text{ m})}{[(0.10)(0.35)^3/12] \text{ m}^4 (0.10) \text{ m}} = 17.31 \text{ kPa} \qquad \text{(u)}$$

The shear stress due to the torque $M_{x'}$ is computed using Eq. (4.24) of Chapter 4.

$$\tau_{x'y'} = \frac{M_{y'}}{k_1 h b^2}$$

Figure 4.17 shows that $h = 350$ mm is the long side of the cross section; k_1 is obtained from Table 4.1 for $h/b = 350/100 = 3.5$ as approximately 0.275.

$$\tau_{x'y'} = \frac{(309.4 \text{ N} \cdot \text{m})}{(0.275)(0.35)(0.1)^2 \text{ m}^3} = 321.5 \text{ kPa} \qquad \text{(v)}$$

$$\sigma_b = \begin{pmatrix} -930.4 & 338.8 \\ 338.8 & 0 \end{pmatrix} \text{kPa} \qquad \text{(w)}$$

The stress at point b is shown in Fig. 8.12g. ∎

8.5 THIN-WALLED PRESSURE VESSELS

Thin-walled pressure vessel theory is an appropriate topic for study at this time. A thin-walled pressure vessel can be analyzed using the elementary methods of mechanics of materials. Certain simplifying assumptions can be made with regard to the stress distribution that result in rather simple formulations for computing stress at a point. At the same time the formulas are accurate and give results that are acceptable for engineering design. The previous studies of stress at a point give a normal stress on opposite faces of the element and zero normal stress on the other two faces. Normal stresses act on all four sides of an element when thin-walled pressure vessels are analyzed. Hence, a practical problem can be defined that leads to a stress element of a more general nature.

Pressure vessels have numerous practical uses: water tanks, pipelines, and steam boilers, among others. A vessel under pressure must qualify as either thin-walled or thick-walled. The actual thickness is not the criterion, but rather the comparison of the thickness to the radius of the vessel. A balloon is an extreme example of a thin-walled vessel. A gun barrel, where the thickness of the barrel is of the same order of magnitude as the bore of the barrel, would be termed a thick-walled cylinder. The radial stress along the radius of a thick-walled member is not constant and should be computed. However, for thin-walled vessels the stress in the radial direction can be neglected.

There are two types of thin-walled pressure vessels that can be analyzed in an elementary way. Cylindrical and spherical vessels are very popular three-dimensional geometric shapes that are used as pressure vessels.

A cylindrical pressure vessel may have flat ends, so that geometrically it is a right circular cylinder, or it may have spherical caps attached to its ends. These configurations are illustrated in Fig. 8.13. The analysis for the cylindrical portion of the tank is the same in either case. Prior to deriving stress equations for thin-walled pressure vessels a review of pressure distributions on curved surfaces is in order.

Consider the curved circular surface of Fig. 8.14. A uniform pressure p acts normal to the surface. The circular surface is defined by the angle α and the length L. An elemental strip is isolated as shown in Fig. 8.14b. The force on the element is dP and acts normal to the element. The resultant dP is equivalent to $pLr\,d\theta$, the internal pressure times the area over which it acts. The horizontal component of the force is

$$dP_H = pLr \cos \theta \, d\theta$$

$$P_H = \int_0^\alpha pLr \cos \theta \, d\theta = pLr \sin \alpha$$

The horizontal component of the pressure force is merely the projected area of the curved surface shown in Fig. 8.14a times the pressure. Using

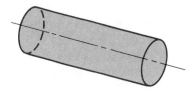

Flat ends

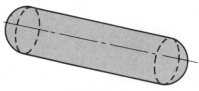

Spherical ends

FIGURE 8.13
Cylindrical tanks.

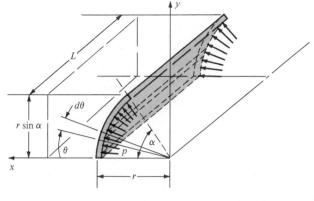

F!GURE 8.14
(a) Pressure distribution
acting on a curved surface;
(b) force acting on the strip
element.

this concept it is a fairly simple matter to derive the stress equation for a thin-walled pressure vessel.

The cylindrical vessel of Fig. 8.15 has a radius of r and wall thickness t. Since t is much smaller than r, no distinction need be made between the inside radius and the outside radius of the cylinder. Since t is small compared to r, it is in order to assume that the normal stress is uniform through the thickness. A circular slice of length ΔL is removed from the cylinder and shown in Fig. 8.15b. The slice is cut in half and the free body is loaded internally with pressure p and is equivalent to a single force P. The force P is equal to p times the projected area of the curved surface.

$$P = 2pr\,\Delta L$$

The balance of forces gives

$$P = 2F = 2\sigma_h\,\Delta L\,t$$

and F is related to the stress σ_h since σ_h is the force divided by the area. Solving for stress gives

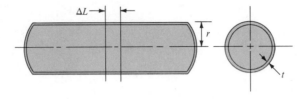

(a)

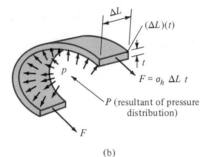

(b)

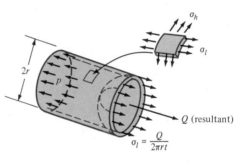

FIGURE 8.15

(c)

$$\sigma_h = \frac{pr}{t} \tag{8.8}$$

The subscript h indicates hoop stress since the slice of the cylinder resembles a hoop.

The stress along the longitudinal axis of the cylinder is σ_l and is obtained using the free body of Fig. 8.15c. The derivation is independent of the length of the free body. Summation of forces along the axis of the vessel gives the desired result. The force acting on the end of the cylinder is given by the pressure acting on the area of the cylinder. The balancing force is distributed around the cylinder wall of the cut section and is given by the longitudinal stress times the area. Again, since $t \ll r$,

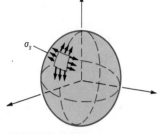

FIGURE 8.16
Spherical pressure vessel.

the area can be accurately evaluated as the circumference of the cylinder times the wall thickness.

$$2\pi rt\sigma_l = \pi r^2 p$$

$$\sigma_l = \frac{pr}{2t} \tag{8.9}$$

For a cylindrical pressure vessel the longitudinal stress is one-half the hoop stress.

A spherical pressure vessel can be analyzed in a similar manner. A free body of the vessel of Fig. 8.16 would be the same along any diameter due to the symmetry of the vessel. The dashed lines of Fig. 8.15c would represent a free body of a spherical pressure vessel. The analysis for the spherical vessel is analogous to that for the longitudinal stress in the cylinder. The projected area of the sphere is πr^2, and therefore

$$\sigma_s = \frac{pr}{2t} \tag{8.10}$$

the same as σ_l for the cylinder.

It becomes obvious that spherical ends are equivalent to flat ends for a cylinder as far as stress due to internal pressure is concerned. There are other considerations when designing cylindrical pressure vessels. The joints where the ends are connected to the cylinder usually require special attention because of the abrupt change in the geometry of the vessel. Visualize an element situated along the line where a spherical end is connected to a cylindrical tank as shown in Fig. 8.17. The hoop stress for the cylinder portion would be given by Eq. (8.8), yet the same stress, assuming that the spherical end is being analyzed, would be given by Eq. (8.10). This indicates that along the connection there is an abrupt change in stress by a factor of 1/2. There is a change in stress, but the actual analysis is beyond the scope of this text. It will suffice merely to point out the problem area and note that standard design codes offer additional information. The elementary formulas are extremely accurate for the design of pressure vessels except where there is some change in geometry that would violate the assumed geometry of Figs. 8.15 and 8.16. For instance, if a hole needed to be cut in the side of a pressure vessel and then repaired with a covering so that the vessel could again be used as a pressure vessel, the stresses around the hole and at the juncture of the covering with the original vessel would not be obtainable using the elementary theory.

A single example should illustrate the use of the thin-walled pressure vessel equations for problems involving combined stresses.

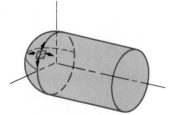

FIGURE 8.17
Abrupt change from a cylinder to a sphere.

EXAMPLE 8.10

A cylindrical thin-walled pressure vessel of Fig. 8.18 is supported as a simple beam and weighs 120 lb/ft. The inside radius is 18 in. and the wall thickness is 0.25 in. For an internal pressure of 170 psi compute the maximum tensile and compressive normal stress at the midpoint of the span.

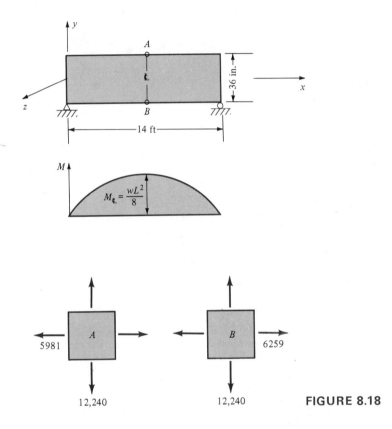

FIGURE 8.18

Solution:
The moment diagram for the uniform load is shown in Fig. 8.18. The moment at the center is given by

$$M = \frac{(120)(12)(168)^2}{8} = 35{,}280 \text{ in.·lb}$$

Normal stress due to bending for points A and B is

$$\sigma_{A,B} = \frac{(35{,}280)(18)}{\pi(18)^3(0.25)} = 139 \text{ psi}$$

Stresses due to internal pressure are given by Eqs. (8.8) and (8.9).

$$\sigma_h = \frac{(170)(18)}{0.25} = 12{,}240 \text{ psi}$$

$$\sigma_l = \frac{(170)(18)}{2(0.25)} = 6120 \text{ psi}$$

In this example the stress due to the weight of the tank is practically negligible compared to the stress due to internal pressure. The results are tabulated as follows and shown in Fig. 8.18.

$$\sigma_A = \begin{pmatrix} 5981 & 0 \\ 0 & 12{,}240 \end{pmatrix} \qquad \sigma_B = \begin{pmatrix} 6259 & 0 \\ 0 & 12{,}240 \end{pmatrix} \qquad \blacksquare$$

8.6 BEAMS WITH INITIAL CURVATURE

Beams with initial curvature were discussed in Section 6.5 of Chapter 6, and problems that involved combined stresses were purposely avoided. Beams with initial curvature are often subject to combined stresses due to the manner in which they are used. Again, the C-clamp mentioned in Chapter 6 can serve as an example of a curved beam subject to combined stress.

An illustrative example should serve to demonstrate the analysis of a beam with initial curvature subject to combined stresses. Equation (6.30) and Table 6.4 of Chapter 6 will serve to compute flexure stresses.

EXAMPLE 8.11

A split ring is made of a circular steel member 18 mm in diameter as illustrated in Fig. 8.19. The inside radius of the split ring is 9 mm. The applied loads are each 3 N as shown in the figure. Compute the state of stress as points A and B.

Solution:
A free body that includes the point A is shown in Fig. 8.19b. The moment with respect to the centroid of the section is

$$M = (36 \text{ mm})(3 \text{ N}) = 108 \text{ N·mm}$$

Summing forces gives a tensile force of 3 N acting at the cut section. The normal stress at point A is

$$\sigma_A = K_i \frac{Mc}{I} + \frac{P}{A} \qquad \text{(tension)}$$

where K_i is obtained from Table 6.4 of Chapter 6. The quantity

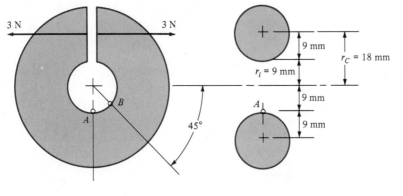

(a)

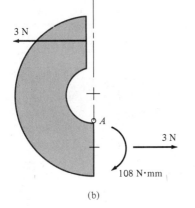

(b)

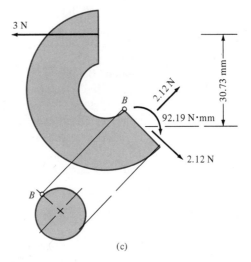

(c)

FIGURE 8.19

$$\frac{r_C}{r_C - r_i} = \frac{18}{9} = 2$$

and hence K_i is found as 1.62.

$$\sigma_A = \frac{(1.62)(108 \text{ N·mm})(9 \text{ mm})}{\pi(9 \text{ mm})^4/4} + \frac{3\text{N}}{\pi(9 \text{ mm})^2}$$

$$\sigma_A = 0.306 \text{ N/mm}^2 + 0.012 \text{ N/mm}^2 = 0.318 \text{ MPa}$$

$$\sigma_A = \begin{pmatrix} 0.318 & 0 \\ 0 & 0 \end{pmatrix} \text{ MPa}$$

The section at point B shown in Fig. 8.19c has a reactive moment of

$$M = (30.73 \text{ mm})(3 \text{ N}) = 92.19 \text{ N·mm}$$

The reactive force is 3N and has components of (3)(cos 45°) N, as shown in Fig. 8.18c. The component forces act on the section to produce a shear and an axial tension.

$$\sigma_B = \frac{(1.62)(92.19 \text{ N·mm})(9 \text{ mm})}{\pi(9 \text{ mm})^4/4} + \frac{2.12 \text{ N}}{\pi(9 \text{ mm})^2}$$

$$= 0.269 \text{ MPa} \qquad \text{(tension)}$$

$$\sigma_B = \begin{pmatrix} 0.269 & 0 \\ 0 & 0 \end{pmatrix} \text{MPa}$$

∎

8.7 SUMMARY

Combined stresses are the result of applying the stress equations of the previous chapters to compute the state of stress at a point. The state of stress at a point has been represented using the notation

$$\sigma = \begin{pmatrix} \sigma_x & \tau_{xy} \\ \tau_{xy} & \sigma_y \end{pmatrix} \tag{8.6}$$

and an element to represent a two-dimensional state of stress.

The use of vector analysis has been fully exploited and appears as a computational tool in most of the examples presented. The interpretation of vector forces and vector moments acting on a cut section has been emphasized and their relation to the beam sign convention has been demonstrated.

The vector transformation for two dimensions has been introduced and formally derived in the appendix of this chapter. The vector transformation has been employed to compute stresses on a cut section that is not normal to a global coordinate axis. An important application of the vector transformation allows for combined shear stresses to be computed at locations on a circular cross section other than the coordinate axes.

Beam and frame structures with both two- and three-dimensional loading have been illustrated. An important concept concerning free-body sections has been illustrated and should be stated again. Free-body sections have been analyzed using an equivalent-force system and also by balancing the actions on a structure with its reactions. Both methods are equivalent. Examples 8.8 and 8.9 illustrate the methodology of the concepts. In Example 8.9 a free-body section is cut and the applied loading is located in its acting position and is not on the free body. The method has been applied primarily to fixed-end structures to avoid computing reactions. The second method is illustrated in Example 8.8. A free body is isolated and the reactions at a cut section are computed such that the free body is in equilibrium.

Thin-wall pressure vessel theory has been developed for cylindrical pressure vessels. The circumferential stress or hoop stress is

$$\sigma_h = \frac{pr}{t} \tag{8.8}$$

and the longitudinal stress is

$$\sigma_l = \frac{pr}{2t} \tag{8.9}$$

Spherical pressure vessels have equal stresses in any direction and are given by

$$\sigma_s = \frac{pr}{2t} \tag{8.10}$$

The beam with initial curvature that was discussed in Chapter 6 is shown to be an example of combined stresses. In almost any situation a curved beam will have normal stress due to bending superposed on normal stress caused by an axial force.

8.8 APPENDIX: VECTOR TRANSFORMATIONS

This appendix is intended to be a brief discussion of a specialized application of matrix theory. It is obvious that the use of matrix multiplication is not necessary to achieve the vector transformation. Examples 8.6 and 8.7 were concerned with determining the components of a force vector in a rotated coordinate system, and it was accomplished without mention of a matrix. On the other hand, a course in mechanics of materials offers an excellent opportunity to introduce the transformation matrix. In more advanced studies, applied theory of elasticity, theory of structures, or dynamics of machinery, the formulation of problems using matrix concepts has become an indispensable computational tool.

The object is to demonstrate that if the unit vectors defining the new coordinate system are written in the proper form, then a matrix multiplication with the original force vector yields the components of the transformed vector. The reader unfamiliar with matrix multiplications should refer to Appendix D for a discussion of elementary matrix theory.

The vector **F** of Fig. A8.1 lies in the x-y coordinate plane. The transformation to be illustrated is termed a rotation about the z axis. The transformed vector will lie in the x'-y' system. The transformation is of the components of **F** in the x-y system into components of **F** in the x'-y' system. The resultant vector remains unchanged. The rotation is through an angle θ measured counterclockwise from the positive x axis.

The transformation is for the components of **F**; hence, **F** is written in component form, $\mathbf{F} = F_x\mathbf{i} + F_y\mathbf{j}$. The components are written as a column matrix, sometimes referred to as a column vector.

$$\mathbf{F} \Rightarrow \begin{Bmatrix} F_x \\ F_y \end{Bmatrix} \Rightarrow \{F\} \tag{A8.1}$$

where the notation $\Rightarrow$ stands for "is the same as."

The component F_x is shown in Fig. A8.1b as a vector along the x axis. The computations to find the *components* of F_x in the x'-y' system are straightforward. The components are shown in Fig. A8.1b.

$$F_{x'} = F_x \cos \theta \tag{A8.2}$$

$$F_{y'} = -F_x \sin \theta \tag{A8.3}$$

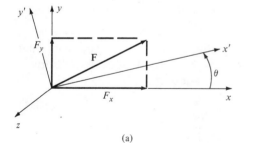

(a)

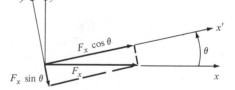

(b)

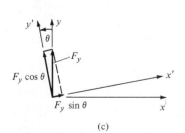

(c)

(d)

FIGURE A8.1

F_y, the y component of **F**, is shown in Fig. A8.1c. It has components

$$F_{x'} = F_y \sin \theta \tag{A8.4}$$

$$F_{y'} = F \cos \theta \tag{A8.5}$$

The total component of $F_{x'}$ is obtained by adding Eqs. (A8.2) and (A8.4); that is

$$F_{x'} = F_x \cos \theta + F_y \sin \theta \tag{A8.6}$$

Similarly for $F_{y'}$,

$$F_{y'} = -F_x \sin \theta + F_y \cos \theta \tag{A8.7}$$

In matrix notation, Eqs. (A8.6) and (A8.7) are combined as

$$\begin{Bmatrix} F_{x'} \\ F_{y'} \end{Bmatrix} = \begin{Bmatrix} F_x \cos \theta + F_y \sin \theta \\ -F_x \sin \theta + F_y \cos \theta \end{Bmatrix} \tag{A8.8}$$

The right-hand side of Eq. (A8.8) is the result of a matrix multiplication, and the matrix equation is written as

$$\begin{Bmatrix} F_{x'} \\ F_{y'} \end{Bmatrix} = \begin{bmatrix} \cos \theta & \sin \theta \\ -\sin \theta & \cos \theta \end{bmatrix} \begin{Bmatrix} F_x \\ F_y \end{Bmatrix} = \begin{Bmatrix} F_x \cos \theta + F_y \sin \theta \\ -F_x \sin \theta + F_y \cos \theta \end{Bmatrix} \tag{A8.9}$$

The square matrix of Eq. (A8.9) is classically termed the transformation matrix and will be referred to as $[T]$.

$$[T] = \begin{bmatrix} \cos \theta & \sin \theta \\ -\sin \theta & \cos \theta \end{bmatrix} \tag{A8.10}$$

In the future any vector transformation for a vector lying in a coordinate plane and rotated through an angle θ can be obtained by substituting into Eq. (A8.9), which in matrix notation is

$$\{F'\} = [T]\{F\} \tag{A8.11}$$

For Further Study

ASME Boiler and Pressure Vessel Code, American Society of Mechanical Engineers, New York, 1974.

Budynas, R. G., *Advanced Strength and Applied Stress Analysis*, McGraw-Hill, New York, 1977.

Curtis, C. W., *Linear Algebra*, Allyn and Bacon, Boston, 1974.

Spiegel, M. R., *Vector Analysis*, Schaum Publishing Company, New York, 1959.

Ugural, A. C., and S. K. Fenster, *Advanced Strength and Applied Elasticity*, 2nd ed., Elsevier, New York, 1986.

8.9 PROBLEMS

8.1 A beam 75 by 100 mm is loaded with a single 40-kN force as shown in Fig. P8.1. Compute the stress at points A, B, and C of section a-a located 1 m from the fixed end. Show your results as stresses acting on a stress block and write the results using notation for stress at a point.

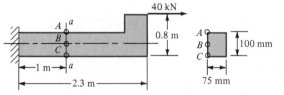

FIGURE P8.1

8.2 The beam of Fig. P8.2 is 2 by 10 in. in cross section and is loaded as shown. Compute the stress at points A, B, and C at the midpoint of the beam. Present your results using notation for stress at a point and sketch the stress elements for points A, B, and C.

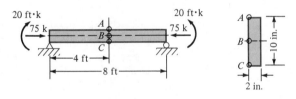

FIGURE P8.2

8.3 A wooden member is to be fabricated using 2- by 10-in. members. It has been determined that the loading will produce a negative bending moment of 36 ft·k and an axial force of 9 k tension. The depth of the member must be 10 in. How many 2- by 10-in. members must be used so that the normal stress will not exceed 5000 psi in tension or 4200 psi in compression?

8.4 The hollow shaft of Fig. P8.4 is 1.2 m long, with a cross section as illustrated in the figure. The loading consists of an axial force, a transverse force, and a torque. Compute the state of stress at points A and B located at the fixed end. Sketch your results on an element.

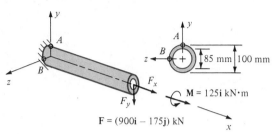

FIGURE P8.4

8.5 The beam of Fig. P8.5 is loaded as illustrated. Compute the width of the beam to the nearest 0.5 in. if the allowable normal stress is not to exceed 4500 psi in tension or compression.

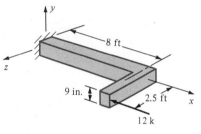

FIGURE P8.5

8.6 A fixed-free beam with a square cross section is loaded at the free end with a force $\mathbf{F} = (6.0\mathbf{i} + 7.5\mathbf{k})$ kN. Compute the stress at the fixed end for points A, B, and C.

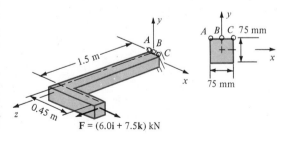

FIGURE P8.6

8.7 The fixed-free beam of Fig. P8.7 is loaded as illustrated. Determine the state of stress at points A and B.

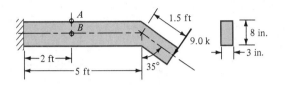

FIGURE P8.7

8.8 The beam of Fig. P8.8 is fabricated using 5-in. standard-weight pipe. Determine the state of stress at points A and B.

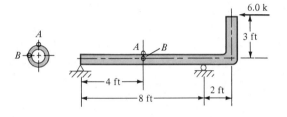

FIGURE P8.8

8.9 A square beam 12 ft long is loaded as shown in Fig. P8.9. Compute the state of stress at points A and B located at midspan and at the midpoints of the sides. Sketch the elements and give answers using notation for stress at a point.

8.10 A simply supported beam is loaded in both transverse directions as shown in Fig. P8.10. Compute the stress at points A and B on a section 1 m from the right

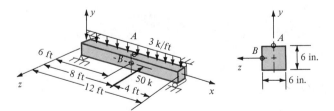

FIGURE P8.9

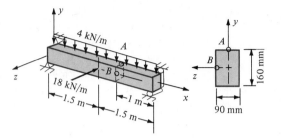

FIGURE P8.10

end. Sketch the element showing the stresses acting on an element.

8.11 The fixed-free beam of square cross section 6 by 6 in. shown in Fig. P8.11 is loaded with a single concentrated force. Compute the state of stress at points A and B. (See Section 4.7 of Chapter 4.)

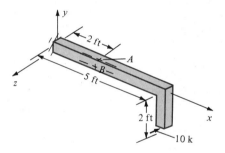

FIGURE P8.11

8.12 The pipe of Fig. P8.12 has an outside diameter of 250 mm and the wall is 25 mm thick. A single force of $\mathbf{F} = (40\mathbf{i} + 60\mathbf{j} - 30\mathbf{k})$ kN is applied at the free end

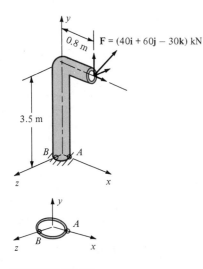

F = (40i + 60j − 30k) kN

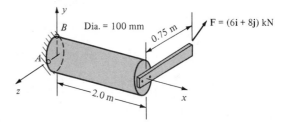

FIGURE P8.14

FIGURE P8.12

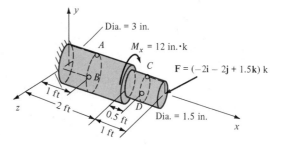

FIGURE P8.15

through the centroid of the section. Compute the state of stress at points A and B at the fixed end. Show your results on a sketch of the element.

8.13 Determine the state of stress at points A and B for the beam of Fig. P8.13. The beam is fixed at the left end and carries a uniform load of 2 k/ft. (See Section 4.7 of Chapter 4.)

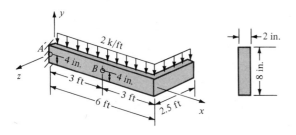

FIGURE P8.13

8.14 A circular shaft 100 mm in diameter is fixed at one end and loaded as shown in Fig. P8.14. Compute the state of stress at points A and B located at the fixed end. Present your analysis as the state of stress acting on an element.

8.15 A stepped shaft is loaded as shown in Fig. P8.15. Compute the state of stress at points A and B on a section located 1 ft from the fixed end and at points C

and D located 2.5 ft from the fixed end. Sketch your results on elements representing the points A, B, C, and D.

8.16 Compute the state of stress at points A, B, C, D, E, and F at sections a, b, and c, respectively, for the structure of Fig. P8.16. Sketch the stress elements and write your answers using the notation for stress at a point.

8.17 The circular shaft of Fig. P8.17 is fixed at one end and is bent such that the free end projects outward as shown. Assume that the axial vector is a force vector with a magnitude of 500 lb and compute the state of stress at points A and B. Section C is 6 in. from the fixed end.

8.18 Repeat Problem 8.17 assuming that the axial vector is an axial moment (torque) of 1200 in·lb.

8.19 A 1-in.-diameter circular beam is fixed at one end and is curved in the x, z-plane to form a quarter circle as shown in Fig. P8.19. Compute the maximum normal and shear stresses on a section of the beam defined by θ = 30 degrees. Assume a = 11 in. and P = 1000 lb. (*Note:* It may be beneficial to refer to Chapter 5, Example 5.17.)

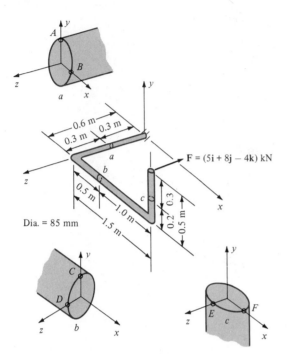

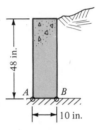

$\mathbf{F} = (5\mathbf{i} + 8\mathbf{j} - 4\mathbf{k})$ kN

Dia. = 85 mm

FIGURE P8.16

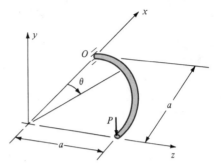

FIGURE P8.19

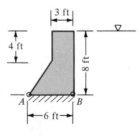

FIGURE P8.20

8.21 A small gravity dam with the cross section shown in Fig. P8.21 is to be constructed from reinforced concrete and is intended to retain water. Determine the state of stress at the toe, point A, and the heel, point B. Locate the neutral axis.

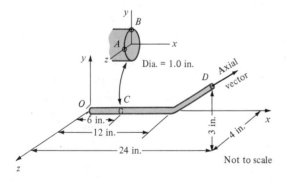

FIGURE P8.17

FIGURE P8.21

8.20 A retaining wall is required to withstand soil pressure as shown in Fig. P8.20. Assume the unit weight of the soil is 95 lb/ft^3 and the wall is concrete. Compute the normal stresses at points A and B.

8.22 A concrete dam has the dimensions shown in Fig. P8.22 and is intended to contain water. Compute the stresses at points A and B and locate the neutral axis. $\gamma_{\text{conc}} = 23.6$ kN/m^3; $\gamma_w = 9.81$ kN/m^3.

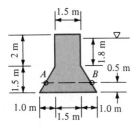

FIGURE P8.22

8.23 A retaining wall 8 ft high is constructed using wide-flange structural members, W10x45, driven vertically into the ground every 4 ft. The structural members act as fixed-free beams illustrated in Fig. P8.23. Assume the wall that spans the distance between the wide-flange members is of sufficient strength to hold the soil mass without being overstressed. Compute the maximum normal and shear stresses acting on the wide-flange members. Assume the unit weight of soil as 105 lb/ft³.

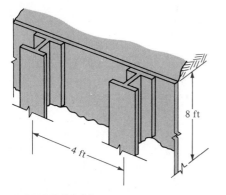

FIGURE P8.23

8.24 A short column, shown in Fig. P8.24, is loaded with a single axial force. Compute the normal stresses at the extreme corners and locate the neutral axis.

8.25 A short column supports two axial forces as shown in Fig. P8.25. Compute the normal stresses at each corner and locate the neutral axis.

8.26 A short column is loaded with two concentrated forces as illustrated in Fig. P8.26. Compute the state of stress at each corner on a section 24 in. below the top of the column.

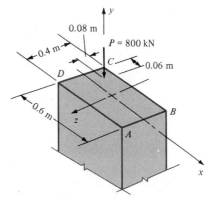

FIGURE P8.24

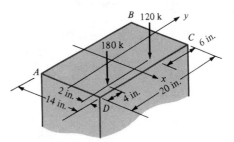

FIGURE P8.25

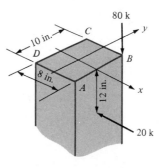

FIGURE P8.26

8.27 A 4-in. standard pipe (nominal diameter) is fixed-free, as shown in Fig. P8.27. The axis of the pipe is oriented at 35 degrees with the x axis and lies in the x-z plane. A moment vector, $\mathbf{M} = 12\mathbf{i} + 18\mathbf{j} - 10\mathbf{k}$ ft·lb,

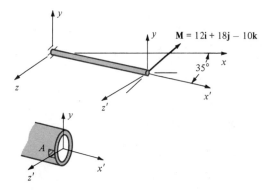

FIGURE P8.27

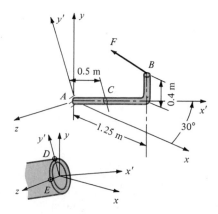

FIGURE P8.29

acts at the free end. Compute the state of stress at point *A* located at the fixed end.

8.28 The solid circular beam of Fig. P8.28 is loaded with a single force, $\mathbf{F} = 20\mathbf{k}$ kN. Compute the state of stress at point *A* at the fixed end. The diameter of the pipe is 120 mm.

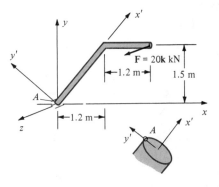

FIGURE P8.28

8.29 A circular pipe lies in the *x-y* plane and extends upward 30 degrees above the *x* axis. The outside diameter of the pipe is 60 mm and the wall thickness is 10 mm. A force *F* with components $F_x = -16$, $F_y = 30$, and $F_z = 24$ kN is applied at the free end *B*. Compute the state of stress at points *D* and *E* of section *C* located 0.5 m from the fixed end along the *x'* axis of Fig. P8.29.

8.30 A solid circular rod 0.1 m in diameter is subject to a shear force of 6.5**j** kN as shown in Fig. P8.30. Use

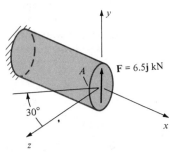

FIGURE P8.30

the method of Example 8.7 to compute the shear stress at point *A*.

8.31 A solid brass rod 1.25 in. in diameter carries a force of 2000 lb applied as shown at point *B* in Fig. P8.31. Compute the state of stress at point *A* located at the fixed end. See Example 8.7.

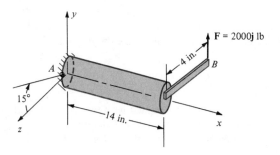

FIGURE P8.31

8.32 A solid circular rod 30 mm in diameter is loaded with 7 kN at the free end as shown in Fig. P8.32. Compute the state of stress at point A located on a section 0.25 m from the fixed end.

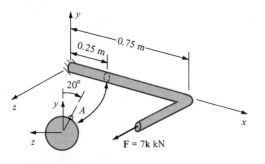

FIGURE P8.32

8.33 The frame of Fig. P8.33 is loaded as illustrated. The support at C is a pin and B rests on a roller-type support. Determine the state of stress at section a-a for points D and E as shown in the detail view. The cross section is 5 by 2.5 in.

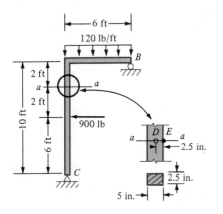

FIGURE P8.33

8.34 The structure of Fig. P8.34 is loaded as shown. Compute the normal force, shear, and bending moment acting on section a-a as follows: **a.** Use section AC as a free body; **b.** use section BC as a free body.

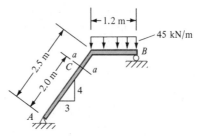

FIGURE P8.34

8.35 A rectangular bar 2 by 4 in. in cross section is positioned as shown in Fig. P8.35. End A is pin-connected and B rests on a roller support. A 300-lb force acts 2 ft from A. Determine the state of stress at a section 2 ft from B. Compute results for points C and D, at the center of the cross section and at the top surface, respectively.

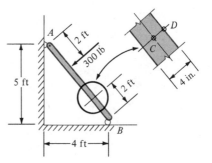

FIGURE P8.35

8.36 Compute the state of stress at points C and D, section a-a, for the structure of Fig. P8.36.

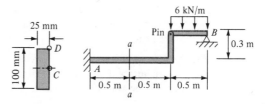

FIGURE P8.36

8.37 The circular structure of Fig. P8.37 is square in cross section, 3 by 3 in. Compute the normal force, shear, and bending moment acting on section *a-a*.

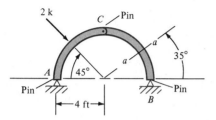

FIGURE P8.37

8.38 The structure of Fig. P8.38 is loaded and supported as illustrated. The members are square in cross section, 70 by 70 mm. Compute the maximum normal stress and shear stress at section *a-a*.

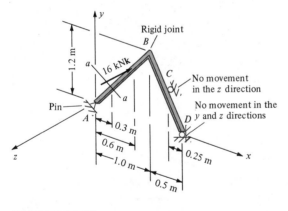

FIGURE P8.38

8.39 The structure of Fig. P8.39 is pinned at *A* and *B* and fixed at *C*. Compute the maximum normal and shear stresses at the fixed end.

8.40 A cylindrical tank in the form of a right circular cylinder is subject to an internal pressure of 90 psi. Compute the thickness of the wall if the stress is not to exceed 6000 psi.

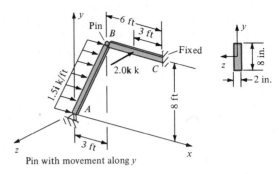

FIGURE P8.39

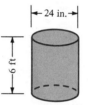

FIGURE P8.40

8.41 A cylindrical pressure vessel illustrated in Fig. P8.41 has an internal pressure of 65 kPa, a torque of $T = 90$ kN·m, and a force $F = 60$ kN. Compute the maximum normal stress and shear stress at the base of the tank for a wall thickness of 10 mm.

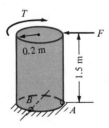

FIGURE P8.41

8.42 A cylindrical pressure vessel is supported as shown in Fig. P8.42. The tank and its contents weigh 750 lb/ft. The tank is subject to an internal pressure of 26 psi and has wall thickness of 0.1 in. Compute the normal stress at points *A* and *B*.

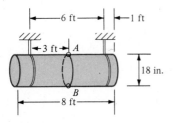

FIGURE P8.42

8.43 A cylindrical storage tank in the shape of a right circular cylinder is loaded as shown in Fig. P8.43. The internal pressure is 75 lb/ft² and the wall thickness is 0.2 in. Assume the tank is fixed at the base and compute the state of stress at points A and B.

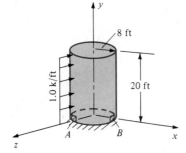

FIGURE P8.43

8.44 The curved bar of Fig. P8.44 has a rectangular cross section 0.5 by 0.75 in. Compute the normal stress

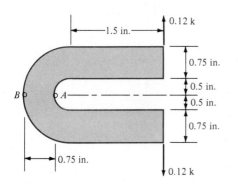

FIGURE P8.44

at points A and B. Use the correction factor method to compute flexural stresses.

8.45 A split ring illustrated in Fig. P8.45 is loaded with 40-lb forces applied at the centroid of the cross section. Compute the normal stresses at points A and B.

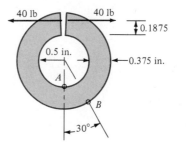

FIGURE P8.45

8.46 A crane hook is fabricated using a circular steel rod 50 mm in diameter as shown in Fig. P8.46. Compute the maximum load P that can be applied if the normal stress at section A-B is not to exceed 120 MPa.

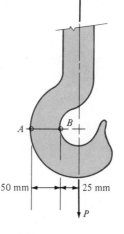

FIGURE P8.46

CHAPTER 9

STRESS TRANSFORMATION

9.1 INTRODUCTION

The concept of stress transformation is the culmination of the preceding seven chapters. The equations of Chapter 8 have been applied to a variety of physical problems. Stress at a point has been studied by visualizing the point as an element. The element has always been oriented in a plane cartesian coordinate system. In this chapter the element is assumed to be oriented in a rotated coordinate system while the stress remains in the original system. Hence, the objective is to define the original stresses in the new coordinate system. A primary goal of this chapter is to present a thorough discussion of the transformation of stress at a point.

Throughout the text the difference between force and stress has been emphasized. Finally, the difference is illustrated by showing that mathematically they transform differently. In fact, it is shown that stress transformation is a combined transformation of area and force, two different vector quantities. A graphical solution referred to as Mohr's circle of stress transformation is developed and related to the transformation equations.

A brief discussion of strain transformation is included. The point is made that stress transformation and strain transformation are similar.

The relation between engineering design and the specification of stress at a point is discussed. Several examples that illustrate engineering analysis and design are presented.

9.2   ELEMENTARY APPLICATION OF STRESS TRANSFORMATION

The axially loaded short column of Chapter 2 can be considered to be an elementary problem in stress transformation due to the elementary manner in which the column is loaded. The column of Fig. 2.13 is illustrated in Fig. 9.1 and the methods of Chapter 8 will be used to investigate the stress at a point. Choose a point anywhere on a cut section, such as point A. The normal stress is $\sigma = P/A = P/bh$. The state of stress at point A is shown in Fig. 9.1b as

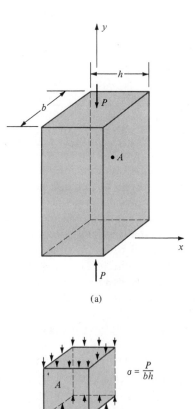

(a)

(b)

FIGURE 9.1
Axially loaded short column.

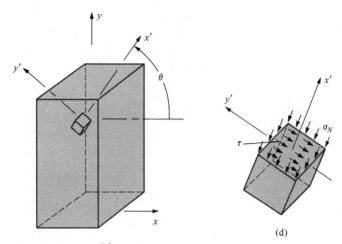

(c)

(d)

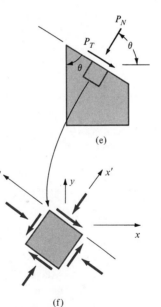

(e)

FIGURE 9.1
(continued)

(f)

$$\boldsymbol{\sigma}_A = \begin{pmatrix} 0 & 0 \\ 0 & -\dfrac{P}{bh} \end{pmatrix}$$

As was illustrated in Chapter 2, the failure actually occurred along the plane illustrated in Fig. 9.1c. Hence, the problem is as follows: Given the state of stress as shown in Fig. 9.1b, determine the state of stress on a rotated element as shown in Fig. 9.1d. The solution of this problem was demonstrated in a very informal manner in Chapter 2. The inclined area of Fig. 9.1c is $A' = A/\sin\theta = bh/\sin\theta$. The force components shown in Fig. 9.1e acting in a normal and tangential direction to the inclined surface are

$$P_N = P\sin\theta$$

$$P_T = P\cos\theta$$

The normal and tangential stresses are

$$\sigma_N = \frac{P_N}{A'} = \frac{P\sin^2\theta}{bh}$$

$$\tau = \frac{P_T}{A'} = \frac{P\sin\theta\cos\theta}{bh}$$

The stresses acting on the element of Fig. 9.1d are

$$\sigma_{A'} = \begin{pmatrix} \dfrac{-P\sin^2\theta}{bh} & -\dfrac{P\sin\theta\cos\theta}{bh} \\[3mm] -\dfrac{P\sin\theta\cos\theta}{bh} & -\dfrac{P\cos^2\theta}{bh} \end{pmatrix}$$

as shown in Fig. 9.1f.

The primary purpose of this chapter is to organize the preceding process of transforming stress at a point. To facilitate the process the vector transformation of Chapter 8 will be employed.

9.3   STRESS TRANSFORMATION USING THE WEDGE METHOD

The wedge method derives its name from the process involved in stress transformation. A wedge is sliced out of the stress element and treated as a free body. Let the element of Fig. 9.2a represent the state of stress at a point. Recall that the element is actually the face of an elemental cube with normal and shear stresses drawn as they act at a point.

The element lies in a plane of x and y. The problem is to compute the stresses on an element that is rotated θ degrees with respect to the x axis. It is important to understand that the element of Fig. 9.2a is *not* rotated to a new position. A new element is defined that is in a plane rotated θ degrees with respect to the x-y axis. The new element is shown using solid lines. In order to transform the stress from the x-y system to the x'-y' system, the stress must first be written as a force. The reason for changing stress to force is to facilitate the summation of force components.

The area of the new element in the x'-y' system is first defined in the original coordinates. To accomplish this, view the element of Fig.

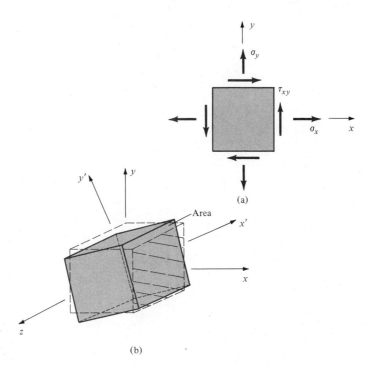

(a)

(b)

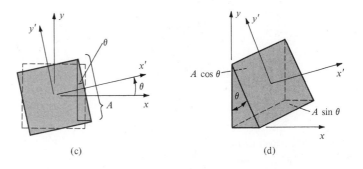

FIGURE 9.2   (c)   (d)

9.2*b* along the *z* axis, as shown in Fig. 9.2*c*. The inclined area will be referred to as *A*. Then the area component in the *x* direction is *A* cos θ and in the *y* direction is *A* sin θ. The wedge of Fig. 9.2*c* is illustrated in Fig. 9.2*d*.

The wedge method employs the concept discussed above. The element of Fig. 9.2*a* is shown again in Fig. 9.3*a*. The angle θ is defined as shown and the problem is to define the stresses in the *x-y* coordinates acting on the face of a rotated element that is normal to the *x'* axis. The area normal to the *x'* axis is defined as *A*, and the areas normal to the *x* and *y* axes are *A* cos θ and *A* sin θ as shown in Fig.

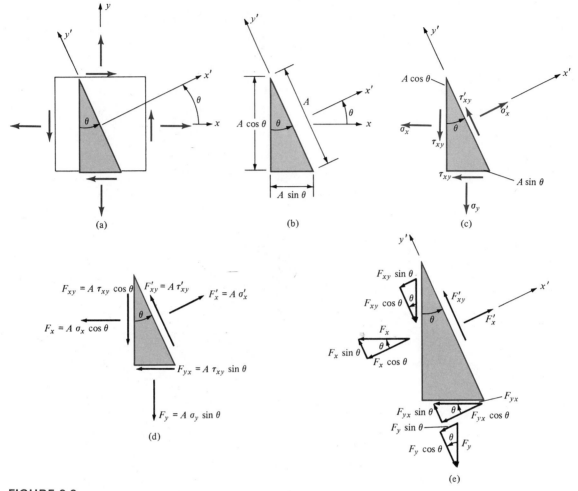

(a)

(b)

(c)

(d)

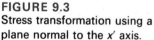

(e)

FIGURE 9.3
Stress transformation using a
plane normal to the x' axis.

9.3*b*. The wedge is visualized as shown in Fig. 9.3*c*. The stress acting
on the face of the wedge normal to the x' axis is σ'_x and the tangential stress is τ'_{xy}, both acting in a positive direction. The original
stresses are illustrated acting on the element. They remain unchanged
even though they are now acting on smaller areas, $A\cos\theta$ and $A\sin\theta$.
Recall, stress is force per unit area and is now distributed over the
partial faces of the cube. The stresses are changed to forces by multiplying them by their respective areas as shown in Fig. 9.3*d*.

The forces of Fig. 9.3*d* are now converted into components along
the x'-y' coordinate directions as shown in Fig. 9.3*e*. Summation of the
forces in the x' direction gives the normal F'_x.

$$F'_x - F_x \cos \theta - F_y \sin \theta - F_{yx} \cos \theta - F_{xy} \sin \theta = 0 \qquad \textbf{(9.1)}$$

Substituting the forces of Fig. 9.3d, dividing by A, and solving for σ'_x gives

$$\sigma'_x = \sigma_x \cos^2 \theta + \sigma_y \sin^2 \theta + 2\tau_{xy} \sin \theta \cos \theta \qquad \textbf{(9.2)}$$

Summing forces along y' in Fig. 9.3e gives

$$F'_y - F_y \cos \theta + F_x \sin \theta + F_{yx} \sin \theta - F_{xy} \cos \theta = 0 \qquad \textbf{(9.3)}$$

Again, the forces are converted to stresses to yield an expression for τ'_{xy}, as follows:

$$\tau'_{xy} = (\sigma_y - \sigma_x) \sin \theta \cos \theta + \tau_{xy}(\cos^2 \theta - \sin^2 \theta) \qquad \textbf{(9.4)}$$

Equations (9.2) and (9.4) are the stresses on the plane normal to the x' axis. The stresses on the plane normal to the y' axis, defined by the same rotation angle θ, are shown in Fig. 9.4 (next page). Summing forces on the element of Fig. 9.4 yields σ'_y as

$$\sigma'_y = \sigma_x \sin^2 \theta + \sigma_y \cos^2 \theta - 2\tau_{xy} \sin \theta \cos \theta \qquad \textbf{(9.5)}$$

and τ'_{xy} can be verified to be the same as Eq. (9.4).

Equations (9.2), (9.4), and (9.5) represent the transformation equations for a two-dimensional state of stress defined in cartesian coordinates. The example of Section 9.2 will be solved again using these equations.

EXAMPLE 9.1

Given the state of stress at a point defined in Fig. 9.1, compute the stresses on a plane rotated an angle θ counterclockwise from the x axis.

Solution:
The element is shown in Fig. 9.5 oriented with respect to the x-y axes. The rotated element is shown inside the original element in its rotated position. The stress acting on the element is given as

$$\sigma_A = \begin{pmatrix} 0 & 0 \\ 0 & -\dfrac{P}{bh} \end{pmatrix} = \begin{pmatrix} \sigma_x & \tau_{xy} \\ \tau_{xy} & \sigma_y \end{pmatrix}$$

Substituting into Eqs. (9.2), (9.5), and (9.4) gives the stresses on the new element.

$$\sigma'_x = -\frac{P}{bh} \sin^2 \theta$$

$$\sigma'_y = -\frac{P}{bh} \cos^2 \theta$$

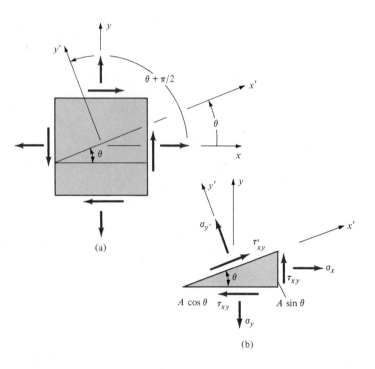

(a)

(b)

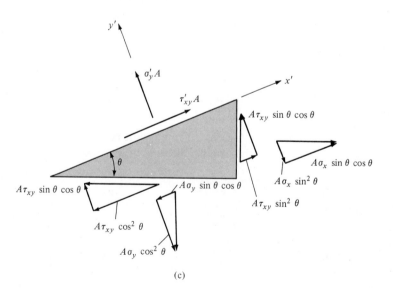

FIGURE 9.4
Stress transformation using a
plane normal to the y' axis.

(c)

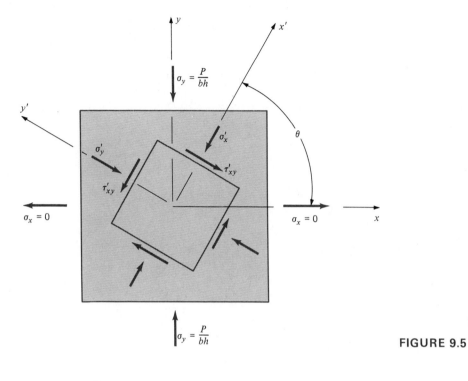

FIGURE 9.5

$$\tau'_{xy} = -\frac{P}{bh} \cos \theta \sin \theta$$

The positive x'-y' axes are shown in Fig. 9.5. The stresses acting on the new element are all negative and act as shown in Fig. 9.5, which is similar to Fig. 9.2d.   ■

EXAMPLE 9.2

Assume that the state of stress shown in Fig. 9.6 has been computed using the formulas of the previous chapters.

a. Compute the state of stress on an element rotated 35 degrees counterclockwise with respect to the original element.

b. Compute the state of stress on an element rotated 35 degrees clockwise with respect to the original element.

Solution:

a. The element of Fig. 9.6a is oriented in a positive x-y coordinate system. The stresses can be written in stress notation as

$$\sigma = \begin{pmatrix} 6 & -4 \\ -4 & -5 \end{pmatrix} \text{ksi}$$

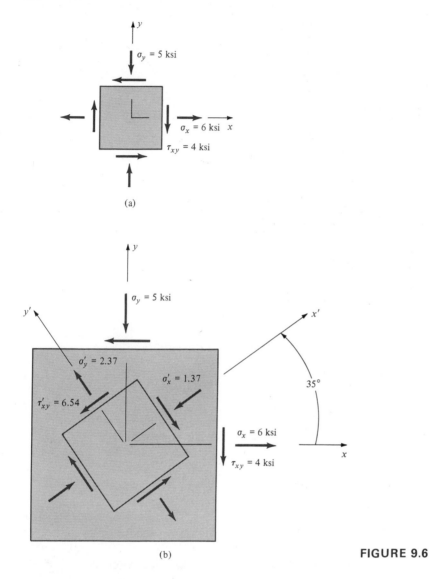

(a)

(b)

FIGURE 9.6

noting that σ_y and τ_{xy} are negative according to the definition for stress sign convention given in the preceding chapter.

The new element is shown in its rotated position oriented 35 degrees counterclockwise from the x axis. According to the derivation, this would be positive $\theta = 35$ degrees. Substituting into Eq. (9.2) gives σ_x'.

$$\sigma_x' = 6 \cos^2 35° - 5 \sin^2 35° + (2)(-4) \sin 35° \cos 35°$$
$$= 4.03 - 1.64 - 3.76 = -1.37 \text{ ksi}$$

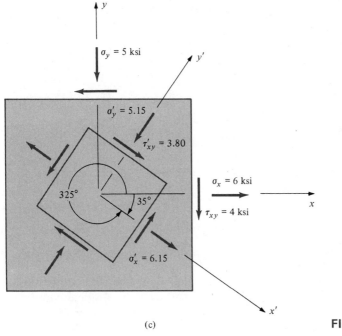

(c)

FIGURE 9.6 (continued)

Similarly, Eqs. (9.5) and (9.4) give σ'_y and τ'_{xy}, respectively.

$$\sigma'_y = 6 \sin^2 35° - 5 \cos^2 35° - (2)(-4) \sin 35° \cos 35°$$
$$= 1.97 - 3.36 + 3.76 = 2.37 \text{ ksi}$$

$$\tau'_{xy} = (-5 - 6) \sin 35° \cos 35° - (4)(\cos^2 35° - \sin^2 35°)$$
$$= -5.17 - 1.37 = -6.54$$

The stresses on an element rotated 35 degrees counterclockwise are written as

$$\boldsymbol{\sigma'} = \begin{pmatrix} -1.37 & -6.54 \\ -6.54 & 2.37 \end{pmatrix} \text{ksi}$$

The element is shown in Fig. 9.6b.

b. A rotation of the element 35 degrees clockwise means that θ would be either 325 degrees or -35 degrees, and the solution would follow the previous analysis. The rotated element is shown in Fig. 9.6c. The positive x'-y' axes are shown superposed upon the x-y axes. The stress transformation operations give the stresses in the new coordinates.

$$\sigma'_x = 6 \cos^2 325° - 5 \sin^2 325° + (2)(-4) \sin 325° \cos 325°$$
$$= 4.03 - 1.64 + 3.76 = 6.15 \text{ ksi}$$

Note: The only difference between σ'_x of part **a** and σ'_x of part **b** is the sign in the third term. Recall the trigonometric relations

$$\sin(-\theta) = -\sin\theta \quad \text{and} \quad \cos(-\theta) = \cos\theta$$

and the reason for the similarities in the analysis becomes obvious.

$$\sigma'_y = 6\sin^2 325° - 5\cos^2 325° - (2)(-4)\sin 325° \cos 325°$$
$$= 1.97 - 3.36 - 3.76 = -5.15 \text{ ksi}$$

$$\tau'_{xy} = (-5-6)\sin 325° \cos 325° - 4(\cos^2 325° - \sin^2 325°)$$
$$= 5.17 - 1.37 = 3.8 \text{ ksi}$$

Hence,

$$\sigma' = \begin{pmatrix} 6.15 & 3.8 \\ 3.8 & -5.15 \end{pmatrix} \text{ksi}$$

as illustrated in Fig. 9.6c. ∎

The sense or direction of the stresses in the rotated coordinate system are interpreted in the same way as in the original coordinate system. A positive σ'_x is tension and acts in the positive x' direction. A positive shear, τ'_{xy}, acts adjacent to the x' face of the element in the direction of the positive y' axis.

The angle θ as shown in Fig. 9.3 or 9.6b defines a surface of the element rather than defining the direction of σ'_x. The sketch of Fig. 9.7 illustrates the concept. Visualize a unit vector drawn normal to the surface. The normal vector defines the orientation of the surface of the rotated element and θ defines the angle between the x axis and the normal vector. Figure 9.3 illustrates the use of θ. The angle is de-

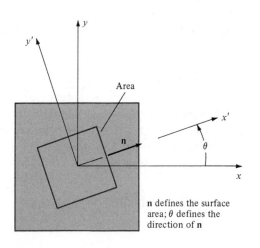

FIGURE 9.7
The unit vector **n** defines the surface; θ defines the direction of **n**.

n defines the surface area; θ defines the direction of **n**

fined in Fig. 9.3*b*, but the wedge-free body of Fig. 9.3*c* defines stresses acting on the surface normal to the *x'* axis. Hence, the analyst should think of θ as defining a surface area as opposed to a stress direction.

The stress σ'_y given by Eq. (9.5) is obtainable by substituting $(\theta + \pi/2)$ for θ of Eq. (9.2). Recall that numerically

$$\sin\left(\theta + \frac{\pi}{2}\right) = \cos\theta \quad\text{and}\quad \cos\left(\theta + \frac{\pi}{2}\right) = -\sin\theta$$

9.4   STRESS TRANSFORMATION MATRIX

In the previous chapter a vector transformation equation was established as Eq. (A8.9). The transformation of a vector quantity such as force or area is written according to Eq. (A8.9) as

$$\begin{Bmatrix} F'_x \\ F'_y \end{Bmatrix} = \begin{bmatrix} \cos\theta & \sin\theta \\ -\sin\theta & \cos\theta \end{bmatrix} \begin{Bmatrix} F_x \\ F_y \end{Bmatrix} \tag{9.6}$$

The complete transformation of stress requires two transformations of the form of Eq. (9.6). An area transformation must occur as shown in Figs. 9.3*b* and 9.4*b*. The force transformation illustrated in Figs. 9.3*e* and 9.4*c* is the second transformation. It has been established that both force and area can be considered vector quantities. A complete matrix transformation using Eq. (9.6) can be used to transform all stress components in the *x-y* system into the *x'-y'* system in one operation. The mathematics that justifies the transformation of stress is usually not required in an undergraduate engineering curriculum. The concept is discussed here for two reasons:

1. To emphasize the basic difference between force and stress transformation

2. To illustrate a method of stress transformation that can be extended to three dimensions

The complete transformation for a two-dimensional state of stress, transforming the stress components from an element in the *x-y* system to an element defined by a rotation of θ into a *x'-y'* system,[1] is given as

$$\begin{bmatrix} \sigma_{x'} & \tau_{xy'} \\ \tau_{xy'} & \sigma_{y'} \end{bmatrix} = \begin{bmatrix} \cos\theta & \sin\theta \\ -\sin\theta & \cos\theta \end{bmatrix}$$
$$\times \begin{bmatrix} \sigma_x & \tau_{xy} \\ \tau_{xy} & \sigma_y \end{bmatrix} \begin{bmatrix} \cos\theta & -\sin\theta \\ \sin\theta & \cos\theta \end{bmatrix} \tag{9.7}$$

---

[1] S. F. Borg, *Matrix-Tensor Methods in Continuum Mechanics*, Van Nostrand, Princeton, N.J., 1963, p. 42.

The stresses at a point, written in matrix format, in the original system are premultiplied by the transformation matrix and postmultiplied by the transpose of the transformation matrix. The rules of matrix multiplication are applied to Eq. (9.7).

The first two matrices on the right-hand side of Eq. (9.7) are multiplied together to give

$$\begin{bmatrix} \sigma_{x'} & \tau_{xy'} \\ \tau_{xy'} & \sigma_{y'} \end{bmatrix} = \begin{bmatrix} \sigma_x \cos\theta + \tau_{xy}\sin\theta & \tau_{xy}\cos\theta + \sigma_y\sin\theta \\ -\sigma_x\sin\theta + \tau_{xy}\cos\theta & -\tau_{xy}\sin\theta + \sigma_y\cos\theta \end{bmatrix}$$
$$\times \begin{bmatrix} \cos\theta & -\sin\theta \\ \sin\theta & \cos\theta \end{bmatrix} \tag{9.8}$$

Equation (9.8) represents a part of the transformation and can be visualized when compared to the wedge method of the previous article. If the terms of the matrix were multiplied by A they would correspond to the forces of Fig. 9.3d or it would correspond to the process of going from Fig. 9.3b to 9.3c to 9.3d. The term $\sigma_x \cos\theta + \tau_{xy}\sin\theta$ corresponds to the forces $F_x = \sigma A \cos\theta$ and $F_{yx} = \tau_{xy}A\sin\theta$, the force components acting in the x direction of Fig. 9.3d. Similarly, $\tau_{xy}\cos\theta + \sigma_y\sin\theta$ correspond to the y components illustrated in Fig. 9.3d. The terms on the second row of the matrix correspond to the x and y force components of Fig. 9.4.

Completion of the matrix multiplication of Eq. (9.8) corresponds to the summation process between Figs. 9.3d and 9.3e, and similarly for Fig. 9.4.

$$\begin{bmatrix} \sigma_{x'} & \tau_{xy'} \\ \tau_{xy'} & \sigma_{y'} \end{bmatrix} =$$
$$\begin{bmatrix} \sigma_x \cos^2\theta + \sigma_y\sin^2\theta + 2\tau_{xy}\sin\theta\cos\theta & (\sigma_y - \sigma_x)\sin\theta\cos\theta + \tau_{xy}(\cos^2\theta - \sin^2\theta) \\ (\sigma_y - \sigma_x)\sin\theta\cos\theta + \tau_{xy}(\cos^2\theta - \sin^2\theta) & \sigma_x\sin^2\theta + \sigma_y\cos^2\theta - 2\tau_{xy}\sin\theta\cos\theta \end{bmatrix} \tag{9.9}$$

Equation (9.9) represents the total stress transformation and can be compared to Eqs. (9.2)–(9.4). The matrix notation of Appendix 8.1 of Chapter 8 and Appendix D at the end of the book can be employed to write Eq. (9.7) in more concise form, as follows:

$$[\sigma] = [T][\sigma][T]^T \tag{9.10}$$

The transpose of $[T]$ is written $[T]^T$.

9.5 PRINCIPAL STRESSES AND MAXIMUM SHEAR STRESS

The magnitude of stress is a major criterion in the design of structures. Equations for computing stresses have been derived and illustrated. Methods of combining the stresses at a point have been demonstrated.

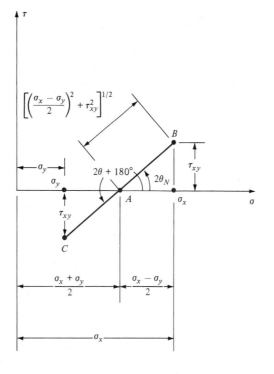

FIGURE 9.9
An interpretation of
Eqs. (9.11) and (9.12).

Substituting Eqs. (a) and (b) into Eq. (9.11) gives the value of the *principal stresses* that act on the *principal planes* of stress given by θ of Eq. (9.12).

$$\sigma_{\substack{max \\ min}} = \sigma_{1,2} = \frac{\sigma_x + \sigma_y}{2} \pm \left[\left(\frac{\sigma_x - \sigma_y}{2} \right)^2 + \tau_{xy}^2 \right]^{1/2} \tag{9.13}$$

The algebraically larger stress of Eq. (9.13) is designated σ_1, and σ_2 is the lesser stress.

The plane of maximum shearing stress is obtained from Eq. (9.4), which again is rearranged as

$$\tau'_{xy} = -\frac{(\sigma_x - \sigma_y)}{2} \sin 2\theta + \tau_{xy} \cos 2\theta \tag{9.14}$$

Equation (9.14) is differentiated with respect to θ and the result set equal to zero. Hence, an expression for θ_s corresponding to maximum and minimum shear stress is obtained.

$$\tan 2\theta_s = -\frac{(\sigma_x - \sigma_y)/2}{\tau_{xy}} \tag{9.15}$$

Equation (9.15) may be compared to Eq. (9.12). The angle defined by Eq. (9.15) is the negative reciprocal of Eq. (9.12), and it is concluded that $2\theta_N$ defines a plane that is normal to the plane defined by $2\theta_s$. The plane of the principal stress given by θ_N is always oriented at 45 degrees to the plane of the maximum shearing stress. Again, Eq. (9.15) has two roots and, when substituted into Eq. (9.14), will yield two values for shear stress. It turns out that the maximum and minimum shear stresses are numerically equal with opposite signs. The maximum shear stress is given by

$$\tau_{\substack{max \\ min}} = \pm \left[\left(\frac{\sigma_x - \sigma_y}{2} \right)^2 + \tau_{xy}^2 \right]^{1/2} \tag{9.16}$$

The plane defined by θ_N of Eq. (9.12) locates the plane of the principal stresses. If θ_N is substituted into Eq. (9.14) the result is $\tau'_{xy} = 0$. Hence, the shear stress is zero, corresponding to the orientation of the element for principal stresses. The reverse is not true. There are normal stresses acting on the element in the orientation corresponding to the maximum shear stress.

EXAMPLE 9.3

Compute the principal stresses, planes of principal stress, maximum shear stress, and plane of maximum shear stress for the state of stress of Example 9.2.

Solution:

The principal stresses are computed by substituting directly into Eq. (9.13) as follows:

$$\sigma_{1,2} = \frac{(6-5)}{2} \pm \left[\frac{(6+5)^2}{2} + (-4)^2 \right]^{1/2} = 0.50 \pm 6.80$$

$$\sigma_1 = 7.30 \text{ ksi} \qquad \sigma_2 = -6.30 \text{ ksi}$$

Using Eq. (9.12),

$$\tan 2\theta_N = \frac{-4}{[(6+5)/2]} = -0.73$$

$$2\theta_N = -36° \qquad \text{or} \qquad 2\theta_N = -36° + 180° = 144°$$

then

$$\theta_N = -18° \qquad \text{or} \qquad \theta_N = 72° \qquad \text{(measured from the } x \text{ axis)}$$

It is not a good practice to guess which value of θ_N corresponds to σ_1. However, substituting $\theta_N = -18°$ into Eq. (9.2) will yield either σ_1 or σ_2.

$$\sigma'_x = 6 \cos^2(-18°) - 5 \sin^2(-18°) + (2)(-4) \sin(-18°) \cos(-18°)$$
$$= 5.43 - 0.48 + 2.35 = 7.30 \text{ ksi}$$

It turns out that $\theta_N = -18°$ corresponds to the plane of σ_1. This result is illustrated in Fig. 9.10a. The result is also shown in the plot of Fig. 9.8. For $\theta_N = 360° - 18° =$

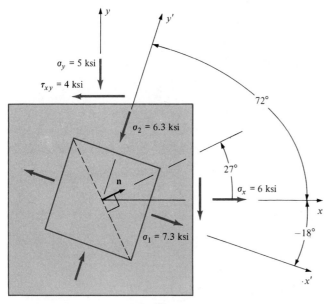

(a)

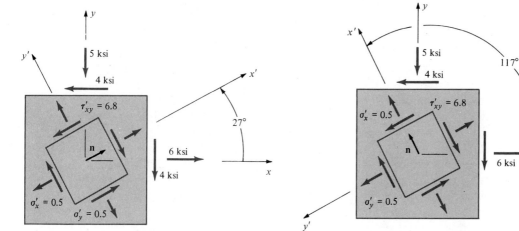

(b)

(c)

FIGURE 9.10

$342°$, the curve of σ'_x indicates a maximum value of 7.3 ksi. The minimum stress is $\sigma_2 = -6.3$ ksi and the angle defining a vector drawn normal to that plane is $\theta_N = 72°$. The shear stress associated with the rotated element of Fig. 9.10a is zero. This fact can again be verified by substituting into Eq. (9.4). The plot of Eq. (9.4), shown in Fig. 9.8, indicates that $\tau_{xy} = 0$ at both $\theta_N = -18°$ and $\theta_N = 72°$.

The maximum shear is given by Eq. (9.16); that is,

$$\tau_{max} = \pm\left[\left(\frac{6+5}{2}\right)^2 + 4^2\right]^{1/2} = \pm 6.80 \text{ ksi}$$

The plane of maximum shear is determined using Eq. (9.15).

$$\tan 2\theta_s = -\frac{(6+5)/2}{-4} = 1.38$$

$$2\theta_s = 54° \quad \text{or} \quad 2\theta_s = 234°$$

$$\theta_s = 27° \quad \text{or} \quad \theta_s = 117°$$

There remains the question of which sign to use for τ_{max} and which value of θ_s to use. Note that an angle of 27 degrees would bisect the element of Fig. 9.10a as shown. The angle θ_s defines the unit normal of the plane, not an extension of the plane. This result is shown in Fig. 9.10b. Substitute $\theta_s = 27°$ in Eq. (9.4) to obtain

$$\tau'_{xy} = (-5 - 6)\sin 27° \cos 27° - 4(\cos^2 27° - \sin^2 27°)$$
$$= -4.45 - 2.35 = -6.8 \text{ ksi}$$

The angle θ_s corresponds to a negative shear stress of 6.8 ksi, as illustrated in Fig. 9.10b. In order to verify that either value of θ_s will lead to the same result let $\theta_s = 117°$. Equation (9.4) yields

$$\tau'_{xy} = 4.45 + 2.35 = 6.8 \text{ ksi}$$

This result is shown in Fig. 9.10c. Note that the x' axis is in the direction of a unit normal vector drawn with respect to the surface in question. A positive shear for the rotation of Fig. 9.10c is the same as a negative shear for the rotation of Fig. 9.10b. Hence, the plus and minus answers for τ'_{xy} have the same physical meaning.

All of the foregoing results are shown in the plot of Fig. 9.8. The results for principal stresses and maximum shear stress that have been computed appear in Fig. 9.8 at the computed values of θ and again at 180 degrees removed. These answers correspond to an 180-degree revolution of the elements of Figs. 9.10b and c.

The normal stress associated with the maximum shear stress is obtained by substituting $\theta_s = 27°$ into Eq. (9.2) to yield

$$\sigma'_x = 0.5$$

Similarly,

$$\sigma'_y = 0.5$$

These results are shown in Fig. 9.10b. The normal stresses for Fig. 9.10c are obtained in a similar manner.

The value of θ that corresponds to σ_1 can be anticipated without formally substituting θ_1 into Eq. (9.2). The following rule can be applied to identify the plane of σ_1. If the magnitude of θ_1 is within plus or minus 45°, $(-45° \le \theta_1 \le 45°)$, the algebraically larger principal stress will be within plus or minus 45° of the algebraically larger normal stress. ■

A useful property pertains to the sum of the normal stresses. For any angle of rotation of the element, the sum of the normal stresses is invariant or remains constant.

For the plane elements that have been discussed thus far the invariance can be shown by adding Eqs. (9.2) and (9.5) to give

$$\sigma'_x + \sigma'_y = \sigma_x + \sigma_y$$

The invariance can be verified for the two previous examples. In Example 9.1 the given normal stress is $\sigma_y = -P/bh$, and the principal stresses add to give

$$\sigma_1 + \sigma_2 = -\left(\frac{P}{bh}\right)(\sin^2 \theta + \cos^2 \theta) = \frac{-P}{bh}$$

It can be verified that the sum of the normal stresses of all three element configurations are equal for Example 9.2. It follows that

$$\sigma_x + \sigma_y = \sigma'_x + \sigma'_y = \sigma_1 + \sigma_2 \tag{9.17}$$

An additional useful relation between maximum shear stress and the principal stresses can be obtained by taking the difference of the two principal stresses of Eq. (9.13) and substituting the maximum shear stress of Eq. (9.16). The resulting relation is

$$\tau_{max} = \frac{\sigma_1 - \sigma_2}{2} \tag{9.18}$$

9.6   MOHR'S CIRCLE FOR STRESS TRANSFORMATION

In 1895 a German scientist, Otto Mohr, developed a graphical representation for the equations of stress transformation. This graphical construction is referred to as Mohr's circle for stress transformation. Today, with the availability of inexpensive calculators, Mohr's circle has lost much of its classical usefulness. However, for the engineer who masters the basic concepts of construction and interpretation of the circle it offers an immediate approximate check on the solution for principal stresses, maximum shear stress, or the standard stress transformation equations.

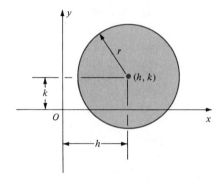

FIGURE 9.11

The general equation of a circle in cartesian coordinates is written

$$(x - h)^2 + (y - k)^2 = r^2 \qquad \textbf{(a)}$$

and appears as in Fig. 9.11.

Equation (9.11) may be rearranged as follows:

$$\sigma_x' - \frac{(\sigma_x - \sigma_y)}{2} = \frac{(\sigma_x - \sigma_y)}{2} \cos 2\theta + \tau_{xy} \sin 2\theta \qquad \textbf{(b)}$$

Squaring both sides of this equation gives

$$\left[\sigma_x' - \frac{(\sigma_x - \sigma_y)}{2}\right]^2 = \left[\frac{(\sigma_x - \sigma_y)}{2}\right]^2 \cos^2 2\theta$$

$$+ (\sigma_x - \sigma_y)\tau_{xy} \sin 2\theta \cos 2\theta + \tau_{xy}^2 \sin^2 2\theta \qquad \textbf{(c)}$$

Both sides of Eq. (9.14) may be squared and written in the form

$$\tau_{xy}'^2 = \left[\frac{(\sigma_x - \sigma_y)}{2}\right]^2 \sin^2 2\theta - (\sigma_x - \sigma_y)\tau_{xy} \sin 2\theta \cos 2\theta$$

$$+ \tau_{xy}^2 \cos^2 2\theta \qquad \textbf{(d)}$$

Adding Eqs. (c) and (d) gives

$$\left[\sigma_x' - \frac{(\sigma_x - \sigma_y)}{2}\right]^2 + \tau_{xy}'^2 = \left[\frac{(\sigma_x - \sigma_y)}{2}\right]^2 + \tau_{xy}^2 \qquad \textbf{(e)}$$

Equation (e) is in the form of Eq. (a) with

$$h = \frac{\sigma_x - \sigma_y}{2} \qquad k = 0 \qquad r^2 = \left(\frac{\sigma_x - \sigma_y}{2}\right)^2 + \tau_{xy}^2$$

A circle can be constructed using the analogy between Eqs. (a) and (e), from which the stress **transformation** can be determined.

Consider the element of Fig. 9.12 and the corresponding coordinate axis. Normal stresses are plotted along the abscissa and shear stresses

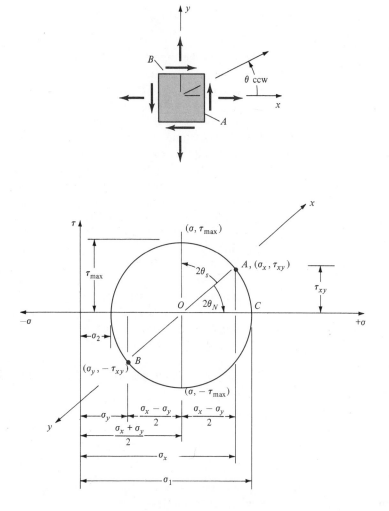

FIGURE 9.12
Mohr's circle.

along the ordinate. For purposes of illustration assume that all stresses are positive and σ_x is greater than σ_y. The stresses acting on the surface of the positive element normal to the x axis, side A, are plotted as point A on Mohr's circle. Hence, a sign convention is established for plotting the shear stress. If the shear stress is positive on side A of the element, it is plotted upward on Mohr's circle. The stresses on side B are plotted with positive σ_y plotted to the right of the origin and positive τ_{xy} plotted downward on Mohr's circle. Points A and B are connected, and the intersection with the σ axis, point O, has coordinates $(\sigma_x + \sigma_y)/2$ and zero as shown in Fig. 9.12. This point is the center of the circle defined by Eq. (e). The length OA or OB is given by

$$OA = \left\{ \left[\frac{\sigma_x - \sigma_y}{2} \right]^2 + \tau_{xy}^2 \right\}^{1/2}$$

and is the radius of the circle of Eq. (e). The circle with the center at point O and radius OA of Fig. 9.12 is the graphical representation of Eq. (e) and is referred to as Mohr's circle of stress.

The principal stresses σ_1 and σ_2, if scaled from Mohr's circle, are equivalent to the values computed using Eq. (9.13). The magnitude of the maximum shear stress is the radius of the circle and is equivalent to results obtained using Eq. (9.16). The angle $2\theta_N$ or angle AOC defines the double angle of the angle between the x axis and x' axis. The interpretation of the angle θ is not straightforward. The sign convention for Mohr's circle does not follow any sign convention developed thus far; in fact, the sign convention is peculiar only to Mohr's circle. Visualize the extension of the line OA as the x axis and the extension of the line OB as the y axis. For principal stresses the extension of the line OC is the x' axis. Then half the double angle on Mohr's circle measured clockwise is the counterclockwise angle on the element between the x axis and the x' axis. This rather awkward interpretation of the direction of rotation of the element could be corrected if positive shear stresses acting on the face of the original element were plotted in the negative τ direction. But plotting positive stress in the negative direction is as awkward as measuring in the clockwise direction. The following example will aid in understanding the idiosyncrasies of Mohr's circle.

EXAMPLE 9.4

a. Use Mohr's circle to compute the principal stresses, principal planes, and maximum shear stress for the element of Example 9.3.
b. Use Mohr's circle and solve the problem stated in Example 9.2.

Solution:
a. The element is shown in Fig. 9.13a. Mohr's circle is constructed using the following procedure.

(1) Locate the point (σ_x, τ_{xy}) associated with face A.

(2) Locate the point (σ_y, τ_{xy}). Note that τ_{xy} in this step is plotted in the opposite direction as in step (1). (In this example σ_y is negative and is plotted to the left of the origin.)

(3) Connect points A and B. The intersection of the σ axis and the line AB is $(\sigma_x + \sigma_y)/2 = (6 - 5)/2 = 0.5$ ksi, the center of Mohr's circle. Construct a circle with the center at point O and passing through points A and B.

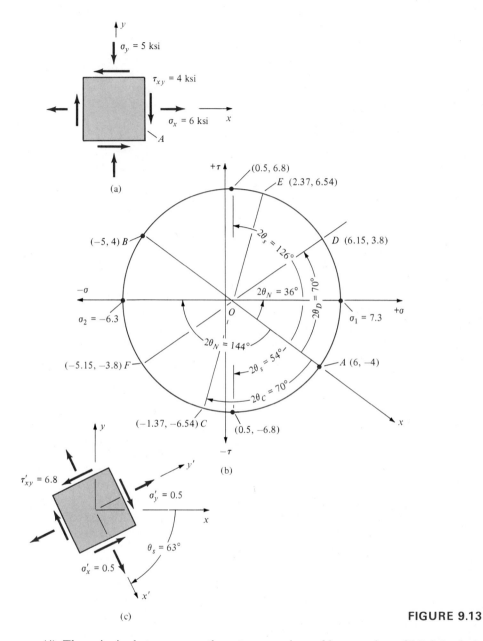

FIGURE 9.13

(4) The principal stresses are the extreme points of intersection of Mohr's circle with the σ axis. The maximum shear stress is the maximum value of τ_{xy} corresponding to points directly above and below the point O. Note that $\tau_{max} = \pm 6.8$ ksi.

(5) The angle $2\theta_N$ measured from the line AB to the horizontal axis is 36 degrees ccw. The angle of rotation on the element would be 18 degrees cw and would corre-

spond to the face of the element containing $\sigma_1 = 7.3$ ksi. The second angle $2\theta_N$ measured from line AB to the σ_2 axis is 144 degrees cw and corresponds to an angle of 72 degrees ccw, as shown in Fig. 9.10a.

(6) The plane of the maximum shear stress is given by $2\theta_s = 54$ degrees cw and corresponds to an angle of 27 degrees ccw, a negative shear, and a positive normal stress of 0.5 ksi, as shown in Fig. 9.10b. The angle $2\theta_s = 126$ degrees ccw would be a rotation of 63 degrees cw. This rotation is shown in Fig. 9.13c and can be verified to correspond to $\theta_s = 117$ degrees ccw of Fig. 9.10c. Compare the orientation of the x'-y' axes of Figs. 9.10c and 9.13c; the elements are oriented identically. However, the axes are oriented 90 degrees to each other, but a positive maximum shear in either coordinate system yields the same result on the figure.

b. The problem of Example 9.2 is based upon the element of Fig. 9.13a. Stresses on a plane defined by a rotation of 35 degrees ccw are to be determined. A 35-degree ccw rotation would be represented by a 70-degree cw rotation from the line AB. The double angle is defined by $2\theta_C = 70°$ and defines a point C on the circle of $\sigma' = -1.37$ ksi, $\tau' = -6.54$. These results are shown in Fig. 9.6b. A 35-degree cw rotation of the element is defined as $2\theta_D = 70$ degrees ccw on Mohr's circle. The stress state on the plane is $\sigma' = 6.15$ and $\tau' = 3.8$. The results are shown on the element of Fig. 9.6c.

Thus far only the σ_x' stresses have been obtained from Mohr's circle. The σ_y' stresses of Figs. 9.6b and c are on a plane rotated an additional 90 degrees from the σ_x' stresses. A 90-degree angle on the element is 180 degrees on Mohr's circle. Hence, for the 35-degree ccw rotation of Fig. 9.6b the stresses on the y' face are obtained by extending the line CO through the origin to intersect the circle on the opposite side, point E, giving $\sigma' = 2.37$. Note that the sign of the shear stress has been previously determined to be negative. An extension of the line DO to point F gives $\sigma' = -5.15$. If the coordinates drawn in Figs. 9.6b and c were rotated an additional 90 degrees, the shear stresses at points E and F would be in the proper direction even though the signs appear to be in disagreement. ■

> The transformation equations are most convenient for computing stresses on a rotated element. Mohr's circle is best employed only for principal stresses and maximum shear stress.

EXAMPLE 9.5

Compute the principal stresses, planes of principal stress, and maximum shear stress for the element of Fig. 9.14a. Sketch the elements showing the principal stresses and maximum shear stresses.

Solution:
Mohr's circle for the element of Fig. 9.14a is shown in Fig. 9.14b. The principal stresses are $\sigma_1 = -1.76$ and $\sigma_2 = -6.24$ and can be verified using Eq. (9.13). The

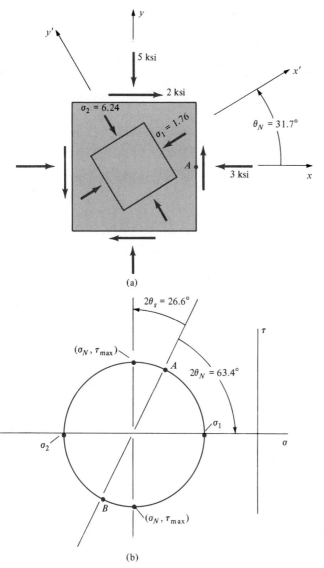

(a)

(b)

FIGURE 9.14

orientation is the angle between the x axis (the extension of line AB on Mohr's circle) and the horizontal axis. Then $2\theta_N = 63.4$ degrees cw on Mohr's circle and would be 31.7 degrees ccw for the element of Fig. 9.14a. For the x' axis defined in Fig. 9.14a, $\sigma_1 = -1.76$ ksi. The stress on the adjacent face is $\sigma_2 = -6.24$ ksi.

The maximum shear stress is shown in Fig. 9.14c. The orientation is obtained by measuring the angle $2\theta_s = 26.6$ degrees in Fig. 9.14b. The angle θ_s is 13.3 degrees

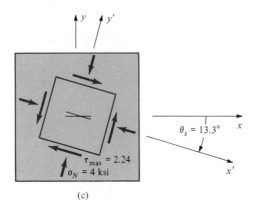

(c)

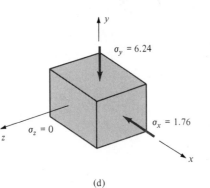

(d)

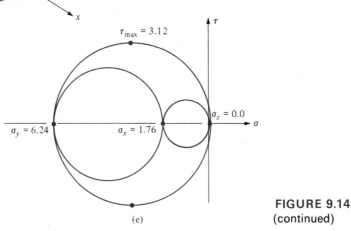

(e)

FIGURE 9.14
(continued)

ccw and defines a coordinate axis normal to the element. The values of stress on the element correspond to the stress at the point defined by $2\theta_s$ on Mohr's circle.   ■

Example 9.5 will serve to illustrate an important result when both principal stresses have the same sign and the problem can be classified as plane stress. A review of plane stress, as defined in Chapter 3, indicates that the stress in the third orthogonal direction is zero. A plane stress element with the stresses of Example 9.5 would have the result: $\sigma_1 = 0$, $\sigma_2 = -1.76$, and $\sigma_3 = -6.24$.

Let the principal stresses correspond to the x-y-z axis of Fig. 9.14d. Note that these are principal stresses and that the shear stress on each face is zero. Figure 9.14e shows a sketch of Mohr's circle for the three-dimensional stress state. The circle for σ_x and σ_y is the same as that shown in Fig. 9.14b. The circle for σ_y and σ_z corresponds to the plane element that would be seen normal to the x axis in Fig. 9.14d. The significant result is the effect on the maximum shear stress. From Mohr's circle, or using Eq. (9.18),

$$\tau_{max} = \frac{\sigma_1 - \sigma_3}{2} = \frac{0 + 6.24}{2} = 3.12 \text{ ksi}$$

This special case is discussed in more detail in Chapter 14.

The next example problem illustrates, using Mohr's circle, a state of uniaxial stress and a state of pure shear.

EXAMPLE 9.6

---

a. Given the state of uniaxial stress shown on the element of Fig. 9.15, use Mohr's circle to determine the maximum shear stress and the plane of maximum shear stress.

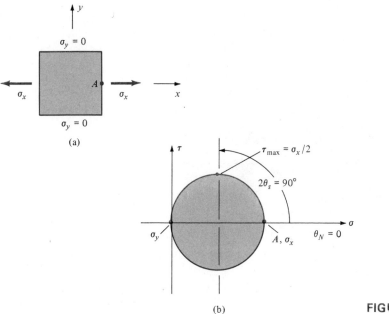

(a)

(b)

FIGURE 9.15

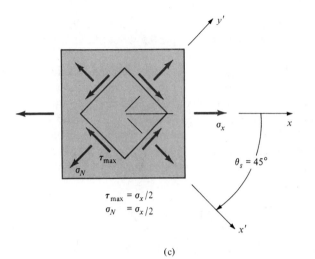

$$\tau_{max} = \sigma_x/2$$
$$\sigma_N = \sigma_x/2$$

(c)

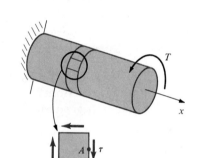

(d)

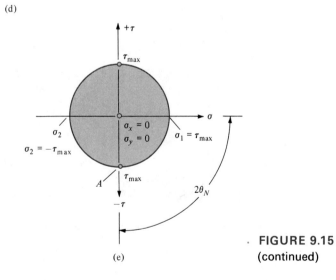

(e)

FIGURE 9.15
(continued)

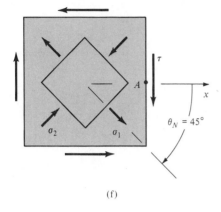

(f)

FIGURE 9.15 (continued)

b. A circular rod subject to a torque is shown in Fig. 9.15d. Investigate the state of stress for a plane element that is subject to a state of pure shear.

Solution:

a. Mohr's circle for the element of Fig. 9.15a is shown in Fig. 9.15b. The result is that $\tau_{max} = \pm \sigma_x/2$ and acts at an angle of $\pm 45°$ with respect to the x axis. Note that the normal stress that acts on the plane of maximum shear stress is $\sigma_N = \sigma_x/2$.

b. A free body of the element is illustrated in Fig. 9.15d. The element pictured in Fig. 9.15d is enlarged to illustrate the state of stress at a point, and the pure shear situation is shown in Fig. 9.15e. Substituting into principal stress equations gives

$$\sigma_{1,2} = \pm \tau_{xy}$$

The sketch of Mohr's circle indicates similar results. The element face A of Fig. 9.15d is a point on the negative τ axis of Mohr's circle. The face containing the positive principal stress is 90 degrees ccw from the vertical axis; hence, the element is rotated 45 degrees cw to show the principal stresses. The principal stress element is shown in Fig. 9.15f. If the rod were made of material that is weaker in tension than shear, a failure could be anticipated along a spiral lying at 45 degrees to the axial direction. ■

EXAMPLE 9.7

A circular fixed-free rod of 2 in. diameter is made of a hypothetical elastic material that is stronger in shear than in tension, but is stronger in compression than in shear. Assume that the limiting compressive and shear stresses are 20 ksi and 10 ksi, respectively, and the allowable tension is 50 percent of the allowable shear stress. A torque of 8000 in.·lb and a compressive force P are applied at the free end of the rod as shown in Fig. 9.16. Compute the limiting values of P to ensure that the allowable stresses are not exceeded.

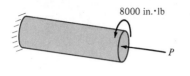

8000 in.·lb

P

(a)

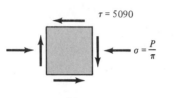

$\tau = 5090$

$\sigma = \dfrac{P}{\pi}$

(b)

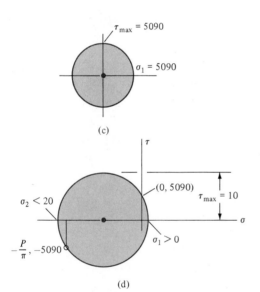

$\tau_{max} = 5090$

$\sigma_1 = 5090$

(c)

τ

$(0, 5090)$

$\tau_{max} = 10$

$\sigma_2 < 20$

$\sigma_1 > 0$

σ

$-\dfrac{P}{\pi}, -5090$

(d)

FIGURE 9.16

Solution:
The state of stress is shown in Fig. 9.16b. The applied shear stress is

$$\tau = \frac{(8000 \text{ in.·lb})(1 \text{ in.})}{\pi(1)^4 \text{ in.}^4/2} = 5090 \text{ psi}$$

and the normal stress is

$$\sigma = \frac{P}{\pi}$$

Assume $P = 0$ and sketch the state of stress using Mohr's circle as shown in Fig. 9.16c. Based upon the discussion of Example 9.15 it is obvious that if $P = 0$, the material is overstressed in tension. P must be at least large enough to shift the origin of the circle of Fig. 9.16c by 90 psi to the left. Consider the sketch of Fig. 9.16d and assume the shear stress is maximum, $\tau = 10$ ksi. If P is greater than zero, which it must be, the maximum compressive stress must be less than the diameter of Mohr's circle. Therefore, when $\tau = 10$ ksi, the compressive stress is less than 20 ksi.

The actual value of P is computed using the principal stress equation. Assume $\tau_{max} = 10$ ksi.

$$10 = \pm\left[\left(\frac{P}{2\pi}\right)^2 + (5.09)^2\right]^{1/2}$$

Square both sides of the equation and solve for P.

$$100 - 25.91 = \frac{P^2}{4\pi^2}$$

$$P = 54.1 \text{ k}$$

Compute the corresponding principal stresses.

$$\sigma_1 = \frac{-54.1}{2\pi} + 10 = 1.4 \text{ ksi}$$

$$\sigma_2 = \frac{-54.1}{2\pi} - 10 = -18.6 \text{ ksi}$$

For the limiting value of P based upon $\sigma_1 = 5$ ksi,

$$5 = \frac{-P}{2\pi} \pm \left[\left(\frac{P}{2\pi}\right)^2 + (5.09)^2\right]^{1/2}$$

$$\left(5 + \frac{P}{2\pi}\right)^2 = \left(\frac{P}{2\pi}\right)^2 + 25.91$$

$$P = \frac{0.91\pi}{5} = 0.57 \text{ k}$$

Therefore,

$$0.57 \text{ k} \leq P \leq 54.1 \text{ k} \qquad \text{(compression)} \qquad \blacksquare$$

9.7 STRAIN TRANSFORMATION

Strain transformation is mathematically identical to stress transformation. In this section the equations for strain transformation will be developed using the mechanics of materials approach. The analogy between stress and strain transformation will be demonstrated.

Axial strain was defined in Chapter 2 using the concept that strain was the deformation in a given length divided by the length. The deri-

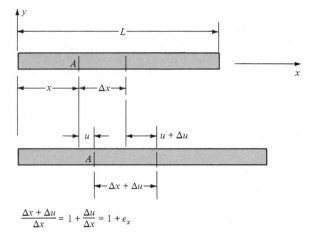

FIGURE 9.17
Axial strain.

vation will be reviewed using the following arguments. Consider the free body of Fig. 9.17. An element of length Δx is allowed to elongate as a tensile force is imagined to act on the bar. The deformation of the rod at point A is u. The Δx element deforms an amount Δu, where Δu can be considered to be a small change in u. The strain is the change in length divided by the original length, $\varepsilon_x = \Delta u / \Delta x$. The deformed length divided by the original length is

$$\frac{\Delta x + \Delta u}{\Delta x} = 1 + \varepsilon_x \tag{9.19}$$

Then $1 + \varepsilon_x$ represents a dimensionless quantity that when multiplied by any length yields the final deformed length.

The deformation of a small cube could be visualized as in Fig. 9.18a. For the purpose of illustrating strain transformation the element will be viewed as a plane element in an x-y coordinate system as shown in Fig. 9.18b.

A shear strain is visualized as shown in Fig. 9.18c. This configuration can be considered as a state of pure shear strain. Note, Δu is produced solely by the action of shear. The deformation in the y direction is Δv. The shear strain is given by

$$\gamma_1 = \frac{\Delta v}{\Delta x} \qquad \gamma_2 = \frac{\Delta u}{\Delta y} \tag{a}$$

The total shear strain is

$$\gamma_1 + \gamma_2 = \gamma_T$$

or

$$\gamma_T = \frac{\Delta v}{\Delta x} + \frac{\Delta u}{\Delta y} = \gamma_1 + \gamma_2 = \gamma_{xy} \tag{9.20}$$

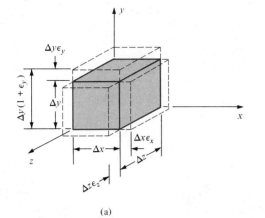

(a)

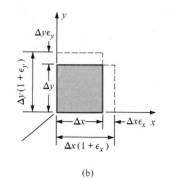

(b)

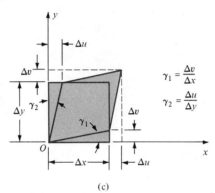

$$\gamma_1 = \frac{\Delta v}{\Delta x}$$

$$\gamma_2 = \frac{\Delta u}{\Delta y}$$

(c)

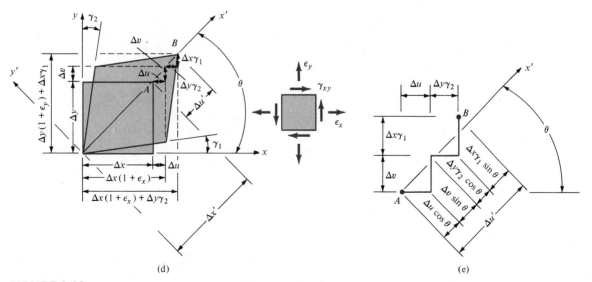

(d)

(e)

FIGURE 9.18
(a) Normal strains in three dimensions; (b) normal strains in two dimensions; (c) pure shear strain; (d) normal strain superposed on pure shear strain; (e) detail view of (d).

where γ_T can be replaced with the notation γ_{xy}.

A general state of strain at a point is a superposition of Figs. 9.18*b* and *c* as shown in Fig. 9.18*d*. The deformation of the strained element is idealized as point *A* located on the undeformed element as it deforms to a corresponding point *B* on the deformed element. The deformation is composed of four parts; two normal strains and two shear strains. Imagine that the same element is now defined in a new *x'-y'* coordinate system. The rotation is also shown in Fig. 9.18*e*, where the separate deformations comprising the distance from *A* to *B* are labeled and shown separately. It should be kept in mind that the order in which these deformations occur is arbitrary. The deformation along the *x'* axis is $\Delta u'$ and is the sum of the four components illustrated in Fig. 9.18*e*. The components can be added to give $\Delta u'$ in terms of deformations in the *x-y* coordinate system.

$$\Delta u' = \Delta u \cos \theta + \Delta v \sin \theta + \Delta y \gamma_2 \cos \theta + \Delta x \gamma_1 \sin \theta \qquad \textbf{(b)}$$

Recall that $\Delta u = \Delta x \varepsilon_x$ and $\Delta v = \Delta y \varepsilon_y$. Substitution into Eq. (b) gives

$$\Delta u' = \Delta x \varepsilon_x \cos \theta + \Delta y \varepsilon_y \sin \theta + \Delta y \gamma_2 \cos \theta + \Delta x \gamma_1 \sin \theta \qquad \textbf{(c)}$$

Divide Eq. (c) by $\Delta x'$ and note that $\Delta x / \Delta x' = \cos \theta$ and $\Delta y / \Delta x' = \sin \theta$ while $\Delta u' / \Delta x' = \varepsilon_x'$, the normal strain along the *x'* axis; then

$$\varepsilon_x' = \varepsilon_x \cos^2 \theta + \varepsilon_y \sin^2 \theta + \gamma_2 \cos \theta \sin \theta + \gamma_1 \sin \theta \cos \theta \qquad \textbf{(d)}$$

Substituting Eq. (9.20) gives the final form of the transformation:

$$\varepsilon_x' = \varepsilon_x \cos^2 \theta + \varepsilon_y \sin^2 \theta + \gamma_{xy} \cos \theta \sin \theta \qquad \textbf{(9.21)}$$

There is an analogy between Eqs. (9.2) and (9.21). The conclusion is that strains transform in exactly the same way as stresses. They have similar mathematical properties, which implies that the transformation matrix of Eq. (9.6) is applicable to strains in the same way as stresses. The strain matrix must be symmetrical in the same manner as the stress matrix; recall that $\tau_{xy} = \tau_{yx}$. For strains $\gamma_T = \gamma_1 + \gamma_2$, but $\gamma_1 \neq \gamma_2$. Hence, with no loss of generality the strain matrix or expression for strain at a point can be written

$$\varepsilon = \begin{pmatrix} \varepsilon_x & \dfrac{\gamma_1 + \gamma_2}{2} \\ \dfrac{\gamma_1 + \gamma_2}{2} & \varepsilon_y \end{pmatrix} \Rightarrow \begin{pmatrix} \varepsilon_x & \varepsilon_{xy} \\ \varepsilon_{yx} & \varepsilon_y \end{pmatrix} \qquad \textbf{(9.22)}$$

for the purpose of strain transformation. The total shear strain using an engineering definition is

$$\gamma_T = \varepsilon_{xy} + \varepsilon_{yx} = \gamma_{xy} \qquad \text{where } \varepsilon_{xy} = \varepsilon_{yx} \qquad \textbf{(e)}$$

Hence, $\gamma_T = \gamma_{xy} = 2\varepsilon_{xy}$. The strain transformation can be written in matrix format as

$$\begin{bmatrix} \varepsilon'_x & \varepsilon'_{xy} \\ \varepsilon'_{yx} & \varepsilon'_y \end{bmatrix} = \begin{bmatrix} \cos\theta & \sin\theta \\ -\sin\theta & \cos\theta \end{bmatrix}$$

$$\times \begin{bmatrix} \varepsilon_x & \varepsilon_{xy} \\ \varepsilon_{yx} & \varepsilon_y \end{bmatrix} \begin{bmatrix} \cos\theta & -\sin\theta \\ \sin\theta & \cos\theta \end{bmatrix} \quad \textbf{(9.23)}$$

Completing the matrix multiplication gives the transformation equations with $\varepsilon_{xy} = \varepsilon_{yx}$ as

$$\begin{bmatrix} \varepsilon'_x & \varepsilon'_{xy} \\ \varepsilon'_{xy} & \varepsilon'_y \end{bmatrix} =$$

$$\begin{bmatrix} \varepsilon_x \cos^2\theta + \varepsilon_y \sin^2\theta + 2\varepsilon_{xy}\cos\theta\sin\theta & (-\varepsilon_x + \varepsilon_y)\sin\theta\cos\theta + \varepsilon_{xy}(\cos^2\theta - \sin^2\theta) \\ (-\varepsilon_x + \varepsilon_y)\sin\theta\cos\theta + \varepsilon_{xy}(\cos^2\theta - \sin^2\theta) & \varepsilon_x \sin^2\theta + \varepsilon_y \cos^2\theta - 2\varepsilon_{xy}\sin\theta\cos\theta \end{bmatrix} \quad \textbf{(9.24)}$$

It follows that in the transformed system

$$\gamma'_1 + \gamma'_2 = \gamma'_T = \gamma'_{xy} \qquad \text{or} \qquad \gamma'_T = \gamma'_{xy} = 2\varepsilon'_{xy} \qquad \textbf{(f)}$$

The analogy between stress and strain thus far indicates that all previous theory and concepts such as principal stresses and Mohr's circle are equally applicable to strains. The transformation for shear strains using Eq. (9.24) is

$$\gamma'_{xy} = 2(\varepsilon_y - \varepsilon_x)\sin\theta\cos\theta + \gamma_{xy}(\cos^2\theta - \sin^2\theta) \qquad \textbf{(9.25)}$$

Additional discussion of the use of γ_{xy} versus ε_{xy} for the definition of strain is included in Section 9.8.

A noteworthy application of strain transformation occurs in the reduction of experimental data obtained from electric strain gages. Without elaborating on the technical details of electric strain gages, it will suffice to comment that they are extremely accurate. A thin wire of known electrical resistance is attached to a structural member that is to be loaded and deformed. As the member deforms, the wire deforms; hence, its electrical resistance changes. The change in electrical resistance is directly proportional to the strain.

A strain gage may be used to measure strain along a single axis or along three intersecting axes. A gage that measures strain along three directions is referred to as a strain rosette. A popular configuration is illustrated in Fig. 9.19. The strains would be measured along the x and y axes and at 45 degrees between the axes. The direction that the strains would be measured would be along the x axis, $\theta_0 = 0°$, the y axis, $\theta_{90} = 90°$, and at $\theta_{45} = 45°$. The three known normal strains can be used to compute the normal and shear strains at point O. Sub-

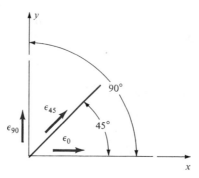

FIGURE 9.19
Graphical representation of a strain gage rosette.

stituting into Eq. (9.21) using the measured strain and their corresponding directions gives three equations and three unknowns.

$$\varepsilon_0 = \varepsilon_x \cos^2 0° + \varepsilon_y \sin^2 0° + \gamma_{xy} \sin 0° \cos 0°$$

$$\varepsilon_{45} = \varepsilon_x \cos^2 45° + \varepsilon_y \sin^2 45° + \gamma_{xy} \sin 45° \cos 45°$$

$$\varepsilon_{90} = \varepsilon_x \cos^2 90° + \varepsilon_y \sin^2 90° + \gamma_{xy} \sin 90° \cos 90°$$

Solving these equations gives the complete description of the strain at point O.

$$\varepsilon_x = \varepsilon_0 \qquad \varepsilon_y = \varepsilon_{90}$$

$$\gamma_{xy} = 2\varepsilon_{45} - \varepsilon_0 - \varepsilon_{90} \tag{9.26}$$

The strains along any other direction can be found using the strain transformation equations. Stress is usually a more significant design parameter. Equations (9.26) are usually used to compute strains, then the stress-strain relations are used to compute stresses. The stress data are used in the manner outlined in the first part of this chapter.

Equation (3.6) can be used to give additional applications for Eqs. (9.26). For a plane element oriented in an x-y coordinate system,

$$\varepsilon_x = \frac{\sigma_x - v\sigma_y}{E} \qquad \varepsilon_y = \frac{\sigma_y - v\sigma_x}{E} \tag{9.27}$$

In terms of stress these equations give

$$\sigma_x = \frac{E(\varepsilon_x + v\varepsilon_y)}{1 - v^2} \qquad \sigma_y = \frac{E(\varepsilon_y + v\varepsilon_x)}{1 - v^2} \tag{9.28}$$

The shear stress is related to the shear strain by

$$\tau_{xy} = G\gamma_T = G\gamma_{xy} \tag{9.29}$$

Given the strain rosette data, Eqs. (9.26) will yield the strains in an x-y coordinate system and Eqs. (9.28) and (9.29) give the corresponding stresses.

EXAMPLE 9.8

A strain gage rosette such as the one of Fig. 9.19 is attached to an aluminum casting. The state of strain obtained from the gage reading is $\varepsilon_0 = 0.00075$, $\varepsilon_{45} = 0.00015$, and $\varepsilon_{90} = -0.0009$. Compute the state of stress for an element oriented along the x-y axis. Compute the principal stresses, principal planes of stress, and maximum shear stress.

Solution:

Material properties for aluminum are $10(10^6)$ psi for the modulus of elasticity, $3.8(10^6)$ psi for the shear modulus, and 0.3 for Poisson's ratio. The strain at the point is obtained from Eqs. (9.26).

$$\varepsilon_x = \varepsilon_0 = 0.00075$$

$$\varepsilon_y = \varepsilon_{90} = -0.00090$$

and

$$\gamma_{xy} = 2(0.00015) - 0.00075 - (-0.00090) = 0.00045$$

The state of stress is determined using Eq. (9.28).

$$\sigma_x = \frac{10(10^6)[7.5(10^{-4}) + (0.3)(-9)(10^{-4})]}{1 - 0.3^2}$$

$$= 5.27(10^3) \text{ psi}$$

$$\sigma_y = \frac{10(10^6)[-9(10^{-4}) + (0.3)(7.5)(10^{-4})]}{1 - 0.3^2}$$

$$= -7.42(10^3) \text{ psi}$$

Also,

$$\tau_{xy} = 3.8(10^6)(4.5)(10^{-4}) = 1.71(10^3) \text{ psi}$$

$$\sigma = \begin{pmatrix} 5.27 & 1.71 \\ 1.71 & -7.42 \end{pmatrix} (10^3) \text{ psi}$$

The principal stresses are given by Eq. (9.13)

$$\sigma_{1,2} = \frac{(5.27 - 7.42)(10^3)}{2} \pm \left[\left(\frac{5.27 + 7.24}{2} \right)^2 + (1.71)^2 \right]^{1/2} (10^3)$$

$$\sigma_1 = -1.07(10^3) + 6.57(10^3) = 5.50(10^3) \text{ psi}$$

$$\sigma_2 = -1.07(10^3) - 6.57(10^3) = -7.64(10^3) \text{ psi}$$

Also,

$$\tau_{max} = \pm 6.57(10^3) \text{ psi}$$

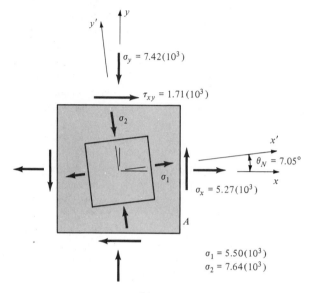

(a)

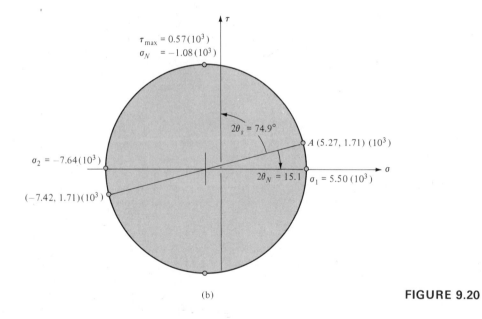

(b)

FIGURE 9.20

$$2\theta_N = \tan^{-1}\frac{1.71}{(5.27+7.42)/2} = 15.1°$$

The proper orientation of the element is obtained from a sketch of Mohr's circle as shown in Fig. 9.20.

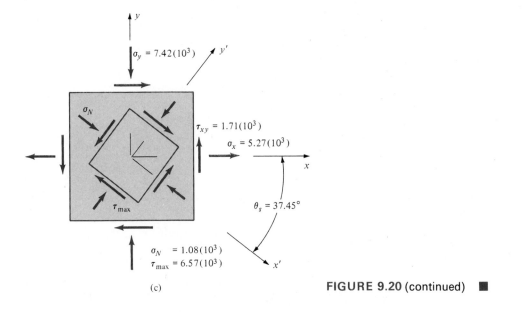

$\sigma_y = 7.42(10^3)$

σ_N

$\tau_{xy} = 1.71(10^3)$

$\sigma_x = 5.27(10^3)$

x

$\theta_s = 37.45°$

τ_{max}

$\sigma_N \quad = 1.08(10^3)$
$\tau_{max} = 6.57(10^3)$

x'

(c)   **FIGURE 9.20** (continued) ■

9.8 THE ISOTROPIC ELASTIC PARAMETERS E, G, v, AND K, AND SHEAR STRAIN

In Chapter 3, Section 3.4, a fundamental relation between Young's modulus E, Poisson's ratio v, and the shear modulus G, was given in Eq. (3.9) as

$$G = \frac{E}{2(1 + v)}$$

Equation (3.9) can be derived using the results of Example 9.11b and Eqs. (9.21), (9.27), and (9.29). A state of pure shear is shown in Fig. 9.15d, and principal stresses are computed as

$$\sigma_1 = \tau_{max} = \tau_{xy} \tag{a}$$

and

$$\sigma_2 = -\tau_{max} = -\tau_{xy} \tag{b}$$

The principal stresses are shown in Fig. 9.15f acting on the planes defined by $\theta = 45°$ and $45° + \pi/2$, respectively.

The strain transformation is given by Eq. (9.21) in terms of the total shear strain, γ_{xy}. Substituting $\varepsilon_x = \varepsilon_y = 0$ and $\theta = 45°$ into Eq. (9.21) gives

$$\varepsilon'_x = \frac{\gamma_{xy}}{2} = \frac{\tau_{xy}}{2G} \tag{c}$$

using Eq. (9.29). Write the first of Eqs. (9.27) in terms of principal strain and stress as

$$\varepsilon_1 = \frac{\sigma_1 - v\sigma_2}{E} \qquad \text{(d)}$$

Substitute Eqs. (a) and (b),

$$\varepsilon_1 = \frac{\tau_{xy}(1 + v)}{E} \qquad \text{(e)}$$

Equating Eqs. (c) and (e) gives Eq. (3.9) from Chapter 3; that is,

$$G = \frac{E}{2(1 + v)}$$

It is important to recognize the two different definitions of shear strain. The so-called engineering definition of shear strain has been denoted as γ_{xy} or γ_T in the previous section. Equation (3.9) is valid for the engineering definition. The definition dictated by the mathematical theory of elasticity is referred to as ε_{xy} and ε_{yx} in Eqs. (9.22), (9.23), and (9.24). The two different definitions are obviously related. The notation of the mathematical theory of elasticity is necessary so that the theory can be written within a particular formal mathematical framework. For additional study see the references at the end of this chapter.

The bulk modulus was defined by Eq. (3.10) of Chapter 3. The expression can be derived by visualizing a cube of material that is loaded with a hydrostatic pressure, p. All six faces are loaded with the compressive pressure, p. Use Eqs. (3.3), (3.4), and (3.5) and recognize that $\sigma_x = \sigma_y = \sigma_z = -p$. The three equations are

$$\varepsilon_x = \frac{-p + v(2p)}{E}$$

$$\varepsilon_y = \frac{-p + v(2p)}{E} \qquad \text{(f)}$$

$$\varepsilon_z = \frac{-p + v(2p)}{E}$$

Adding Eqs. (f)

$$\varepsilon_x + \varepsilon_y + \varepsilon_z = \frac{-3p + 6vp}{E}$$

or

$$\varepsilon_x + \varepsilon_y + \varepsilon_z = \frac{-3(1 - 2v)p}{E}$$

The bulk modulus was defined in Eq. (3.10) as

$$K = \frac{E}{3(1 - 2v)} = \frac{-p}{e}$$

where

$$e = \varepsilon_x + \varepsilon_y + \varepsilon_z$$

The bulk modulus is the ratio of the hydrostatic pressure to the volume change of the elemental cube.

9.9 TRANSFORMATION PROPERTIES OF THE SECOND MOMENT OF THE AREA AND UNSYMMETRICAL BENDING

In Chapter 5 a plane area was considered as a vector. The moment of area was obtained using the vector product, and the result was also considered to have vector properties. The moment of the moment of an area, or so called moment of inertia, does not have vector properties in the same way as area or moment of area. It turns out that stress, strain, and the second moment of the area all have similar mathematical properties and transform using the same transformation laws. In more advanced studies they are termed the stress tensor, the strain tensor, and the inertia tensor. The transformation of the second moment of the area is actually a special case of the classical transformation of the inertia tensor. The second moment of the area is, by its definition, a two-dimensional quantity. The inertia tensor, which is encountered primarily in dynamics, is the second moment of the mass and is basically a three-dimensional quantity. The transformation of the second moment of the area is identical to the transformation of plane stress. The second moment of the area is defined using the plane area of Fig. 9.21 as

$$I_x = \int_A y^2 \, dA \qquad I_y = \int_A x^2 \, dA \qquad I_{xy} = \int_A xy \, da \qquad \textbf{(9.30)}$$

I_{xy} is classically referred to as the product of inertia. The second moment of the area will be written in a manner similar to stress at a point.

$$\mathbf{M} = \begin{pmatrix} I_y & I_{xy} \\ I_{xy} & I_x \end{pmatrix} \qquad \textbf{(9.31)}$$

The transformation, modeled after the stress transformation, in matrix form is

$$[M'] = [T][M][T]^T \qquad \textbf{(9.32)}$$

Note that the subscript notation along the diagonal of Eq. (9.31) is reversed with Eq. (9.7). The transformation for stress corresponds to

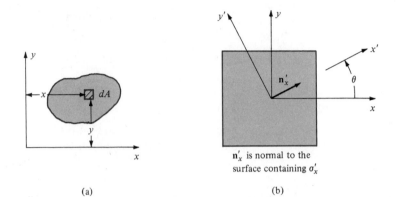

(a)

$\mathbf{n}'_x$ is normal to the
surface containing σ'_x

(b)

$I'_y = \int x'^2 \, dA$

$\mathbf{n}'_x$ is normal to the
axis of I'_y

(c)

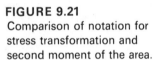

FIGURE 9.21
Comparison of notation for
stress transformation and
second moment of the area.

the plane containing σ_x defined by the unit normal of Fig. 9.21b. The same transformation for the second moment of area corresponds to a plane defined by $\mathbf{n}_{x'}$ of Fig. 9.21c. The unit normal of Fig. 9.21c is normal to the y' axis and defines the second moment of the area about the y' axis. Hence, the notation of Eq. (9.31) properly defines the orientation of second area moments.[2] Performing the operations indi-

---

[2] The inertia tensor occurs in several topics of mechanics. In dynamics the mass moment of inertia occurs in the conservation of momentum equations. Equation (9.31) is often written in the following form.

$$\mathbf{M} = \begin{pmatrix} I_x & -I_{xy} \\ -I_{xy} & I_y \end{pmatrix}$$

The transformation of this equation will give exactly the same results as Eq. (9.33).

cated in Eq. (9.32) gives the following transformation equations for the second moment of the area.

$$I'_y = I_y \cos^2 \theta + I_x \sin^2 \theta + 2I_{xy} \sin \theta \cos \theta$$

$$I'_x = I_y \sin^2 \theta + I_x \cos^2 \theta - 2I_{xy} \sin \theta \cos \theta \tag{9.33}$$

$$I'_{xy} = (I_x - I_y) \sin \theta \cos \theta + I_{xy}(\cos^2 \theta - \sin^2 \theta)$$

It follows that principal axes of moments of area can be computed in exactly the same manner as principal planes of stress.

$$I_{\substack{max \\ min}} = I_{1,2} = \frac{I_y + I_x}{2} \pm \left[\left(\frac{I_y - I_x}{2} \right)^2 + I_{xy}^2 \right]^{1/2} \tag{9.34}$$

$$I_{xy_{max}} = \pm \left[\left(\frac{I_y - I_x}{2} \right)^2 + I_{xy}^2 \right]^{1/2} \tag{9.35}$$

$$\tan 2\theta_N = \frac{I_{xy}}{(I_y - I_x)/2} \tag{9.36}$$

Equations (9.34)–(9.36) can be compared to Eqs. (9.13), (9.16), and (9.12). There is a correspondence between I_y and σ_x, I_x and σ_y, and I_{xy} and τ_{xy}.

A primary use of the principal second moments of area is in unsymmetrical bending of an unsymmetrical cross section. Previous examples have dealt with bending moments that act as components about each of the centroid axes of a symmetrical cross section. If the beam cross section is unsymmetrical, meaning that the principal axes of the second moment of the area are not axes of symmetry and cannot be found by inspection, the principal axes must be located using Eqs. (9.34)–(9.36). Once the principal axes are found, the bending moment can be resolved into components about the principal axes and the basic formula $\sigma = Mc/I$ can be used with respect to each principal axis.

Numerous examples and problems in previous chapters dealt with beams subject to loading in both transverse directions. In all cases the beam cross section was symmetrical. In the following example an unsymmetrical cross section will be loaded in such a way that bending occurs with respect to the principal axes of the second moment of the area. To accomplish this type of loading the loads must act through the shear center. The shear center would be computed as illustrated in Chapter 7. In the following example it will be assumed that the location of the shear center has been computed. Recall that if the beam has at least one axis of symmetry, a principal second moment of the area will coincide with the axis of symmetry, which is obvious by inspection of Eq. (9.34). If there is an axis of symmetry, I_{xy} is zero and $I_{1,2}$ will be I_x or I_y. The following example will illustrate bending stress computations for an unsymmetrical cross section.

EXAMPLE 9.9

A simply supported beam 3 m long with the cross section shown in Fig. 9.22a is loaded as shown in Fig. 9.22b. Assume the loading passes through the shear center; hence, there will be no torsional warping of the section. Compute the bending stresses at the center of the span.

Solution:
The centroid is located with respect to point A on the section.

$$\bar{x} = \frac{(0.02)(0.2)(0.01) + (0.02)(0.3)(0.29) + (0.26)(0.04)(0.15)}{(0.02)(0.2) + (0.02)(0.3) + (0.26)(0.04)}$$

$$= \frac{3.34(10^{-3})}{2.04(10^{-2})} = 0.1637 \text{ m}$$

$$\bar{y} = \frac{(0.02)(0.2)(0.1) + (0.02)(0.3)(0.15) + (0.26)(0.04)(0.02)}{2.04(10^{-2})}$$

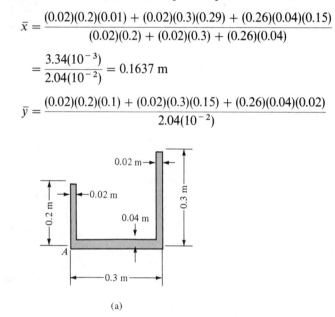

(a)

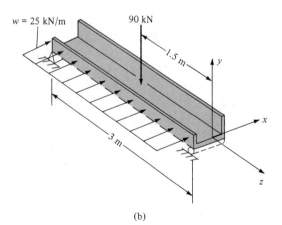

(b)

FIGURE 9.22

$$\bar{y} = \frac{1.508(10^{-3})}{2.04(10^{-2})} = 0.0739 \text{ m}$$

The second moments of the area with respect to the centroid axis are computed next.

$$I_x = \frac{1}{3}(0.02)(0.1261)^3 + \frac{2}{3}(0.02)(0.0739)^3 + \frac{1}{3}(0.02)(0.2261)^3$$

$$+ \frac{1}{12}(0.26)(0.04)^3 + (0.26)(0.04)(0.0539)^2$$

$$= 1.274(10^{-4}) \text{ m}^4$$

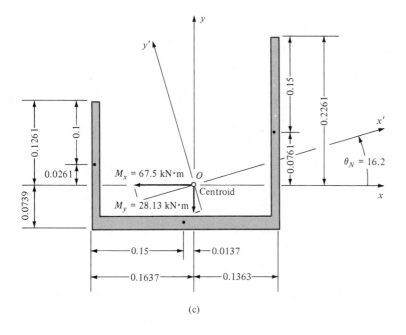

(c)

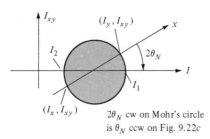

$2\theta_N$ cw on Mohr's circle
is θ_N ccw on Fig. 9.22c

(d)

FIGURE 9.22
(continued)

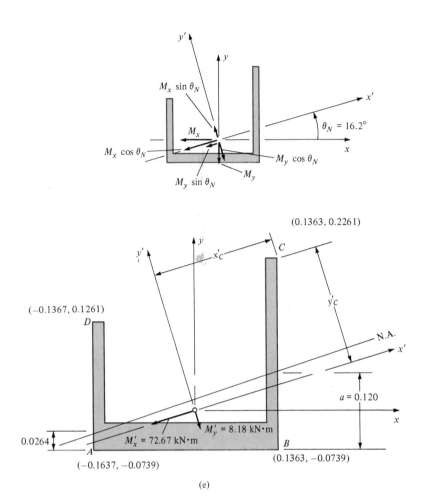

(−0.1367, 0.1261)

(0.1363, 0.2261)

$M_x \sin \theta_N$

M_x

$M_x \cos \theta_N$

$M_y \cos \theta_N$

$M_y \sin \theta_N$

M_y

$\theta_N = 16.2°$

x'_C

y'_C

N.A.

$a = 0.120$

$M'_y = 8.18$ kN·m

$M'_x = 72.67$ kN·m

0.0264

(−0.1637, −0.0739)

(0.1363, −0.0739)

(e)

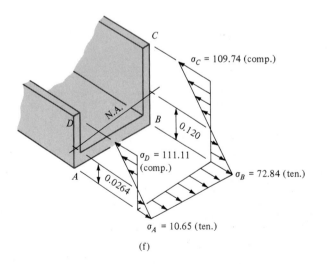

$\sigma_C = 109.74$ (comp.)

N.A.

0.120

$\sigma_D = 111.11$ (comp.)

$\sigma_B = 72.84$ (ten.)

0.0264

$\sigma_A = 10.65$ (ten.)

(f)

FIGURE 9.22
(continued)

$$I_y = \frac{1}{3}(0.04)(0.1637)^3 + \frac{1}{3}(0.04)(0.1363)^3 + \frac{1}{12}(0.2)(0.02)^3 + (0.2)(0.02)(0.1537)^2$$

$$+ \frac{1}{12}(0.3)(0.02)^3 + (0.3)(0.02)(0.1263)^2$$

$$= 2.828(10^{-4})\,\text{m}^4$$

$$I_{xy} = (0.02)(0.2)(-0.1537)(0.0261) + (0.02)(0.3)(0.1263)(0.0761)$$
$$+ (0.04)(0.26)(-0.0137)(-0.0539)$$

$$= +0.493(10^{-4})\,\text{m}^4$$

The principal moments of inertia are obtained from Eq. (9.34).

$$I_{1,2} = \frac{2.828(10^{-4}) + 1.274(10^{-4})}{2}$$

$$\pm \left[\frac{(2.828 - 1.274)^2(10^{-4})^2}{4} + (0.493)^2(10^{-4}) \right]^{1/2}$$

$$= 2.051(10^{-4}) \pm 0.920(10^{-4})$$

$$I_1 = 2.971(10^{-4})\,\text{m}^4 \qquad I_2 = 1.131(10^{-4})\,\text{m}^4$$

$$\tan 2\theta_N = \frac{0.493(10^{-4})}{(2.828 - 1.274)(10^{-4})/2} = 0.635$$

$$2\theta_N = 32.4°$$

$$\theta_N = 16.2°$$

A sketch of Mohr's circle will indicate the value, I_1 or I_2, associated with $\theta_N = 16.45°$. Mohr's circle for moment of inertia is shown in Fig. 9.22d. Since I_{xy} is positive it is plotted in the positive direction locating the point (I_y, I_{xy}) on point A. As mentioned previously, I_y is the moment of inertia about the y axis of Fig. 9.22c. The plane of the y axis is defined by a unit normal vector acting in the x direction. The angle θ_N is counterclockwise on Mohr's circle and is clockwise on the beam cross section. θ_N defines the direction of a unit normal to the axis about which I_1 would be computed. Hence, I_1 is with respect to the y' axis and I_2 is with respect to the x' axis. The principal axes are x' and y', $I_1 = I'_y$ and $I_2 = I'_x$.

The maximum bending moments at the center of the beam of Fig. 9.22c are

$$M_x = (45\ \text{kN})(1.5)\ \text{m} = 67.5\ \text{kN}\cdot\text{m}$$

$$M_y = (37.5\ \text{kN})(1.5)\ \text{m} - \frac{(25\ \text{kN/m})(1.5)^2}{2} = 28.13\ \text{kN}\cdot\text{m}$$

These moments are indicated as vectors on the beam cross section of Fig. 9.22c. The M'_x and M'_y moment vectors are computed using the vector transformation of Eq. (8A.9).

$$M'_x = -67.5 \cos 16.2° - 28.13 \sin 16.2° = -72.67 \text{ kN·m}$$

$$M'_y = 67.5 \sin 16.2° - 28.13 \cos 16.2° = -8.18 \text{ kN·m}$$

To compute bending stresses the points A through D of Fig. 9.22e must be located relative to the x'-y' system. The coordinates relative to the x-y system are obtained from Fig. 9.22c and shown on Fig. 9.22e. Once again the vector transformation is used with $\cos 16.2° = 0.960$ and $\sin 16.2° = 0.279$.

$$x'_A = (-0.1637)(0.960) - (0.0739)(0.279) = -0.1778 \text{ m}$$

$$y'_A = -(-0.1637)(0.279) - (0.0739)(0.960) = -0.0242 \text{ m}$$

$$x'_B = (0.1363)(0.960) - (0.0739)(0.279) = 0.1046 \text{ m}$$

$$y'_B = -(0.1363)(0.279) - (0.0739)(0.960) = -0.1089 \text{ m}$$

$$x'_C = (0.1363)(0.960) + (0.2261)(0.279) = 0.1939 \text{ m}$$

$$y'_C = -(0.1363)(0.279) + (0.2261)(0.960) = 0.1791 \text{ m}$$

$$x'_D = (-0.1637)(0.960) + (0.1261)(0.279) = -0.1220 \text{ m}$$

$$y'_D = -(-0.1637)(0.279) + (0.1261)(0.960) = 0.1677 \text{ m}$$

The flexure formula can be used with respect to the x'-y' axis.

$$\sigma_A = \frac{M'_x y'_A}{I'_x} + \frac{M'_y x'_A}{I'_y} = \frac{(72.67)(0.0242)}{(1.131)(10^{-4})} - \frac{(8.18)(0.1778)}{(2.971)(10^{-4})} = 10.65 \text{ kPa}$$

$$\sigma_B = \frac{(72.67)(0.1089)}{(1.131)(10^{-4})} + \frac{(8.18)(0.1046)}{(2.971)(10^{-4})} = 72.84 \text{ kPa}$$

$$\sigma_C = -\frac{(72.67)(0.1791)}{(1.131)(10^{-4})} + \frac{(8.18)(0.1939)}{(2.971)(10^{-4})} = -109.74 \text{ kPa}$$

$$\sigma_D = -\frac{(72.67)(0.1677)}{(1.131)(10^{-4})} - \frac{(8.18)(0.1220)}{(2.971)(10^{-4})} = -111.11 \text{ kPa}$$

A sketch of the stress distribution is shown in Fig. 9.22f. The neutral axis is located using similar triangles; for example,

$$\frac{a}{72.84} = \frac{0.3}{109.74 + 72.84} \qquad a = 0.120$$

This example is intended to illustrate the use of the transformation equations. Structural shapes with no axis of symmetry are sometimes analyzed using the foregoing procedure. ∎

9.10 SUMMARY

Stress transformation has been discussed and its importance in engineering design has been emphasized. Stress transformation has been limited to two-dimensional stress states. A large majority of engineering problems are adequately solved in two dimensions. The approach in this chapter has been to attempt to connect the physical and mathematical concepts such that the reader understands physically what happens when stress is transformed. With such an understanding the reader should be confident in the mathematical structure of the transformation. The extension to strain transformation is demonstrated, and transformation of the second moment of the area is shown to be analogous. General three-dimensional stress transformations have not been discussed. The extension of the two-dimensional transformation would be based upon the matrix transformations.

The important stress transformation equations were introduced early in the chapter.

$$\sigma'_x = \sigma_x \cos^2 \theta + \sigma_y \sin^2 \theta + 2\tau_{xy} \sin \theta \cos \theta \qquad \textbf{(9.2)}$$

$$\sigma'_y = \sigma_x \sin^2 \theta + \sigma_y \cos^2 \theta - 2\tau_{xy} \sin \theta \cos \theta \qquad \textbf{(9.5)}$$

$$\tau'_{xy} = (\sigma_y - \sigma_x) \sin \theta \cos \theta + \tau_{xy}(\cos^2 \theta - \sin^2 \theta) \qquad \textbf{(9.4)}$$

The two-dimensional tensor transformation was given as Eq. (9.7).

The derivation of the principal stress equations is the most significant application of the stress transformation equations.

$$\sigma_{1,2} = \frac{\sigma_x + \sigma_y}{2} \pm \left[\left(\frac{\sigma_x - \sigma_y}{2} \right)^2 + \tau_{xy}^2 \right]^{1/2} \qquad \textbf{(9.13)}$$

$$\tau_{\text{max}} = \pm \left[\left(\frac{\sigma_x - \sigma_y}{2} \right)^2 + \tau_{xy}^2 \right]^{1/2} \qquad \textbf{(9.16)}$$

The planes of principal stress and maximum shear stress are

$$\tan 2\theta_N = \frac{\tau_{xy}}{(\sigma_x - \sigma_y)/2} \qquad \textbf{(9.12)}$$

and

$$\tan 2\theta_s = -\frac{(\sigma_x - \sigma_y)/2}{\tau_{xy}} \qquad \textbf{(9.15)}$$

Mohr's circle for equations of the type of Eq. (9.13) can be a powerful ally when locating the correct solution of Eq. (9.12) corresponding to σ_1 and σ_2. Mohr's circle should be viewed as a graphical aid to be used in conjunction with Eqs. (9.13) and (9.16). If an analysis for stress on a plane other than the principal plane is desired, Eqs. (9.2), (9.4), and

(9.5) should be used. Mohr's circle results in additional confusion and, since it is a graphical method, is not as accurate as the equations.

Strain transformation has been introduced; however, the physical concepts have not been discussed in as much detail as stress transformation.

Transformation of the second moment of the area, which is sometimes introduced in engineering mechanics (statics) has been shown to be applicable for problems of unsymmetrical bending.

For Further Study

Boresi, A. P. and O. M. Sidebottom, *Advanced Mechanics of Materials*, 4th ed., Wiley, New York, 1985.

Borg, S. F., *Matrix-Tensor Methods in Continuum Mechanics*, Van Nostrand, Princeton, N.J., 1963.

Dally, J. W. and W. F. Riley, *Experimental Stress Analysis*, 2nd ed., McGraw-Hill, New York, 1978.

Sokolnikoff, I. S., *Mathematical Theory of Elasticity*, 2nd ed., McGraw-Hill, New York, 1956.

9.11   PROBLEMS

Note: Some of the problems in this section specify parameters for hypothetical materials. The reader should keep in mind that these problems are intended to illustrate problem solving and do not imply that such hypothetical materials exist.

9.1 A test specimen 2 by 2 in. in cross section is subjected to a compressive load of 7500 lb.
a. Use the wedge method to compute the normal and shearing stresses on a plane that is rotated 55 degrees counterclockwise from a horizontal axis as shown in Fig. P9.1.
b. Use the stress transformation equations to determine the state of stress on an element rotated 55 degrees counterclockwise with respect to an x axis.

9.2 A circular post 6 in. in diameter is subjected to a compressive force of 40 k. Compute the state of stress on an element rotated 20 degrees clockwise with respect to a horizontal axis. Use the wedge method and compare your answers to the stress transformation equations.

9.3 A 50- by 75-mm rectangular block shown in Fig. P9.3 is subject to a compressive load. The normal

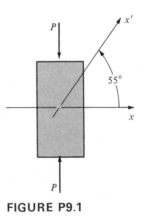

FIGURE P9.1

stress and shearing stress on the plane defined by $\theta = 35°$ are not to exceed 4 MPa and 2 MPa, respectively. Compute the maximum load that can be applied to the block.

9.4–9.7 Compute the stresses acting on an element

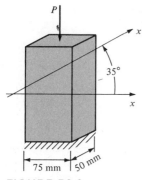

FIGURE P9.3

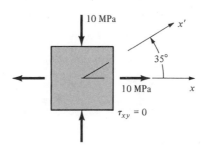

FIGURE P9.9

rotated θ degrees counterclockwise from the x axis for the given state of stress.

9.4 $\theta = 45°$ ccw

$$\sigma = \begin{pmatrix} 6 & 0 \\ 0 & 6 \end{pmatrix} \text{MPa}$$

9.5 $\theta = 45°$ ccw

$$\sigma = \begin{pmatrix} -5 & 0 \\ 0 & 5 \end{pmatrix} \text{MPa}$$

9.6 $\theta = 72°$ ccw

$$\sigma = \begin{pmatrix} 30 & 18 \\ 18 & -25 \end{pmatrix} \text{ksi}$$

9.7 $\theta = 50°$ cw

$$\sigma = \begin{pmatrix} 0 & 5 \\ 5 & 2 \end{pmatrix} \text{MPa}$$

9.8–9.13 The elements represent the stress at a point for various loading conditions in Figs. P9.8–P9.13.

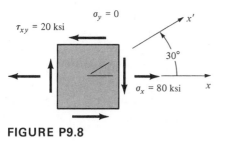

FIGURE P9.8

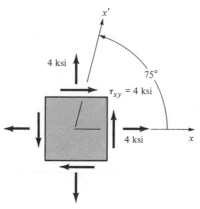

FIGURE P9.10

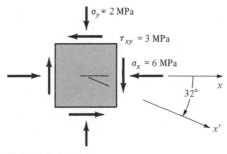

FIGURE P9.11

a. Compute the principal stresses, maximum shear stress and planes of principal stress.
b. Compute the stresses acting on an element oriented with respect to the indicated x'-y' axis.
Use the equations developed in this chapter.

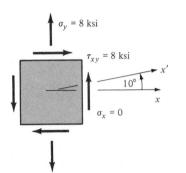

FIGURE P9.12

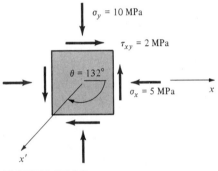

FIGURE P9.13

9.14–9.19 Compute the principal stresses, maximum shear stress, and principal planes using Mohr's circle for the elements of Problems 9.8–9.13.

9.20 A concrete test specimen is cylindrical in shape, 6 in. in diameter and 12 in. long. An axial force P is applied and increased until the cylinder fails. Investigate the stress at a point on the cylinder. Compare the relative magnitudes of the tensile, compressive, and shear stresses.

9.21 A simply supported beam of length L carries a concentrated force of P at its midpoint, as shown in Fig. P9.21. The allowable normal stress in tension and compression is 35 MPa. The allowable shear stress is 14 MPa. Compute the allowable load P using the stress element at point A. Show that the maximum shear stress caused by the transverse shear is insignificant. Assume $L = 3$ m.

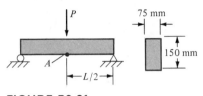

FIGURE P9.21

9.22 A circular rod is fixed free and subject to a torque T in.·lb. The radius of the rod is 2 in. The allowable tensile stress for the material is 80 percent of the allowable shear stress. Compute the maximum torque that can be applied to the rod if the allowable shear stress is 1400 psi.

9.23 For the rod of Problem 9.22 assume the torque is 2 in.·k and compute the magnitude of a compressive axial force that would allow the rod to be stressed to its maximum allowable shear.

9.24 A solid circular shaft 75 mm in diameter is made of a material that must be designed so that the normal stress for tension cannot exceed 30 percent of the normal stress for compression and the shearing stress cannot exceed 45 percent of normal stress in compression. Assume an allowable compressive normal stress of 15 MPa and compute the maximum torque that can be applied to the shaft.

9.25 A machine element is fabricated by butt welding two circular shafts together along the inclined surface illustrated in Fig. P9.25. The shaft is subjected to a compressive axial force and a torque such that the shear stress at the section is twice the magnitude of the compressive normal stress at the same section. However, the tensile stress normal to the weld cannot exceed 300 psi. Compute the allowable magnitude of the axial force and applied torque. The torque causes negative shear stress.

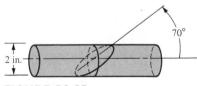

FIGURE P9.25

9.26 A 100-mm-diameter shaft is subjected to combined bending and torsion as illustrated in Fig. P9.26. Compute the principal stresses and maximum shear stresses at points A and B.

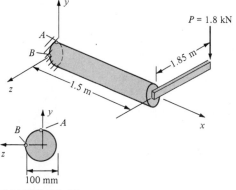

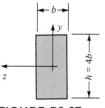

FIGURE P9.26

9.27 A beam with the cross section of Fig. P9.27 is subjected to a bending moment $M_z = +72 \ \text{ft·k}$ and a positive vertical shear of 12 k. Assume a design criterion that the shear stress must not exceed 10 percent of the normal stress (tension or compression) and compute the value of b. $\sigma_{\text{allow}} = 1800$ psi.

FIGURE P9.27

9.28 A cylindrical pressure vessel is subjected to an internal pressure of 50 psi and an external torque of 16,000 in.·k acting to cause negative shear stress on the vessel. The pressure vessel is 100 in. in diameter with a wall thickness of 0.25 in. Determine the principal stresses, planes of principal stress, and maximum shear stress.

9.29 Assume that the pressure vessel of Problem 9.28 has spiral welded seams making an angle of 75° with the longitudinal axis of the vessel, as illustrated in Fig. P9.29. Compute the normal and shearing stresses acting on the weld.

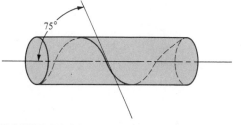

FIGURE P9.29

9.30 For the pressure vessel of Problem 9.29, compute an angle of the spiral welded seams such that the shear stress acting on the weld is minimum. Compute the angle of the welded seams to make the normal stress acting on the weld a minimum.

9.31 Assume a design criterion for welded seams that requires that shearing stresses be no greater than one-third the maximum normal stresses for the weld. Compute the angle of the spiral welded seam in Problem 9.29 that would cause a shear stress equal to one-third the normal stress on the weld.

9.32 A spherical pressure vessel 3.5 m in diameter is subjected to an internal pressure of 3 MPa. Compute the wall thickness of the vessel if the normal stress is not to exceed 100 MPa.

9.33 The vessel of Problem 9.32 is to be fabricated by welding sections of sheet metal together. Discuss the effect of shear stresses on the weld. Support your discussion with computations.

9.34 A cylindrical closed storage tank is loaded as shown in Fig. P9.34. The internal pressure is 75 psi. Compute the wall thickness if the allowable normal stress is 15,000 psi and the allowable shear stress is 8500 psi. Assume that the tank is fixed at the base.

9.35 The beam of Fig. P9.35 has a strain gage attached at point A. For a longitudinal strain measurement of $+0.00075$ compute the magnitude of P. What strain would occur at point B? $E = 8(10^6)$ psi.

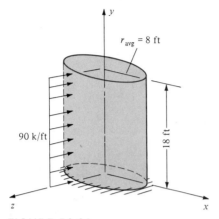

FIGURE P9.34

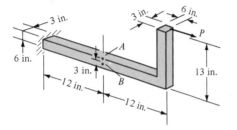

FIGURE P9.35

9.36 A strain gage rosette such as the one in Fig. 9.19 is attached to a structural member. The state of strain obtained from the gage readings is $\varepsilon_0 = -0.000825$, $\varepsilon_{45} = -0.00025$, and $\varepsilon_{90} = 0.00065$. Compute the state of stress for an element oriented along the x-y axis. Compute the principal stresses, principal planes of stress and maximum shear stress. $E = 10(10^6)$; $v = 0.33$; $G = 3.76(10^6)$.

9.37 A cylindrical pressure vessel has two single-axis strain gages attached at right angles, as illustrated in Fig. P9.37. Gage A reads 0.002475 and gage B reads

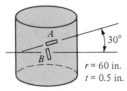

FIGURE P9.37

0.001305. Compute the internal pressure assuming $E = 10(10^6)$ psi and $v = 0.3$. (*Hint:* Convert strain to stress and use the relation for σ_h and σ_l.)

9.38 A delta strain gage rosette has three strain gages oriented as shown in Fig. P9.38. Using Eq. (9.21), compute ε_x, ε_y, and γ_{xy}.

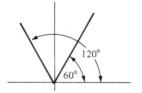

ε_0 (x axis)
ε_{60} at 60°
ε_{120} at 120°

FIGURE P9.38

9.39 Derive a parallel-axis theorem for the product of inertia. *Hint:* Refer to your engineering mechanics (statics) textbook.

9.40 Compute the second moments of area I'_x, I'_y and the product of inertia I'_{xy} for any angle θ using the cross section of Fig. P9.40.

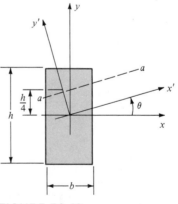

FIGURE P9.40

9.41 Compute the second moments of area and product of inertia with respect to the axis a-a of Fig. P9.40. Line a-a is parallel to the x' axis.

9.42 Locate the centroid of the angle of Fig. P9.42. Compute I'_x, I'_y, and I'_{xy} with respect to the centroid. Compute I'_x, I'_y, and I'_{xy} with respect to an axis rotated 75 degrees counterclockwise from the x axis. Compute the principal area moments and the corresponding principal planes.

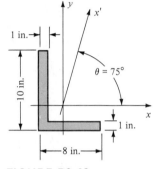

FIGURE P9.42

9.43 The section of Fig. P9.43 has vector moments of $M_x = 120 \text{ kN·m}$ and $M_y = 85 \text{ kN·m}$ acting about the centroid. Locate the principal axes for the second moment of the area and compute the stresses at points A, B, and C. Locate the neutral axis.

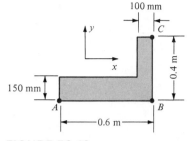

FIGURE P9.43

9.44 The section of Fig. P9.44 has vector moments of $M_x = 50 \text{ k·ft}$ and $M_y = 80 \text{ k·ft}$ acting about the centroid. Locate the principal axes for the second moment of the area and compute the stresses at points A through D.

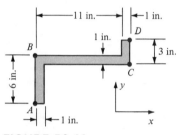

FIGURE P9.44

CHAPTER

DEFLECTION OF BEAMS

10.1   INTRODUCTION

Deflections of beams due to transverse loading is a topic of utmost importance in the engineering analysis of structures. In this chapter systematic methods are presented for computing beam deflections. The emphasis is placed upon understanding the physical problem and solution of its mathematical counterpart.

In Chapter 2 axially loaded members were analyzed for deflection and in almost every loading situation one algebraic equation was sufficient to solve the problem. A similar observation can be made for the rotation of circular members as studied in Chapter 4. The deflection of transversely loaded members presents a different situation. There is a relatively simple equation that governs the deflection analysis, but it is a differential equation, and its solution is dependent upon load conditions and boundary conditions.

A method for computing beam deflections that is fundamental in mechanics of materials is dependent upon writing a differential equation describing either the load function or the moment function. This method of analysis is thoroughly discussed in the first few sections of this chapter. Additional practice with writing moment equations using various coordinate origins for the same problem occurs naturally.

The method of superposition is discussed. Superposition of beam deflection problems, from an engineering design viewpoint, is a practical design aid.

The use of singularity functions is discussed in this chapter. The method is included as a separate section at the end of the chapter.

An extensive discussion of virtual work is included in this chapter. Energy methods, the concept of the balance of energy in a physical system, are equally important as deflection analysis using the basic differential equation. The so-called unit load method is developed for axially loaded members, torsionally loaded members, and transversely loaded beams. The unit load method is an interpretation of virtual work, while virtual work is a basic interpretation of the balance of energy. Virtual work specialized as the unit load method lends itself, in a natural way, to a vector analysis type of formulation. Once again, vector analysis is used to aid in problem solving.

Finally, a brief discussion of moment area theorems is presented. The moment area method is a classical interpretation of the differential relationship between moment, slope, and deflection.

10.2 DIFFERENTIAL EQUATIONS GOVERNING THE DEFLECTION OF BEAMS

Differential equations governing beam deflections due to transverse loading are based upon the shear and moment relations of Chapter 5, the relation between moment and curvature, and the geometry of the deflected beam. The deflected axis of the beam is the elastic curve. Beam deflection problems are solved by determining the equation defining the elastic curve.

Consider the beam element of Fig. 10.1. A small section of beam, ΔS in length, is sliced out of the beam. In the deformed position, ΔS remains constant along the neutral axis. The angle $\Delta\theta$ is defined in Figs. 10.1a and b as the change in slope along the length Δx, and also is the angle defining the deformed section with respect to a center of curvature. The radius of curvature, ρ, is the distance from the center of curvature to the neutral axis of the beam. The geometry of the deformed section is used to obtain a relation between the angle $\Delta\theta$ and the axial deformation Δu, as shown in Fig. 10.1b. The line a'-b' is constructed parallel to line a-b, thereby forming an angle $\Delta\theta$ with the line c-d. The deformation Δu occurs at some arbitrary distance y above the neutral axis.

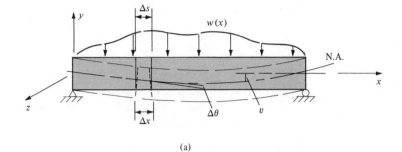

(a)

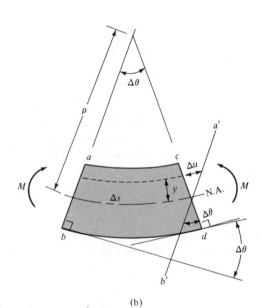

FIGURE 10.1
Deformed beam element.

(b)

$$\Delta u = y \, \Delta \theta \tag{10.1}$$

Similarly, ΔS is related to $\Delta \theta$ as

$$\Delta S = \rho \, \Delta \theta \tag{10.2}$$

Combining Eqs. (10.1) and (10.2), eliminating $\Delta \theta$, gives

$$\frac{\Delta u}{\Delta S} = \frac{y}{\rho} \quad \text{or} \quad \lim_{\Delta S \to 0} \frac{\Delta u}{\Delta S} = \frac{du}{dS} = \frac{y}{\rho} \tag{10.3}$$

Since Δu represents the deformation at y above the neutral axis, the strain at the same location y above the neutral axis is

$$\varepsilon = \frac{du}{dS} \tag{10.4}$$

and

$$\frac{1}{\rho} = \frac{\varepsilon}{y} \tag{10.5}$$

Recognize that ε is the strain along an axial fiber and that transverse strain ε_y or ε_z is assumed to be zero. The relation between stress and strain, $\varepsilon = \sigma/E$, can be used. The axial or normal stress in the beam is $\sigma = My/I$. Substituting into Eq. (10.5),

$$\frac{1}{\rho} = \frac{\sigma}{yE} \tag{10.6}$$

or

$$\frac{1}{\rho} = \frac{M_z}{EI_z} \tag{10.7}$$

and is the basic moment curvature relation for bending of a beam due to transverse loading. It is assumed that bending occurs about the z axis and that z is a principal axis of the cross section.

Equation (10.6) can be used to compute the stress for situations in which a structure is deformed with a given radius of curvature.

EXAMPLE 10.1

A thin, flat strip of aluminum 4 mm thick and 15 mm wide is used as a strap-type support for a circular tank 1.2 m in diameter. Compute the bending stress in the strip. Assume $E = 70$ GPa.

Solution:

Using Eq. (10.6), y is replaced with $\pm c$ since extreme fiber stress is to be computed.

$$\sigma = \frac{Ec}{\rho} = \frac{70(10^9) \text{ N/m}^2 \, 2(10^{-3}) \text{ m}}{0.6 \text{ m}} = 2.33(10^8) \text{ N/m} = 233 \text{ MPa} \qquad \blacksquare$$

A general differential equation relating beam deflection to bending moment is obtained from Eq. (10.7) and the curvature expression as it is presented in any calculus text. The curvature equation is

$$\frac{1}{\rho} = \frac{d^2v/dx^2}{[1 + (dv/dx)^2]^{3/2}} \tag{10.8}$$

where v is the deflection of the beam. Note, as shown in Fig. 10.1, that v is the deviation of the elastic curve from the undeformed x axis and represents deflection. The vertical coordinate is y, and v is the deflection in the y direction. Equation (10.8) is valid for any curve where v is a function of x. The deflection gradient of the beam of Fig. 10.1 can be considered to be very small, meaning that dv/dx is small. The term $(dv/dx)^2$ can be neglected for the theory of beam deflections, thus simplifying Eq. (10.8). Combining Eqs. (10.8) and (10.7) gives the differential equation governing the deflection of beams caused by bending as

$$\frac{d^2v}{dx^2} = \frac{M}{EI} \tag{10.9}$$

The subscript z has been dropped. It is understood that M and I are with respect to an axis on the cross section that is normal to v.

Equation (10.1) is significant since it leads to a geometrical relation between strain and change in slope. Divide both sides of Eq. (10.1) by ΔS and take the limit as $\Delta S \to 0$.

$$\lim_{\Delta S \to 0} \frac{\Delta u}{\Delta S} = y \lim_{\Delta S \to 0} \frac{\Delta \theta}{\Delta S} \tag{10.10}$$

Comparing Eq. (10.10) to Eqs. (10.5) and (10.6) indicates that the change in slope is related directly to the bending moment.

$$\frac{d\theta}{dS} = \frac{M_z}{EI_z} \tag{10.11}$$

The assumption that provides for neglecting the square of the slope implies that the change in length of the beam is negligible as the beam deflects transversely. For all practical purposes the length dS in Fig. 10.1 is the same as dx. The term $d\theta/dS$ of Eq. (10.11) can be replaced with $d\theta/dx$. The term θ is the slope of the beam, the change in v with respect to x.

$$\theta = \frac{dv}{dx} \quad \text{and} \quad \frac{d\theta}{dx} = \frac{d^2v}{dx^2} \tag{10.12}$$

Equation (10.12) also implies Eq. (10.9) subject to the assumption that θ is small. Small θ implies that

$$\frac{dv}{dx} = \tan \theta = \theta$$

It follows that Eq. (10.9) represents an approximate theory for beam deflections and implies what is termed the elementary theory of beam deflections.

Equation (10.9) is probably the most useful form of the differential equation. The differential equation can be written in terms of beam loading using the load, shear, and moment relations of Chapter 5, $dV/dx = w$, and $dM/dx = V$. The following basic relations exist for computing beam deflections:

v, the beam deflection, is the deviation of the elastic curve from the x axis.

$dv/dx = v' = \theta$ is the slope as defined by Eq. (10.12).

$d^2v/dx^2 = v'' = M/EI$ is the bending moment as defined by Eq. (10.9).

$$\frac{d}{dx}\left(EI\,\frac{d^2v}{dx^2}\right) = (EIv'')' = \frac{dM}{dx} = V \qquad \text{(the shear)} \qquad \textbf{(10.13)}$$

$$\frac{d^2}{dx^2}\left(EI\,\frac{d^2v}{dx^2}\right) = (EIv'')'' = \frac{dV}{dx} = w \qquad \text{(the load)} \qquad \textbf{(10.14)}$$

Equations (10.9) and (10.14) are the most often used differential equations. Equation (10.14) is used only when the load can be written as a continuous function of x. Sometimes the beam load can be described by a continuous function that is unwieldy when attempting to write a moment equation. In that case, Eq. (10.14) is the obvious choice. An example problem, after some discussion of boundary conditions, will illustrate the use of the equations.

10.3   BOUNDARY CONDITIONS FOR BEAM DEFLECTIONS

The differential equations of the preceding section are simple to solve; merely integrate both sides of the equation. Each integration yields a constant of integration that can only be evaluated by considering the conditions upon v, v', v'' or v'''. Boundary conditions are a combination of applied mathematics and physical reasoning.

Support conditions and types of loading are the same as described in Chapter 5. A review and discussion of their application for solving beam deflection problems is in order. The elastic curve can be imagined, as shown in Fig. 10.2, to be a line that lies in a x-y coordinate system with v measuring the deviation of the line from the x axis. The type of load and boundary conditions, together with Eqs. (10.9), (10.12), (10.13) and (10.14), completely describe the elastic curve in some interval, which for convenience will be $\overline{OL}$, the *total* length of the beam.

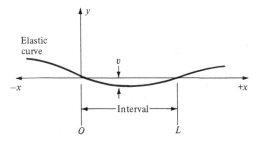

FIGURE 10.2
The elastic curve is defined in an interval (*OL*).

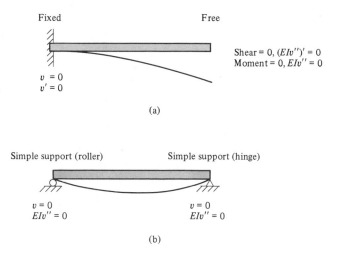

FIGURE 10.3
Boundary conditions.

Boundary conditions can be put into two categories, static boundary conditions that are dependent upon force and moment and kinematic boundary conditions that are dependent upon slope and deflection. Boundary conditions are identified physically and interpreted mathematically. Figure 10.3 illustrates the boundary conditions that occur most frequently for structural members. The fixed support of Fig. 10.3a is characterized by zero deflection ($v = 0$) and zero slope ($v' = 0$). A free end has zero shear and zero bending moment. Shear is given by Eq. (10.13); hence, zero shear implies $(EIv'')' = 0$. Zero moment is represented by $EIv'' = 0$. A simple support, shown in Fig. 10.3b, is represented by zero deflection and zero moment. Additional boundary conditions can be specified by forcing any one of the parameters of Eqs. (10.9), (10.12), (10.13), and (10.14) to have a value other than zero.

10.4 DEFLECTION OF BEAMS USING DIFFERENTIAL EQUATIONS

An example problem will best illustrate the use of the differential equations and boundary conditions. Consider the following problem that will be solved using both the second-order and fourth-order differential equations.

EXAMPLE 10.2

A simply supported beam of length L is subjected to a positive moment M_0 applied at the left end. Determine the equation of the elastic curve using the second-order and fourth-order differential equations. Assume EI is constant.

Solution:

The beam is shown in Fig. 10.4. The boundary conditions correspond directly to Fig. 10.3b with the exception that $v''(x = 0) = M_0$. Note the use of the beam sign convention of Chapter 5. The fourth-order equation is solved as follows assuming that EI is constant:

$$EIv^{IV} = w = 0 \tag{a}$$

Integrating both sides of the equation gives

$$EIv''' = C_1 \tag{b}$$

where C_1 is the constant of integration. Additional integrations give

$$EIv'' = C_1 x + C_2 \tag{c}$$

$$EIv' = \frac{C_1 x^2}{2} + C_2 x + C_3 \tag{d}$$

$$EIv = \frac{C_1 x^3}{6} + \frac{C_2 x^2}{2} + C_3 x + C_4 \tag{e}$$

The four boundary conditions are used to solve for the four constants of integration. The boundary conditions are on v and v''; hence, only Eqs. (c) and (e) can be used to solve for all four constants. Substituting $v''(x = 0) = M_0$ into Eq. (c),

$$M_0 = 0 + C_2 \quad \text{or} \quad C_2 = M_0$$

Substituting $v''(x = L) = 0$ and C_2 into Eq. (c),

$$0 = C_1 L + M_0 \quad \text{or} \quad C_1 = -\frac{M_0}{L}$$

The curvature equation is obtained by substituting both C_1 and C_2 into Eq. (c).

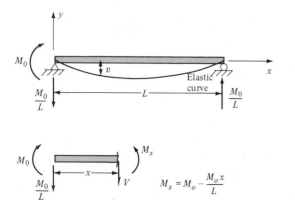

Boundary conditions

$v \ (x = 0) = 0$
$v \ (x = L) = 0$
$v'' (x = 0) = M_o$
$v'' (x = L) = 0$

FIGURE 10.4

$$EIv'' = -\frac{M_0 x}{L} + M_0 \tag{f}$$

The boundary condition $v(x = 0) = 0$ is substituted into Eq. (e)

$$0 = 0 + C_4 \quad \text{or} \quad C_4 = 0$$

Substituting $v(x = L) = 0$ into Eq. (e) along with C_1 and C_2 completes the solution for the constants.

$$0 = -\left(\frac{M_0}{L}\right)\left(\frac{L^3}{6}\right) + \left(\frac{M_0 L^2}{2}\right) + C_3 L \quad \text{or} \quad C_3 = -\left(\frac{M_0 L}{3}\right)$$

The equation of the elastic curve is obtained by substituting all constants into Eq. (e).

$$EIv = -\left(\frac{M_0}{6L}\right)(x^3) + \left(\frac{M_0 x^2}{2}\right) - \left(\frac{M_0 L x}{3}\right) \tag{g}$$

The same solution can be developed using the second-order equation. The moment must be evaluated by writing a moment equation using the free body of Fig. 10.4b.

$$EIv'' = M = M_0 - \left(\frac{M_0 x}{L}\right) \tag{h}$$

This result is the same as Eq. (f); the first two integrations have been bypassed.

Integrating Eq. (h) twice will give results similar to Eqs. (d) and (e). The solution for the constants of integration is dependent upon the deflection boundary conditions $v(x = 0) = 0$ and $v(x = L) = 0$. The final result will be the same as Eq. (g) above. ∎

Usually, the second-order equation leads to a solution involving less algebraic manipulation. Occasionally, a beam loading is described by a simple mathematical function, but the construction of a moment equation using the free-body approach becomes somewhat complicated. In these cases, the fourth-order equation is the proper choice. Concentrated loadings represent an extra difficulty using the fourth-order equation; but a moment equation using the free-body method is straightforward, which suggests that the second-order equation is more suitable for concentrated forces.

The next example should give additional insight concerning the selection of the best equation for a given loading.

EXAMPLE 10.3

The fixed-free beam of Fig. 10.5 has a continuous loading that varies as a second-order curve. Develop the equation of the elastic curve. Assume EI is constant.

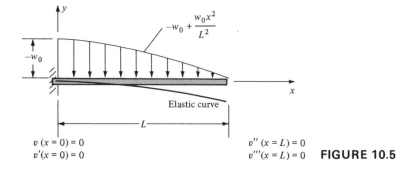

$$v\,(x = 0) = 0$$
$$v'(x = 0) = 0$$

$$v''\,(x = L) = 0$$
$$v'''(x = L) = 0 \quad \textbf{FIGURE 10.5}$$

Solution:
The boundary conditions are shown in Fig. 10.5. The load function is given and can be verified to be $w = -w_0$ at $x = 0$ and $w = 0$ at $x = L$. A moment equation for the coordinate origin at the left end of the beam is not a straightforward computation; hence, choose the fourth-order differential equation.

$$EIv^{IV} = -w_0 + \frac{w_0 x^2}{L^2} \tag{a}$$

Two integrations yield equations for shear and moment, or curvature.

$$EIv'' = -w_0 x + \frac{w_0 x^3}{3L^2} + C_1 \tag{b}$$

$$EIv'' = -\frac{w_0 x^2}{2} + \frac{w_0 x^4}{12L^2} + C_1 x + C_2 \tag{c}$$

The boundary conditions at the free end can be used to determine C_1 and C_2. Substituting into Eq. (b),

$$v'''(x = L) = 0 = -w_0 L + \frac{w_0 L}{3} + C_1$$

or

$$C_1 = \frac{2w_0 L}{3}$$

Substituting C_1 into Eq. (c) and applying the boundary condition

$$v''(x = L) = 0 = -\frac{w_0 L^2}{2} + \frac{w_0 L^2}{12} + \frac{2w_0 L^2}{3} + C_2$$

or

$$C_2 = -\frac{w_0 L^2}{4}$$

Substitute the constants into Eq. (c) and continue the integration.

$$EIv' = -\frac{w_0x^3}{6} + \frac{w_0x^5}{60L^2} + \frac{w_0Lx^2}{3} - \frac{w_0L^2x}{4} + C_3 \tag{d}$$

$$EIv = -\frac{w_0x^4}{24} + \frac{w_0x^6}{360L^2} + \frac{w_0Lx^3}{9} - \frac{w_0L^2x^2}{8} + C_3x + C_4 \tag{e}$$

The boundary conditions on v and v' at $x = 0$ are used to evaluate C_3 and C_4. From Eq. (d),

$$EIv'(x = 0) = 0 \quad \text{or} \quad C_3 = 0$$

and Eq. (e) gives

$$EIv(x = 0) = 0 \quad \text{or} \quad C_4 = 0$$

Equation (e) is the equation of the elastic curve

$$EIv = -\frac{w_0x^4}{24} + \frac{w_0x^6}{360L^2} + \frac{w_0Lx^3}{9} - \frac{w_0L^2x^2}{8} \tag{f}$$

The same result can be obtained using Eq. (10.9), the curvature equation. A free-body diagram must be used to write a moment equation and would be the same as Eq. (c) after the constants of integration have been determined. It should be obvious that the fourth-order equation offers a more direct solution for this beam loading. ■

10.5 BEAMS WITH DISCONTINUOUS LOADING

An abrupt change in the load represents a discontinuity. A concentrated load or the point where a uniform load begins or ends is an example of a discontinuous load. These load conditions are not describable using simple mathematical functions. New conditions, in addition to the boundary conditions, must be introduced in order to use the integration method. An example of a beam with a single concentrated load will illustrate the concept.

EXAMPLE 10.4

The beam of Fig. 10.6 carries a single concentrated force. Write the moment equation and derive the equation for the slope and the elastic curve.

Solution:
The moment diagram and moment equations are shown in Fig. 10.6. Two moment equations are required and will be written with respect to an origin at the left end.

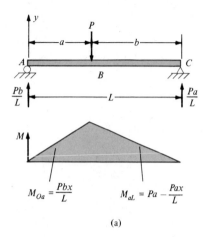

$$M_{Oa} = \frac{Pbx}{L} \qquad\qquad M_{aL} = Pa - \frac{Pax}{L}$$

(a)

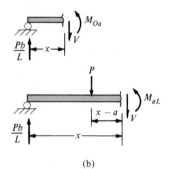

(b)

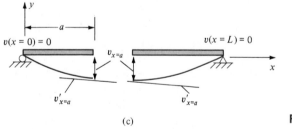

(c)

FIGURE 10.6

The free-body sections are shown in Fig. 10.6b, and summing moments at the cut section gives the proper moment equations.

$$M_{Oa} = \frac{Pbx}{L} \tag{a}$$

$$M_{aL} = \frac{Pbx}{L} - P(x - a) \tag{b}$$

Substituting $b = L - a$ into Eq. (b) and rearranging gives a more suitable form for M_{aL}.

$$M_{aL} = Pa - \frac{Pax}{L} \tag{c}$$

Section AB:

$$EIv'' = \frac{Pbx}{L}$$

$$EIv' = \frac{Pbx^2}{2L} + C_1 \tag{d}$$

$$EIv = \frac{Pbx^3}{6L} + C_1 x + C_2 \tag{e}$$

For $v(x = 0) = 0$, it follows that $C_2 = 0$.

Section BC:

$$EIv'' = Pa - \frac{Pax}{L}$$

$$EIv' = Pax - \frac{Pax^2}{2L} + C_3 \tag{f}$$

$$EIv = \frac{Pax^2}{2} - \frac{Pax^3}{6L} + C_3 x + C_4 \tag{g}$$

For $v(x = L) = 0$, Eq. (g) gives

$$0 = \frac{PaL^2}{2} - \frac{PaL^2}{6} + C_3 L + C_4 \tag{h}$$

There are three unknown constants. Two additional equations are required that can be combined with Eq. (h) for evaluating the constants. These are obtained from Eqs. (i) and (j). The moment for the beam is described by two separate equations; it follows that the slope and deflection are each described by two sets of equations. The boundary conditions are illustrated in Fig. 10.6c. The condition $v(x = 0) = 0$ can apply only to the left-hand section while $v(x = L) = 0$ can apply only to the right-hand section. Additional conditions, referred to as continuity conditions, are available at the point $x = a$, where the left-hand section must literally fit together with the right-hand section. The conditions are that the slope and deflection must be continuous. These conditions can be written as

$$v'\Big|_{\substack{x=a \\ \text{left}}} = v'\Big|_{\substack{x=a \\ \text{right}}} \tag{i}$$

$$v\Big|_{\substack{x=a \\ \text{left}}} = v\Big|_{\substack{x=a \\ \text{right}}} \tag{j}$$

Continuing with the solution will illustrate the use of these conditions. Equation (i) is evaluated by substituting $x = a$ into Eqs. (d) and (f) and equating the slope at $x = a$.

$$\frac{Pba^2}{2L} + C_1 = Pa^2 - \frac{Pa^3}{2L} + C_3 \tag{k}$$

Similarly, Eq. (j) is evaluated using Eq. (e) and (g) with $x = a$.

$$\frac{Pba^3}{6L} + C_1 a = \frac{Pa^3}{2} - \frac{Pa^4}{6L} + C_3 a + C_4 \tag{l}$$

Simultaneous solution of Eqs. (h), (k), and (l) gives the constants C_1, C_3, and C_4.

$$C_1 = -\frac{Pb}{6L}(L^2 - b^2)$$

$$C_3 = -\frac{Pa}{6L}(2L^2 - a^2)$$

$$C_4 = \frac{Pa^3}{6}$$

The equations for the elastic curve are obtained by substituting the constants in Eqs. (e) and (g).

Section AB:

$$EIv_{Oa} = \frac{Pb}{6L}(x^3 - L^2 x + b^2 x) \tag{m}$$

Section BC:

$$EIv_{aL} = \frac{Pa}{6L}(-x^3 + 3Lx^2 - 2L^2 x - a^2 x + La^2) \tag{n}$$

Equations for the slope are obtained by substituting the proper constants into Eqs. (d) and (f). ∎

It becomes obvious that the analysis involving continuity conditions can be lengthy and tedious. The following example will demonstrate a method that will simplify the procedure and provide experience using a coordinate origin at two different locations in the same problem.

EXAMPLE 10.5

Repeat the analysis for the beam of Example 10.4 using the coordinates of Fig. 10.7. Section *BC* has its origin at the right end of the beam while section *AB* remains unchanged. Note the change in the slope continuity condition.

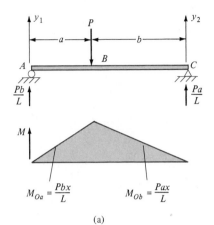

$$M_{Oa} = \frac{Pbx}{L} \qquad M_{Ob} = \frac{Pax}{L}$$

(a)

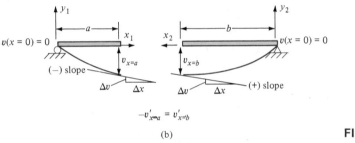

(b)

FIGURE 10.7

Solution:

Physically the boundary and continuity conditions remain unchanged.

$$v(x_1 = 0) = 0 \qquad v(x_2 = 0) = 0$$

Note that the mathematical description of the condition at point C has changed.

$$v\big|_{\substack{x=a \\ \text{left}}} = v\big|_{\substack{x=b \\ \text{right}}} \tag{a}$$

The deflection continuity condition is evaluated at $x = b$ on the right due to the change in origin for section BC.

$$-v'\big|_{\substack{x=a \\ \text{left}}} = v'\big|_{\substack{x=b \\ \text{right}}} \tag{b}$$

A minus sign occurs on one side of the slope continuity condition. The reason is illustrated in Fig. 10.7b. A negative slope in the x_1-y_1 coordinate system is positive in the x_2-y_2 coordinate system. The analysis for section AB is identical to Example 10.4, Eqs. (a), (d), and (e).

$$EIv'' = \frac{Pbx}{L} \tag{c}$$

$$EIv' = \frac{Pbx^2}{2L} + C_1 \qquad\qquad\qquad \text{(d)}$$

$$EIv = \frac{Pbx^3}{6L} + C_1 x + C_2 \qquad\qquad\qquad \text{(e)}$$

The condition $v(x_1 = 0) = 0$ gives $C_2 = 0$.

The moment equation for section CB, with origin at C, is shown in Fig. 10.7a.

$$EIv'' = \frac{Pax}{L} \qquad\qquad\qquad \text{(f)}$$

$$EIv' = \frac{Pax^2}{2L} + C_3 \qquad\qquad\qquad \text{(g)}$$

$$EIv = \frac{Pax^3}{6L} + C_3 x + C_4 \qquad\qquad\qquad \text{(h)}$$

The condition $v(x_2 = 0) = 0$ gives $C_4 = 0$.

Evaluating Eqs. (a) and (b) gives two equations with two unknowns.

$$\frac{Pba^3}{6L} + C_1 a = \frac{Pab^3}{6L} + C_3 b$$

$$-\frac{Pba^2}{2L} - C_1 = \frac{Pab^2}{2L} + C_3$$

Solving for C_1 and C_3 gives, after some algebraic manipulation,

$$C_1 = \frac{-Pb}{6L}(L^2 - b^2) \qquad C_2 = 0$$

$$C_3 = \frac{Pa}{6L}(L^2 - a^2) \qquad C_4 = 0$$

Substituting C_1 and C_3 into Eqs. (e) and (h) gives solutions for the elastic curve that are equivalent to the previous example. It is important to become accustomed to shifting the coordinate axis to different locations in the same problem. At first there may be some additional confusion in formulating the problem, but invariably a judicious choice of coordinates will simplify the computational work.   ■

The use of singularity functions for beam deflections leads to an organized formulation of the problem. Singularity functions are discussed in Section 10.11. It is in order, if so desired, to include that material for the analysis of beam deflections at this time.

10.6   BEAMS WITH VARIABLE CROSS-SECTIONAL PROPERTIES

It was shown in Chapter 6 that bending stress is dependent upon both bending moment and second moment of the area. It is good design practice to optimize the use of materials in some cases and allow the second moment of the area to vary. Two examples are illustrated in Fig. 10.8. The simply supported beam of Fig. 10.8a has an abrupt change in the second moment of the area. The bending moment is larger near the center; hence, the second moment of the area is increased in the

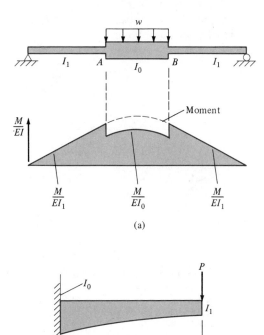

(a)

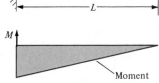

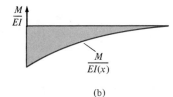

(b)

FIGURE 10.8
Beams with variable second
moment of the area.

center portion of the beam. Equation (10.9) clearly indicates that if I is a function of x, the solution of the differential equation will depend upon the functional form of both M and I. An M/EI diagram is shown in Fig. 10.8a indicating that three separate continuous functions must be integrated to completely define the elastic curve. Continuity conditions must be used at points A and B to connect the three functions.

The second moment of the area of the fixed-free beam of Fig. 10.8b varies continuously along the length of the beam. The moment diagram is linear, but the M/EI diagram varies nonlinearly. If the variation in the second moment of the area can be described using a mathematical function, valid between $0 \leq x \leq L$, Eq. (10.9) can be integrated directly. Consider the following example.

EXAMPLE 10.6

A fixed-free beam of length L has a concentrated force P applied at the free end. The depth of the beam varies such that the second moment of the area varies linearly from I_0 at the fixed end to $I_0/2$ at the free end. The beam is sketched in Fig. 10.9 along with a plot of the second moment of the area. Determine the equation of the elastic curve using the curvature differential equation.

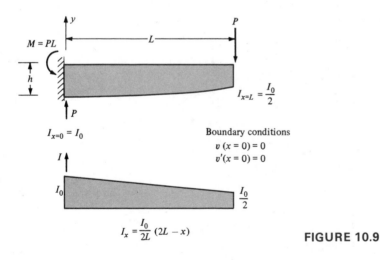

FIGURE 10.9

Solution:
The depth of the beam varies as the cube root of the linear variation in the second moment of the area. The equation defining the change in the second moment of the area is obtained using elementary methods.

$$I = \frac{I_0(2L - x)}{2L}$$

Equation (10.9) is written

$$v'' = \frac{2LM}{EI_0(2L - x)} = \frac{2PL}{EI_0}(x - L)/(2L - x)$$

$$= \frac{2PL}{EI_0}[x/(2L - x) - L/(2L - x)] \tag{a}$$

Equation (a) can be integrated term by term.

$$v' = \frac{2PL}{EI_0}[-x - 2L\ln(2L - x) + L\ln(2L - x)] + C_1$$

$$= \frac{-2PL}{EI_0}[x + L\ln(2L - x)] + C_1 \tag{b}$$

$$v = \frac{-2PL}{EI_0}[x^2/2 + L(x - 2L)\ln(2L - x) - Lx] + C_1x + C_2 \tag{c}$$

The boundary condition $v'(x = 0) = 0$ gives

$$0 = \frac{-2PL}{EI_0}[0 + L\ln(2L)] + C_1$$

or

$$C_1 = \frac{2PL^2\ln(2L)}{EI_0} \tag{d}$$

$v(x = 0) = 0$ substituted in Eq. (c) gives C_2.

$$C_2 = \frac{4PL^3\ln(2L)}{EI_0} \tag{e}$$

Combining Eqs. (c), (d) and (e) gives the equation of the elastic curve.

$$v = \frac{-2PL^2}{EI_0}\left[\frac{x^2}{2L} - x + (x - 2L)\ln\left(\frac{2L - x}{2L}\right)\right] \tag{f}$$

The elastic curve for the same problem, assuming a constant second moment of the area, I_0, is obtained in a straightforward manner.

$$v = \frac{P}{EI_0}\left(\frac{x^3}{6} - \frac{Lx^2}{2}\right) \tag{g}$$

Substituting $x = L$ into Eqs. (f) and (g) gives a comparison of the deflection at the free end. Equation (f) gives

$$v = \frac{-2PL^2}{EI_0}\left[\frac{L}{2} - L - L\ln(0.5)\right] = \frac{-0.386PL^3}{EI_0}$$

compared to

$$v = \frac{-0.333PL^3}{EI_0}$$

for the beam with constant second moment of the area. ∎

The foregoing problem will be solved using an approximation for the variable moment of inertia in a later section.

10.7 BEAM DEFLECTIONS USING SUPERPOSITION

Superposition of stresses was discussed in Chapter 8. Superposition of deflections merely implies that the deflections caused by individual types of loading can be superposed to yield a final deflection equation. In the previous examples the relation between deflection and load was linear. For instance, Eq. (g) of Example 10.2 shows that if the applied moment loading was increased by a constant factor, the deflection would increase by the same factor.

Superposition can be applied to the situation shown in Fig. 10.10. The simply supported beam carries two concentrated forces. The analysis can be divided into two parts. Each part can be evaluated separately and the results then combined. Consider the following example.

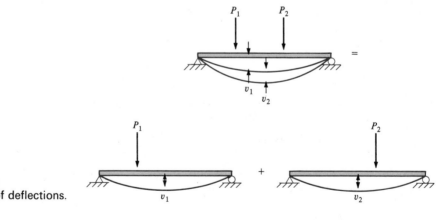

FIGURE 10.10
Superposition of deflections.

EXAMPLE 10.7

The simply supported beam of Fig. 10.11 is subject to a uniform load w_1 and uniformly varying load w_2. Use superposition to determine the equation of the elastic curve and the equation defining the slope of the beam.

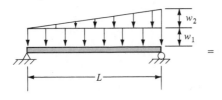

EI is constant

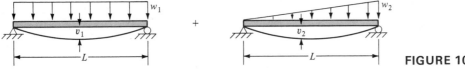

FIGURE 10.11

Solution:

The beam is analyzed in two parts, as illustrated in Fig. 10.11. Superposition analysis is usually accomplished using a set of beam tables or superposition tables, such as Appendix E, where numerous solutions for elementary beam loadings are tabulated. The equation of the elastic curve for the beam of Fig. 10.11 is obtained by adding the equation given by Cases C3 and C7 as found in the beam tables.

$$v = v_1 + v_2 = \frac{w_1 x}{24EI}(L^3 - 2Lx^2 + x^3) + \frac{w_2 x}{360LEI}(7L^4 - 10L^2 x^2 + 3x^4)$$

$$= \frac{w_1 x}{12EI}\left[\frac{L^3}{30}\left(15 + \frac{7w_2}{w_1}\right) - Lx^2\left(1 + \frac{w_2}{3w_1}\right) + \frac{x^3}{2}\left(1 + \frac{w_2 x}{5w_1 L}\right)\right] \tag{a}$$

The point of maximum deflection occurs at the point where the slope is zero for the simply supported beam. The slope equation can be obtained by adding the slope equations as they are given in Appendix E or by differentiating Eq. (a). Differentiating Eq. (a) gives

$$v' = \frac{w_1}{360EI}\left[L^3\left(15 + \frac{7w_2}{w_1}\right) - 30L^2 x\left(3 + \frac{w_2}{w_1}\right) + 5x^3\left(12 + \frac{3w_2 x}{w_1 L}\right)\right] \tag{b}$$

The point of maximum deflection is obtained by letting $v' = 0$ in Eq. (b) and solving for x. Equation (b) is a fourth-order equation; hence, there should be four roots or four values of x that would satisfy the condition $v' = 0$. A closed-form analysis of the equation is a rather formidable task. Both Eqs. (a) and (b) lend themselves to computer solution. In fact it is very straightforward to program the equations using a minicomputer with graphics capability and directly view the elastic curve and slope diagram. ∎

EXAMPLE 10.8

Use superposition and Case E9 of Appendix E to compute the deflection at the free end of the fixed-free beam of Fig. 10.12.

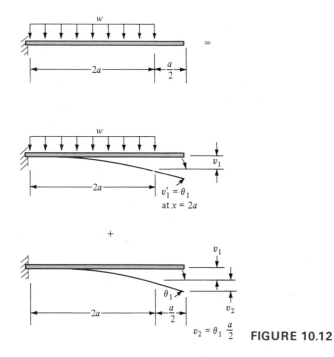

FIGURE 10.12

Solution:

The method of analysis is illustrated in Fig. 10.12. The total deflection v is made up of two parts, v_1 and v_2, as illustrated. The uniform load causes deflection v_1 and is evaluated using Case E9 and a beam of length $2a$.

$$v_1 = \frac{w(2a)^4}{8EI} = \frac{2wa^4}{EI} \tag{a}$$

The deflection v_1 corresponds to the deflection at the free end of a fixed-free beam of length $2a$. The deflection v_2 is computed as

$$v_2 = \theta_1 \left(\frac{a}{2}\right)$$

The unloaded portion of the beam extends downward with constant slope. For the unloaded portion of the beam, the moment is zero; hence, the curvature is zero. The slope θ_1 corresponds to the slope at the free end of a fixed-free beam of length $2a$. Use Case E9 again.

$$v_2 = \frac{w(2a)^3}{6EI}\left(\frac{a}{2}\right) = \frac{2}{3}\left(\frac{wa^4}{EI}\right)$$
\hfill **(b)**

$$v = v_1 + v_2 = \frac{8}{3}\left(\frac{wa^4}{EI}\right)$$

The same result can be obtained by substituting $a = 2a$, $b = a/2$, and $L = 5a/2$ for Case E10. ∎

10.8 VIRTUAL WORK AND THE UNIT LOAD METHOD

The unit load method is an application of the methods of virtual work. The development of the final equations is achieved using Eq. (3.20) for linear elastic materials.

$$\Delta W = \Delta U \tag{10.15}$$

The change in external work is equal to the change in strain energy.

This chapter has thus far dealt with the analysis of the basic beam structure using the solution of the differential equation exclusively. Energy methods are equally significant. The basic concept of virtual work is a powerful analysis tool for continued study in mechanics of materials.

The intent of this chapter has been to demonstrate the analysis of beam deflections and has been limited to two-dimensional problems subject to flexural loading. The method of virtual work lends itself to investigating the deflections of beams subject to loadings that produce deflections due to bending and shear, rotation due to torsion, and axial deformations. The vector methods of Chapter 5 for writing shear and moment equations find significant application in constructing the unit load method equations.

The unit load method equations for axial deformation will be derived first. The derivation centers about finding the deflection or movement of a particular joint in a truss. A truss represents a structure in which both external and internal forces can be easily visualized. In addition, the deflection analysis for a truss is a worthwhile problem and deserves some discussion for its own sake.

Virtual work is distinguished from real work by the fact that physically it does not exist. Virtual work is an imaginary concept that enables the analyst to effectively use Eq. (10.15) to develop a method for the analysis of structures. Virtual work can be formulated using either a virtual force or a virtual displacement. An extensive derivation of the principles of virtual work has been given by J. T. Oden.[1] A virtual

[1] J. T. Oden, *Mechanics of Elastic Structures*, McGraw-Hill, New York, 1967, chap. 8.

work principle can be formulated either as *principle of virtual displacements* or as a *principle of virtual forces*. The choice of formulation depends upon the application. For the purpose of deriving the equations for the unit load method, the principles will be stated without proof.

The principle of virtual displacements is as follows: A deformable structural system is in equilibrium if the total external virtual work is equal to the total internal virtual work for every virtual displacement consistent with the constraints.

A similar statement can be referred to as the principle of virtual forces: The strains and displacements in a deformable structural system are compatible and consistent with the constraints if the total external complementary virtual work is equal to the total internal complementary virtual work for every system of virtual forces and stresses that satisfy the equations of equilibrium.

The mathematical statement of either principle appears in the form of Eq. (10.15). The difference in the two principles will not be obvious to the reader who has no previous experience with energy methods. In this application we wish to use the principle of virtual forces. The motivation arises from the fact that we wish to solve for a real, non-virtual displacement. It must be emphasized that the material property has been specified to be linear elastic, which means that complementary energy can be taken equal to the strain energy. Strain energy was defined in Chapter 3 and complementary energy will be defined in Chapter 13. In this chapter we will merely state that the assumption of a linear elastic material allows us to restate a portion of the principle of virtual forces as "the total external virtual work is equal to the total virtual strain energy."

The truss of Fig. 10.13a will serve as a model for deriving the basic equation. The members of a truss are assumed to be pin-connected and hence carry axial force only, and the corresponding member deformations are axial. Assume that the deflection of point A is to be computed for the truss of Fig. 10.13a. The deflection at A is a combination of the individual truss members shortening or elongating and rotating about their pin connections. For the purpose of the derivation the truss of Fig. 10.13a is assumed to be unloaded. To compute the deflection at A, apply a virtual force δF at A recognizing that δF has vector properties. The truss deflects an amount $\Delta_{\delta F}$ due to the application of the virtual force as shown in Fig. 10.13b. The structure is in equilibrium and Eq. (10.15) must be satisfied, or the external work due to the virtual force must balance the internal strain energy due to the virtual force.

$$(1/2)\delta F \cdot \Delta_{\delta F} = U_{\delta F} \tag{a}$$

These terms are written symbolically and do not need to be evaluated

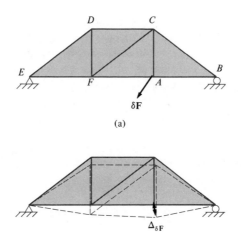

(a)

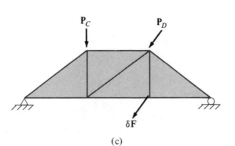

(b)

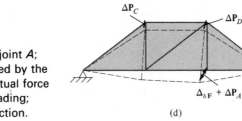

(c)

FIGURE 10.13
(a) Virtual force at joint *A*;
(b) deflection caused by the
virtual force; (c) virtual force
plus actual truss loading;
(d) total truss deflection.

(d)

since they will cancel later. They are included in order to make the derivation complete. The factor of 1/2 appears because of the assumption of a linear elastic material, as was discussed in Chapter 3.

The actual truss loading, assumed as $\mathbf{P}_C$ and $\mathbf{P}_D$ in Fig. 10.13c, is added to the truss of Fig. 10.13a causing additional joint movement of the truss $\Delta\mathbf{P}_A$, as shown in Fig. 10.13d. The balance of energy for the actual loading can be written as

$$[(1/2)\mathbf{P}_D \cdot \Delta\mathbf{P}_D] + [(1/2)\mathbf{P}_C \cdot \Delta\mathbf{P}_C] = U_P \tag{b}$$

The external work in Eqs. (a) and (b) is the scalar product of one half the force vector with the corresponding displacement vector. The strain energy is the sum of the strain energy of all individual truss members. One additional energy component exists. The virtual force $\delta \mathbf{F}$ displaces an amount $\Delta \mathbf{P}_A$ as shown in Figs. 10.13c and d to produce external virtual work $\delta \mathbf{F} \cdot \Delta \mathbf{P}_A$. Note that the factor of 1/2 is not included since $\delta \mathbf{F}$ is not a function of $\Delta \mathbf{P}_A$. The corresponding strain energy U_{virtual}, is the product of the internal force in each member caused by $\delta \mathbf{F}$ as the internal force moves through the deformation caused by the application of the actual loads such that

$$\delta \mathbf{F} \cdot \Delta \mathbf{P}_A = \text{External virtual work}$$
$$U_{\text{virtual}} = \text{Virtual strain energy} \qquad \textbf{(c)}$$

Adding Eqs. (a) and (b), and including (c), gives the total balance of energy.

$$[(1/2)\delta \mathbf{F} \cdot \Delta_{\delta F}] + [(1/2)\mathbf{P}_D \cdot \Delta \mathbf{P}_D] + [(1/2)\mathbf{P}_C \cdot \Delta \mathbf{P}_C] + [\delta \mathbf{F} \cdot \Delta \mathbf{P}_A]$$
$$= U_{\delta F} + U_P + U_{\text{virtual}} \qquad \textbf{(d)}$$

Obviously, all terms in Eq. (d) can be canceled to leave Eq. (d) as

$$\delta \mathbf{F} \cdot \Delta \mathbf{P}_A = U_{\text{virtual}} \qquad \textbf{(e)}$$

The right-hand side of Eq. (e) is evaluated by formally analyzing the truss for the application of the virtual force $\delta \mathbf{F}$. Let these forces be represented by $\mathbf{p}$. The deformation of each truss member due to the application of the actual load is computed by the formula $u = PL/AE$, which was given in Chapter 2. The force in each truss member is $\mathbf{P}$. Then the internal work or strain energy is the sum of $\mathbf{p} \cdot PL/AE$ for each truss member. Equation (e) is written

$$\delta \mathbf{F} \cdot \Delta \mathbf{P}_A = \sum_{i=1}^{N} \mathbf{p} \cdot \frac{PL}{AE}$$

where N is the number of truss members. The virtual force $\delta \mathbf{F}$ will be assumed to be of unit magnitude. Replacing $\Delta \mathbf{P}_A$ with Δ gives the basic virtual work equation as

$$1 \cdot \Delta = \sum_{i=1}^{N} \mathbf{p} \cdot PL/AE \qquad \textbf{(10.16)}$$

The unknown Δ corresponds to the deflection at the point where the unit load is applied and is always in the direction of the unit load. The scalar product can be implied and the formality of the dot product, $1 \cdot \Delta$, can be replaced with a Δ. Letting $\delta \mathbf{F} = 1$ is referred to as the unit load method, and it follows that the method is merely an application of the principle of virtual forces.

EXAMPLE 10.9

Given the truss illustrated in Fig. 10.14, compute the vertical deflection at joint C and the horizontal deflection at joint D. The horizontal members have an area of 3 in.^{2}, the vertical 2 in.^{2}, and the diagonals 4 in.^{2}. All members are steel. $E = 30(10^6)$ psi.

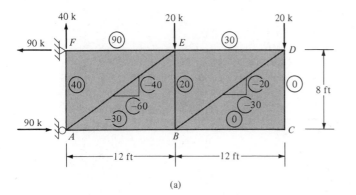

(a)

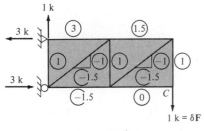

(b)

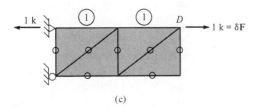

(c)

FIGURE 10.14

Solution:

Since forces are required for all truss members, the method of joints can be used effectively to analyze the truss. The actual member forces are given in Table 10.1 under the heading P_i and are shown circled in Fig. 10.14a. A unit load is applied at point C in the vertical direction since the vertical deflection is desired. The analysis

TABLE 10.1

Member	P_i, kips	L_i, ft	A_i, in.^{2}	Point C p_i, kips	Point C $\dfrac{p_i P_i L_i}{A_i}$	Point D p_i, kips	Point D $\dfrac{p_i P_i L_i}{A_i}$
AB	-30	12	3	-1.50	180	0	0
BC	0	12	3	0	0	0	0
DC	0	8	2	1.0	0	0	0
DE	30	12	3	1.50	180	1.0	120
FE	90	12	3	3.0	1080	1.0	360
AF	40	8	2	1.0	160	0	0
AE	-72.11	14.42	4	-1.8	467.9	0	0
BD	-36.05	14.42	4	-1.80	233.9	0	0
BE	20	8	2	1.0	80	0	0
				Totals	2381.8		480

for internal forces is shown in Fig. 10.14b. Figure 10.14c shows the results for a unit horizontal force applied at joint D. Equation (10.16) is evaluated systematically in Table 10.1.

$$\Delta_C = \sum_{i=1}^{N} \frac{p_i P_i L_i}{A_i 30(10^3)} \,(12 \text{ in./ft}) = \frac{(2381.8)(12)}{30(10^6)}$$

$$= 0.95 \text{ in.}$$

$$\Delta_D = \sum_{i=1}^{N} \frac{p_i P_i L_i}{A_i 30(10^3)} \,(12 \text{ in./ft}) = \frac{(480)(12)}{30(10^6)}$$

$$= 0.19 \text{ in.}$$

A computer program is given in Appendix G that is intended for this type of truss analysis. ∎

When the unit load method is used, the virtual force is assumed to act in the direction of the anticipated deflection. If the unit load is assumed to act in the opposite direction of the deflection, the final result will be negative, indicating that the computed deflection acts in the opposite direction of the assumed unit force.

Virtual work, formulated as the unit load method, for beam deflections or shaft rotations differs from axially loaded members. The internal strain energy for an axially loaded member is represented by force and deformation acting along the axis of the member. A beam subject to

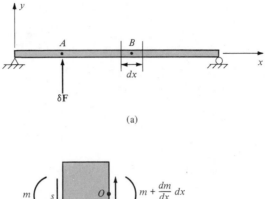

(a)

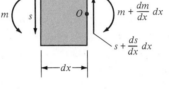

(b)

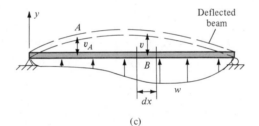

(c)

FIGURE 10.15

bending has a continuously distributed moment and a distributed slope. The strain energy occurs as the moment acts through the beam rotation. The basic relation for beam deflections, corresponding to Eq. (10.16) for axial deformation, can be developed using the theoretical concepts presented in the first sections of this chapter.

Consider the beam of Fig. 10.15a. A positive virtual force is applied at point A, the point where deflection is to be computed. An element at point B, located at any arbitrary section along the beam, is shown in Fig. 10.15b. The internal shear and moment due to the virtual force are designated s and m, respectively. Summing moments about point O, as was done in Chapter 5, gives

$$\frac{dm}{dx} = s \qquad\qquad \textbf{(a)}$$

and can be differentiated

$$\frac{d^2m}{dx^2} = \frac{ds}{dx} \tag{b}$$

The actual beam load w is shown positive in Fig. 10.15c. The deflection at A due to the acting load is v_A. The deflection of the element at B is v. The statement of virtual work, using a virtual force and assuming a linear elastic material, is

$$W_{\text{virtual}} = U_{\text{virtual}} \tag{c}$$

where U_{virtual} is the virtual strain energy and can be evaluated as internal virtual work or *the scalar product of the internal virtual forces on the element of Fig. 10.15b and the actual displacement of the element integrated over the length of the beam.* Equation (c) is evaluated as

$$\delta \mathbf{F} \cdot \mathbf{v}_A = \int_0^L \left(-\mathbf{s} + \mathbf{s} + \frac{d\mathbf{s}}{dx}\,dx \right) \cdot \mathbf{v} \tag{c}$$

Substituting Eq. (b) and rearranging

$$\delta \mathbf{F} \cdot \mathbf{v}_A = \int_0^L \left(\frac{d^2\mathbf{m}}{dx^2} \right) \cdot \mathbf{v}\,dx \tag{d}$$

The integral can be integrated by parts twice.

$$\delta \mathbf{F} \cdot \mathbf{v}_A = \mathbf{v} \cdot \frac{d\mathbf{m}}{dx}\bigg|_0^L - \int_0^L \frac{d\mathbf{m}}{dx} \cdot \frac{d\mathbf{v}}{dx}\,dx$$

$$\delta \mathbf{F} \cdot \mathbf{v}_A = \mathbf{v} \cdot \frac{d\mathbf{m}}{dx}\bigg|_0^L - \frac{d\mathbf{v}}{dx} \cdot \mathbf{m}\bigg|_0^L + \int_0^L \mathbf{m} \cdot \frac{d^2\mathbf{v}}{dx^2}\,dx \tag{e}$$

The boundary conditions for the beam are $v(x = 0) = 0$, $v(x = L) = 0$, which reduces the first term on the right to zero; and $m(x = 0) = 0$, $m(x = L) = 0$, which reduces the second term to zero. There are other boundary conditions such as fixed or free. Assume that the derivation had been based upon a fixed-free beam. At the fixed end $v = dv/dx = 0$ and at the free end $m = dm/dx = 0$, and it can be shown that the first two terms on the right of Eq. (e) will be zero. Substituting Eq. (10.9) for d^2v/dx^2 gives the final form of the unit load method equation.

$$\delta \mathbf{F} \cdot \mathbf{v}_A = \int_0^L \mathbf{m} \cdot \frac{M\,dx}{EI} \tag{10.17}$$

It is customary to let $\delta \mathbf{F} = \mathbf{1}$ in Eq. (10.17).

Equation (10.17) can be derived using different models of strain energy. A straightforward derivation, but less convincing for the inexperienced reader, is based on Eq. (3.19). In Chapter 3 strain energy was shown to be one-half the volume integral of stress times strain. The virtual strain energy would be the scalar product of stress due to the application of the virtual force and strain due to the application of the actual load. The one-half does not appear since the quantities are not functionally dependent. Using the previous notation and the flexural stress formula

$$\sigma_{\text{virtual}} = \frac{my}{I} \tag{f}$$

$$\varepsilon_{\text{actual}} = \frac{\sigma}{E} = \frac{My}{EI} \tag{g}$$

Equating virtual external work and virtual strain energy gives

$$\delta \mathbf{F} \cdot \mathbf{v}_A = \int_V \sigma_{\text{virtual}} \varepsilon_{\text{actual}} \, dV \tag{h}$$

$$\delta \mathbf{F} \cdot \mathbf{v}_A = \int_A y^2 \int_0^L \frac{m}{I} \frac{M}{EI} \, dx \, dA = \int_0^L \mathbf{m} \cdot \frac{\mathbf{M}}{EI} \, dx \tag{10.17[2]}$$

This method will be used to derive the virtual work equations for a circular shaft subject to an external torque.

The method of virtual work for beam deflections involves applying a unit force at the point where the deflection is to be computed. The unit force must be in the direction of the desired deflection. The moment m due to the unit force is written in equation form and multiplied by M, the moment due to the actual load, also in equation form. The integration is performed for the total length of the beam. Consider the following example.

EXAMPLE 10.10

Compute the deflection of the free end of the fixed-free beam of Fig. 10.16a. Assume EI is a constant.

Solution:
Equation (10.17) will be used with $\delta F = 1$ and v is the deflection at the free end of the beam. The virtual load is applied at the free end as shown in Fig. 10.16b. Before

---

[2] The "dot" indicating the scalar product has merely been inserted. The mathematics defining the scalar product of the quantities in Eq. (h) is usually not presented to undergraduate engineers.

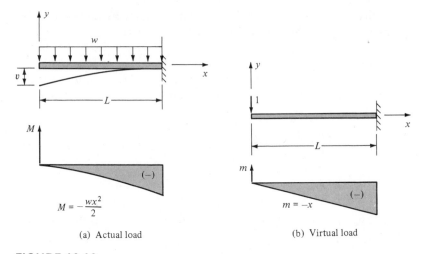

$$M = -\frac{wx^2}{2}$$

(a) Actual load

$$m = -x$$

(b) Virtual load

FIGURE 10.16

the integral can be evaluated, the functional form of the moments must be substituted into the equation. Moment diagrams for M and m are sketched in Fig. 10.16. The moment equations can be verified to be

$$M = \frac{-wx^2}{2} \qquad m = -(1)(x)$$

Equation (10.17) gives the deflection as

$$(1)(v) = \int_0^L (x)\left(\frac{wx^2}{2EI}\right) dx$$

$$v = \frac{wL^4}{8EI}$$

∎

The unit load method, as an application of virtual work, is straightforward and easily applicable to beam deflection problems. Compared to the differential equation method the unit load method gives the deflection at a point. On the other hand, the solution of the differential equation gives the equation of the elastic curve and is valid at any point along the beam.

The following example will illustrate the manner in which different coordinate locations along the beam axis can be used to advantage.

EXAMPLE 10.11

Compute the deflection at the center of the simply supported beam of Fig. 10.17a. Assume EI is constant.

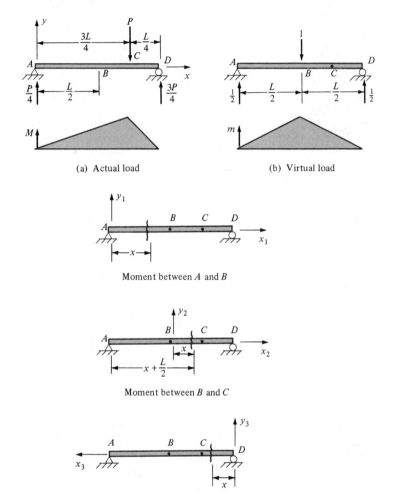

(a) Actual load
(b) Virtual load

Moment between A and B

Moment between B and C

Moment between C and D

(c)

FIGURE 10.17

Solution:
The actual load moment diagram and the virtual load moment diagram corresponding to a unit load at midspan are shown in Fig. 10.17. The integration along the beam must correspond to continuous moment intervals. The interval from A to B is

continuous since the unit load moment diagram changes abruptly at B. The interval from B to C is continuous since the actual moment diagram changes abruptly at C. The interval from C to D is continuous. The integration of Eq. (10.17) must be accomplished using three separate integrals. In this example the origin for the moment equations will be chosen such that the lower limit of each integral will be zero. This approach often lessens the algebraic work. Three coordinate locations are shown in Fig. 10.17c. The equations for M and m of Eq. (10.17) must correspond to the same coordinate locations for each beam interval. The moment equations for each interval of Fig. 10.17c are written using the standard procedure.

$$M_{AB} = \frac{Px}{4} \qquad M_{BC} = \frac{P(x + L/2)}{4} \qquad M_{DC} = \frac{3Px}{4}$$

$$m_{AB} = \frac{x}{2} \qquad m_{BC} = \frac{L}{4} - \frac{x}{2} \qquad m_{DC} = \frac{x}{2}$$

$$EIv_B = \int_0^{L/2} \frac{Px^2\, dx}{8} + \int_0^{L/4} \frac{P(-x^2 + L^2/4)\, dx}{8} + \int_0^{L/4} \frac{3Px^2\, dx}{8}$$

$$v_B = \frac{11PL^3}{768EI}$$

The coordinate origin can be located at any point that is convenient and expedites the use of the unit load method. The rules for choosing free bodies and writing moment equations as given in Chapter 5 should be followed.   ∎

Virtual work methods are especially useful for problems that combine axial deformation and flexural deflection. As an example, consider the following problem.

EXAMPLE 10.12

The uniformly loaded beam of Fig. 10.18 is hinged at the left end and supported by a cable that passes over a frictionless pulley at the right end. Compute the deflection at midspan and at the cable-supported end. Assume the following parameters: $E = 30(10^3)$ ksi for the beam and cable, I for the beam $= 60$ in.^{4}, A for the cable $= 0.75$ in.^{2}, L for the beam $= 8$ ft, $w = 2.5$ k/ft, and the length of the cable is given in Fig. 10.18.

Solution:
The reactions are computed and shown in Fig. 10.18a to be 10 k at each end. The force in the cable is also 10 k. A unit force is applied at point B, the center of the beam, as shown in Fig. 10.18b. The internal work is obtained by integrating from A to B, then C to B, and summing along the axially loaded cable from C to D and D to E.

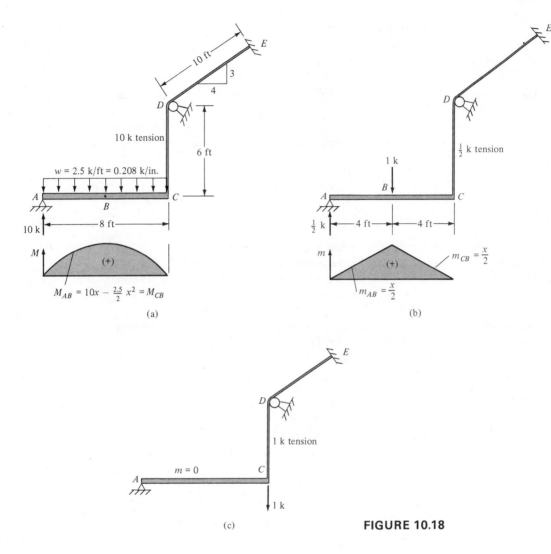

(a)

(b)

(c)

FIGURE 10.18

$$1 \text{ k } v_B \text{ in.} = 2 \int_0^{48} \frac{(10x - 0.208x/2)(x/2)}{(60)(30)(10^3)} \, dx$$

$$+ \left(\frac{1}{2}\right) \frac{(10)(6)(12)}{(0.75)(30)(10^3)} + \left(\frac{1}{2}\right) \frac{(10)(10)(12)}{(0.75)(30)(10^3)}$$

$$v_B = 0.128 + 0.016 + 0.027 = 0.171 \text{ in.}$$

A unit load is placed at C as shown in Fig. 10.18c to compute the deflection at C. Note that $m = 0$ and the only contribution to the internal energy is due to the cable.

$$v_C = \frac{(1)(10)(6)(12)}{(0.75)(30)(10^3)} + \frac{(1)(10)(10)(12)}{(0.75)(30)(10^3)}$$

$$= 0.032 + 0.053 = 0.085 \text{ in.}$$

The deflection at point C is due to the elongation of the cable. ∎

An additional example should illustrate the application of the method of virtual work for a beam with a variable second moment of the area.

EXAMPLE 10.13

Compute the deflection at the free end of the beam of Example 10.6, shown in Fig. 10.9. Divide the beam into four segments and assume the second moment of the area for each section to be equal to the average second moment of the area of the section.

Solution:
The beam is shown in Fig. 10.19. The average second moment of the area in the four sections of the beam is shown in Fig. 10.19b. The coordinate origin is taken at the free end. Due to the simplicity of the moment equations, the coordinate system will remain at the free end for evaluating the integrals.

$$M = -Px \qquad m = -x$$

$$v = \int_0^{L/4} \frac{16}{9} \frac{Px^2}{I_0 E} \, dx + \int_{L/4}^{L/2} \frac{16}{11} \frac{Px^2}{I_0 E} \, dx + \int_{L/2}^{3L/4} \frac{16}{13} \frac{Px^2}{I_0 E} \, dx + \int_{3L/4}^{L} \frac{15}{16} \frac{Px^2}{I_0 E} \, dx$$

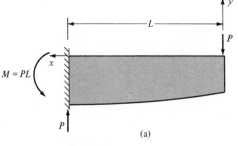

(a)

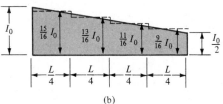

(b)

FIGURE 10.19

$$v = \frac{16PL}{3I_0E}\left|\frac{1}{(9)(64)} + \frac{1}{11}\left(\frac{1}{8} + \frac{1}{64}\right) + \frac{1}{13}\left(\frac{27}{64} - \frac{1}{8}\right) + \frac{1}{16}\left(1 - \frac{27}{64}\right)\right|$$

$$= \frac{0.377PL^3}{EI_0}$$

This result compares very well with the exact answer of

$$v = \frac{0.386PL^3}{EI_0}$$

given in Example 10.6.   ∎

The unit load method is applicable for computing the slope of a beam subject to transverse loading. In this case the unit load is a unit couple applied at the point where the slope is to be computed. The external virtual work is the unit couple multiplied with the actual slope.

$$(1) \cdot \theta = \int_0^L \frac{\mathbf{C} \cdot \mathbf{M}}{EI} \, dx \tag{10.18}$$

where $\mathbf{C}$ is the virtual internal moment caused by the application of the unit couple. $\mathbf{M}$ is the moment due to the actual load.

EXAMPLE 10.14

Compute the slope at the left end of the beam of Fig. 10.20a.

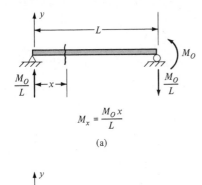

$$M_x = \frac{M_O x}{L}$$

(a)

$$C_x = 1 - \frac{x}{L}$$

(b)   **FIGURE 10.20**

Solution:

The origin is assumed at the left end. A unit couple is applied at the left end as shown in Fig. 10.20b. The moment equations are obtained in the usual manner.

$$M = \frac{M_0 x}{L} \qquad C = 1 - \frac{x}{L} = \frac{L - x}{L}$$

$$(1)(\theta) = \int_0^L \frac{M_0 x}{EIL} \frac{(L - x)}{L} \, dx = \frac{M_0}{EIL^2} \int_0^L (xL - x^2) \, dx$$

$$\theta = \frac{M_0 L}{6EI}$$

■

The rotation of a circular shaft subject to the action of an applied torque can be computed using virtual work. To derive the basic equation, imagine a shaft subject to a virtual torque δT that acts through the actual rotation, ϕ. The internal virtual work is the integral of the virtual stress times the actual strain taken over the volume.

$$\delta T \phi = \int_v \tau_{\text{virtual}} \gamma_{\text{actual}} \, dV \qquad \text{(i)}$$

$$\tau_{\text{virtual}} = \frac{tr}{J}$$

$$\gamma_{\text{actual}} = \frac{\tau}{G} = \frac{Tr}{JG}$$

where J is the polar second moment of the area and G is the shear modulus. Substituting into Eq. (i) and recognizing that $J = \int_A r^2 \, dA$ gives the proper form of the equation:

$$\delta T \cdot \phi = \int_A \int_0^L \frac{t \cdot T}{GJ^2} \, dx \, r^2 \, dA = \int_0^L \frac{t \cdot T}{JG} \, dx \qquad \text{(10.19)}$$

It can be verified that Eq. (10.19) gives the correct value for the rotation of a shaft subject to a torque T.

EXAMPLE 10.15

Compute the rotation at the free end of a shaft subject to a torque applied at the free end.

Solution:

The shaft is shown in Fig. 10.21. A unit torque is applied, which from statics gives $t = 1$. Substituting into Eq. (10.19) gives

$$\phi = \int_0^L 1 \cdot \frac{T \, dx}{JG} = \frac{TL}{JG}$$

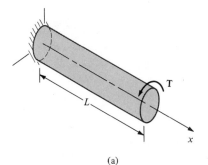

(a)

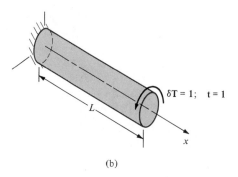

(b)

FIGURE 10.21
(a) Actual torque; (b) virtual
torque.

The illustration is almost trivial. Virtual work can be used effectively for problems involving variable applied torques. However, the primary purpose for introducing the subject will be demonstrated in the next few paragraphs.

The subject of shear deformation due to transverse loading will not be discussed at this time. It will suffice to say that shear deformations are extremely small compared to flexural deformations. The exception to the rule occurs for very short, deep beams where the length and depth of a beam approach being equal.

The next example will illustrate problems where deformations due to axial force, flexural loads, and torsional moments can be combined. Vector analysis will be used to organize the computations.

EXAMPLE 10.16

Compute the deflection at point C for the beam of Fig. 10.22. The beam is fixed at A and carries a uniform load along the length BC. Assume E, G, I, and J are constant along the entire beam.

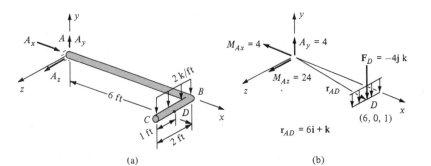

(a)

(b)

$\mathbf{r}_{AD} = 6\mathbf{i} + \mathbf{k}$

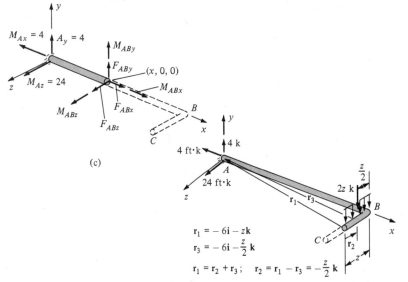

(c)

$\mathbf{r}_1 = -6\mathbf{i} - z\mathbf{k}$

$\mathbf{r}_3 = -6\mathbf{i} - \dfrac{z}{2}\mathbf{k}$

$\mathbf{r}_1 = \mathbf{r}_2 + \mathbf{r}_3; \quad \mathbf{r}_2 = \mathbf{r}_1 - \mathbf{r}_3 = -\dfrac{z}{2}\mathbf{k}$

(d)

(e)

(f)

FIGURE 10.22

Solution:

Assume positive reaction forces at A, given by (A_x, A_y, A_z).

$$\Sigma\mathbf{F} = 0 \qquad A_x\mathbf{i} + A_y\mathbf{j} + A_z\mathbf{k} - (2\text{ k/ft})(2\text{ ft})\mathbf{j} = 0$$

$$A_x = 0 \qquad A_y = 4\text{ k} \qquad A_z = 0$$

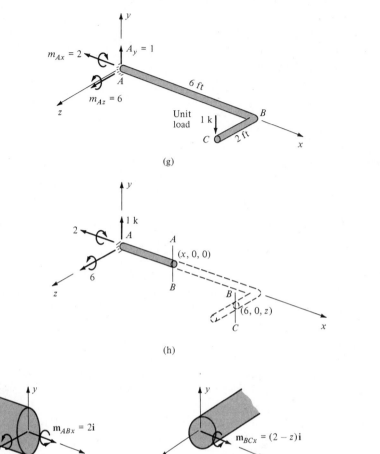

(g)

(h)

Section AB Section BC

(i)

FIGURE 10.22
(continued)

Assume positive reaction moments at A, given by (M_{Ax}, M_{Ay}, M_{Az}).

$$\Sigma \mathbf{M}_A = 0 \qquad M_{Ax}\mathbf{i} + M_{Ay}\mathbf{j} + M_{Az}\mathbf{k} + (6\mathbf{i} + \mathbf{k}) \times (-4\mathbf{j}) = 0$$

$$M_{Ax} = -4 \text{ ft} \cdot \text{k} \qquad M_{Ay} = 0 \qquad M_{Az} = 24 \text{ ft} \cdot \text{k}$$

Reactions are shown in Fig. 10.22b.

Moment equations will be referenced to an origin at A even though in some cases there could be a more convenient location. Cut a section between A and B, and the free body is as shown in Fig. 10.22c. Assume positive forces and moments at the cut section and write equilibrium equations to obtain moment and force equations valid for the interval A-B.

$$\Sigma \mathbf{F}_{AB} = 0 \qquad F_{ABx}\mathbf{i} + F_{ABy}\mathbf{j} + F_{ABz}\mathbf{k} + 4\mathbf{j} = 0$$

$$F_{ABx} = 0 \qquad F_{ABy}\mathbf{j} = -4\mathbf{j}\,\mathbf{k} \qquad F_{ABz} = 0$$

$$\Sigma \mathbf{M}_{AB} = 0 \qquad M_{ABx}\mathbf{i} + M_{ABy}\mathbf{j} + M_{ABz}\mathbf{k} + (x\mathbf{i}) \times (4\mathbf{j}) - 4\mathbf{i} + 24\mathbf{k} = 0$$

$$M_{ABx}\mathbf{i} = 4\mathbf{i} \qquad M_{ABy} = 0 \qquad M_{ABz}\mathbf{k} = (-24 + 4x)\mathbf{k}$$

A free body cut between B and C is shown in Fig. 10.22d. The reactions at A and the distributed load between B and the free-body cut must be included in the moment equation. As shown in Fig. 10.22d, the loading is distributed over a variable length z and its centroid is located at $-z/2$ from the cut. Assume a positive force system acting at the cut section; it follows that

$$F_{BCx}\mathbf{i} + F_{BCy}\mathbf{j} + F_{BCz}\mathbf{k} + 4\mathbf{j} - 2z\mathbf{j} = 0$$

$$F_{BCx} = 0 \qquad F_{BCy}\mathbf{j} = (2z - 4)\mathbf{j} \qquad F_{BCz} = 0$$

$$M_{BCx}\mathbf{i} + M_{BCy}\mathbf{j} + M_{BCz}\mathbf{k} + (-6\mathbf{i} - z\mathbf{k}) \times (4\mathbf{j}) + (-z/2\mathbf{k}) \times (-2z\mathbf{j}) - 4\mathbf{i} + 24\mathbf{j} = 0$$

$$M_{BCx}\mathbf{i} = (z^2 - 4z + 4)\mathbf{i} \qquad M_{BCy}\mathbf{j} = (24 - 24)\mathbf{j} = 0 \qquad M_{BCz} = 0$$

The vector results for the two free bodies are illustrated in Figs. 10.22e and f. Both of the force vectors produce shear and, as explained previously, shear deformations will be neglected. The moment vectors of Fig. 10.22e represent a bending moment and a torque. The moment vector of Fig. 10.22f is a bending moment about the x axis. These moment vectors represent the internal effects due to the actual loading.

The vertical deflection at C is computed by placing a unit load at C. By inspection the deflection will be downward; hence, the unit load will be $-1\mathbf{j}$ in kip units. The structure is shown in Fig. 10.22g. Again, reactions are computed by assuming positive forces and moments at A and applying the equilibrium equations. The results will be virtual internal force and moment. In keeping with the previous notation, force will be denoted as p and moment as m. The reactions at A will be symbolized with an A.

$$A_x\mathbf{i} + A_y\mathbf{j} + A_z\mathbf{k} - (1)\mathbf{j} = 0$$

$$A_x = 0 \qquad A_y = 1 \qquad A_z = 0$$

$$m_{Ax}\mathbf{i} + m_{Ay}\mathbf{j} + m_{Az}\mathbf{k} + (6\mathbf{i} + 2\mathbf{k}) \times (-\mathbf{j}) = 0$$

$$m_{Ax}\mathbf{i} = -2\mathbf{i} \qquad m_{Ay} = 0 \qquad m_{Az}\mathbf{k} = 6\mathbf{k}$$

The reactions are shown in Fig. 10.22g.

The free body of Fig. 10.22h is used to compute the virtual shear and moment equations along AB.

$$p_{ABx} = 0 \qquad p_{ABy}\mathbf{j} = -(1)\mathbf{j} \qquad p_{ABz} = 0$$

$$m_{ABx}\mathbf{i} + m_{ABy}\mathbf{j} + m_{ABz}\mathbf{k} + (-x\mathbf{i}) \times (1\mathbf{j}) - 2\mathbf{i} + 6\mathbf{k} = 0$$

$$m_{ABx}\mathbf{i} = 2\mathbf{i} \qquad m_{ABy} = 0 \qquad m_{ABz}\mathbf{k} = (x - 6)\mathbf{k}$$

The free-body cut BC is also shown in Fig. 10.22h.

$$p_{BCx} = 0 \qquad p_{BCy}\mathbf{j} = -1\mathbf{j} \qquad p_{BCz} = 0$$

$$m_{BCx}\mathbf{i} + m_{BCy}\mathbf{j} + m_{BCz}\mathbf{k} + (-6\mathbf{i} - z\mathbf{k}) \times (1\mathbf{j}) - 2\mathbf{i} + 6\mathbf{k} = 0$$

$$m_{BCx}\mathbf{i} = (2 - z)\mathbf{i} \qquad m_{BCy} = 0 \qquad m_{BCz} = 0$$

Again, p_{BCy} represents a shear and will be neglected.

The moment vectors acting on the cut sections are shown in Fig. 10.22*i*. The statement of virtual work given by Eq. (10.17) for deformation due to bending moment is combined with Eq. (10.19) for deformation due to torsional moment. Along the interval AB the actual moments represent the deformation of the section when divided by the appropriate constants—EI for bending and JG for torque. Terms multiplied by $\mathbf{i}$ are torques and terms multiplied by $\mathbf{k}$ represent bending moments. Along section BC, moments multiplied by $\mathbf{k}$ represent torques and terms multiplied by $\mathbf{i}$ represent bending moments. The virtual work equation can be written in a general form:

$$(1) \cdot v_c = \int_A^B \frac{\mathbf{m}_{AB} \cdot \mathbf{M}_{AB}\, dx}{\text{(beam constants)}} + \int_B^C \frac{\mathbf{m}_{BC} \cdot \mathbf{M}_{BC}\, dz}{\text{(beam constants)}}$$

$$v_c = \int_0^6 [2\mathbf{i} - (x - 6)\mathbf{k}] \cdot \left[\frac{4}{JG}\mathbf{i} + \frac{(-24 + 4x)\mathbf{k}}{EI} \right] dx$$

$$+ \int_0^2 [(2 - z)\mathbf{i}] \cdot \left[\frac{(z^2 - 4z + 4)\mathbf{i}}{EI} \right] dz$$

The limits of integration correspond to an origin at point A. Since AB lies along the x axis, the limits are 0 and 6; and since CD lies along the z axis and emanates from the x axis the limits are 0 and 2. The scalar product indicated in the equation above gives the following.

$$v_c = \int_0^6 \left[\frac{8}{JG} + \frac{(-4x^2 + 48x - 144)}{EI} \right] dx + \int_0^2 \frac{(-z^3 + 6z^2 - 12z + 8)}{EI}\, dx$$

$$= \left[\frac{8x}{JG} + \frac{1}{EI} \left(\frac{-4x^3}{3} + \frac{48x^2}{2} - 144x \right) \right]_0^6 + \frac{1}{EI} \left[\frac{-z^4}{4} + \frac{6z^3}{3} - \frac{12z^2}{2} + 8z \right]_0^2$$

$$= \left(\frac{48}{JG} + \frac{288}{EI} + \frac{4}{EI} \right)(12)^3 \text{ in.}$$

The solution v_c is in three terms. It is possible to identify the separate parts of the solution in terms of their contribution to the total deflection. Consider the beam again as illustrated in Fig. 10.23. The method of superposition can be used to compute the total reaction. The beam can be viewed as in Fig. 10.23*b* as lying in the x-y plane. The deflection component v_{AB} is computed as

$$v_{AB} = \frac{PL^3}{3EI} = \frac{(4)(6^3)}{3EI} = \frac{288}{EI}$$

and is the second term of the virtual work solution. Viewing the beam along the x axis as shown in Fig. 10.23*c* illustrates the origin of the third term. The 2-ft section

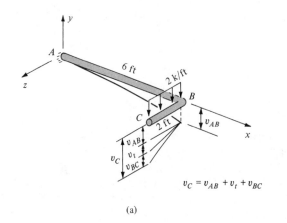

$$v_C = v_{AB} + v_t + v_{BC}$$

(a)

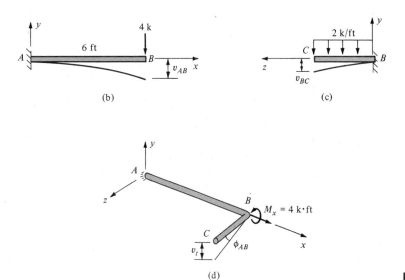

(b)

(c)

(d)

FIGURE 10.23

behaves as a uniformly loaded fixed-free beam.

$$v_{BC} = \frac{wL^4}{8EI} = \frac{(2 \text{ k/ft})(2 \text{ ft})^4}{8EI} = \frac{4}{EI}$$

The torque acting on member AB causes a downward rotation of point C. The angle of rotation, shown in Fig. 10.23d, is computed using the formula of the previous example.

$$\phi_{AB} = \frac{TL}{JG} = \frac{(4 \text{ k} \cdot \text{ft})(6 \text{ ft})}{JG} = \frac{24}{JG}$$

The deflection component v_t is the arc length computed as ϕ_{AB} times the length of BC.

$$v_t = \frac{(2)(24)}{JG} = \frac{48}{JG}$$

$$v_c = v_t + v_{AB} + v_{BC} = \left(\frac{48}{JG} + \frac{288}{EI} + \frac{4}{EI}\right)(12)^3 \text{ in.}$$

Note that the dimensional units in this example have not been converted to inches. The final answer is multiplied by $(12 \text{ in./ft})^3$ to indicate a result in inches. ■

The solution for this problem is of course much shorter using superposition. However, superposition is not always a viable method of analysis since it is dependent upon visualizing the component deflections that add together to produce the final deflection. The example was intended to illustrate the basic method of virtual work and the organizational power of vector analysis.

10.9   MOMENT DIAGRAMS BY PARTS

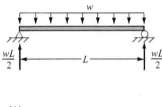

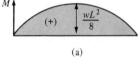

(a)

The construction of moment diagrams by parts represents an intermediate step toward efficient use of moment area theorems. The concept of superposition provides the basis for justifying moment diagrams by parts. Consider the uniformly loaded beam of Fig. 10.24a. The beam reactions can be computed as $wL/2$ and the moment at midspan of the beam is $M_{\textbf{\textcentoldstyle}} = wL^2/8$. A moment diagram by parts drawn with respect to end B appears in Fig. 10.24b. The beam is imagined to be fixed at point B. The simple beam reactions are considered as forces acting on the beam. A moment diagram is drawn for each individual load. The uniform load moment diagram is shown negative below the x axis. The moment diagram due to the left reaction is positive. If the diagrams are superposed, the result will be as shown in Fig. 10.24a. For instance, computing the moment at midspan for the diagrams of Fig. 10.24b gives

$$M = \frac{wL^2}{4} - \frac{wL^2}{8} = \frac{wL^2}{8}$$

the same as in Fig. 10.24a.

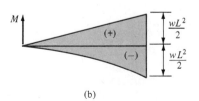

(b)

FIGURE 10.24
Moment diagram by parts.

EXAMPLE 10.17

Construct the moment diagram for the beam of Fig. 10.25 with respect to point B.

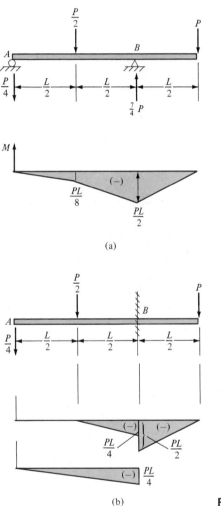

(a)

(b)

FIGURE 10.25

Solution:

Reactions are computed as shown in Fig. 10.25a. The structure, fixed at point B, is shown in Fig. 10.25b. The moment diagram is composed of three separate parts—one for each concentrated force. Note that the moment is balanced with respect to point B. Summing the moments on the left gives $-PL/2$ and is equal to the moment to the right of B. ∎

EXAMPLE 10.18

Construct the moment diagram for the beam of Fig. 10.26a with respect to the mid-point, point C.

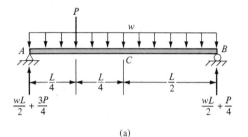

(a)

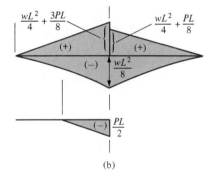

(b)

FIGURE 10.26

Solution:
Again, reactions are computed as illustrated in Fig. 10.26a. The moment diagram is shown in Fig. 10.26b. There are five components for the diagram—one for each separate load component. ∎

10.10 MOMENT AREA THEOREMS

Moment area theorems are included in the text because the topic is classical. The basic method appears in many structural analysis and machine design textbooks. There is also some insight to be gained in

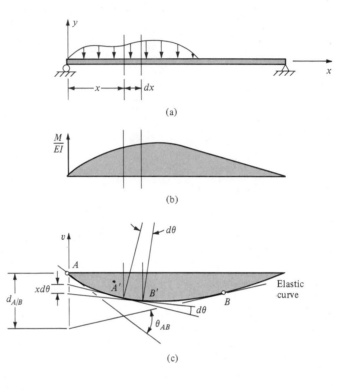

FIGURE 10.27

the relation between moment, slope, and deflection. The theory can be developed using Eq. (10.11), where dS is replaced with dx. Recall that, for small deflection gradients, dS is replaced with dx because the change in length of the beam is negligible.

$$d\theta = \frac{M\,dx}{EI} \tag{10.20}$$

An interpretation of the equation is given in Fig. 10.27. The vertical lines defining the elemental length dx are extended to intersect the elastic curve, points A' and B' on Fig. 10.27c. Tangent lines to the elastic curve at the intersection, points A' and B', define the angle $d\theta$. If Eq. (10.20) is integrated over some finite length of beam, the result would be the change in slope over that length. If A and B represent two points on the beam, the result would be as follows:

$$\int_{\theta_A}^{\theta_B} d\theta = \int_{x_A}^{x_B} \frac{M}{EI}\,dx$$

$$\theta_B - \theta_A = \theta_{AB} = \int_{x_A}^{x_B} \frac{M}{EI}\,dx \tag{10.21}$$

Consider the following example.

EXAMPLE 10.18

Compute the slope at the free end of a uniformly loaded fixed-free beam of length L.

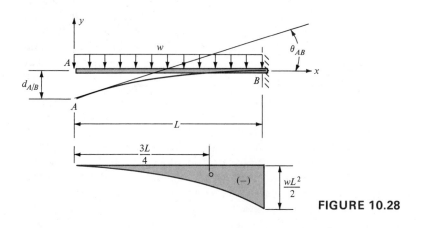

FIGURE 10.28

Solution:

The beam is shown in Fig. 10.28. A line is drawn tangent to the elastic curve at both points A and B. The angle included between the two tangent lines is θ_{AB}, as given by Eq. (10.21). It turns out that the change in slope, given by the area of the M/EI diagram, is the slope at the free end of the beam. This result occurs only because the slope at the fixed end is zero.

$$\theta_{AB} = \theta_B - \theta_A = 0 - \theta_A = \text{Area } \frac{M}{EI} \text{ diagram}$$

$$= \frac{1}{EI}\left(\frac{-wL^2}{2}\right)(L)\left(\frac{1}{3}\right) = \frac{-wL^3}{6EI}$$

$$\theta_B = 0 \qquad \theta_A = \frac{wL^3}{6EI}$$ ■

A sign convention for the moment area method is at best somewhat tedious. The coordinate system of Fig. 10.27 is considered positive. Integration from A to B is in the positive direction. Hence, if point A, the lower limit of integration, is always on the negative side of point B and the sign of the moment is included in the computation, the proper sign for the slope will be obtained.

The second moment area theorem is somewhat more difficult to visualize, and the reader should pay close attention to the following discussion. The deflection is not computed directly; rather, the deviation of a tangent line from the elastic curve is computed. Consider again Fig. 10.27. The tangent lines at A' and B' are extended to the left until they intersect the vertical y axis. The arc length, $x\,d\theta$ shown in the figure, is the vertical distance between the two tangent lines. The integration of $x\,d\theta$ between A and B would be the vertical distance shown in Fig. 10.27 labeled $d_{A/B}$. The second moment area theorem yields the tangential deviation between a point A on the elastic curve and a tangent line drawn with respect to any other point B on the elastic curve. Note the sign convention that was described previously.

$$\int_{0}^{-d_{A/B}} d(d_{A'/B'}) = -d_{A/B} = \int_{x_A}^{x_B} x\,d\theta = \int_{x_A}^{x_B} \frac{xM}{EI}\,dx \qquad (10.22)$$

Equation (10.22) is interpreted as the moment of the M/EI diagram between points A and B with respect to the point A. The negative sign in the limits can be explained by referring to Fig. 10.27c. For the derivation a negative downward load was assumed. Hence, the deflection is downward, but the bending moment and curvature are positive. Positive curvature will create a situation where the deviation from the elastic curve to a tangent drawn at any point on the elastic curve is always downward. The opposite is true for negative curvature. Hence, the sign convention is to multiply by a minus one when the curvature is positive, but leave negative curvature unmodified. Remember to include the proper sign for the M/EI diagrams.

EXAMPLE 10.19

Compute the deflection at the free end for the beam of Fig. 10.28.

Solution:
According to the second moment area theorem, Eq. (10.22), the deviation of point A on the elastic curve with respect to the tangent at B would give the deflection at A. Note, however, that this result is true only because the tangent at B lies along the original undeformed axis. Summing moments about point A gives the deviation at point A. The curvature is negative; hence, the minus sign on the left of Eq. (10.22) is neglected. The proper sign is obtained by simply carrying the sign of the M/EI diagram through the computations.

$$d_{A/B} = v_A = \frac{-wL^2}{2EI}\left(\frac{L}{3}\right)\left(\frac{3L}{4}\right) = \frac{-wL^4}{8EI} \qquad \blacksquare$$

Fixed-free beams are easily analyzed using moment area theorems. Consider the following more general problem.

EXAMPLE 10.20

The simply supported beam of Fig. 10.29 carries a single concentrated force. Compute the deflection at the point of application of the force, the center of the beam, and the maximum deflection.

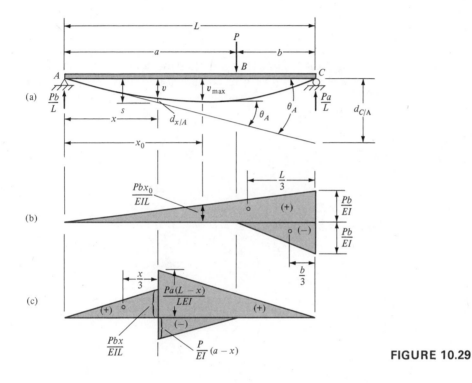

FIGURE 10.29

Solution:

For this problem a general solution for v, with $(0 \leq x \leq a)$, will be derived. The deflections can be obtained by merely substituting into the general equation. The procedure for computing v will be outlined and referenced to Fig. 10.29. Deflections cannot be computed directly for this problem as in the previous example. Only the deviation of a point on the elastic curve with respect to a tangent drawn at a second point on the elastic curve can be computed.

1. Construct a tangent at A and compute $d_{C/A}$ by summing the moment of the M/EI diagram of Fig. 10.29b with respect to C.

2. Use similar triangles to evaluate s of Fig. 10.29a.

3. Compute $d_{x/A}$ by summing moments of M/EI using Fig. 10.29c.

4. The final result is $v = s - d_{x/A}$.

Step 1: The minus sign of Eq. (10.22) is transposed to the right.

$$d_{C/A} = \frac{-(Pb)(L/2)(L/3)}{EI} + \frac{Pb(b/2)(b/3)}{EI} = \frac{-Pb(L^2 - b^2)}{6EI}$$

Step 2:

$$s = d_{C/A}\left(\frac{x}{L}\right) = \frac{-Pbx(L^2 - b^2)}{6EIL}$$

Step 3: It is necessary to understand that the M/EI diagram for computing the deviation is only that part between the point where deviation is to be computed and the point where the tangent is constructed. The portion of the M/EI diagram to the right of length x in Fig. 10.29c is not to be used. It is shown merely to make the diagram complete.

$$d_{x/A} = \frac{-Pb}{EI}\left(\frac{x}{L}\right)\left(\frac{x}{2}\right)\left(\frac{x}{3}\right) = \frac{-Pbx^3}{6EI}$$

Step 4:

$$v = s - d_{x/A} = \frac{-Pbx(L^2 - b^2)}{6EIL} + \frac{Pbx^3}{6EIL} = \frac{-Pbx(L^2 - b^2 - x^2)}{6EIL}$$

At the point of application of the load ($x = a$),

$$v(x = a) = \frac{-Pab(L^2 - b^2 - a^2)}{6EIL} = \frac{-Pa^2b^2}{3EIL}$$

At the center ($x = L/2$),

$$v(x = L/2) = \frac{-Pb(3L^2 - 4b^2)}{48EI}$$

The maximum deflection occurs at the point where the slope is zero or where the tangent to the elastic curve is horizontal. The location of the point of maximum slope is x_0 in Fig. 10.29a. The first moment area theorem can be used to find the change in slope as the slope changes from θ_A to 0. The area of the M/EI diagram of Fig. 10.29b would be only that part between $x = 0$ and $x = x_0$.

$$0 - \theta_A = \frac{Pbx_0}{EIL}(12x_0)$$

The angle θ_A can also be written in terms of $d_{C/A}$ as

$$\theta_A = \frac{d_{C/A}}{L} = \frac{-Pb(L^2 - b^2)}{6EIL}$$

Equating the previous two equations permits the solution for x_0 as follows:

$$\frac{Pbx_0^2}{2EIL} = \frac{Pb(L^2 - b^2)}{6EIL}$$

$$x_0 = [(L^2 - b^2)/3]^{1/2}$$

which, if substituted into the equation for v gives the maximum deflection.

$$v_{max} = \frac{-Pb}{3EIL}\left(\frac{L^2 - b^2}{3}\right)^{3/2}$$ ∎

10.11 SINGULARITY FUNCTIONS FOR BEAM DEFLECTIONS

Singularity functions are similar to the unit step function that is used extensively in electrical circuit theory and they are used in conjunction with the integration method. Singularity functions offer convenience and elegance when writing moment equations. As a mathematical tool they tend to aid in organizing the mathematical manipulations. On the other hand, singularity functions, when used without first understanding the use of the continuity conditions of the integration method, can obscure the physical understanding of the process of analysis.

The singularity function will be defined as

$$y(x) = \langle x - a \rangle^n \tag{10.23}$$

There are several rules to be followed when using the function defined by Eq. (10.23)

1. x is the coordinate measured from the origin;

2. a is some value of x; and

3. n is an integer, $n \geq 0$.

When $x < a$, the quantity in angle brackets vanishes. When $x \geq a$, the function becomes a regular mathematical function $(x - a)^n$. The function of Eq. (10.23) can be integrated or differentiated.

$$\int \langle x - a \rangle^n \, dx = \frac{\langle x - a \rangle^{n+1}}{n+1} + c \qquad (n \geq 0) \tag{10.24}$$

$$\frac{d}{dx}\langle x - a \rangle^n = n\langle x - a \rangle^{n-1} \qquad (n \geq 1) \tag{10.25}$$

The singularity function is used to write a moment equation for the beam when the integration method is being used. Consider the following elementary example.

EXAMPLE 10.21

Write the equation of the elastic curve for the simply supported beam of Fig. 10.30.

Solution:

$$M = \frac{P}{2}\langle x - 0\rangle^1 - P\left\langle x - \frac{L}{2}\right\rangle^1 \qquad (10.26)$$

For $x < L/2$ the second term vanishes. The value of n is 1 since the moment is a linear function of x. For $x \geq L/2$ the moment equation is given by the sum of both terms.

$$M = \frac{Px}{2} - P\left(x - \frac{L}{2}\right); \qquad \frac{L}{2} \leq 0 \leq L$$

The differential equation relating curvature and moment is written using Eq. (10.26).

$$EIv'' = \frac{P}{2}\langle x\rangle^1 - P\left\langle x - \frac{L}{2}\right\rangle^1 \qquad \textbf{(a)}$$

$$EIv' = \frac{P}{4}x^2 - \frac{P}{2}\left\langle x - \frac{L}{2}\right\rangle^2 + C_1 \qquad \textbf{(b)}$$

$$EIv = \frac{Px^3}{12} - \frac{P}{6}\left\langle x - \frac{L}{2}\right\rangle^3 + C_1 x + C_2 \qquad \textbf{(c)}$$

The boundary condition is $v = 0$ when $x = 0$. Note, for $x = 0$, the second term on the right-hand side of Eq. (c) would vanish because $L/2 > 0$.

Substituting $v = 0$, $x = 0$, gives $C_2 = 0$. The second boundary condition is $v = 0$, $x = L$. For $x = L$ the second term becomes a regular mathematical function.

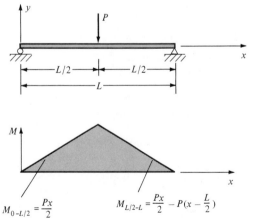

FIGURE 10.30

$$0 = \frac{PL^3}{12} - \frac{P}{6}\left(L - \frac{L}{2}\right)^3 + C_1 L \qquad C_1 = \frac{-PL^2}{16}$$

$$EIv = \frac{Px^3}{12} - \frac{P}{6}\left\langle x - \frac{L}{2}\right\rangle^3 - \frac{PL^2}{16}x \qquad \textbf{(d)}$$

Equation (d) can be compared to Case E1 of Appendix E and even though the deflection equations appear to be different, they give the same value of v for corresponding values of x. ∎

A discontinuous uniform or uniformly varying load cannot be represented using a single singularity function. A singularity function, once written into the moment equation, can only be removed by subtracting the singularity function. This leads to a superposition type of formulation for the moment equation. The next two examples should illustrate the concept.

EXAMPLE 10.22

Write the equation of the elastic curve for the simply supported beam of Fig. 10.31.

Solution:
The three beam loadings of Figs. 10.31b–d add up to give the loading of Fig. 10.31a. Using Fig. 10.31b.

$$M = \frac{wL}{3}\langle x - 0\rangle^1 - \frac{w}{2}\langle x - 0\rangle^1\langle x - 0\rangle^1$$

or

$$M = \frac{wL}{3} - \frac{wx^2}{2} \qquad (\textit{Note: The exponents add to give } n = 2.)$$

Using Fig. 10.31b plus Fig. 10.32c,

$$M = \frac{wLx}{3} - \frac{wx^2}{2} + \frac{w}{2}\left\langle x - \frac{L}{3}\right\rangle^2$$

For Fig. 10.31b plus Fig. 10.31c plus Fig. 10.31d,

$$EIv'' = M = \frac{wLx}{3} - \frac{wx^2}{2} + \frac{w}{2}\left\langle x - \frac{L}{3}\right\rangle^2 - \frac{w}{2}\left\langle x - \frac{2L}{3}\right\rangle^2 \qquad \textbf{(a)}$$

Equation (a) is the complete moment equation using singularity functions.

$$EIv' = \frac{wLx^2}{6} - \frac{wx^3}{6} + \frac{w}{6}\left\langle x - \frac{L}{3}\right\rangle^3 - \frac{w}{6}\left\langle x - \frac{2L}{3}\right\rangle^3 + C_1$$

$$EIv = \frac{wLx^3}{18} - \frac{wx^4}{24} + \frac{w}{24}\left\langle x - \frac{L}{3}\right\rangle^4 - \frac{w}{24}\left\langle x - \frac{2L}{3}\right\rangle^4 + C_1 x + C_2 \qquad \textbf{(b)}$$

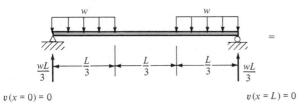

$v(x = 0) = 0$ $v(x = L) = 0$

(a)

$=$

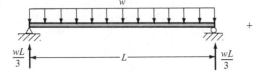

(b)

$+$

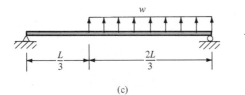

(c)

$+$

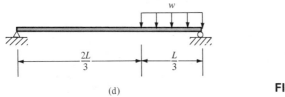

(d)

FIGURE 10.31

The boundary conditions are $v(x = 0)$ and $v(x = L) = 0$. Substituting $x = 0$, $v = 0$ gives $C_2 = 0$. Substituting $x = L$, $v = 0$ gives C_1.

$$0 = \frac{wL^4}{18} - \frac{wL^4}{24} + \frac{w}{24}\left(L - \frac{L}{3}\right)^4 - \frac{w}{24}\left(L - \frac{2L}{3}\right)^4 + C_1 L$$

$$C_1 = \frac{7wL^3}{324}$$

Substituting C_1 into Eq. (b) gives the equation of the elastic curve. ∎

EXAMPLE 10.23

Write the equation of the elastic curve for the beam of Fig. 10.32.

Solution:
The beam loadings of Figs. 10.32b–d add to give the loading of Fig. 10.32a. Using Fig. 10.32b,

$$M = \frac{wL}{6}\langle x - 0\rangle^1 - \frac{w}{2L}\langle x - 0\rangle^1\langle x - 0\rangle^1\frac{\langle x - 0\rangle^1}{3} + \frac{wL}{3}\langle x - L\rangle^1$$

$$M = \frac{wL}{6}x - \frac{wx^3}{6L} + \frac{wL}{3}\langle x - L\rangle^1 \tag{a}$$

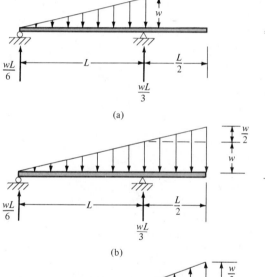

(a)

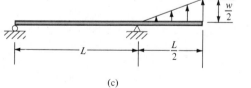

(b)

=

+

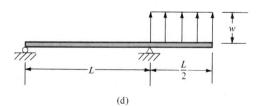

(c)

+

(d)

FIGURE 10.32

For Fig. 10.32c,

$$M = \frac{w}{2L}\langle x - L\rangle^1 \langle x - L\rangle^1 \frac{\langle x - L\rangle^1}{3} = \frac{w}{6L}\langle x - L\rangle^3 \qquad \textbf{(b)}$$

For Fig. 10.32d,

$$M = w\langle x - L\rangle^1 \frac{\langle x - L\rangle^1}{2} = \frac{w}{2}\langle x - L\rangle^2 \qquad \textbf{(c)}$$

The moment equation is the sum of Eqs. (a)–(c).

$$EIv'' = M = \frac{wLx}{6} - \frac{wx^3}{6L} + \frac{wL}{3}\langle x - L\rangle^1 + \frac{w}{6L}\langle x - L\rangle^3 + \frac{w}{2}\langle x - L\rangle^2$$

$$EIv' = \frac{wLx^2}{12} - \frac{wx^4}{24L} + \frac{wL}{6}\langle x - L\rangle^2 + \frac{w}{24L}\langle x - L\rangle^4 + \frac{w}{6}\langle x - L\rangle^3 + C_1$$

$$EIv = \frac{wLx^3}{36} - \frac{wx^5}{120L} + \frac{wL}{18}\langle x - L\rangle^3 + \frac{w}{120L}\langle x - L\rangle^5$$

$$+ \frac{w}{24}\langle x - L\rangle^4 + C_1 x + C_2 \qquad \textbf{(d)}$$

The boundary conditions are $v(x = 0) = 0$ and $v(x = L) = 0$. Substituting $v = 0$ when $x = 0$ gives $C_2 = 0$. Substituting $v = 0$ when $x = L$,

$$0 = \frac{wL^4}{36} - \frac{wL^4}{120} + C_1 L$$

$$C_1 = -\frac{7wL^3}{360}$$

Substituting C_1 and C_2 into Eq. (d) gives the equation of the elastic curve. ∎

10.12 SUMMARY

In this chapter the basic theory associated with the deflection of beams has been presented. Historically, the differential equation relating deflection to applied load or bending moment has been the basis for the theory of beam deflections. The idea of balancing the external and internal mechanical energy has usually found lesser application in mechanics of materials. The introduction of the unit load method in this chapter is intended to illustrate a general method for computing deflections of structures and to initiate the reader to the concept of virtual work.

The use of the differential equation is limited because of the mushrooming algebraic work involved for problems with complicated loading patterns.

The differential equations were derived in Section 10.2.

$$\frac{dv}{dx} = v' = \theta \tag{10.12}$$

$$\frac{d^2v}{dx^2} = v'' = \frac{M}{EI} \tag{10.9}$$

$$\frac{d}{dx}\left(EI\,\frac{d^2v}{dx^2}\right) = (EIv'')' = \frac{dM}{dx} = V \tag{10.13}$$

$$\frac{d^2}{dx^2}\left(EI\,\frac{d^2v}{dx^2}\right) = (EIv'')'' = \frac{dV}{dx} = w \tag{10.14}$$

Equations (10.9) and (10.14) were demonstrated to be the most useful.

Beams with discontinuous properties such as the loading or the second moment of the area were discussed, and continuity conditions on slope and deflection were illustrated.

The method of superposition was introduced and briefly discussed. The method is further illustrated in Chapter 11 for the analysis of statically indeterminate structures and, as presented in this chapter, involves breaking a structure with complicated loading down into a series of simpler basic problems and then superposing their solutions.

The concept of virtual work was cast in a usable form by introducing the unit load method. The deformation equation for axially loaded members is

$$1 \cdot \Delta = \sum_{i=1}^{N} \frac{\mathbf{p}_i \mathbf{P}_i L_i}{A_i E_i} \tag{10.16}$$

The deflection equation for beams is

$$1 \cdot \mathbf{v}_A = \int_0^L \frac{\mathbf{m} \cdot \mathbf{M}\,dx}{EI} \tag{10.17}$$

where $\mathbf{m}$, $\mathbf{M}$, and I are possible functions of x.

The equation for computing the torsional rotation of a bar subject to a twisting moment was shown to be

$$1 \cdot \phi = \int_0^L \frac{\mathbf{t} \cdot \mathbf{T}\,dx}{JG} \tag{10.19}$$

The deformation due to transverse shear was purposefully neglected.

The unit load method equations were combined to analyze a simple three-dimensional problem. The use of the scalar product that occurs in Eqs. (10.16), (10.17), and (10.19) was demonstrated.

The moment area theorems were derived and applications were illustrated. The first moment area theorem pertains to the computation of slope.

$$\theta_B - \theta_A = \int_{x_A}^{x_B} \frac{M}{EI}\, dx \tag{10.21}$$

The theorem may be stated as:

> The change in slope between two points A and B is equal to the area of the M/EI diagram between points A and B.

The second moment area theorem pertains to the computation of deflection; however, deflections are usually not computed directly.

$$d_{A/B} = -\int_{x_A}^{x_B} \frac{xM}{EI}\, dx \tag{10.22}$$

The theorem may be stated as:

> The deviation between a point A on the elastic curve and a tangent drawn to the elastic curve at point B is equal to the moment of the M/EI diagram between points A and B. The moment is with respect to the point A.

For Further Study

Anton, H., *Calculus*, Wiley, New York, 1983 (see p. 784 for the basic curvature relation).

Chadwick, P., *Continuum Mechanics*, George Allen and Unwin, London, 1976.

Laible, J.P., *Structural Analysis*, Holt, Rinehart and Winston, New York, 1985.

Oden, J. T., *Mechanics of Elastic Structures*, McGraw-Hill, New York, 1967.

Spiegel, M. R., *Vector Analysis*, Schaum, New York, 1959.

10.13 PROBLEMS

10.1 A steel wire 2 mm in diameter and 1 m in length has equal bending moments applied at each end. Compute the radius of curvature corresponding to a stress of 110 MPa. Assume $E = 200$ GPa. Compute the magnitude of the bending moments.

10.2 A standard-weight steel pipe 2 in. in diameter is subject to a constant bending moment. Compute the radius of curvature if the extreme fiber stress is 18,000 psi. $E = 30(10^6)$ psi.

10.3 A flat steel strip 3/8 in. wide is used as a hoop for a keg that has an outside diameter of 36 in. Compute the thickness of the hoop if the stress is not to exceed 20 ksi when the keg is empty.

10.4 Given the conditions of Problem 10.3, assume that the keg is filled with a material causing an internal pressure of $p = 0.5$ psi and hoops are positioned at 3 in. center to center. Compute the thickness of the hoop if the stress is not to exceed 20 ksi.

10.5–10.18 Use the differential equation method to determine the equation of the elastic curve for Figs. P10.5 to P10.18. Use the fourth-order equation for Problems 10.10 and 10.14. Assume EI to be constant. Use the coordinate system when indicated.

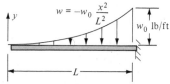

$$w = -w_0 \frac{x^2}{L^2}$$

FIGURE P10.10

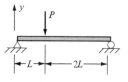

FIGURE P10.11

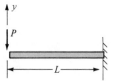

FIGURE P10.5

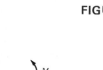

FIGURE P10.6

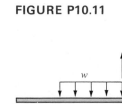

FIGURE P10.12

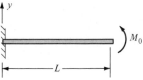

FIGURE P10.7

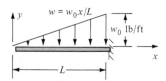

$$w = w_0 x/L$$

FIGURE P10.8

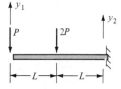

FIGURE P10.13

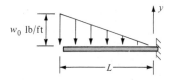

FIGURE P10.9

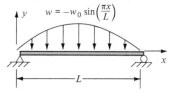

$$w = -w_0 \sin\left(\frac{\pi x}{L}\right)$$

FIGURE P10.14

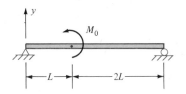

FIGURE P10.15

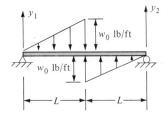

FIGURE P10.16

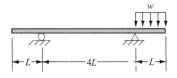

FIGURE P10.17

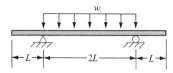

FIGURE P10.18

10.19 Repeat Problem 10.15 using a coordinate y_1 at the left end and y_2 at the right end.

10.20 Compute the equation of the elastic curve for the beam of Fig. P10.20. The left end undergoes a small deflection v_l.

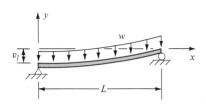

FIGURE P10.20

10.21 Compute the equation of the elastic curve for the beam of Fig. P10.21. The left end sustains a counter-clockwise rotation of $PL^2/20EI$.

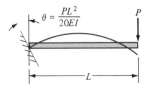

FIGURE P10.21

10.22 Compute the magnitude of the moment M_0 that will prevent rotation at the right end of the beam of Fig. P10.22. Note, $v'(x = L) = 0$.

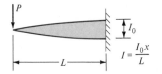

FIGURE P10.22

10.23 The second moment of the area of the beam of Fig. P10.23 varies linearly as $I = I_0 x/L$. Determine the equation of the elastic curve in terms of P, L, E, and I_0.

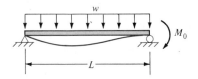

FIGURE P10.23

10.24 The beam of Fig. P10.24 is fabricated by joining two different materials. The 0.25-m section is steel, $E = 200$ GPa and the 0.35-m section is bronze, $E = 100$ GPa. The cross section is 50 mm by 50 mm. Determine the equation of the elastic curve and the magnitude of the deflection at the free end.

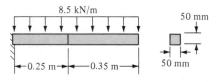

FIGURE P10.24

10.25 Determine the equation of the elastic curve for the beam of Fig. P10.25. Assume E is constant.

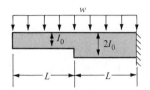

FIGURE P10.25

10.26 Determine the equation of the elastic curve for the beam of Fig. P10.26. Compute the deflection at the center of the span. Assume $E = 20$ GPa.

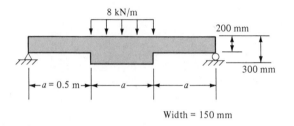

Width = 150 mm

FIGURE P10.26

10.27–10.31 Use superposition to compute the deflection at point A in Figs. P10.27 to P10.31.

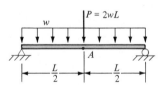

FIGURE P10.27

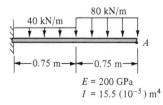

$E = 200$ GPa
$I = 15.5\,(10^{-5})\,\text{m}^4$

FIGURE P10.28

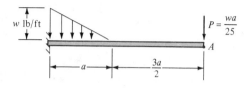

FIGURE P10.29

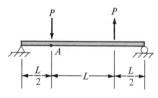

FIGURE P10.30

FIGURE P10.31

10.32 Compute the vertical and horizontal deflection of joint A for the truss of Fig. P10.32. Assume the area of all members to be $2.5(10^{-3})\,\text{m}^2$. $E = 200$ GPa.

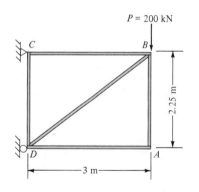

FIGURE P10.32

10.33 Compute the vertical deflection of joints D and E in terms of l, A, and E for the truss of Fig. P10.33. Assume that A and E are the same for all members.

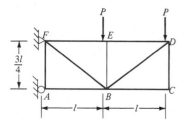

FIGURE P10.33

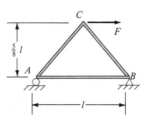

FIGURE P10.36

10.34 Compute the vertical deflection of joint B for the truss of Fig. P10.34. Assume AE is constant.

10.37 For the truss of Fig. P10.37 compute the vertical deflection of joint C and the horizontal deflection of joint D. Assume the area of all members is $4(10^{-3})$ m^2 and E is 70 GPa.

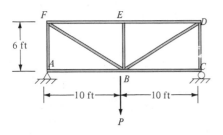

FIGURE P10.34

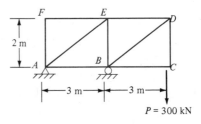

FIGURE P10.37

10.38 Use virtual work (the unit load method) to find the deflection at C and the slope at A for the beam of Fig. P10.38.

10.35 Compute the vertical deflection at joint B for the truss of Fig. P10.35. The area of the top chord and bottom chord members is 16 in.^{2}, twice the area of the interior truss members. Assume $E = 2(10^6)$ psi.

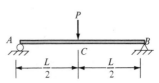

FIGURE P10.38

10.39 Compute the deflection at B and the slope at C using the unit load method for the beam of Fig. P10.39.

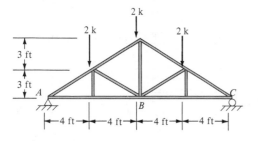

FIGURE P10.35

10.36 Compute the horizontal deflection of the joints A and C in terms of F, l, A, and E for the truss of Fig. P10.36. Assume AE is the same for all members.

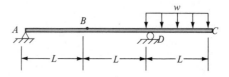

FIGURE P10.39

10.40 Use the unit load method to find the deflection and slope at the free end of the beam of Fig. P10.40.

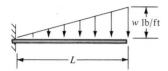

FIGURE P10.40

10.41 Use the unit load method to find the deflection and slope at the midpoint of the beam of Fig. P10.41.

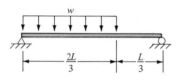

FIGURE P10.41

10.42 Compute the deflection at the center of the span and the slope at the right support using the unit load method for the beam of Fig. P10.42.

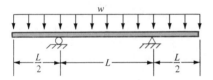

FIGURE P10.42

10.43 Use the unit load method to find the deflection at the free end and the slope at the center of the beam of Fig. P10.43.

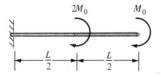

FIGURE P10.43

10.44 Compute the deflection and slope at the free end of the beam using the unit load method for the beam of Fig. P10.44.

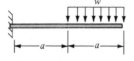

FIGURE P10.44

10.45 Use the unit load method to compute the deflection and slope at the free end of the beam of Fig. P10.45.

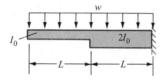

FIGURE P10.45

10.46 Use the unit load method to compute the deflection at the center of the span for the beam of Fig. P10.46. Assume E is constant for the structure.

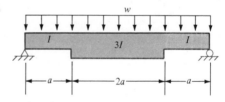

FIGURE P10.46

10.47 The second moment of the area for the beam of Fig. P10.47 varies from I_0 to $3I_0/2$ at the fixed end. Divide the beam into two sections and use an average value of the second moment of the area for each section to compute the deflection at the free end. Use the unit load method and assume E is constant.

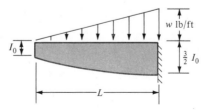

FIGURE P10.47

10.48 The structure of Fig. P10.48 is a solid aluminum bar with a cross section 25 mm by 75 mm. Compute the deflection normal to the beam at point A and the axial movement at point B. $E = 70$ GPa.

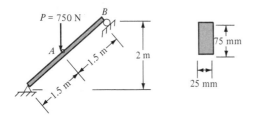

FIGURE P10.48

10.49 Use the unit load method to compute the deflection at midspan and at point B of Fig. P10.49. Assume $EI = 1.2(10^8)$ lb·in.² for the beam and $AE = 3.75(10^5)$ lb for the rod.

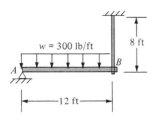

FIGURE P10.49

10.50 Use the unit load method to compute the deflection at point C of Fig. P10.50 in terms of P, L, A, E, and I.

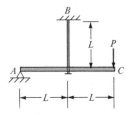

FIGURE P10.50

10.51 Compute the deflection at C and the elongation of the wire support for the beam structure of Fig. P10.51.

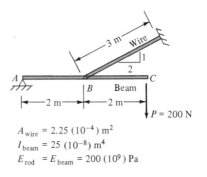

$A_{\text{wire}} = 2.25\,(10^{-4})\text{ m}^2$
$I_{\text{beam}} = 25\,(10^{-8})\text{ m}^4$
$E_{\text{rod}} = E_{\text{beam}} = 200\,(10^9)\text{ Pa}$

FIGURE P10.51

10.52 Use the unit load method to compute the angle of rotation at points A and B of the circular shaft of Fig. P10.52. Assume that JG is constant.

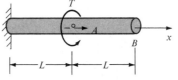

FIGURE P10.52

10.53 The circular shaft of Fig. P10.53 is subjected to a torque, T, as illustrated. Assume that I, J, E, and G are constant for the shaft and compute the deflection of the free end using the unit load method.

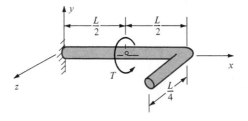

FIGURE P10.53

10.54 Compute the vertical deflection at point A for the frame of Fig. P10.54. Use the unit load method and assume that EI is constant for the entire structure.

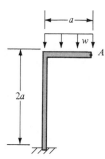

FIGURE P10.54

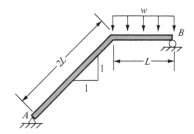

FIGURE P10.57

10.55 Use the unit load method to compute the horizontal deflection at point A for the frame of Fig. P10.55. Assume that EI is constant for the structure.

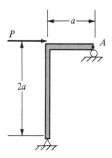

FIGURE P10.55

10.56 Use the unit load method to compute the vertical and horizontal deflection at point A for the beam of Fig. P10.56.

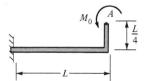

FIGURE P10.56

10.57 Use the unit load method to compute the horizontal movement at support B for the structure of Fig. P10.57. Assume that EI is constant for the structure.

10.58 Use the unit load method to compute the vertical deflection at point A and the horizontal deflection at D for the frame of Fig. P10.58.

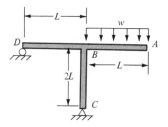

FIGURE P10.58

10.59 Compute the deflection at the free end in the direction of the 6-k force for the structure of Fig. P10.59. Assume that the area is 16 in.^{2}, $E = 30(10^6)$ psi, $G = 12(10^6)$ psi, and $J = 20$ in.^{4}. Neglect deformation due to shear.

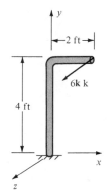

FIGURE P10.59

10.60 A moment of magnitude $M = (2\mathbf{i} + 4\mathbf{j} + 3\mathbf{k})\,\text{ft}\cdot\text{k}$ is applied at the free end of the beam of Fig. P10.60. Compute the deflection in each coordinate direction at point A. Assume E, I, G, and J are constant for the entire beam.

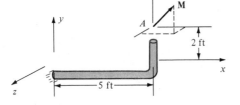

FIGURE P10.60

10.61 Compute the rotation in each coordinate direction for the beam of Fig. P10.60.

10.62 Compute the deflection in the z direction for the structure of Fig. P10.62 in terms of P, E, G, J, and I. Neglect deformation due to shear.

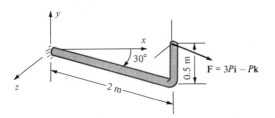

FIGURE P10.62

10.63–10.74 Construct moment diagrams by parts with respect to point A for Figs. P10.63–P10.74.

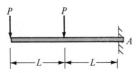

FIGURE P10.63

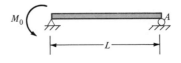

FIGURE P10.64

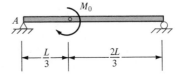

FIGURE P10.65

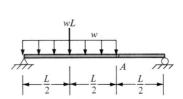

FIGURE P10.66

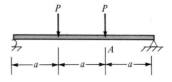

FIGURE P10.67

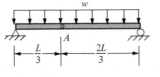

FIGURE P10.68

FIGURE P10.69

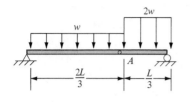

FIGURE P10.70

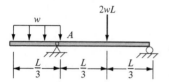

FIGURE P10.71

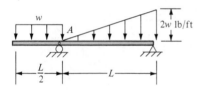

FIGURE P10.72

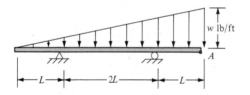

FIGURE P10.73

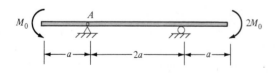

FIGURE P10.74

10.75 Use the moment area method to compute the slope at A of Fig. P10.75.

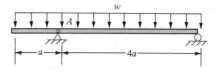

FIGURE P10.75

10.76 Use the moment area method to compute the deflection at A of Fig. P10.76.

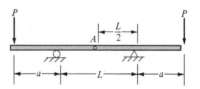

FIGURE P10.76

10.77 Use the moment area method to compute the slope and deflection at A of Fig. P10.77.

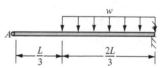

FIGURE P10.77

10.78 Use the moment area method to compute the deflection at A and the maximum deflection between the supports of Fig. P10.78.

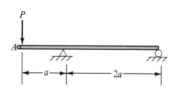

FIGURE P10.78

10.79 Use the moment area method to compute the deflection at the center of the span of Fig. P10.79. Assume the cross section of the beam is 4 in. by 12 in. and $E = 2(10^6)$ psi.

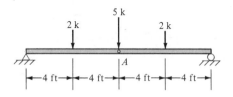

FIGURE P10.79

10.80 Use the moment area method to compute the slope and deflection at the free end of Fig. P10.80.

FIGURE P10.80

10.81 Use the moment area method to compute the deflection at A of Fig. P10.81.

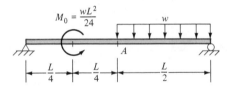

FIGURE P10.81

10.82 Use the moment area method to compute the deflection at A and maximum deflection between the supports of Fig. P10.82.

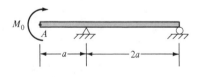

FIGURE P10.82

10.83 Use the moment area method to compute the slope and deflection at A of Fig. P10.83.

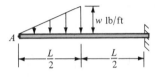

FIGURE P10.83

10.84 Use the moment area method to compute deflections at A and B and the maximum deflection between the supports of Fig. P10.84.

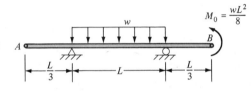

FIGURE P10.84

Use singularity functions to determine the equation of elastic curve for the following problems.

10.85 Chapter 10, Problem 10.13.

10.86 Chapter 10, Problem 10.12.

10.87 Chapter 10, Problem 10.16.

10.88 Chapter 10, Problem 10.15.

10.89 Chapter 10, Problem 10.18.

10.90 Chapter 10, Problem 10.28.

10.91 Chapter 10, Problem 10.30. Assume the origin at the right end of the beam.

10.92 Chapter 10, Problem 10.41.

10.93 Chapter 10, Problem 10.43.

CHAPTER 11

STATICALLY INDETERMINATE STRUCTURES

11.1 INTRODUCTION

A statically determinate structure can be analyzed for external reactions using the equilibrium equations and conditions that are discussed in statics. A structural system is termed statically indeterminate when there are more unknown external reactions than there are equilibrium equations and conditions of statics. The structure is statically indeterminate because of additional reaction constraints. These constraints are actually additional boundary conditions, which are used to formulate equations that combine with the equations of statics. The additional boundary conditions will be called deformation conditions. In any analysis the total number of unknown reactions is equal to the number of equations of statics plus the deformation conditions. The purpose of this chapter is to illustrate basic methods for interpreting the deformation conditions.

The discussion begins with axially loaded rods and torsional loads applied to circular rods because these problems are conceptually simpler than beam problems. There are numerous techniques for analyzing indeterminate beams. The discussion in this chapter is limited to the force method, the displacement method, and the integration method.

477

Thermal effects on structures were discussed briefly in Chapter 3. Thermal effects are more pronounced for indeterminate structures and a thorough discussion is presented illustrating different types of thermal effects.

The analysis of a statically indeterminate truss is a straightforward application of the force method. The truss is an example of a structure that can be indeterminate either externally or internally, or both. Finally, frames and three-dimensional structures are analyzed using the unit load method of the previous chapter in conjunction with the force method.

11.2   ELEMENTARY STATICALLY INDETERMINATE STRUCTURES

Structural members that are basically loaded axially or torsionally will be termed elementary. Axially loaded members are characterized by the axial deformation equation

$$u = \frac{PL}{AE} \qquad\qquad (11.1)$$

Structural members subject to torque deform or rotate according to the equation

$$\phi = \frac{TL}{JG} \qquad\qquad (11.2)$$

Beams subject to transverse loading are characterized by a variety of deformation relations. The basic concept for analyzing statically indeterminate structures will be developed for axially loaded members.

Visualize the rod of Fig. 11.1 that is rigidly fixed at both ends. An axial force P is applied at a distance a from the left end. The problem is to compute the reactions P_A and P_B. The equations of equilibrium are always valid for determinate or indeterminate structures. Summing forces along the axis of the beam gives one equation with two unknowns:

$$P_A + P_B = P \qquad\qquad (a)$$

The reactions have arbitrarily been assumed to act toward the left. A

FIGURE 11.1
Statically indeterminate axially loaded rod.

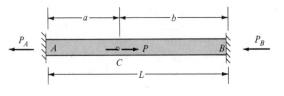

second equation is obtained by considering the deformation of the rod. The method is dependent upon assuming a free-body cut, computing the deformation associated with the free bodies, and relating that information to the physical boundary conditions. The solution for the rod will be continued as an example problem.

EXAMPLE 11.1

Compute the reactions P_A and P_B for the rod of Fig. 11.1 by assuming a free-body cut at B. Assume the area and modulus of elasticity are constant.

Solution:

The rod is drawn again in Fig. 11.2a. A free body resulting from a cut at B is shown in Fig. 11.2b. The reaction at the right end is computed from statics as $(P - P_A)$. An

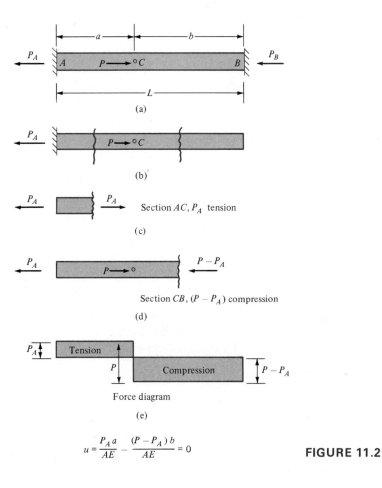

Section AC, P_A tension

(c)

Section CB, $(P - P_A)$ compression

(d)

Force diagram

(e)

$$u = \frac{P_A a}{AE} - \frac{(P - P_A) b}{AE} = 0$$

FIGURE 11.2

axial-force diagram is shown in Fig. 11.2e. The axial force in the section AC is obtained from Fig. 11.2c as P_A tension. Figure 11.2d illustrates a free body for section CB. The axial-force diagram shows that the section from A to C is in tension and the section CB is in compression. Substituting into Eq. (11.1) gives an expression for the deformation of the free body of Fig. 11.2b.

$$u = \frac{P_A a}{AE} - \frac{(P - P_A)b}{AE} \tag{b}$$

The deformation condition for the actual structure is that the axial displacement at point B is zero. Hence, Eq. (b) should be set equal to zero.

$$P_A a - Pb + P_A b = 0$$

or

$$P_A L = Pb \qquad P_A = \frac{Pb}{L} \tag{c}$$

The deformation condition yields a second equation to use with Eq. (a) of the previous discussion.

$$P_A + P_B = P$$

Substituting P_A gives

$$P_B = P - \frac{Pb}{L} = \frac{P(L - b)}{L} = \frac{Pa}{L} \tag{d}$$

Both reactions act to the left, as assumed in Fig. 11.2a.   ■

EXAMPLE 11.2

Compute the reactions P_A and P_B for the rod of Fig. 11.1 by assuming a free-body cut at C. Assume AE is constant for the beam.

Solution:
The beam is cut at point C and gives the free bodies of Figs. 11.3b and c. The deformation condition is not zero displacement as in the previous example since the cut section is not at the support where the deformation can be specified to be zero. Rather, at point C the elongation of section AC must be equal to the shortening of section CB.

$$u_A = u_B$$

According to the free body of Fig. 11.3b,

$$u_A = \frac{P_A a}{AE}$$

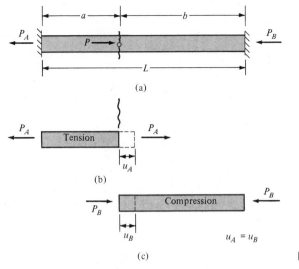

(a)

(b)

(c)

FIGURE 11.3

Similarly, using Fig. 11.3c,

$$u_B = \frac{P_B b}{AE}$$

and

$$\frac{P_A a}{AE} = \frac{P_B b}{AE} \qquad P_A = P_B(b/a)$$

Substituting into the equilibrium equation,

$$P_B\left(\frac{b}{a}\right) + P_B = P \qquad P_B = \frac{Pa}{L} \qquad \text{and} \qquad P_A = \frac{Pb}{L}$$

Obviously, the results are the same as in the previous example. The significant point is that the manner in which the deformation condition is formulated is dependent upon the choice of free body. A poor choice of free-body diagrams can lead to a more complicated analysis. Practice is an essential factor in the successful analysis of statically indeterminate structures. ∎

Consider a circular rod subject to a torque as shown in Fig. 11.4. The ends of the rod are fixed against rotation. The reactive torques are named T_A and T_B. An equation of equilibrium gives $T_A + T_B = T$. The deformation condition is illustrated in Fig. 11.4d as a plot of ϕ. The rotation at the point where the torque T is applied can be computed

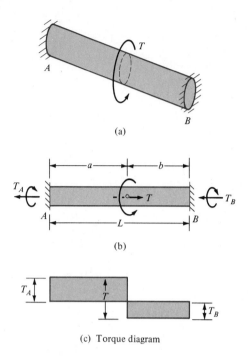

(a)

(b)

(c) Torque diagram

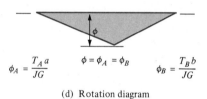

FIGURE 11.4
Statically indeterminate rod
subjected to a torsional load.

$$\phi_A = \frac{T_A a}{JG} \qquad \phi = \phi_A = \phi_B \qquad \phi_B = \frac{T_B b}{JG}$$

(d) Rotation diagram

using either section of the rod. Hence, a deformation condition is obtained: $T_A a/JG = T_B b/JG$. The deformation condition is combined with the equilibrium equation to complete the solution.

EXAMPLE 11.3

Compute the reactions for the circular rod of Fig. 11.5. Assume JG is a constant. The torque T_1 is related to T_2 as $T_1 = 4T_2$.

Solution:
The magnitude of T_A and T_B are unknown; however, a sketch such as Fig. 11.5b can serve to illustrate the relation between the various torques. The rotation diagram can be used to visualize a relation that can be written with respect to ϕ_C. The rotation using

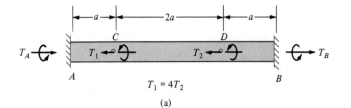

$$T_1 = 4T_2$$

(a)

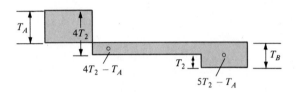

(b) Torque diagram (sketch)

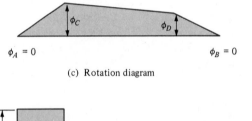

(c) Rotation diagram

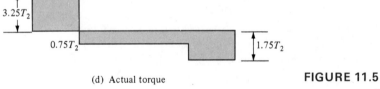

(d) Actual torque

FIGURE 11.5

the section to the left of point C should equal the rotation at C as found using the right-hand section.

$$\phi_C|_{\text{left}} = \phi_C|_{\text{right}}$$

$$\frac{T_A a}{JG} = \frac{(4T_2 - T_A)2a}{JG} + \frac{(5T_2 - T_A)a}{JG}$$

$$T_A = \frac{13T_2}{4} \tag{a}$$

The equilibrium equation can be written as

$$T_A + T_B - T_1 - T_2 = 0$$

or

$$T_A + T_B - 5T_2 = 0 \qquad \qquad \textbf{(b)}$$

Substituting Eq. (a),

$$\frac{13T_2}{4} + T_B - 5T_2 = 0$$

$$T_B = \frac{7T_2}{4}$$

The actual torque diagram is shown in Fig. 11.5d. ■

Examples 11.1–11.3 are straightforward, and with very little practice their analysis can be mastered. The moment equation that leads to Eq. (a) of Example 11.3 could cause some difficulty. It is always appropriate to cut a free-body section between points C and D and write an equilibrium equation that should lead to a correctly posed expression for torque.

A statically indeterminate problem that occurs often in engineering design can be thought of as a composite axially loaded structure. The structure is composed of two materials that are loaded in such a way that both materials deform together. The following example will illustrate the use of the deformation equation for this type of problem.

EXAMPLE 11.4

A concrete test cylinder 6 in. in diameter and 12 in. high is reinforced with six 1-in.-diameter steel bars placed symmetrically in the cylinder. Compute the percentage of a force P that is carried separately by the concrete and steel. $E_{st} = 30(10^6)$ psi; $E_{conc} = 4(10^6)$ psi.

Solution:
The test specimen is shown in Fig. 11.6. It is assumed that the force P is applied evenly through a rigid plate. The area of the test cylinder is computed as

$$A_{cyl} = \frac{\pi d^2}{4} = \frac{\pi(6)^2}{4} = 28.27 \text{ in.}^2$$

The area of the steel bars is computed as

$$A_{st} = \frac{6\pi(1)^2}{4} = 4.71 \text{ in.}^2$$

The area of the concrete is

$$A_{conc} = A_{cyl} - A_{st} = 28.27 - 4.71 = 23.56 \text{ in.}^2$$

Consider the six reinforcing bars to act together and, as illustrated in the figure, the concrete and steel deform an equal amount.

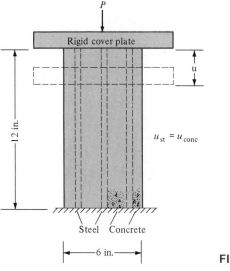

FIGURE 11.6

$$u_{st} = u_{conc}$$

$$\frac{P_{st}(12 \text{ in.})}{(4.71 \text{ in.}^2)(30)(10^6) \text{ psi}} = \frac{P_{conc}(12 \text{ in.})}{(23.56 \text{ in.}^2)(4)(10^6) \text{ psi}}$$

$$P_{st} = 1.50P_{conc}$$

The equilibrium equation gives a second relationship between P_{st} and P_{conc}.

$$P = P_{st} + P_{conc} = 1.50P_{conc} + P_{conc}$$

$P_{conc} = 0.40P$, or 40 percent of P

$P_{st} = 0.60P$, or 60 percent of P ∎

EXAMPLE 11.5

The rigid rod of Fig. 11.7 is pinned at the left end and supported by wires at points B and C. Compute the reactions due to the force applied at point D. Assume the wires have equal areas and are of the same material.

Solution:

It is important to recognize that the rigid bar rotates to a deflected position without bending. The diagram of Fig. 11.7b illustrates the deformation. The relation between the elongation of wire B and wire C is obtained using similar triangles.

$$\frac{u_B}{a} = \frac{u_C}{3a} \qquad 3u_B = u_C$$

Evaluating u_B and u_C gives a relation between R_B and R_C.

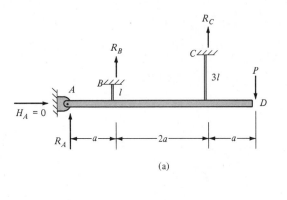

(a)

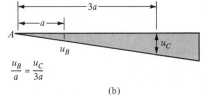

$$\frac{u_B}{a} = \frac{u_C}{3a}$$

(b)   **FIGURE 11.7**

$$\frac{3R_B l}{AE} = \frac{R_C 3l}{AE} \qquad \text{or} \qquad R_B = R_C$$

The equilibrium equations give two additional equations.

$$P - R_A - R_B - R_C = 0$$

$$\Sigma M_A = 0 \qquad aR_B + 3aR_C - 4aP = 0$$

Solving the three equations yields the desired results.

$$R_B = P \qquad R_A = -P$$

$$R_C = P \qquad H_A = 0$$

The negative sign for R_A indicates that R_A is downward rather than upward as assumed in Fig. 11.7a.   ∎

11.3 STATICALLY INDETERMINATE BEAMS

The basic concept for determining the reactions for a statically indeterminate beam structure will be developed using the force or flexibility method. The force method is easier to visualize than the displacement or stiffness method. However, modern-day highly indeterminate structural problems are analyzed using the stiffness method because it is more applicable to computer analysis of structures.

The Force Method

The force method consists of removing support conditions until the structure is statically determinate. The supports that are removed are referred to as redundants. The statically determinate structure is referred to as the primary structure. After removing redundants the deflection is computed at the location of each redundant. The magnitude is computed for each redundant reaction that will return the primary deflected structure back to its original undeflected position. The magnitude of the redundant reaction is the unknown reaction at the point of redundancy. After all redundants are replaced, the equilibrium equations are used to complete the analysis.

The procedure is similar to the previous section. The equilibrium equations are combined with deformation conditions to analyze a statically indeterminate beam problem.

EXAMPLE 11.6

Determine the reactions for the statically indeterminate beam of Fig. 11.8. Use the superposition tables, Appendix E, to compute deflections for each statically determinate structure.

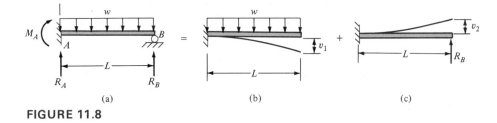

(a)   (b)   (c)

FIGURE 11.8

Solution:

The reaction R_B at B is removed and the primary structure is shown in Fig. 11.8b. The redundant is returned as shown in Fig. 11.8c. The original structure can be visualized as the superposition of the structures of Figs. 11.8b and c if the proper deformation relation is satisfied; namely,

$$v_1 - v_2 = 0 \qquad v_1 = v_2$$

The deflections are obtained using the superposition tables, Appendix E, Case E9 and Case E8.

$$v_1 = \frac{wL^4}{8EI} \qquad \text{and} \qquad v_2 = \frac{R_BL^3}{3EI}$$

The deflection condition gives R_B in terms of w.

$$\frac{R_B L^3}{3EI} = \frac{wL^4}{8EI} \qquad R_B = \frac{3wL}{8}$$

Summing moments about A of Fig. 11.8a gives

$$M_A + \frac{wL^2}{2} - R_B L = 0$$

$$M_A = \frac{3wL^2}{8} - \frac{wL^2}{2} = \frac{-wL^2}{8}$$

where the negative sign indicates that the M_A should be in the opposite direction to that shown in Fig. 11.8a. Summing forces will give R_A.

$$R_A + R_B - wL = 0$$

$$R_A = wL - \frac{3wL}{8} = \frac{5wL}{8}$$ ■

EXAMPLE 11.7

Determine the reactions for the beam of Fig. 11.8a assuming that M_A is redundant.

Solution:
The beam is shown in Fig. 11.9a. The moment at A is shown as negative moment using the beam sign convention. The moment at A is released and the primary struc-ture is shown in Fig. 11.9b as a simple beam. The deformation condition is that the rotation at A is zero. Hence, compute θ_A and find the moment M_A of Fig. 11.9c that would rotate the beam back to zero rotation. Using the superposition tables, Cases E3 and E5.

$$\theta_A = \frac{wL^3}{24EI}$$

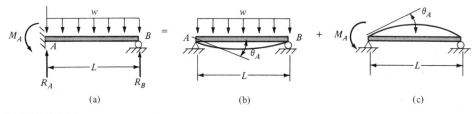

(a)   (b)   (c)

FIGURE 11.9

For the beam of Fig. 11.9c,

$$\theta_A = \frac{M_A L}{3EI}$$

$$\frac{M_A L}{3EI} = \frac{wL^3}{24EI} \qquad M_A = \frac{wL^2}{8}$$

The reactions can be computed using statics

$$R_A = \frac{5wL}{8} \qquad R_B = \frac{3wL}{8} \qquad\qquad\qquad ■$$

To further illustrate the dependence of the analysis on the deformation condition, consider the following problem in which the support has been allowed to settle by a small amount.

EXAMPLE 11.8

The beam of Fig. 11.10 is statically indeterminate to the first degree; that is, there is one more unknown reaction than there are equations of equilibrium.

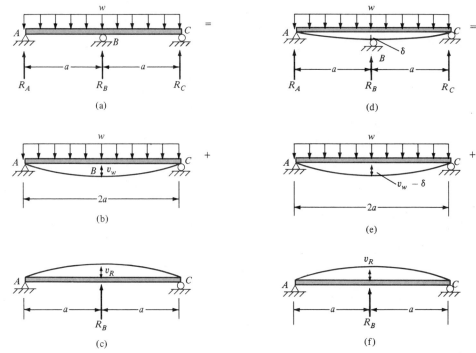

FIGURE 11.10

a. Compute the reactions in terms of w and L.

b. Assume the center reaction R_B settles by a small amount, δ. Compute the reactions in terms of δ.

c. Assume a steel S8x23 beam, $w = 4$ k/ft, and $a = 10$ ft, and compare the solutions of parts **a** and **b** for $\delta = 0.5$ in. $E = 30(10^3)$ ksi.

Solution:

a. The center support is removed as shown in Fig. 11.10b. Case E3, Appendix E, is used to compute v_w for $L = 2a$.

$$v_w = \frac{5w(2a)^4}{384EI} = \frac{5wa^4}{24EI} \tag{a}$$

$$v_R = \frac{R_B(2a)^3}{48EI} = \frac{R_B a^3}{6EI} \quad \text{(case 1)} \tag{b}$$

The deflection condition is $v_w = v_R$.

$$R_B = \frac{5wa}{4}$$

$$\Sigma M_A = 0 \qquad R_B a + R_C(2a) - \frac{w(2a)^2}{2} = 0$$

$$R_C = \frac{3wa}{8}$$

$$\Sigma F = 0 \qquad R_A + R_B + R_C - 2wa = 0$$

$$R_A = \frac{3wa}{8}$$

b. The beam of Fig. 11.10d illustrates the deflection of the center support. When the center support is removed, the beam appears to deflect the same amount as shown in Fig. 11.10b for part **a**. Since the support was allowed to settle an amount δ prior to removing the support, the total deflection is actually $v_w - \delta$, as shown in Fig. 11.10e. The deformation condition is

$$v_w - \delta = v_R \tag{c}$$

The reaction R_B returns the support only to the position shown in Fig. 11.10d. Substituting Eqs. (a) and (b) into (c) gives the reaction R_B.

$$\frac{5wa^4}{24EI} - \delta = \frac{R_B a^3}{6EI}$$

$$R_B = \frac{5wa}{4} - \frac{6EI\delta}{a^3} \tag{d}$$

$$R_C = \frac{3wa}{8} + \frac{3EI\delta}{a^3} = R_A \tag{e}$$

c. The second moment of the area for a S8x23 structural member is obtained from Appendix C as $I = 64.9$ in.^{4}. For the beam of part *a*, $w = 4$ k/ft $= (4/12)$ k/in., and $a = 120$ in.

$$R_A = R_C = \frac{(3)(4/12)(120)}{8} = 15 \text{ k}$$

$$R_B = \frac{(5)(4/12)(120)}{4} = 50 \text{ k}$$

The reactions for the beam of part *b* are obtained from Eqs. (d) and (e).

$$R_B = 50 \text{ k} - \frac{(6)(30)(10^3) \text{ ksi } (64.9 \text{ in.}^4)}{(120 \text{ in.})^3} (0.5 \text{ in.}) = 50 \text{ k} - 3.38 \text{ k}$$

$$= 46.62 \text{ k}$$

$$R_A = R_C = 15 \text{ k} + \frac{(3)(30)(10^3)(64.9)}{(120)^3} (0.5) = 15 + 1.69 = 16.69 \text{ k}$$

This problem illustrates that what might appear to be a small settlement of the center support has a significant effect on the reactions. ∎

The problems thus far have been indeterminate to the first degree. Beams that are indeterminate to the second degree or higher orders tend to increase the computational effort but strengthen the understanding of the basic force method. Consider the following problem, which is statically indeterminate to the second degree.

EXAMPLE 11.9

Compute the reactions for the fixed-fixed beam of Fig. 11.11.

Solution:
The beam is statically indeterminate to the second degree. Two redundant reactions must be removed to make the beam statically determinate. Removing the moment and shear reaction at *B* will create a primary structure corresponding to Case E10 of Appendix E. The redundant shear reaction is applied as shown in 11.11*c* to cause both a deflection and rotation. Similarly, the redundant moment reaction is applied as shown in Fig. 11.11*d*. The direction of these redundants is not arbitrary. They have been applied in the same direction that they were assumed to act in Fig. 11.11*a*. There are two deformation conditions. The deflection and slope at *B* are zero. Adding the deformations as they are shown in Figs. 11.11*b–d* gives the conditions

$$v_B - v_{BR} + v_{BM} = 0 \tag{a}$$

$$\theta_B - \theta_{BR} + \theta_{BM} = 0 \tag{b}$$

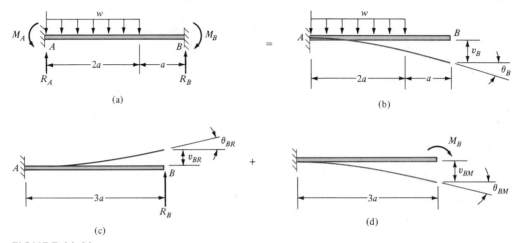

FIGURE 11.11

Note that as each redundant is applied it has an effect on the other redundant. Equations (a) and (b) are evaluated using Cases E8, E10, and E13 of Appendix E. The following analogy must be used:

Nomenclature		
Example 11.9		Appendix E
$3a$	$=$	L
$2a$	$=$	a
a	$=$	b

$$\frac{w(2a)^3}{24}\left[4(3a) - 2a\right] - \frac{R_B(3a)^3}{3} + \frac{M_B(3a)^2}{2} = 0 \qquad \textbf{(a)}$$

$$\frac{w(2a)^3}{6} - \frac{R_B(3a)^2}{2} + M_B(3a) = 0 \qquad \textbf{(b)}$$

or

$$\frac{10wa^4}{3} - 9R_B a^3 + \frac{9M_B a^2}{2} = 0 \qquad \textbf{(a)}$$

$$\frac{4wa^3}{3} - \frac{9R_B a^2}{2} + 3M_B a = 0 \qquad \textbf{(b)}$$

Solving Eqs. (a) and (b) simultaneously gives R_B and M_B.

$$M_B = \frac{4wa^2}{9} \qquad R_B = \frac{16wa}{27}$$

Summing moments about A gives an equation for M_A.

$$M_A - \frac{4wa^2}{9} - 2wa^2 + \frac{(3a)(16wa)}{27} = 0$$

$$M_A = \frac{2wa^2}{3}$$

and

$$R_A + \frac{16wa}{27} - 2wa = 0$$

$$R_A = \frac{38wa}{27}$$

■

It follows that if the structure is statically indeterminate to the third degree, there will be three deformation equations.

The Integration Method

The integration method is the same as determining the equation of the elastic curve as discussed in Chapter 10. The intent, however, is to compute reactions for a statically indeterminate structure. It will be shown that the additional boundary conditions produced by additional reactions can be used in conjunction with equations defining the slope and elastic curve to compute reactions for indeterminate structures.

The integration method is useful when a particular superposition solution is not available or when it is convenient to pose the problem in terms of the load function or when reaction forces and moments are produced by support movement.

The following example is statically indeterminate and the load function is expressed in terms of a sine function. A straightforward method of solution is to use the fourth-order differential equation that relates deflection and load.

EXAMPLE 11.10

Compute reactions for the beam of Fig. 11.12. Use the fourth-order differential equation that relates deflection and load.

Solution:
The differential equation is written for constant EI.

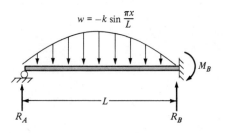

Boundary conditions:
$v(0) = v(L) = 0$
$v'(L) = 0$
$v''(0) = 0$
$EIv'''(0) = R_A$

FIGURE 11.12

$$EIv^{IV} = w(x) = -k \sin \frac{\pi x}{L}$$

$$EIv''' = \frac{kL}{\pi} \cos \frac{\pi x}{L} + C_1$$

$$EIv'' = \frac{kL^2}{\pi^2} \sin \frac{\pi x}{L} + C_1 x + C_2$$

The boundary conditions can be used to compute C_1 and C_2.

$$EIv'''(0) = R_A = \frac{kL}{\pi} + C_1 \qquad EIv''(0) = 0$$

$$C_1 = -\frac{kL}{\pi} + R_A \qquad\qquad C_2 = 0$$

Then

$$EIv' = -\frac{kL^3}{\pi^3} \cos \frac{\pi x}{L} - \frac{kL}{2\pi} x^2 + \frac{R_A x^2}{2} + C_3$$

$$EIv = -\frac{kL^4}{\pi^4} \sin \frac{\pi x}{L} - \frac{kLx^3}{6\pi} + \frac{R_A x^3}{6} + C_3 x + C_4$$

$v(0) = 0$ yields $C_4 = 0$

$v(L) = 0$ yields $0 = -\frac{kL^4}{6\pi} + \frac{R_A L^3}{6} + C_3 L$

$$C_3 = \frac{kL^3}{6\pi} - \frac{R_A L^2}{6}$$

The final boundary condition is $v'(x = L) = 0$. The constant C_3 is substituted into the slope equation.

$$EIv' = -\frac{kL^3}{\pi^3}\cos\frac{\pi x}{L} + \frac{kL}{6\pi}(L^2 - 3x^2) + \frac{R_A}{6}(3x^2 - L^2)$$

$$0 = \frac{kL^3}{\pi^3} - \frac{kL^3}{3\pi} + \frac{R_A L}{3}$$

$$R_A = \frac{kL}{\pi^3}(\pi^2 - 3) \qquad\qquad\qquad \textbf{(a)}$$

$$\Sigma M_B = 0 \qquad M_B - R_A L + \frac{L}{2}\int_0^L -k\sin\frac{\pi x}{L}\,dx = 0$$

The integral term represents the total load acting on the beam. Because of the symmetry of the load, the centroid lies at $L/2$, and hence the term preceding the integral. Formally, the moment of the load would be

$$\int_0^L -xk\sin\frac{\pi x}{L}\,dx$$

$$M_B = \frac{kL^2}{\pi^3}(\pi^2 - 3) + \left[\frac{kL^2}{2\pi}\cos\frac{\pi x}{L}\right]_0^L = \frac{kL^2}{\pi} - \frac{3kL^3}{\pi^3} - \frac{kL^2}{\pi}$$

$$M_B = -\frac{3kL^2}{\pi^3} \qquad \text{in the direction shown in Fig. 11.12} \qquad \textbf{(b)}$$

$$\Sigma F = 0 \qquad R_A + R_B + \int_0^L -k\sin\frac{\pi x}{L}\,dx = 0$$

$$R_B = \frac{kL}{\pi^3}(\pi^2 + 3) \qquad\qquad\qquad \textbf{(c)}$$

The desired results are given by Eqs. (a), (b), and (c).   ■

The integration method is useful for any beam that has a continuous load function. Concentrated loads and discontinuous loading, when a superposition case is not readily available, are probably best solved using the force method. The primary structure and redundant structures can be solved using the unit load method of Chapter 10.

When the integration method is used for structures that are indeterminant to higher degrees, the additional reactions provide additional boundary conditions.

EXAMPLE 11.11

Develop the differential equations that would be required for computing the reactions for the beam of Fig. 11.13 beginning with the fourth-order equation. Formulate

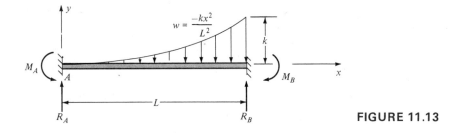

FIGURE 11.13

the boundary conditions that would be used to determine constants of integration and reactions.

Solution:

The moment reactions have been assumed negative. Beginning with the fourth-order differential equation, we will generate four constants of integration.

$$EIv^{IV} = -\frac{kx^2}{L^2}$$

$$EIv''' = -\frac{kx^3}{3L^2} + A$$

$$EIv'' = -\frac{kx^4}{12L^2} + Ax + B$$

The conditions on shear and moment can be used to evaluate A and B. Obviously, $x = 0$ is more desirable than $x = L$. Hence,

$$EIv'''(0) = R_A$$

$$EIv''(0) = -M_A \qquad \text{(as assumed in Fig. 11.13)}$$

The constants A and B will be determined in terms of unknown reactions R_A and M_A. Continuing the integration gives

$$EIv' = \frac{-kx^5}{60L^2} + \frac{Ax^2}{2} + Bx + C$$

$$EIv = \frac{-kx^6}{360L^2} + \frac{Ax^3}{6} + \frac{Bx^2}{2} + Cx + D$$

The constants C and D are evaluated using

$$v'(0) = 0 \qquad v(0) = 0$$

After the constants of integration are evaluated, it is necessary to write two equations that can be used to determine R_A and M_A. The slope and deflection conditions at $x = L$ are still available.

$$v'(x = L) = 0 \qquad v(x = L) = 0$$

The equilibrium equations can be used to compute R_B and M_B.

The integration method is well suited for developing a relation between reactions and joint movements for problems that involve joint displacements or rotations of unloaded beams.   ■

EXAMPLE 11.12

The reaction at B displaces an amount δ for the beam of Fig. 11.14. Compute the reactions and draw shear and moment diagrams.

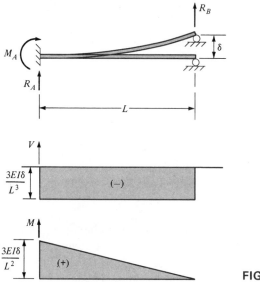

FIGURE 11.14

Solution:

The reactions are assumed positive as shown. Since the load is zero, begin with the fourth-order differential equation as follows:

$$EIv^{IV} = 0 \tag{a}$$

$$EIv''' = A \tag{b}$$

$$EIv'' = Ax + B \tag{c}$$

Equation (b) and $EIv'''(0) = R_A$ give

$$R_A = A$$

Equation (c) and $EIv''(x = L) = 0$ give

$$0 = R_A L + B \qquad B = -R_A L$$

$$EIv' = \frac{R_A x^2}{2} - R_A L x + C \tag{d}$$

$$EIv = \frac{R_A x^3}{6} - \frac{R_A L x^2}{2} + Cx + D \tag{e}$$

Equation (d) and $v'(x = 0) = 0$ give $C = 0$. Equation (e) and $v(x = 0) = 0$ give $D = 0$. The condition $v(x = L) = \delta$ and Eq. (e) give an equation for R_A.

$$EI\delta = \frac{R_A L^3}{6} - \frac{R_A L^3}{2} = -\frac{R_A L^3}{3}$$

$$R_A = -\frac{3EI\delta}{L^3}$$

The equilibrium equations give

$$R_B = \frac{3EI\delta}{L^3} \qquad M = \frac{3EI\delta}{L^3}$$

The shear and moment diagrams are shown in Fig. 11.14.

Obviously, the second-order equation could have been used for this problem since a moment equation is easily written for the beam of Fig. 11.14.   ∎

Displacement Method

The displacement method is somewhat opposite to the force method. Support reactions, either force or moment, are the unknown quantities in the force method. The joint displacements, either translation or rotation, are the unknowns in the displacement method. The first step in the displacement method is to assume that all joints are fixed, and every joint or support is assumed to have zero displacement and rotation. Equilibrium equations for each joint constitutes the set of equations to be solved to determine displacements. Displacement effects at each support are superposed to compute the final support reactions.

The displacement method is sometimes referred to as the stiffness method and uses terminology pertinent to more advanced courses such as matrix analysis of structures. The set of equations that relate joint actions, forces, and moments to joint displacements is written such that the joint displacements are unknowns. The coefficients of the unknowns can be written in the form of a matrix referred to as the stiffness matrix.

Using the displacement method, every indeterminate structure problem is analyzed by assuming that all joints are fixed. This procedure lends itself to computer analysis, whereas the force method requires the analyst to make logical decisions concerning which support reactions should be redundant in order to produce a primary structure—hence the previous statement that the stiffness method lends itself to computer analysis.

The foregoing discussion can be illustrated using an example problem. The problem of Example 11.6 will be solved again using the displacement method. Appendix F will be used as a superposition table for fixed-fixed beams. The next examples illustrate the use of the superposition tables.

EXAMPLE 11.13

Compute the reactions for the beam of Fig. 11.15 using the displacement method.

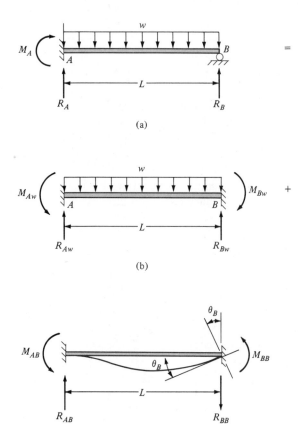

(a)

(b)

(c)

FIGURE 11.15

Solution:

The beam is already fixed against translation and rotation at A. Joint B is free to rotate; hence, assume that joint B is fixed against rotation as shown in Fig. 11.15b. A uniformly loaded fixed-fixed beam can be analyzed using Appendix F. The moment and force reactions due to the uniform load are

$$M_{Aw} = M_{Bw} = \frac{wL^2}{12}$$

where the moments are assumed negative, as shown in Fig. 11.15b.

$$R_{Aw} = R_{Bw} = \frac{wL}{2}$$

In Fig. 11.15c a rotation is applied at joint B. The moment required at B to sustain this rotation is $M_{BB} = 4EI\theta_B/L$ as given by Case F8 of Appendix F. The rotation also causes a moment M_{AB} at point A. The notation can be interpreted as the moment at A due to an action at B. The reactions are obtained using superposition.

$$M_{AB} = \frac{2EI\theta_B}{L} \qquad M_{BB} = \frac{4EI\theta_B}{L}$$

$$R_{AB} = \frac{6EI\theta_B}{L^2} \qquad R_{BB} = \frac{6EI\theta_B}{L^2}$$

The boundary condition is that the moment at joint B is zero. The moment of Fig. 11.15b, M_{Bw} combines with M_{BB}, the moment of Fig. 11.15c, to yield the zero moment of Fig. 11.15a.

$$M_{BB} - M_{Bw} = 0$$

The displacement method uses the boundary condition as an equilibrium equation whereas the force method of Example 11.6 or 11.7 used the boundary condition as a condition on joint movement.

Substituting into the equilibrium equation gives θ_B.

$$\frac{4EI\theta_B}{L} = \frac{wL^2}{12}$$

$$\theta_B = \frac{wL^3}{48EI} \qquad \text{(in the direction shown in Fig. 11.15c)}$$

The actual reactions are given by adding the three reactions of Figs. 11.15a–c, taking into account the equal sign in the figure and solving for M_A.

$$M_A = -M_{Aw} - M_{AB}$$

The moment of Fig. 11.15a was arbitrarily assumed positive, while the moments of Figs. 11.15b and c are negative.

$$M_A = -\frac{wL^2}{12} - \frac{2EI}{L} \cdot \frac{wL^3}{48EI} = -\frac{wL^2}{8}$$

$$R_A = R_{Aw} + R_{AB} = \frac{wL}{2} + \frac{6EI}{L^2} \cdot \frac{wL^3}{48EI} = \frac{5}{8}wL$$

$$R_B = R_{Bw} - R_{BB} = \frac{wL}{2} - \frac{6EI}{L^2} \cdot \frac{wL^3}{48EI} = \frac{3}{8}wL$$

This is the same result obtained in Example 11.6. ■

The displacement method is characterized by fixing all supports as opposed to releasing supports. Every span between supports of a continuous beam is assumed to be a separate fixed-fixed beam.

The following example will improve the understanding of basic concepts.

EXAMPLE 11.14

Compute the reactions for the structure of Fig. 11.16. Construct the shear and moment diagrams.

Solution:

Supports A and C are fixed. There remains to fix B. The spans are visualized to be separated as shown in Fig. 11.16b. The fixed-end moments and shears are obtained from Appendix F.

$$M_{Aw} = M_{Bw} = \frac{wL^2}{12} \qquad \text{(acting as shown in the figure)}$$

$$M_{BP} = M_{CP} = \frac{P2L}{8} = \frac{wL^2}{8}$$

$$R_{Aw} = R_{Bw} = \frac{wL}{2}$$

$$R_{BP} = R_{CP} = \frac{P}{2} = \frac{wL}{4}$$

Support B, which is only temporarily fixed, is given a rotation θ_B counterclockwise as shown in Fig. 11.16c. The reactions are drawn in the direction they would occur for the counterclockwise rotation of joint B. The reactions due to the rotation at B are obtained from Appendix F.

$$M_{AB} = \frac{2EI\theta_B}{L} \qquad M_{BB1} = \frac{4EI\theta_B}{L}$$

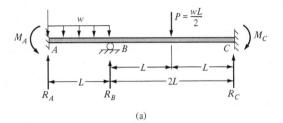

(a)

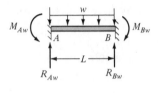
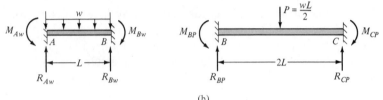

(b)

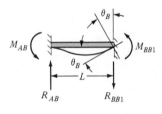
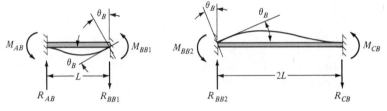

(c)

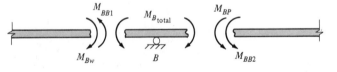

$$-M_{Bw} + M_{BB1} = -M_{B_{total}} = -M_{BP} - M_{BB2}$$

(d)

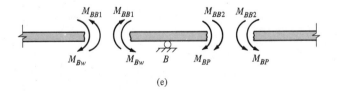

(e)

FIGURE 11.16

$$R_{AB} = \frac{6EI\theta_B}{L^2} \qquad R_{BB1} = \frac{6EI\theta_B}{L^2}$$

$$M_{BB2} = \frac{4EI\theta_B}{2L} = \frac{2EI\theta_B}{L}$$

$$M_{CB} = \frac{2EI\theta_B}{2L} = \frac{EI\theta_B}{L}$$

$$R_{BB2} = \frac{6EI\theta_B}{(2L)^2} = \frac{3EI\theta_B}{2L^2}$$

$$R_{CB} = \frac{6EI\theta_B}{(2L)^2} = \frac{3EI\theta_B}{2L^2}$$

In this example the actual moment at B is not zero, but, it is unknown. The total moment that acts at B on the end of member AB must be equal and opposite to the total moment that acts at B on the end of member BC. Consider the free body of the joint shown in Fig. 11.16d, the moment acting on the joint is $M_{B\,\text{total}}$, the total moment at joint B, and is in equilibrium with the moments acting on each span. Moment equilibrium for joint B gives

$$-M_{Bw} + M_{BB1} = M_{B\,\text{total}} = -M_{BP} - M_{BB2}$$

The boundary condition can be obtained directly from Figs. 11.16b and c by adding the moments acting on joint B.

$$-M_{Bw} + M_{BB1} = -M_{BP} - M_{BB2}$$

A second way to visualize the boundary condition is to let the moments of Figs. 11.16b and c act on the joint as shown in Fig. 11.16e. Let counterclockwise moments be positive and add the four moments acting on joint B.

$$-M_{Bw} + M_{BP} + M_{BB1} + M_{BB2} = 0$$

$$\frac{-wL^2}{12} + \frac{wL^2}{8} + \frac{4EI\theta_B}{L} + \frac{2EI\theta_B}{L} = 0$$

$$\theta_B = \frac{-wL^3}{144EI} \qquad \text{(clockwise rotation)}$$

The moment at B is computed using the moments acting on span AB of Figs. 11.16b and c.

$$M_B = -M_{Bw} + M_{BB1} = \frac{-wL^2}{12} + \frac{4EI}{L}\left(\frac{-wL^3}{144EI}\right) = \frac{-wL^2}{9}$$

The reactions are computed using Figs. 11.16a–c.

$$M_A = M_{Aw} + M_{AB}$$

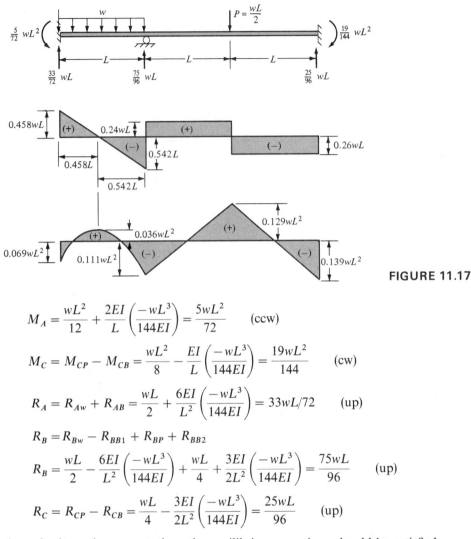

FIGURE 11.17

$$M_A = \frac{wL^2}{12} + \frac{2EI}{L}\left(\frac{-wL^3}{144EI}\right) = \frac{5wL^2}{72} \quad \text{(ccw)}$$

$$M_C = M_{CP} - M_{CB} = \frac{wL^2}{8} - \frac{EI}{L}\left(\frac{-wL^3}{144EI}\right) = \frac{19wL^2}{144} \quad \text{(cw)}$$

$$R_A = R_{Aw} + R_{AB} = \frac{wL}{2} + \frac{6EI}{L^2}\left(\frac{-wL^3}{144EI}\right) = 33wL/72 \quad \text{(up)}$$

$$R_B = R_{Bw} - R_{BB1} + R_{BP} + R_{BB2}$$

$$R_B = \frac{wL}{2} - \frac{6EI}{L^2}\left(\frac{-wL^3}{144EI}\right) + \frac{wL}{4} + \frac{3EI}{2L^2}\left(\frac{-wL^3}{144EI}\right) = \frac{75wL}{96} \quad \text{(up)}$$

$$R_C = R_{CP} - R_{CB} = \frac{wL}{4} - \frac{3EI}{2L^2}\left(\frac{-wL^3}{144EI}\right) = \frac{25wL}{96} \quad \text{(up)}$$

As a check on the computations the equilibrium equations should be satisfied.

$$\Sigma F = 0 \quad \frac{33wL}{72} + \frac{75wL}{96} + \frac{25wL}{96} - wL - \frac{wL}{2} = 0$$

$$\Sigma M_C = 0 \quad M_A - M_C - 3LR_A - 2LR_B + wL\left(\frac{5L}{2}\right) + \frac{wL^2}{2}$$

$$= wL^2\left(\frac{5}{72} - \frac{19}{144} - \frac{99}{72} - \frac{150}{96} + \frac{5}{2} + \frac{1}{2}\right) = 0$$

The equilibrium equations are satisfied; hence, it can be concluded that the analysis is correct. The shear and moment diagrams are constructed in Fig. 11.17. ∎

The displacement method is fundamental to the analysis of structural systems. The displacement method forms the basis for matrix analysis of structures. Elementary matrix algebra is discussed in Appendix D of the textbook. The previous examples involve only one unknown joint rotation. The formulation of the next problem will involve two joint rotations and by example will illustrate the formulation of a stiffness matrix. Understanding the formulation of the stiffness matrix is the key to successfully writing and using computer codes for the analysis and design of complicated structural systems.

EXAMPLE 11.15

The beam of Fig. 11.18 (next page) is statically indeterminate to the first degree. Use the displacement method to compute the reactions.

Solution:
Each support or joint is fixed independently as shown in Fig. 11.18. The interior joint is affected by both spans as will be shown. The interaction of the spans at a joint is probably the most difficult concept to visualize as the structure is assumed fixed at each support and then systematically released.

Figure 11.18b illustrates each fixed span and the corresponding fixed-end moments as obtained from Appendix F. Each joint of each span is given a rotation as shown in Figs. 11.18c–e. All joints are arbitrarily assumed to rotate in a counterclockwise direction. The fixed-end moments and reactions due to joint rotations are shown in Fig. 11.18. An equation is written for each joint by algebraically adding the rotations of Figs. 11.18b–e. Counterclockwise moments are assumed positive.

$$\frac{Pa}{8} + \frac{4EI\theta_A}{a} + \frac{2EI\theta_B}{a} = 0 \tag{a}$$

$$\frac{-Pa}{8} + \frac{Pa}{2} + \frac{2EI\theta_A}{a} + \frac{4EI\theta_B}{a} + \frac{2EI\theta_B}{a} + \frac{EI\theta_C}{a} = 0 \tag{b}$$

$$\frac{-Pa}{2} + \frac{EI\theta_B}{a} + \frac{2EI\theta_C}{a} = 0 \tag{c}$$

The effect of the joint rotation at B on both spans is shown in Eq. (b). The two θ_B terms are added to give $6EI\theta_B/a$. The three simultaneous equations are solved for the rotations.

$$\theta_A = \frac{EIPa^2}{32}$$

$$\theta_B = -\frac{EIPa^2}{8}$$

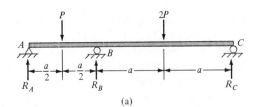

(a)

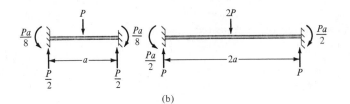

(b)

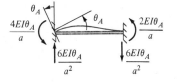

(c)

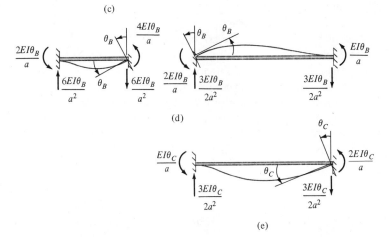

(d)

(e)

FIGURE 11.18

$$\theta_C = \frac{5EIPa^2}{16}$$

The reactions are obtained from Fig. 11.18.

$$R_A = \frac{P}{2} + \frac{6P}{32} - \frac{6P}{8} = \frac{-P}{16}$$

$$R_B = \frac{P}{2} + P - \frac{6P}{32} + \frac{6P}{8} - \frac{3P}{16} + \frac{15P}{32} = \frac{75P}{32}$$

$$R_C = P + \frac{3P}{16} - \frac{15P}{32} = \frac{23P}{32}$$

Using span AB,

$$M_B = \frac{-Pa}{8} + \frac{2Pa}{32} - \frac{4Pa}{8} = \frac{-9Pa}{16}$$

The negative sign indicates that the moment acting at B causes a clockwise rotation, which is a negative moment using the beam sign convention. Shear and moment diagrams are shown in Fig. 11.19.

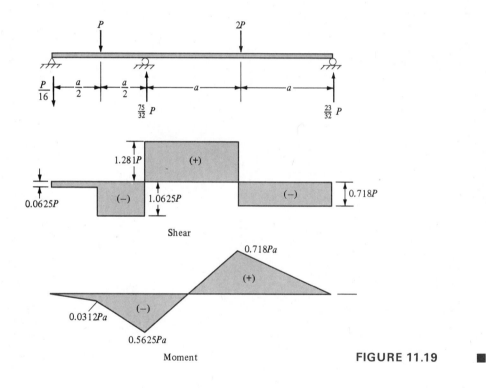

FIGURE 11.19

It is proper to include one more example illustrating the application of the displacement method for a beam with an overhang. An overhang, such as shown in Fig. 11.20, is similar to a fixed-end beam that is allowed to rotate at the fixed end. The fixed-end moment is computed directly and even though the joint may rotate, the moment remains unchanged. The fixed-end moment can be used as a boundary condition. The following example will illustrate the concept.

EXAMPLE 11.16

Use the stiffness method to compute the reactions for the beam of Fig. 11.20.

Solution:

The beam of Fig. 11.20a is separated into two spans in Fig. 11.20b. There is no load on span AB; hence, the fixed-end moments and reactions are zero. The moment acting at B for span BC is PL. Joint B is given a rotation θ_B counterclockwise in Fig. 11.20c.

Since span BC is free at joint C it is allowed to rotate as shown in Fig. 11.20c. The displacement method boundary condition would be for the sum of the moments at joint B. Note that of the four moments shown, two are zero.

$$PL + \frac{4EI\theta_B}{L} = 0 \qquad \theta_B = \frac{-PL^2}{4EI}$$

$$M_A = 0 + \frac{2}{L}\left(\frac{-PL^2}{4}\right) = \frac{-PL}{2} \qquad \text{(clockwise)}$$

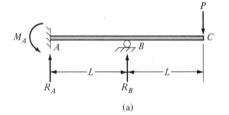

(a)

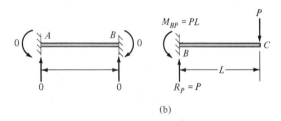

(b)

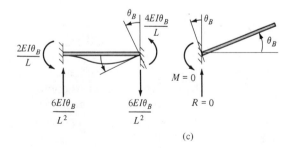

(c)

FIGURE 11.20

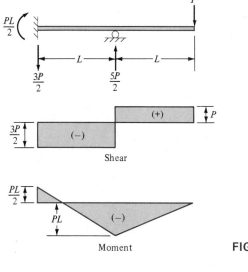

FIGURE 11.21

$$R_A = 0 + \frac{6}{L^2}\left(\frac{-PL^2}{4}\right) = \frac{-3P}{2} \quad \text{(down)}$$

$$R_B = 0 + P - \frac{6}{L^2}\left(\frac{-PL^2}{4}\right) + 0 = \frac{5P}{2} \quad \text{(up)}$$

and, of course,

$$M_B = -PL$$

The shear and moment diagrams are shown in Fig. 11.21. ∎

11.4 COMPLEX STATICALLY INDETERMINATE BEAM STRUCTURES

Occasionally a statically indeterminate structure is a combination of two or more deformable members. The method of analysis is less systematic than for the previous examples. The analyst must visualize the boundary condition or deformation condition and write a corresponding boundary condition equation. The beam of Fig. 11.22 is simply supported with a spring support in the center. A free-body diagram can be used to study the deflection of the beam independent of the spring. The free body of Fig. 11.22b shows that the beam deflects downward due to the uniform load; however, the reaction at C, the force in the spring, causes an upward deflection of the beam. The beam deflection is caused by two separate loads. The spring is in tension and elongates. The boundary condition is that the total deflection of the beam must equal the total deflection of the spring.

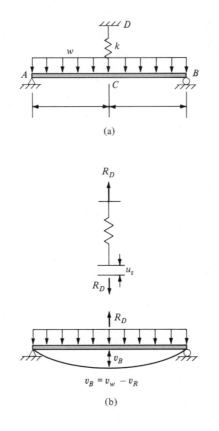

(a)

(b)

FIGURE 11.22

EXAMPLE 11.17

The beam of Fig. 11.22 is simply supported at each end, with a spring support in the center of the span. The spring constant is k and the beam has constant second moment of the area. Compute the reactions.

Solution:
The support at C can be removed to render the structure determinate. The deflection boundary condition is *not* that the deflection at C is zero, but rather that the deflection of the beam must equal the deflection of the spring. The reaction at D is the force in the spring, causing tension. The same force acts upward on the beam at point C. The spring elongates an amount of u_s and is equal to the downward deflection of the beam.

$$u_S = v_B = v_w - v_R \tag{a}$$

The deflection of the spring is $u_s = R_D/k$, where k is the spring constant. The deflection of the beam is obtained from Appendix E.

$v_w = 5wL^4/384EI$ (down)

$$v_R = \frac{R_D L^3}{48EI} \quad \text{up}$$

The uniform load causes a downward deflection while the reaction R_D causes an upward deflection of the beam. Substituting into Eq. (a) gives

$$R_D = \frac{5wL^4/384EI}{(1/k) + (L^3/48EI)}$$

∎

In the example a spring of stiffness k is used as the support. A wire or rod as shown in Fig. 11.23 could be used in place of the spring and the deflection could be computed.

$$u_S = \frac{R_D l}{AE_R}$$

$$R_D = \frac{5wL^4/384E_B I}{(l/AE_R) + (L^3/48EI)}$$

A wire or rod is similar to a spring with spring constant AE/l.

The simply supported beam could be supported by a second beam as shown in Fig. 11.24 that is fixed and supports the simple beam at its free end. Using Appendix E,

$$v_S = \frac{R_D a^3}{3(EI)_2}$$

The beam can be thought of as a spring with spring constant $3(EI)_2/a^3$. Equation (a) gives the relation for computing R_D.

$$\frac{R_D a^3}{3EI_2} = \frac{5wL^4}{384(EI)_1} - \frac{R_D L^3}{48(EI)_1}$$

$$R_D = \frac{(5wL^4/384(EI)_1)}{(a^3/3(EI)_2) + (L^3/48(EI)_1)}$$

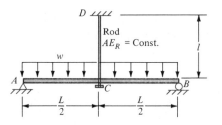

FIGURE 11.23

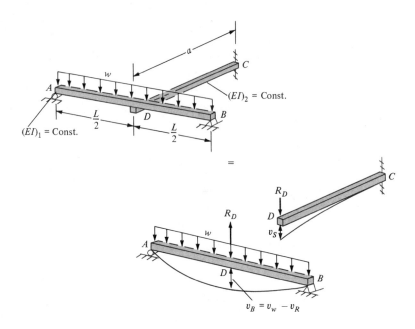

FIGURE 11.24

EXAMPLE 11.18

Two fixed-free beams are arranged as shown in Fig. 11.25. Both beams are steel, $E = 200 \, \text{GPa}$. Beam AB has the second moment of the area, $I_{AB} = 120(10^{-6}) \, \text{m}^4$; and beam CD has the second moment of the area, $I_{CD} = 400(10^{-6}) \, \text{m}^4$. The short post between the beams is 1.2 m in length with $EA = 18 \, \text{GN}$. A small gap of $\delta_B = 0.5 \, \text{mm}$ occurs at B between span AB and the post before the load is applied. Compute the reactions, the axial force in the post, and the deflection at C.

Solution:
The free-body diagram of Fig. 11.25b indicates that the displacement condition would be

$$v_{AB} = 0.5(10^{-3}) \, \text{m} + u_{CD} \qquad \textbf{(a)}$$

The deflection of span AB is made up of two parts—a downward deflection caused by the uniform load and an upward deflection caused by the reactive force R_B.

$$v_{AB} = \frac{wL^4}{8EI} - \frac{R_B L^3}{3EI_{AB}} \qquad \textbf{(b)}$$

A second free body of span CD is shown in Fig. 11.25c. The deflection u_{CD} is made up of two parts; the post is in compression and shortens, and the reaction acts downward on member CD.

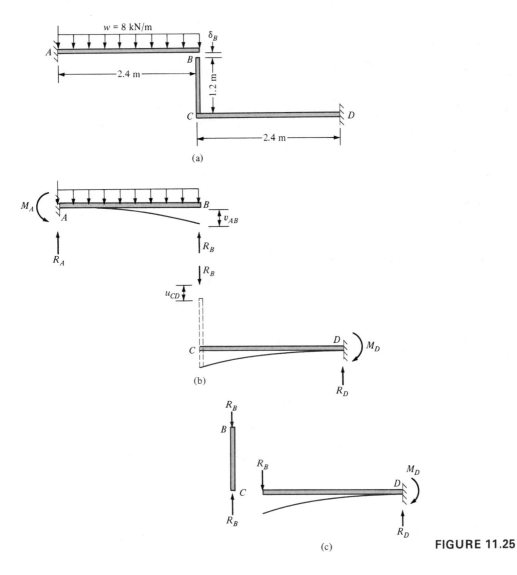

FIGURE 11.25

$$u_{CD} = \frac{R_B L_P}{AE_P} + \frac{R_B L^3}{3EI_{CD}} \tag{c}$$

Substituting the proper values in Eqs. (b) and (c) and then substituting into Eq. (a) gives a single equation for computing R_B:

$$\frac{8000 \text{ N/m } (2.4)^4 \text{ m}^4}{8(200)(10^9) \text{ N/m } (120)(10^{-6}) \text{ m}^4} - \frac{R_B(2.4)^3}{3(200)(10^9)(120)(10^{-6})}$$

$$= 0.5(10^{-3}) + \frac{R_B(1.2)}{18(10^6)} + \frac{R_B(2.4)^3}{3(200)(10^9)(400)(10^{-6})}$$

or

$$138(10^{-3}) - 1.92(10^{-7})R_B = 0.5(10^{-3}) + 0.66(10^{-7})R_B + 0.58(10^{-7})R_B$$

$$R_B = \frac{0.88(10^{-3})}{3.16(10^{-7})} = 2785 \text{ N}$$

The reactions at A are obtained using a free body of span AB.

$$M_A = \frac{(8 \text{ kN/m})(2.4 \text{ m})^2}{2} - (2785 \text{ N})(2.4 \text{ m}) = 16,356 \text{ N} \cdot \text{m}$$

$$R_A = (8 \text{ kN/m})(2.4 \text{ m}) - 2785 \text{ N} = 16,415 \text{ N}$$

Reactions at D are computed using the free body of span CD.

$$M_D = (2785 \text{ N})(2.4 \text{ m}) = 6684 \text{ N} \cdot \text{m}$$

$$R_D = 2785 \text{ N}$$

The axial force in the post is equal to the force R_B, or 2785 N (compression). The deflection at C is the deflection of span CD due to a force at the free end of magnitude 2785 N.

$$v_C = \frac{(2785 \text{ N})(2.4 \text{ m})^3}{(3)(200)(10^9 \text{ N/m})(400)(10^{-6} \text{ m}^4)} = 0.162(10^{-3}) \text{ m} = 0.162 \text{ mm} \qquad \blacksquare$$

11.5 EFFECTS OF TEMPERATURE ON INDETERMINATE STRUCTURES

The effect of temperature change on engineering materials was discussed briefly in Chapter 3. The concept of statically indeterminate structures had not yet been introduced. The deformation caused by temperature change can be accurately computed for elementary structures using the basic equation that was introduced and illustrated in Chapter 3.

$$u = \alpha(\Delta T)L \qquad (11.3)$$

The axial deformation for a rod or wire is u, α is the coefficient of thermal expansion, ΔT is the change in temperature, and L is the original length of the rod or wire.

The use of Eq. (11.3) was illustrated in Example 3.1. When a rod is fixed at each end and subjected to at temperature change, the rod attempts to deform. The deformation is prevented; hence, an axial deformation cannot occur. A stress corresponding to the amount of deformation that would occur in a rod that is free to deform occurs in the rod. The stress is of interest as well as the deformation. If the rod is connected to other deformable structures, the deformation of the rod will produce deformation effects on the entire structure. The computation of these effects is the main topic to be discussed in this section.

The coefficient of thermal expansion varies somewhat for different materials. Several common materials and their approximate coefficients of thermal expansion are given in the following table:

Aluminum	$12.5(10^{-6})/°F$	$22(10^{-6})/°C$
Magnesium	$14.5(10^{-6})/°F$	$26(10^{-6})/°C$
Copper	$10(10^{-6})/°F$	$18(10^{-6})/°C$
Brass	$11(10^{-6})/°F$	$20(10^{-6})/°C$
Steel	$6.5(10^{-6})/°F$	$12(10^{-6})/°C$
Bronze	$10.5(10^{-6})/°F$	$19(10^{-6})/°C$

The force method, as presented in the previous sections, will be used to illustrate the analysis of indeterminate structures subject to temperature change. Consider the following example that should serve as an introduction to the problem and demonstrate a method of analysis.

EXAMPLE 11.19

Two bars are placed end to end, as shown in Fig. 11.26. The bars are rigidly attached at A and B.
a. Compute the stress in each bar if a temperature increase of 50°C occurs.
b. Assume a gap of 0.5 mm between the support and the rod at point B and compute the stress in each rod.

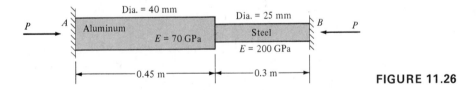

FIGURE 11.26

Solution:
a. The total elongation of the two bars (if they were free) is

$$u_T = u_{st} + u_{al}$$

where u_{st} for the steel member is

$$u_{st} = \alpha(\Delta T)L = [12(10^{-6})/°C](50°C)(0.3 \text{ m}) = 1.8(10^{-4}) \text{ m} \qquad \textbf{(a)}$$

and for the aluminum rod

$$u_{al} = [22(10^{-6})/°C](50°C)(0.45 \text{ m}) = 4.95(10^{-4}) \text{ m}$$

$$u_T = 6.75(10^{-4}) \text{ m} \tag{b}$$

A free-body section cut anywhere in the structure would indicate an axial force of P. The total deformation could be written in terms of axial force.

$$u_T = P\left(\frac{L}{AE}\right)_{st} + P\left(\frac{L}{AE}\right)_{al} \tag{c}$$

$$6.75(10^{-4}) \text{ m} = \frac{4P}{\pi}\left[\frac{0.3 \text{ m}}{(0.04 \text{ m})^2(200)(10^9) \text{ N/m}^2} + \frac{0.45 \text{ m}}{(0.025 \text{ m})^2(70)(10^9) \text{ N/m}^2}\right]$$

$$P = 47.24 \text{ kN} \tag{d}$$

$$\sigma_{st} = \frac{47.24 \text{ kN}}{\pi(0.04 \text{ m})^2/4} = 37.59 \text{ MPa}$$

$$\sigma_{al} = \frac{47.24 \text{ kN}}{\pi(0.025 \text{ m})^2/4} = 96.24 \text{ MPa}$$

The stresses are somewhat high for this example problem and demonstrate the significant effect due to changes in the thermal environment..

A gap of 0.5 mm at B means that the rod can expand prior to making contact with the wall. There is no stress as the rod expands freely. The total expansion of the rod would still be given by Eqs. (a) and (b). Equation (c) is modified to include the effects of the gap.

$$6.75(10^{-4}) \text{ m} - 5.0(10^{-4}) \text{ m} = P\left(\frac{L}{AE}\right)_{st} + P\left(\frac{L}{AE}\right)_{al} \tag{e}$$

$$P = \frac{1.75(10^{-8})}{1.43(10^{-8})} = 12.24 \text{ kN}$$

$$\sigma_{st} = 9.75 \text{ MPa}$$

$$\sigma_{al} = 24.95 \text{ MPa}$$

The gap has significant effect on the stress. ∎

EXAMPLE 11.20

Two steel fixed-free beams are connected at their free ends by a rod as shown in Fig. 11.27. Beam AB is an S4x9.5, 10 ft long; and beam CD is an S5x10, 8 ft long. The rod is magnesium, originally 12 ft in length and with an area of 4 in.^{2}. Compute the axial force in the rod if the temperature of the rod decreases 60°F.

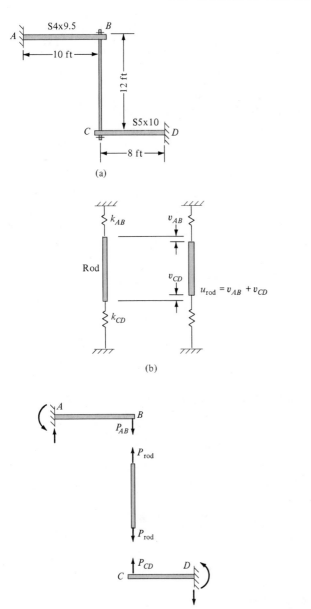

(a)

(b)

(c)

FIGURE 11.27

Solution:

The structure of Fig. 11.27a is idealized in Fig. 11.27b as a rod connected to a spring at each end. The contraction of the rod is seen to be equal to the sum of the spring deformations.

$$u_{\text{rod}} = v_{AB} + v_{CD} \qquad u_{\text{rod}} = u_{\text{temp}} - u_{\text{axial force}} \qquad \textbf{(a)}$$

A free-body section for the spring rod system would indicate that the same axial force acts in the rod and springs.

$$P_{\text{rod}} = P_{AB} = P_{CD} \qquad \textbf{(b)}$$

The idealization is recognized to represent the free bodies of Fig. 11.27c. The deflections of Eq. (a) can be evaluated using the superposition tables and the beam properties of Appendix C:

	Beam	I	L	E
Span AB	S4x9.5	6.79 in.^{4}	120 in.	$30(10^6)$ psi
Span CD	S5x10	12.30 in.^{4}	96 in.	$30(10^6)$ psi
Rod BC	$L = 144$ in.; $A = 4$ in.^{2}; $E = 6.5(10^6)$ psi; $\alpha = 14.5(10^{-6})/°\text{F}$			

Equation (a), after substituting Eq. (b), becomes

$$14.5(10^{-6})(60)(144) = \frac{P_{\text{rod}}}{(3)(30)(10^6)}\left[\frac{(120)^3}{6.79} + \frac{(96)^3}{12.30}\right] + \frac{P_{\text{rod}}(144)}{(4)(6.5)(10^6)}$$

$$P_{\text{rod}} = 34.5 \text{ lb}$$

The axial force is practically negligible. Assuming that the same rod is connected between unyielding supports, the axial force would be tension, computed as

$$u_{\text{rod}} = \frac{P_{\text{rod}}L}{AE} = \alpha(\Delta T)L$$

$$P_{\text{rod}} = \alpha(\Delta T)AE = 14.5(10^{-6})(60)(4)(6.5)(10^6) = 22{,}620 \text{ lb}$$

The yielding supports have significant effect upon the axial force.   ∎

11.6   STATICALLY INDETERMINATE TRUSSES

Trusses can be statically indeterminate externally, internally, or both. In this section, externally indeterminate trusses will be discussed briefly, and an example problem will illustrate a method of analysis for an internally indeterminate truss.

The truss of Fig. 11.28 rests on three simple supports and is indeterminate to the first degree. The truss analysis is similar to beam analysis. The support at B is removed and v_B, the deflection at B, can be computed using the unit load method. A force R_B is applied at B

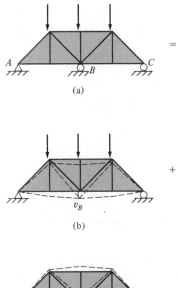

(a)

=

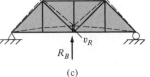

(b)

+

(c)

FIGURE 11.28
Statically indeterminate truss.

as shown in Fig. 11.28c. Again, the deflection v_R caused by a force applied at B can be computed using the unit load method. The unknown reaction is computed using the relation

$$v_R = v_B$$

The remaining external reactions at A and C can be computed using the equations of statics.

An internally statically indeterminate truss is shown in Fig. 11.29. The panel $BCDE$ contains two diagonal members. It is not possible to compute the internal forces using the equations of statics because each joint of the panel contains more unknowns than can be computed using the two equilibrium equations and geometry conditions. The truss analysis follows the concept of the force method of Section 11.2.

Any member of panel $BCDE$ can be removed and the truss will be determinate. The movement between joints connected by the redundant member can be computed. Then the force required to return the joints to their original position will equal the force in the redundant member. The next example will illustrate the method.

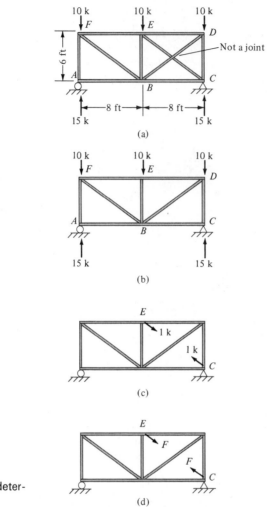

FIGURE 11.29
Internally statically indeterminate truss.

EXAMPLE 11.21

Compute the force in each member of the truss of Fig. 11.29. Assume that all truss members have the same area and modulus of elasticity.

Solution:
The analysis that was outlined briefly will be carried out in detail. Assume member *EC* as the redundant. The choice is not completely arbitrary. Removing member *EC*

TABLE 11.1

	P, k	p, k	L, ft	$p\dfrac{PL}{AE}$	F, k	$p\dfrac{FL}{AE}$	F (actual)
AB	0	0	8	0	0	0	0
BC	0	$-4/5$	8	0	$-4/5$	$5.12F$	7.03
CD	-15	$-3/5$	6	$54/AE$	$-3F/5$	$2.16F$	-9.73
BD	$25/3$	1	10	$83.33/AE$	F	$10F$	-0.45
DE	$-20/3$	$-4/5$	8	$42.67/AE$	$-4F/5$	$5.12F$	0.36
BE	-10	$-3/5$	6	$36/AE$	$-3F/5$	$2.16F$	-4.73
EF	$-20/3$	0	8	0	0	0	-6.67
BF	$25/3$	0	10	0	0	0	8.33
AF	-15	0	6	0	0	0	-15
				$1\cdot u_{CE}=\dfrac{216}{AE}$		$1\cdot u_F=\dfrac{24.56F}{AE}$	$F_{CE}=-8.79$

leaves a primary structure that is symmetrical and hence, easier to analyze. External reactions are computed as 15 k, as shown in Fig. 11.29. The primary structure can be analyzed using the method of joints. The results are given in Table 11.1 under the heading P. A unit load is applied in the direction of the redundant member. Recall that the unit load method requires the solution of the equation.

$$1\cdot \mathbf{u}=\sum_{i=1}^{N}\frac{\mathbf{p}_i\cdot\mathbf{P}_iL_i}{AE} \tag{a}$$

Ideally, a unit load would be placed at joint C and the deflection of joint C computed. Then the unit load would be placed at joint E and the deflection of joint E computed. If both unit loads are placed on the structure at the same time, as shown in Fig. 11.29c, the computation will be simplified. The external reactions are zero and only the members of panel $BCDE$ will have a force. The results are shown in the table under the heading p. Equation (a) is used to compute the relative movement of joints C and E.

$$u_{ce}=\frac{216}{AE}$$

A force F is placed at joints C and D as shown in Fig. 11.29d. The magnitude of the force F should be just enough to return the relative displacements of joints C and E to zero. F is unknown, but the forces in the panel members can be computed in terms of F as illustrated under the heading F in the table. The deflection caused by F can be computed using the unit load method.

$$1\cdot \mathbf{u}_F=\sum_{i=1}^{N}\frac{F_i\cdot P_iL_i}{AE}=\frac{24.56F}{AE} \tag{b}$$

The computation is shown in the table. The deformation condition is

$$u_F = u_{CE}$$

$$F_{CE} = \frac{216}{24.56} = 8.79 \text{ k} \qquad \text{(compression)}$$

The force in member *CE* is 8.79 k. The remaining forces can be computed using the method of joints and are given in the table.   ■

11.7   STATICALLY INDETERMINATE FRAMES AND THREE-DIMENSIONAL STRUCTURES

The force method will be used to illustrate the analysis of frame structures. One of the methods of Chapter 10 for computing deflections of structures must be used in conjunction with the force method. In the examples that follow, the unit load method will be used to compute deflections. Moment equations and moment diagrams for determinate frames were discussed in Chapter 5. A frame structure similar to that of Example 5.13 will be used for an illustrative example. The loading will be less complicated in order to reduce the computational effort.

EXAMPLE 11.22

Compute the reactions for the frame of Fig. 11.30*a*. Use the force method. Assume that *EI* is constant for all members.

Solution:
The basic concept is the same as a beam problem. Choose a redundant—in this case, the horizontal reaction at *D*. Compute the deflection v_D for the primary structure. Apply the reaction as shown in Fig. 11.30*c* and compute v_H in terms of the unknown reaction. The condition for computing H_D is

$$v_H = v_D \tag{a}$$

The unit load method will be used to compute both v_D and v_H. To compute v_D, apply a unit load at *D* and use the theory of the previous chapter.

$$v_D = \int_0^L \frac{mM \, dx}{EI} \tag{b}$$

For convenience let $EI = 1$. Reactions for the primary structure, Fig. 11.31*a*, are computed using the equations of statics and are shown in the figure.

The moment equation along *AB* using the x_1-y_1 axis is zero since there are no horizontal loads.

$$M_{AB} = 0$$

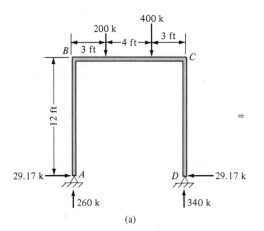

(a)

=

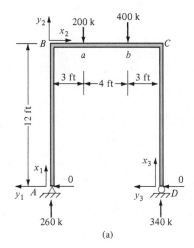

(a)

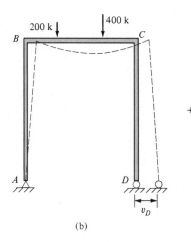

(b)

+

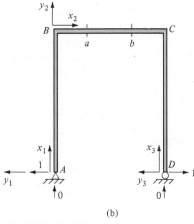

(b)

FIGURE 11.31

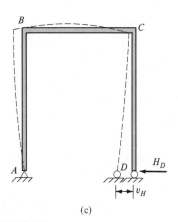

(c)

FIGURE 11.30

Along BC, use the x_2-y_2 axis.

$$M_{Ba} = 260x_2$$

$$M_{ab} = 260x_2 - 200(x_2 - 3) = 60x_2 + 600$$

$$M_{bC} = 260x_2 - 200(x_2 - 3) - 400(x_2 - 7) = -340x_2 + 3400$$

The moment along CD is zero.

$$M_{CD} = 0$$

A unit load is applied at D as shown in Fig. 11.31b. The moment equations for the unit load must be written using the same axis as the previous moments.

$$m_{AB} = (1)(x_1)$$

$$m_{Ba} = m_{ab} = m_{bC} = (1)(12)$$

$$m_{CD} = -(1)(x_3)$$

The deflection of the primary structure is given by the integral of $m \cdot M$ along the total length of the structure.

$$EIv_D = \int_0^{12} (0)\, dx_1 + \int_0^3 (12)(260x_2)\, dx_2 + \int_3^7 (12)(60x_2 + 600)\, dx_2$$

$$+ \int_7^{10} (12)(-340x_2 + 3400)\, dx_2 + \int_0^{12} (0)\, dx_3$$

$$EIv_D = 14{,}040 + 43{,}200 + 18{,}360 = 75{,}600$$

The computation for v_H, as shown in Fig. 11.30c, is accomplished in a similar manner. Assume the horizontal reaction acting to the left on the primary structure as shown in Fig. 11.32a. A unit load is assumed to the left at D. Notice that the moment equations for Fig. 11.32 are the same as Fig. 11.31b except for the sign.

$$M_{AB} = -H_D x_1 \qquad\qquad m_{AB} = -(1)x_1$$

$$M_{BC} = -12H_D \qquad\qquad m_{BC} = -(1)(12)$$

$$M_{CD} = H_D x_3 \qquad\qquad m_{CD} = (1)x_3$$

$$EIv_D = \int_0^{12} H_D x_1\, dx_1 + \int_0^{10} 144H_D\, dx_2 + \int_0^{12} H_D x_3^2\, dx_3$$

$$EIv_D = 576H_D + 1440H_D + 576H_D = 2592H_D$$

Substituting into Eq. (a) gives the desired result

$$2592H_D = 75{,}600$$

$$H_D = 29.17 \text{ k}$$

The remaining reactions are computed using the equations of statics and are shown in Fig. 11.33. Shear and moment diagrams for the frame are shown in Fig. 11.33.

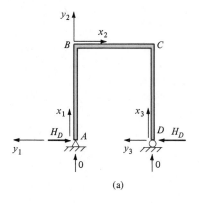

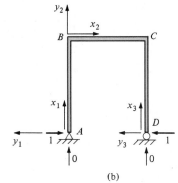

FIGURE 11.32

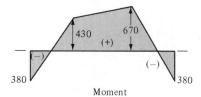

Moment

Shear

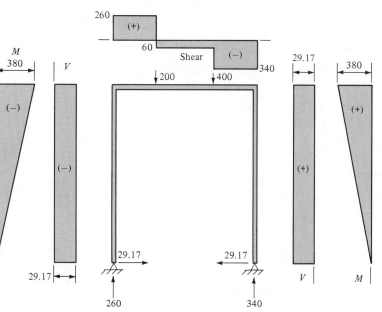

FIGURE 11.33

The stiffness method can be used to analyze frame structures; the analysis would follow that for a beam structure. However, the analysis is complicated by the possible horizontal translation of the frame. The topic should be covered in a course in structural analysis or machine design.

A three-dimensional structure was analyzed in the previous chapter. Vector analysis was used to formulate the moment and force equations and the scalar product was used in conjunction with the application of the unit load method. Again, vector analysis will be used in solving a three-dimensional statically indeterminate structure. The intent is to develop confidence in thinking in terms of vectors and visualizing structures in three dimensions. Again, the force method will be used in conjunction with the unit load method.

EXAMPLE 11.23

Compute the reactions for the frame of Fig. 11.34. The frame is fabricated of circular steel pipe with constant cross-sectional properties along the entire length. Neglect shear and axial deformation when computing deflections.

Solution:
The frame is fixed at A and supported in the y direction at D. Hence, the structure is statically indeterminate to the first degree. Assume D_y as the redundant and proceed to compute v_{Dy}, the deflection at D in the y direction. Moment equations must be written along the entire structure for the actual loading and for a unit load at D applied in the y direction.

The actual load is a moment applied at C. The reactions for the primary structure are shown in Fig. 11.34b. Since there is no force loading on the structure, the force reactions at A are zero.

$$M_A = -6\mathbf{i} - 8\mathbf{j} - 4\mathbf{k}$$

$$\mathbf{F}_A = 0$$

(a)

The moments along each segment of the frame are shown in Figs. 11.34c–e.

$$M_{AB} = 6\mathbf{i} + 8\mathbf{j} + 4\mathbf{k}$$

(b)

$$M_{BC} = 6\mathbf{i} + 8\mathbf{j} + 4\mathbf{k}$$

$$M_{CD} = 0$$

(c)

A unit load is applied upward at D because the reaction was assumed upward, as shown in Fig. 11.35a.

$$\Sigma \mathbf{F} = 0 \quad \mathbf{f} + (1)\mathbf{j} = 0 \quad f_x = 0 \quad f_y = -(1) \quad f_z = 0$$

$$\Sigma \mathbf{m}_A = 0 \quad \mathbf{m}_A + (4\mathbf{i} + 6\mathbf{j} + 2\mathbf{k}) \times (1\mathbf{j}) = 0$$

$$\mathbf{m}_{Ax} = 2\mathbf{i} \quad \mathbf{m}_{Ay} = 0 \quad \mathbf{m}_{Az} = -4\mathbf{k}$$

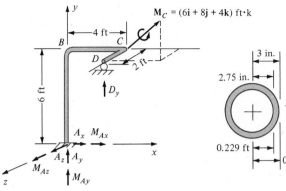

$\mathbf{M}_C = (6\mathbf{i} + 8\mathbf{j} + 4\mathbf{k})$ ft·k

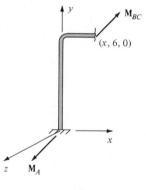

3 in.

2.75 in.

$E = 30(10^3)$ ksi
$G = 12(10^3)$ ksi

0.229 ft

0.25 ft

(a)

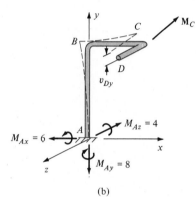

(b)

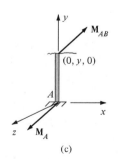

(c)

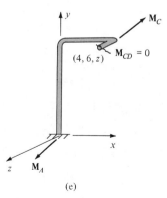

(d)

(e)

FIGURE 11.34

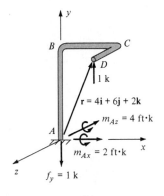

$$\mathbf{r} = 4\mathbf{i} + 6\mathbf{j} + 2\mathbf{k}$$

$m_{Az} = 4 \text{ ft·k}$

$m_{Ax} = 2 \text{ ft·k}$

$f_y = 1 \text{ k}$

(a)

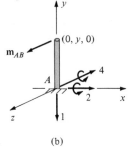

(b)

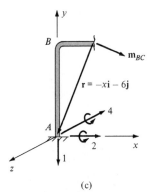

$$\mathbf{r} = -x\mathbf{i} - 6\mathbf{j}$$

(c)

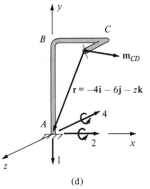

$$\mathbf{r} = -4\mathbf{i} - 6\mathbf{j} - z\mathbf{k}$$

(d)

FIGURE 11.35

Refer to Fig. 11.35*b*.

$$\mathbf{m}_{AB} + 2\mathbf{i} - 4\mathbf{k} + (-y\mathbf{j}) \times (-1\mathbf{j}) = 0 \qquad \textbf{(d)}$$

$$\mathbf{m}_{ABx} = -2\mathbf{i} \qquad \mathbf{m}_{ABy} = 0 \qquad \mathbf{m}_{ABz} = 4\mathbf{k}$$

Refer to Fig. 11.35*c*.

$$\mathbf{m}_{BC} + 2\mathbf{i} - 4\mathbf{k} + (-x\mathbf{i} - 6\mathbf{j}) \times (-1\mathbf{j}) = 0 \qquad \textbf{(e)}$$

$$\mathbf{m}_{BCx} = -2\mathbf{i} \qquad \mathbf{m}_{BCy} = 0 \qquad \mathbf{m}_{BCz} = (4 - x)\mathbf{k}$$

Refer to Fig. 11.35d.

$$\mathbf{m}_{CD} + 2\mathbf{i} - 4\mathbf{k} + (-4\mathbf{i} - 6\mathbf{j} - z\mathbf{k}) \times (-1\mathbf{j}) = 0$$

$$\mathbf{m}_{CDx} = (z - 2)\mathbf{i} \qquad \mathbf{m}_{CDy} = 0 \qquad \mathbf{m}_{CDz} = 0$$

(f)

The deflection is given by evaluating the unit load equation.

$$1 \cdot v_{Dy} = \int_0^L \frac{m \cdot M \, dx}{EI}$$

$$v_{Dy} = \int_0^6 (-2\mathbf{i} + 4\mathbf{k}) \cdot \left(\frac{6\mathbf{i}}{EI} + \frac{8\mathbf{j}}{JG} + \frac{4\mathbf{k}}{EI} \right) dy$$

$$+ \int_0^4 [-2\mathbf{i} + (4 - x)\mathbf{k}] \cdot \left(\frac{6\mathbf{i}}{JG} + \frac{8\mathbf{j}}{EI} + \frac{4\mathbf{k}}{EI} \right) dx + \int_0^2 (0) \, dz$$

The first integral is the scalar product of Eqs. (d) and (b). The moment vector along the y axis is an axial moment or torque and is divided by JG. The other moments are bending moments and divided by EI. The second integral is the scalar product of Eqs. (e) and (c). The torque is acting along the x axis.

$$v_{Dy} = \int_0^6 \frac{(-12 + 16)}{EI} \, dy + \int_0^4 \frac{-12}{JG} + \frac{4(4 - x)}{EI} \, dx$$

$$= \frac{24}{EI} - \frac{48}{JG} + \frac{64}{EI} - \frac{32}{EI} = \frac{56}{EI} - \frac{48}{JG}$$

(g)

Assume the reaction at D is upward (positive) as shown in Fig. 11.36. The reactions and moment equations for D_y turn out to be similar to the previous analysis for the unit load with (1) replaced by D_y. Along AB use the free body of Fig. 11.35b with the reactions of Fig. 11.36.

$$\mathbf{M}_{ABx} = -2D_y\mathbf{i} \qquad \mathbf{M}_{ABy} = 0 \qquad \mathbf{M}_{ABz} = 4D_y\mathbf{k}$$

(h)

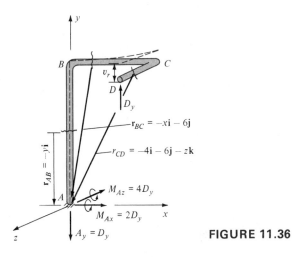

FIGURE 11.36

Along BC,

$$\mathbf{M}_{BC} + 2D_y\mathbf{i} - 4D_y\mathbf{j} + (-x\mathbf{i} - 6\mathbf{j}) \times (-D_y\mathbf{j}) = 0$$

$$\mathbf{M}_{BCx} = -2D_y\mathbf{i} \qquad \mathbf{M}_{BCy} = 0 \qquad \mathbf{M}_{BCz} = (4D_y - D_yx)\mathbf{k}$$

(i)

Along CD,

$$\mathbf{M}_{CD} + 2D_y\mathbf{i} - 4D_y\mathbf{k} + (-4\mathbf{i} - 6\mathbf{j} - z\mathbf{k}) \times (-D_y\mathbf{j}) = 0$$

$$\mathbf{M}_{CDx} = (z - 2)D_y\mathbf{i} \qquad \mathbf{M}_{CDy} = 0 \qquad \mathbf{M}_{CDz} = 0$$

(j)

The deflection caused by the unknown reaction at D requires a unit load at D. The result would be the same as Figs. 11.35a–d given by Eqs. (d)–(f).

$$1 \cdot \mathbf{v}_{\text{reaction}} = \int_0^6 (-2\mathbf{i} + 4\mathbf{k}) \cdot \left(\frac{-2D_y\mathbf{i}}{EI} + \frac{4D_y\mathbf{k}}{EI} \right) dy$$

$$+ \int_0^4 [-2\mathbf{i} + (4 - x)\mathbf{k}] \cdot \left[\frac{-2D_y\mathbf{i}}{JG} + \frac{(4 - x)D_y\mathbf{k}}{EI} \right] dx$$

$$+ \int_0^2 (z - 2)\mathbf{i} \cdot \frac{(z - 2)D_y\mathbf{i}}{EI} dz$$

The first integral is obtained from Eqs. (d) and (h), the second from Eqs. (e) and (i), and the third from Eqs. (f) and (j).

$$v_{\text{reaction}} = \int_0^6 \frac{20}{EI} D_y \, dy + \int_0^4 \left[\frac{4}{JG} + \frac{(16 - 8x + x^2)}{EI} \right] D_y \, dx$$

$$+ \int_0^2 \frac{(4 - 4z + z^2)}{EI} D_y \, dz$$

$$v_{\text{reaction}} = \frac{20}{EI} yD_y \Big|_0^6 + \frac{4x}{JG} + \frac{1}{EI} \left(16x - 4x^2 + \frac{x^3}{3} \right) D_y \Big|_0^4 + \left(4z - 2z^2 + \frac{z^3}{3} \right) \frac{D_y}{EI} \Big|_0^2$$

$$v_{\text{reaction}} = \frac{16}{JG} D_y + \frac{432}{3EI} D_y$$

(k)

Set Eqs. (k) and (g) equal and note that v_{Dy} is assumed upward in Fig. 11.34b and v_{reaction} is also assumed upward in Fig. 11.35. Both solutions correspond to a positive unit load in Fig. 11.35a. The deflection of the primary structure and redundant structure cannot be in the same direction; hence, the following relation contains a minus sign.

$$-v_{\text{reaction}} = v_{Dy}$$

$$-D_y = \frac{(56/EI) - (48/JG)}{(432/3EI) + (16/JG)}$$

(l)

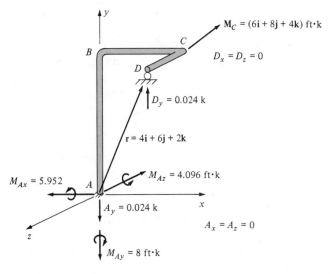

FIGURE 11.37

The constants should be evaluated before proceeding. The units on the right-hand side of Eq. (l) will be kips if $E = 30(10^3)$ ksi (144) in.^{2}/ft^2 = $4.32(10^6)$ k/ft^2; $G = 12(10^3)(144) = 1.73(10^6)$; and I and J are ft^4. The cross section of Fig. 11.34 is used to compute J and I.

$$I = \frac{\pi[(0.25)^4 - (0.229)^4]}{4} = 9.1(10^{-4}) \text{ ft}^4$$

$$J = 2I = 18.2(10^{-4}) \text{ ft}^4$$

$$GJ = 1.73(10^6)18.2(10^{-4}) = 3.15(10^3) \text{ lb·ft}^2$$

$$EI = 4.32(10^6)9.1(10^{-4}) = 3.93(10^3) \text{ lb·ft}^2$$

$$-D_y = \frac{(56/3.93) - (48/3.15)}{(432/3)(3.93) + (16/3.15)}$$

$$D_y = \frac{0.99}{41.7} = 0.024 \text{ k}$$

D_y is upward as shown in Fig. 11.37.

The equations of statics are used to complete the solution for reactions. Sum forces and moments about A. The results are shown in Fig. 11.37.

$$A_x = 0 \qquad A_y = -0.024 \text{ k} \qquad A_z = 0$$

$$\Sigma M_A = 0$$

$$M_{Ax}\mathbf{i} + M_{Ay}\mathbf{j} + M_{Az}\mathbf{k} + (4\mathbf{i} + 6\mathbf{j} + 2\mathbf{k}) \times (0.024\mathbf{j}) + 6\mathbf{i} + 8\mathbf{j} + 4\mathbf{k} = 0$$

$$M_{Ax} = -5.952 \text{ ft·k} \qquad M_{Ay} = -8 \text{ ft·k} \qquad M_{Az} = -4.096 \text{ ft·k} \qquad ■$$

11.8   SUMMARY

The analysis of statically indeterminate structures requires considerably more understanding and experience than determinate structures. Basic concepts have been presented in this chapter. There are numerous specialized techniques of analysis that have emerged throughout the years. Usually, a special technique works well for a special application. With this in mind, the chapter has dealt primarily with the force method.

The stiffness method is conceptually more difficult than the force method. However, the stiffness method is the workhorse of the structural engineering profession. The theory and examples in this chapter should serve as the first step in learning and understanding the stiffness method.

The majority of structural analysis deals with indeterminate structures. The civil engineer deals with structures that have multi-indeterminate properties. These structures are usually of large scale and often must be studied in three dimensions. The mechanical engineer may deal with intricate space mechanisms and linkages. A brief introduction to frames and three-dimensional structures should give readers in these fields a beginning point for further study.

For Further Study

Laible, J. P., *Structural Analysis*, Holt, Rinehart and Winston, New York, 1985.

Sack, R. L., *Structural Analysis*, McGraw-Hill, New York, 1984.

Wang, C. K., *Intermediate Structural Analysis*, McGraw-Hill, New York, 1983.

11.9   PROBLEMS

11.1 Compute the reactions for the rod of Fig. P11.1. Sketch an axial-force diagram. Assume that E is constant for the entire rod.

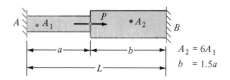

FIGURE P11.1

11.2 For the rod of Fig. P11.1 assume $L = 16$ in. and $A_1 = 1.5$ in.^{2} and compute the allowable force P if the normal stress is not to exceed 16,000 psi tension or compression.

11.3 Compute the reactions for the rod of Fig. P11.3. The applied forces are related such that $P = P_1 = 4P_2$. Assume AE to be constant for the beam.

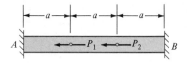

FIGURE P11.3

11.4 Plot the axial-force diagram and deformation diagram for the rod of Problem 11.3.

11.5 A tapered steel rod shown in Fig. P11.5 is fixed at both ends. Compute the reactions at A and B. $E = 200$ GPa. The thickness of the rod is 50 mm.

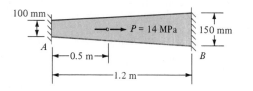

100 mm

$P = 14$ MPa

150 mm

A

$\leftarrow$0.5 m$\rightarrow$

B

$\leftarrow$1.2 m$\rightarrow$

FIGURE P11.5

11.6 The circular shaft of Fig. P11.6 is fixed against rotation at each end. Assume constant JG and compute the reactions at A and B.

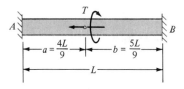

T

A

B

$a = \dfrac{4L}{9}$

$b = \dfrac{5L}{9}$

L

FIGURE P11.6

11.7 The circular shaft of Fig. P11.7 is fabricated using steel for section AC and bronze for section CB. The diameter is 4 in. Compute the reactions at A and B. $G_{bronze} = 6.5(10^6)$ psi; $G_{steel} = 11(10^6)$ psi.

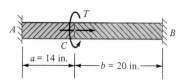

T

A

B

C

$a = 14$ in.

$b = 20$ in.

FIGURE P11.7

11.8 For the shaft and data of Problem 11.7, assume that the reactions are to be equal, $T_A = T_B$. Let the diameter of section AC remain 4 in. and compute the diameter of section CB to satisfy the requirement that $T_A = T_B$.

11.9 The shaft of Fig. P11.9 has constant JG but is subject to torques T_1 and T_2 as shown. For $T_2 = 2T_1$ compute the reactions.

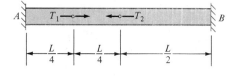

A

T_1

T_2

B

$\dfrac{L}{4}$

$\dfrac{L}{4}$

$\dfrac{L}{2}$

FIGURE P11.9

11.10 A cylindrical test specimen of Fig. P11.10 has an inner core of one material surrounded by a second material. The properties of the test specimen are in the following proportions: $A_{inside} = A_{outside}/2$; $E_{inside} = E_{outside}/2.5$. Determine the percentages of the axial force carried by each material.

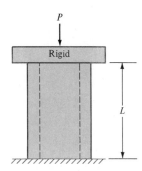

P

Rigid

L

FIGURE P11.10

11.11 The structure shown in Fig. P11.11 is made of two different types of wood. Wood A has $E = 2(10^6)$ psi and wood B has $E = 1.3(10^6)$ psi. The areas are $A_A = 4$ in.^{2} and $A_B = 2.5$ in.^{2}. Note that there are two sections with area A_A. Compute the load carried by each section.

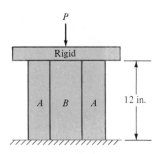

P

Rigid

A B A

12 in.

FIGURE P11.11

11.12 For the structure of Fig. P11.11 assume that each of the three members is to carry an equal force. Assume the modulus of elasticity properties of Problem 11.11 and compute the area required for Wood A in terms of Wood B.

11.13 A square steel container with outside dimensions 100 mm by 100 mm has sides 6 mm thick. The container is filled with a second material, $E = 30$ GPa and leaves a gap of $\delta = 0.025$ mm at the top of the container as shown in Fig. P11.13. If a single force of 275 kN is applied to the container, compute the force carried by each material. $E_{st} = 200$ GPa.

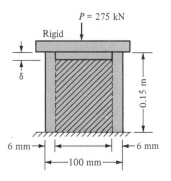

FIGURE P11.13

11.14 A brass rod 0.5 m long is connected to an aluminum rod 0.55 m long as shown in Fig. P11.14. The brass rod has an area of 144 mm^2 and the aluminum 100 mm^2. The gap at point B is 1.5 mm. A load $P = 55$ kN is applied as shown. Compute the reactions at A and B. $E_{brass} = 100$ GPa; $E_{al} = 70$ GPa.

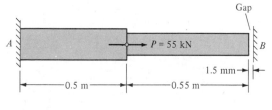

FIGURE P11.14

11.15 A steel 1/2-in.-diameter bolt is snugly attached through an undeformable material 4 in. thick. The bolt has 18 threads per inch. Assume the nut is tightened

one-tenth of a full turn and compute the stress in the bolt. $E_{st} = 30(10^6)$ psi.

11.16 A 0.25-in.-diameter steel bolt is placed inside of a copper tube, $E = 16(10^6)$ psi, with a cross-sectional area of 0.12 in.^{2}. The bolt has 10 threads per inch. Initially the nut fits snugly against the tube. If the nut is tightened one-eighth of a turn, compute the stress in the bolt and the tube of Fig. P11.16.

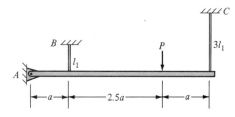

FIGURE P11.16

11.17 The structure of Fig. P11.17 illustrates a rigid bar supported by two wires and a pin support at A. Assume that the wires are made of the same material and have the same cross-sectional area and compute the reactions at A, B, and C.

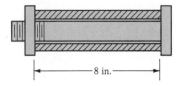

FIGURE P11.17

11.18 For the rigid bar of Fig. P11.17 assume $P = 50$ k, wire B is steel with an area of 0.15 in.^{2}, wire C is bronze with an area of 0.2 in.^{2}, and $l_1 = 10$ in. Compute the reactions at A, B, and C.

11.19 The rigid bar of Fig. P11.19 is pinned at A and supported by a wire at B. A small gap occurs between the bar and the support at C. Assume $P = 4.5$ k, $a = 8$ in., and $b = 16$ in. The length of the wire is 14 in. with an area of 0.05 in.^{2}. The support at C is a wooden post 12 in. long with an area of 2 in.^{2}. Assume that E for the wire is $30(10^3)$ ksi and E for the post is $1.2(10^3)$ ksi. If the gap δ_C is 0.015 in., compute the reactions at A, B, and C.

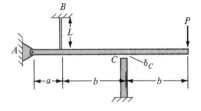

FIGURE P11.19

11.20 Compute the reactions at A, B, and C of Fig. P11.20 in terms of L, K_1, K_2, and P. Assume a rigid bar.

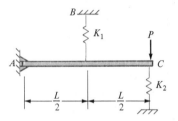

FIGURE P11.20

11.21 Compute the reactions at A, B, and C for the rigid bar of Fig. P11.21. Assume wire B is aluminum with a diameter of 4 mm and wire C is copper with a diameter of 2.5 mm. Assume M_0 is 6 kN·m. See Appendix B.

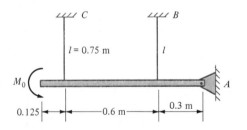

FIGURE P11.21

The force method and superposition tables should be used for Problems 11.22–11.47. However, other methods of analysis are equally applicable. After computing reactions, construct shear and moment diagrams if instructed to do so.

11.22 Compute the reactions in Fig. P11.22.

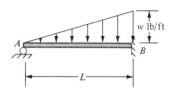

FIGURE P11.22

11.23 Figure P11.23 shows a W8x18 beam. Compute the reactions at A and B, then compute the maximum bending stress in the beam.

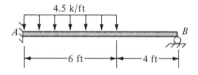

FIGURE P11.23

11.24 Compute the reactions in Fig. P11.24 at A, B, and C and construct shear and moment diagrams.

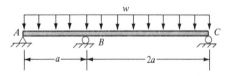

FIGURE P11.24

11.25 Compute the reactions in Fig. P11.25 at A, B, and C.

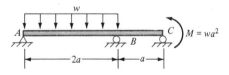

FIGURE P11.25

11.26 Compute the reactions in Fig. P11.26 at A, B, and C.

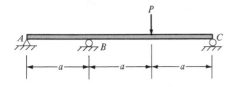

FIGURE P11.26

11.27 Compute the reactions in Fig. P11.27.

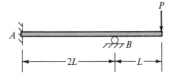

FIGURE P11.27

11.28 Compute the reactions in Fig. P11.28.

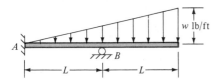

FIGURE P11.28

11.29 Compute the reactions and construct shear and moment diagrams for Fig. P11.29.

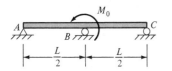

FIGURE P11.29

11.30 Compute the reactions and construct shear and moment diagrams for Fig. P11.30.

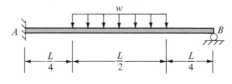

FIGURE P11.30

11.31 Compute the reactions in Fig. P11.31.

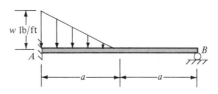

FIGURE P11.31

11.32 Compute the reactions and construct shear and moment diagrams for the beam of Fig. P11.32.

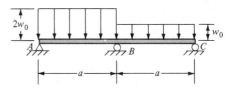

FIGURE P11.32

11.33 Compute the reactions and construct shear and moment diagrams for Fig. P11.33.

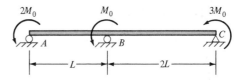

FIGURE P11.33

11.34 Compute the reactions in Fig. P11.34.

11.35 Compute the reactions in terms of δ for the beam of Fig. P11.35.

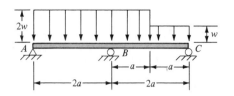

FIGURE P11.34

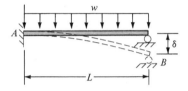

FIGURE P11.35

11.36 Compute the reactions in terms of δ for Fig. P11.36.

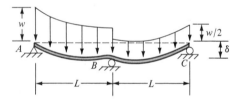

FIGURE P11.36

11.37 Compute the reactions in terms of θ for Fig. P11.37.

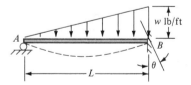

FIGURE P11.37

11.38 Compute the reactions in terms of θ for Fig. P11.38.

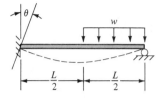

FIGURE P11.38

11.39 Compute the reactions for the beam of Fig. P11.39.

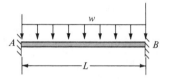

FIGURE P11.39

11.40 Solve Example 11.9 of the text assuming the moment reactions at A and B as redundants.

11.41 Compute the reactions for Fig. P11.41.

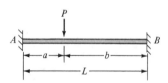

FIGURE P11.41

11.42 Compute the reactions for Fig. P11.42.

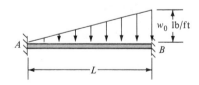

FIGURE P11.42

11.43 Compute the reactions for the structure of Fig. P11.43. Choose the reactions at B and C as redundants and use the superposition tables.

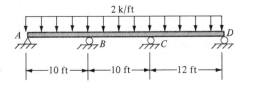

2 k/ft

A B C D

|—10 ft—|—10 ft—|—12 ft—|

FIGURE P11.43

11.44 Compute the reactions at A, B, and C for the beam of Fig. P11.44. Choose the reactions at B and C as redundants and use the superposition tables.

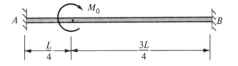

w

A B C

|—L—|—L—|

FIGURE P11.44

11.45 Compute the reactions at A and B for the structure of Fig. P11.45. Choose the moment and force reactions at A as redundants.

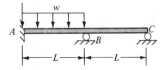

M_0

A B

|—$\frac{L}{4}$—|—$\frac{3L}{4}$—|

FIGURE P11.45

11.46 Compute the reactions at A and B for the structure of Fig. P11.45. Choose the moment reactions at A and B as redundants.

11.47 Compute the reactions for Fig. P11.47.

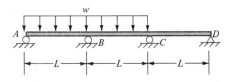

w

A B C D

|—L—|—L—|—L—|

FIGURE P11.47

11.48—11.56 Problems 11.48–11.56 are intended for analysis using the integration method. However, they can be analyzed using other methods. Compute the reactions for the beam of the figure corresponding to the problem number.

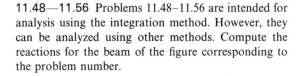

$w = w_0 \dfrac{x^2}{L^2}$

w_0 lb/ft

A B

|————L————|

FIGURE P11.48

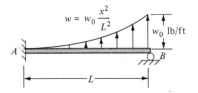

$w = \dfrac{-w_0 x}{L}$

w_0 lb/ft

A B

|————L————|

FIGURE P11.49

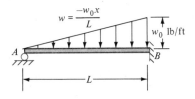

$w = w_0 \left(\dfrac{x^2}{L^2} - 1 \right)$

w_0 lb/ft

A B

|————L————|

FIGURE P11.50

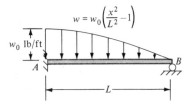

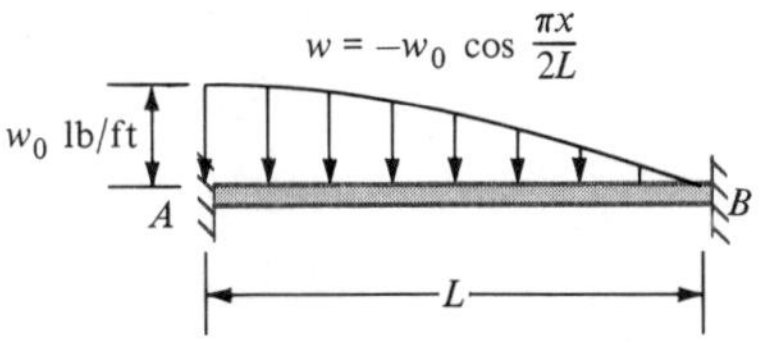

$w = -w_0 \cos \dfrac{\pi x}{2L}$

w_0 lb/ft

A B

|————L————|

FIGURE P11.51

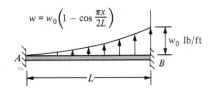

$$w = w_0\left(1 - \cos\frac{\pi x}{2L}\right)$$

w_0 lb/ft

FIGURE P11.52

11.57—11.68 Problems 11.57–11.68 should be analyzed using the displacement method. Compute the force and moment reactions for the beam in each figure and construct shear and moment diagrams. For the structures of Figs. P11.61 and P11.67 divide the load into a uniform load and a uniformly varying load to facilitate using the beam tables.

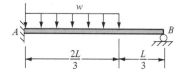

FIGURE P11.57

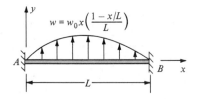

$$w = w_0 x\left(\frac{1 - x/L}{L}\right)$$

FIGURE P11.53

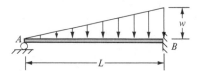

FIGURE P11.58

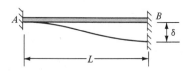

θ_B

FIGURE P11.54

M_0

FIGURE P11.59

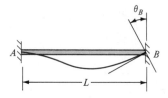

δ

FIGURE P11.55

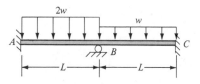

$2w$

w

FIGURE P11.60

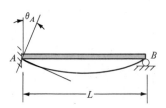

θ_A

FIGURE P11.56

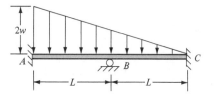

$2w$

FIGURE P11.61

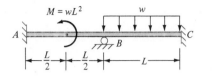

FIGURE P11.62

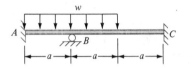

FIGURE P11.63

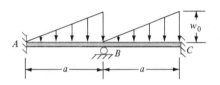

FIGURE P11.64

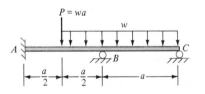

FIGURE P11.65

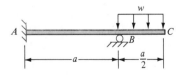

FIGURE P11.66

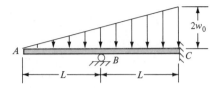

FIGURE P11.67

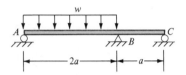

FIGURE P11.68

11.69 The S8x23 beam of Fig. P11.69 carries a partial uniform load of 800 lb/ft. The beam is 14 ft long.
a. Compute the reactions at A and B.
b. Assume support B settles an amount $\delta = 0.25$ in. and compute the reactions.
c. How much would the support at B have to settle for the reaction at B to be one-half of the reaction for no settlement? $E = 30(10^6)$ psi. (See part **a.**)

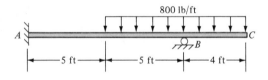

FIGURE P11.69

11.70 Compute the reaction at A and B for the W8x18 beam of Fig. P11.70. Both the beam and the rod are steel.

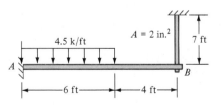

FIGURE P11.70

11.71 The wooden beam of Fig. P11.71 is fixed at the left end and supported by a short wooden post at the right end. Assume E for the beam is $2(10^6)$ psi and for the post is $1.5(10^6)$ psi.

a. Compute the reactions at A and B and find the axial stress in the post.

b. Assume a gap of 0.1 in. between the beam and post prior to loading and compute the axial stress in the post.

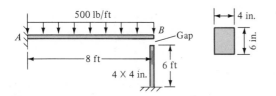

FIGURE P11.71

11.72 Compute the reactions at A and B in terms of k, M_0, L, and EI for the beam of Fig. P11.72.

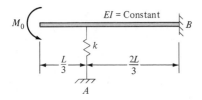

FIGURE P11.72

11.73 A steel rod connects two steel beams as shown in Fig. P11.73. Compute the force in the rod and the reactions at A and D.

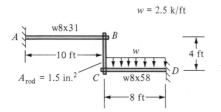

FIGURE P11.73

11.74 The fixed-free beam of Fig. P11.74 is supported by a rod as shown. A small gap exists between the free end of the beam and the rod connection. Compute the axial force in the rod.

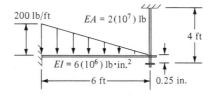

FIGURE P11.74

11.75 Compute the axial force in the rod and the deflection at D for the structure of Fig. P11.75.

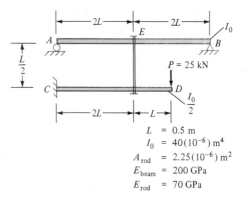

$$
\begin{aligned}
L &= 0.5 \text{ m} \\
I_0 &= 40(10^{-6}) \text{ m}^4 \\
A_{\text{rod}} &= 2.25(10^{-6}) \text{ m}^2 \\
E_{\text{beam}} &= 200 \text{ GPa} \\
E_{\text{rod}} &= 70 \text{ GPa}
\end{aligned}
$$

FIGURE P11.75

11.76 Compute the reaction at D in terms of W, I_0, and E for the beam structure of Fig. P11.76. Assume E is the same for both beams. Both beams are simply supported and beam EC rests on beam AB prior to loading.

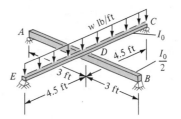

FIGURE P11.76

11.77 Compute the reactions for the beam of Fig. P11.77. *Hint:* Make the support at *C* redundant and use the unit load method to compute the deflection at *C*.

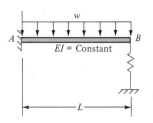

$M_0 = 8(10^3)$ lb·in.

$EI = 4.5(10^6)$ lb·in.^{2}

$k = 8.5(10^3)$ lb/in.

|←——50 in.——→|←——50 in.——→|

FIGURE P11.77

11.78 Compute the stiffness of the spring at *B* to make the reaction at *A* equal 3 times the reaction at *B* for the beam of Fig. P11.78.

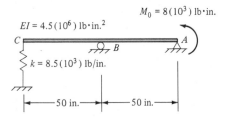

w

EI = Constant

L

FIGURE P11.78

11.79 An aluminum rod 6 mm by 6 mm, 0.30 m in length, is attached between rigid supports. For a temperature increase of 20°C compute the axial stress in the rod.

11.80 Given the rod of Problem 11.79, assume there is a gap of 0.08 mm between the end of the rod and the support at one end. The opposite end remains rigidly attached. For a temperature increase of 50°C, compute the stress in the rod.

11.81 A rod is fabricated from two materials as shown in Fig. P11.81. Section *AB* is bronze and section *BC* is copper. Compute the stress in the rod caused by a 30°C drop in temperature. See Appendix B.

11.82 Given the data and rod of Problem 11.81, compute the magnitude and direction of a force *P* that

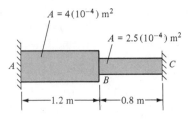

$A = 4(10^{-4})$ m^2

$A = 2.5(10^{-4})$ m^2

|←———1.2 m———→|←——0.8 m——→|

FIGURE P11.81

should be applied at *B* to produce zero stress in member *BC*. Compute the resulting stress in member *AB*.

11.83 Three bars, each of a different material, are connected end to end as shown in Fig. P11.83. The left bar is aluminum, 40 mm in diameter; the center section is brass, 50 mm in diameter; and the section on the right is magnesium, 60 mm in diameter. Compute the axial force in the structure and the stress in each bar for a temperature increase of 20°C. $E_{al} = 70$ GPa; $E_{br} = 60$ GPa; $E_{mag} = 45$ GPa.

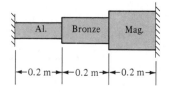

Al.   Bronze   Mag.

|←0.2 m→|←0.2 m→|←0.2 m→|

FIGURE P11.83

11.84 A steel bolt is 1/4 in. in diameter and is encased by a copper tube with an inside diameter of 5/16 in. and wall thickness of 0.025 in. A nut is tightened snugly against the copper tube such that normal stress is not induced in either material. Assume the length of the bolt and tube to be 4 in. Compute the axial force in each member if the temperature decreases 50°F. $E_{st} = 30(10^6)$; $E_{cu} = 18(10^6)$ psi.

11.85 A steel bolt 3/8 in. in diameter fits inside of an aluminum pipe as shown in Fig. P11.85. The aluminum pipe has an inside diameter of 7/16 in. and outside diameter of 1/2 in. Assume the nut is tightened just enough to hold the pipe in place. Compute the stress in each material for a temperature drop of 100°F.

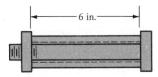

FIGURE P11.85

11.86 Given the data of Problem 11.85 and that the bolt has 12 threads per inch. For a temperature drop of 100°F compute the number of turns required to loosen the nut until the stress returns to zero.

11.87 Given the data for the bolt of Problems 11.85 and 11.86, assume a temperature increase of 100°F and discuss the effect on the system. How many turns of the nut would be required to produce a stress of 10 ksi in the aluminum pipe? Compute the corresponding stress in the steel bolt.

11.88 A 5/8-in. steel rivet is used to connect three 1/2-in.-thick steel plates. The rivet is heated before being placed in the rivet hole. If the rivet cools to a temperature of 70°F, compute the temperature of the rivet when it is inserted to limit the axial stress in the rivet to 15,000 psi.

11.89 The rod of Fig. P11.89 is made of three materials. The section on the left is aluminum with a 2-in. diameter. The section on the right is a 1-in.-diameter solid steel rod with a copper tube covering. The copper tube can be assumed to have an inside diameter of 1 in. and outside diameter of 2 in. Compute the axial force and stress in each material for a temperature increase of 75°F. $E_{cu} = 18(10^6)$ psi; $E_{st} = 30(10^6)$ psi; $E_{al} = 10(10^6)$ psi.

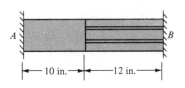

FIGURE P11.89

11.90 The bar of Fig. P11.90 is rigid and not affected by temperature change. Wires AD and BE are aluminum with an area of 0.25 in.². Wire FC is copper with an

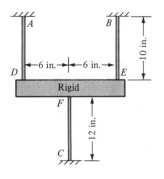

FIGURE P11.90

area of 0.30 in.². Compute the reactions at A, B, and C for a temperature drop of 15°F.

11.91 Given the data of Problem 11.90, assume that the bar is not rigid, but is steel with a cross section of 0.25 in. by 0.4 in. Compute the reactions at A, B, and C and compute the maximum bending stress in the bar for a temperature drop of 15°F.

11.92 Given the data of Problem 11.90, assume the bar is not rigid with properties given in Problem 11.91. Compute the maximum temperature decrease such that the normal stress in the bar or wires does not exceed 18,000 psi.

11.93 Two aluminum wires and a steel wire support a rigid bar as shown in Fig. P11.93. Compute the force in each wire if the temperature drops 75°F.

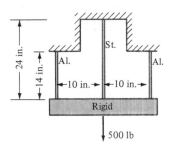

Aluminum wire area = 0.1 in.²

Steel wire area = 0.05 in.²

FIGURE P11.93

11.94 A rigid bar is connected to three rods as shown in Fig. P11.94. The rods can be assumed to be supported such that they can support compressive axial forces. Rods A and C are aluminum and rod B is steel. Assume all rods have the same cross-sectional area. Compute the axial force in each rod due to a temperature decrease of 50°C.

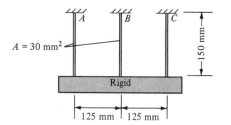

FIGURE P11.94

11.95 Given the data of Problem 11.94, compute the axial force in each rod for a temperature increase of 35°C.

11.96 Given the data of Problem 11.94, assume the bar is not rigid, is 10 mm by 30 mm in cross section, and is made of aluminum. Compute the axial force in each rod and the maximum bending moment in the bar for a temperature drop of 50°C.

11.97 A wooden post 4 in. by 4 in. is fixed at one end and connected to a rod support at the free end as shown in Fig. P11.97. The rod is steel with an area of 1.2 in.². Compute the reactions for a 50°F temperature drop. Assume that the axial deformation of the wooden post can be neglected. Assume $E_{post} = 2(10^6)$ psi.

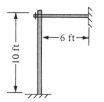

FIGURE P11.97

11.98 The beams of Fig. P11.98 are steel with a cross section 10 mm by 3 mm. A copper wire connects points

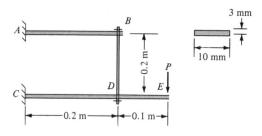

FIGURE P11.98

B and D. The wire is 1 mm in diameter. Assume the force P is zero and compute the axial force in the wire for a temperature drop of 20°C in the wire.

11.99 Given the data of Fig. P11.98, assuming $P = 2.25$ N and a temperature increase of 30°C in the wire, compute the force in the wire and the deflection of point B.

11.100 A circular ring is made of 1/4-in.-diameter aluminum rod. The inside diameter is 24.00 in. The ring must fit snugly around a steel tank of outside diameter 24.06 in.. Compute the temperature differential between the rod and the tank to allow the rod to be slipped in place.

11.101 A thin, circular steel band is used to reinforce a copper pressure vessel. The band is 8 mm by 25 mm and must be heated 120°C to fit snugly over a pressure vessel that is 0.3 m in outside diameter. Compute the axial stress in the band after it cools.

11.102 A circular steel band is to be mounted on a circular steel drum. The drum is 60 in. in diameter. The inside diameter of the circular band is 59.95 in. The band can be heated and placed upon the drum. After cooling there will be a tight fit. This process is called a shrink fit. If the temperature of the band is 75°F before heating, compute the temperature increase to expand the band enough to fit the drum. Assume an extra 0.02 in. in diameter for clearance. Compute the normal stress in the band after cooling. Assume that the drum diameter remains constant.

11.103—11.105 Compute the reactions for the externally statically indetermine trusses of Problems 11.103–11.105. Assume E and A to be constant for all members of each truss in Figs. P11.103—P11.105.

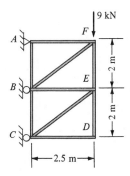

9 kN

FIGURE P11.103

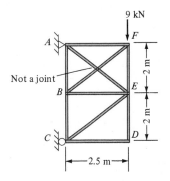

9 kN

Not a joint

FIGURE P11.106

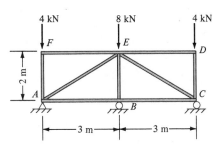

4 kN   8 kN   4 kN

FIGURE P11.104

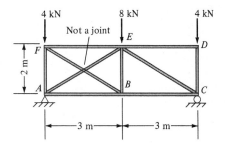

4 kN   8 kN   4 kN

Not a joint

FIGURE P11.107

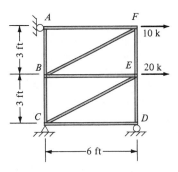

10 k

20 k

FIGURE P11.105

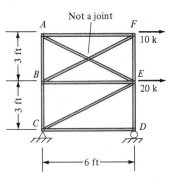

Not a joint

10 k

20 k

FIGURE P11.108

11.106–11.108 Problems 11.106–11.108 are internally statically indeterminate. Assume that all members in the corresponding figures have constant EA and compute the member forces.

11.109 Compute the reactions for the structure of Fig. P11.109. Sketch the shear and moment diagrams.

11.110 Compute the reactions for the structure of Fig. P11.110. Sketch the shear and moment diagrams.

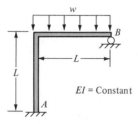

FIGURE P11.109

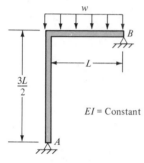

FIGURE P11.110

11.111 Compute the reactions for the structure of Fig. P11.111. Sketch the shear and moment diagrams.

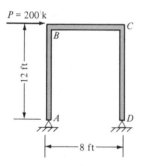

FIGURE P11.111

11.112 Compute the reactions for the frame structure of Fig. P11.112. Contruct shear and moment diagrams. Select a wide-flange structural shape that could be used for the vertical members and not exceed a normal stress of 18,000 psi.

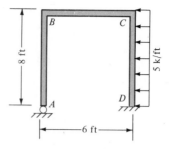

FIGURE P11.112

11.113 Compute reactions for the frame structure of Fig. P11.113. Note that the second moment of the area of the vertical member is twice that of the horizontal member.

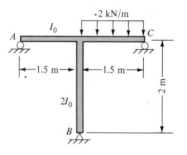

FIGURE P11.113

11.114 Assume the reaction at B as a redundant and compute the reactions for the structure of Fig. P11.114. Assume a square cross section, 100 mm by 100 mm, and compute the maximum normal stress in the structure.

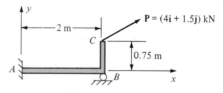

FIGURE P11.114

11.115 Compute the reactions for the structure of Fig. P11.115. Assume the reaction at B as the redundant. The structure is fabricated from standard-weight steel pipe. Assume that each section has the same cross-sectional properties. Note that for a circular cross section $J = 2I$. Let $G = E/2.5$.

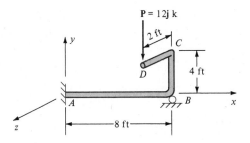

FIGURE P11.115

11.116 Compute the reactions for the structure of Fig. P11.116. The properties of span AB are identical to span BC. Assume that the cross section is circular and the material is aluminum, $G = E/2(1 + v)$, where $v = 0.3$ and $J = 2I$.

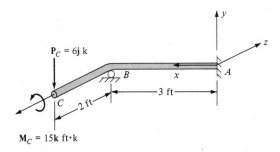

FIGURE P11.116

11.117 The beam of Fig. P11.117 is fixed at A and supported by a rod at B. The beam is circular in cross

section with constant properties throughout its length. A moment loading is applied at C, $M_C = (12\mathbf{i} + 8\mathbf{k})$ ft·k. Assume that all materials are steel, with $E = 30(10^3)$ ksi and $G = 12(10^3)$ ksi, and the area of the rod is 2 in.^{2}. The second moment of the area of the beam is 60 in.^{4}. Compute the force in the rod.

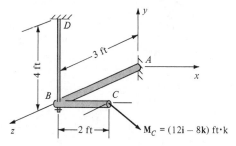

FIGURE P11.117

11.118 The beam of Fig. P11.118 is circular in cross section with diameter of 40 mm. The beam is steel ($E = 200$ GPa, $G = 80$ GPa). The rod support at C is 6 mm in diameter and made of aluminum ($E = 70$ GPa). If the temperature of the wire is lowered 35°C, compute the force in the wire and the reactions at A.

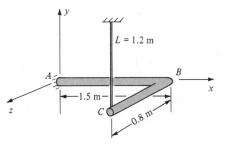

FIGURE P11.118

CHAPTER

COLUMNS

12.1 INTRODUCTION

Columns are predominantly axially loaded compressive members. It is possible to divide the axially loaded members into three categories: short columns, intermediate columns, and long columns. Short columns were essentially covered in Chapter 8. The axially loaded members that were subjected to combined axial force and bending moments were short columns. The failure of a short column is characterized by crushing of the material. Therefore, any column short enough to be analyzed using the methods of Chapter 8 is a short column. A long column fails due to buckling, which means that the column actually bows laterally, suffers a large deformation, and is rendered useless. The failure condition occurs at the instant the column begins to deflect laterally. At the instant of buckling, the material is not overstressed; in fact, buckling of a long column may occur at stresses well below the yield stress for the material. An intermediate column will sustain some degree of axial stress due to both axial load and bending load. The theory that predicts failure by pure buckling will yield an axial force that is far too conservative. The actual failure is sometimes referred to as inelastic buckling. The intermediate column will deflect laterally but will continue to sustain an axial load to resist buckling. As the load increases, the stress will eventually reach the proportional limit. The actual failure is a combined material yielding and buckling. A theoretical method for predicting a priori whether a column is intermediate or long is not available. Hence, column theory, by necessity, is somewhat empirical when compared to other topics in mechanics of materials.

In this chapter very basic design formulas are discussed. The fundamental solution to the theoretically posed column problem is called Euler's formula, which will be derived in the next section. Other design formulas will be used without a discussion of the reasoning that defends their empirical origin.

The secant formula, which is a theoretically correct analysis for a column subject to combined axial load and bending, will be derived. However, the formula is given here because of its historical significance rather than for its value as a design formula.

12.2 LONG COLUMNS AND EULER'S FORMULA

The failure of a long column is predominantly one of buckling. A buckling condition is idealized in Fig. 12.1. The column of Fig. 12.1 is similar to a beam when considering the boundary conditions. The ends are pinned and hence are free to rotate. The load P is increased until the column bows outward. For a long slender column the bending load can be computed with a fair degree of accuracy. Physically the

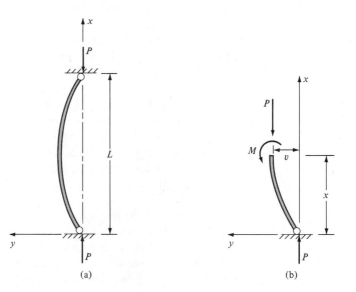

FIGURE 12.1
Buckling of a long column.

column remains straight and is subjected to an axial force. The instant that buckling occurs a bending moment is induced, as illustrated in the free body of Fig. 12.1b. Note that the column is assumed to buckle in the direction of positive y; hence v is positive as shown in the figure. Using the beam sign convention, the moment caused by buckling is negative and is given by $M = -Pv$. Since it is possible to write an expression for M, the integration method can be used to solve the deflection problem.

$$EI \frac{d^2v}{dx^2} = M = -Pv$$

$$\frac{d^2v}{dx^2} + k^2v = 0 \qquad\qquad (12.1)$$

where

$$k^2 = \frac{P}{EI} \qquad\qquad (a)$$

Equation (12.1) is a linear homogeneous differential equation with constant coefficients. Due to the term k^2v, the methods of Chapter 10 cannot be used; the equation cannot be integrated directly. The standard method of solution for Eq. (12.1) is to assume a solution of the form

$$v = Ce^{mx} \qquad\qquad (b)$$

Substituting into Eq. (12.1) gives a *characteristic* equation

$$m^2Ce^{mx} + k^2Ce^{mx} = 0$$

$$m^2 + k^2 = 0 \qquad m = \pm ki \qquad\qquad (c)$$

where i is the imaginary number. Substituting Eq. (c) into Eq. (b) gives the solution

$$vC_1e^{ikx} + C_2e^{-ikx} \qquad\qquad (d)$$

The exponentials of Eq. (d) are rewritten using the Euler formulas.[1]

$$v = A \sin kx + B \cos kx \qquad\qquad (e)$$

The boundary conditions are used to evaluate A and B. Refer to Fig. 12.1 and note that the column is pinned at both ends; hence $v(0) = 0$ and $v(L) = 0$, where L is the length of the column. Substituting into Eq. (e),

---

[1] C. R. Wylie, Jr., *Advanced Engineering Mathematics*, McGraw-Hill, New York, 1960, p. 553.

$$v(0) = 0 = B \qquad \text{therefore, } B = 0$$

$$v(L) = 0 = A \sin kL = 0 \tag{f}$$

Equation (f) must be satisfied at the boundary, $x = L$. The constant A cannot be zero because that would give the solution $v = 0$; hence, $\sin kL = 0$. Sin kL will be zero each time $kL = \pi, 2\pi, 3\pi, \ldots$, or

$$kL = n\pi \qquad n = 1, 2, 3, \ldots$$

$$k = \frac{n\pi}{L} \tag{g}$$

Using Eqs. (a) and (g),

$$\frac{P}{EI} = \frac{n^2\pi^2}{L^2}$$

$$P = \frac{n^2\pi^2 EI}{L^2} \tag{12.2}$$

Equation (12.2) with $n = 1$ is referred to as the Euler column formula, named in honor of the Swiss mathematician Leonhard Euler who is given credit for solving the problem in 1757.

The critical value of P given by Eq. (12.2) occurs when $n = 1$—that is, the least value of P.

$$P_{cr} = \frac{\pi^2 EI}{L^2} \tag{12.3}$$

A logical question is: When can n be greater than 1? Actually, n is never greater than 1 unless the design engineer forces it to be greater than 1 by specifying an intermediate lateral support. This can be shown by substituting Eq. (g) into Eq. (e).

$$v = A \sin\left(\frac{n\pi x}{L}\right) \tag{12.4}$$

Equation (12.4) is the characteristic function that satisfies Eq. (12.1). The constant A is arbitrary and can be assumed to represent the maximum deflection of the buckled column.

Equation (12.4), for $n = 1$, is shown in Fig. 12.2a representing the deflected shape of a buckled column with pinned ends. The column will always deform to this fundamental shape. If the column is supported at its midpoint, the buckled shape will appear as shown in Fig. 12.2b. The critical column load is given by Eq. (12.2) with $n = 2$, $P = 4\pi^2 EI/L^2$. It follows that $n = 3$ corresponds to Fig. 12.2c. A relation exists between Eqs. (12.2) and (12.3). If L is assumed to be $L/2$ in Eq.

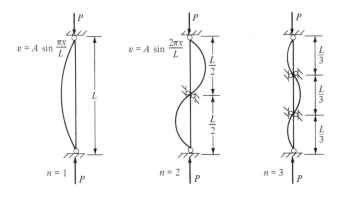

FIGURE 12.2
Buckling modes for a
pinned-pinned column.

(12.3), the critical load will correspond to P of Eq. (12.2) for $n = 2$. This relation leads to the idea of an *effective length*. The effective length of a column is one of the several parameters used in column design and analysis. Columns with various boundary condition combinations—pinned, fixed or free—can be related directly to the Euler formula by modifying the length to correspond to a pinned-pinned column of a given length. Consider the following example that will illustrate the effect of boundary conditions on the fundamental buckling analysis of a column.

EXAMPLE 12.1

---

a. Derive the expression for the fundamental critical buckling load for a fixed-fixed long column. Determine the effective length that should be used to modify the Euler formula to be valid for the fixed-fixed column.
b. Derive a general expression for the critical buckling load for a fixed-fixed column.

Solution:
a. The column is shown in Fig. 12.3*a*. The deflection is assumed to be in the positive y direction. Since the ends are fixed, a moment will develop at each end of the column. A shear reaction is possible at each end also. Consider the deflection shape of the column as shown in Fig. 12.3*b* that illustrates the fundamental buckled shape of a fixed-fixed column. Assuming the fundamental shape simplifies the mathematics of the analysis but limits the solution to be valid only for the fundamental critical load. Due to the assumed symmetry the fixed-end moments should be equal in magnitude but opposite in sign using the beam sign convention. The equations of statics show that $V_A = -V_B = 0$. The governing equation is formulated using the integration method of the previous chapters. The free body of Fig. 12.3*c* will serve to illustrate the derivation of the moment equation.

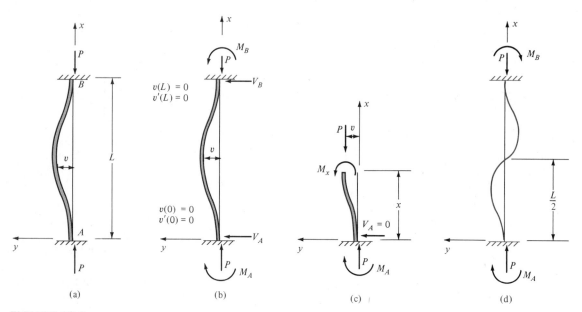

FIGURE 12.3

$$\frac{d^2v}{dx^2} = \frac{M}{EI} = -\frac{Pv}{EI} + \frac{M_A}{EI} \qquad \text{(a)}$$

Let $k^2 = P/EI$; hence,

$$\frac{d^2v}{dx^2} + k^2v = \frac{k^2 M_A}{P} \qquad \text{(b)}$$

The solution of Eq. (b) is obtained in two parts: a homogeneous solution and a particular solution. The homogeneous solution for the left-hand side of Eq. (b) is the same as Eq. (12.1); that is,

$$v_H = A \sin kx + B \cos kx \qquad \text{(c)}$$

The particular solution for the right-hand side can be obtained by assuming a polynomial solution, substituting into the governing equation, and equating coefficients of the polynomial terms. Assume that

$$v_P = Cx^2 + Dx + E$$

$$\frac{dv_p}{dx} = 2Cx + D$$

$$\frac{d^2v_p}{dx^2} = 2C$$

Substitute into Eq. (b):

$$2C + k^2(Cx^2 + Dx + E) = \frac{k^2 M_A}{P}$$

Equating coefficients of powers of x gives

$$C = 0 \qquad D = 0 \qquad E = \frac{M_A}{P} \qquad v_p = \frac{M_A}{P}$$

$$v = v_H + v_P = A \sin kx + B \cos kx + \frac{M_A}{P} \tag{d}$$

$$\frac{dv}{dx} = kA \cos kx - kB \sin kx \tag{e}$$

The boundary conditions at $x = 0$ are $v(0) = v'(0) = 0$ and are substituted into Eqs. (d) and (e):

$$v(0) = 0 = B + \frac{M_A}{P} \qquad B = \frac{-M_A}{P} \tag{f}$$

$$v'(0) = 0 = kA \qquad A = 0 \tag{g}$$

$$v = \frac{M_A}{P}(1 - \cos kx) \tag{h}$$

Equation (h) must satisfy the deflection condition at $x = L$.

$$v(L) = 0 = \frac{M_A}{P}(1 - \cos kL) = 0 \tag{i}$$

The moment M_A is not zero; hence,

$$(1 - \cos kL) = 0$$

or

$$\cos kL = 1$$

The cosine is $+1$ at $2\pi, 4\pi, 6\pi, \ldots$. However, our solution is valid only for the fundamental shape or only for $\cos kL = 2\pi$.

$$kL = 2\pi \tag{j}$$

$$\frac{P}{EI} = k^2 = \frac{4\pi^2}{L^2}$$

$$P = \frac{4\pi^2 EI}{L^2} \tag{k}$$

The result is shown in Fig. 12.4.

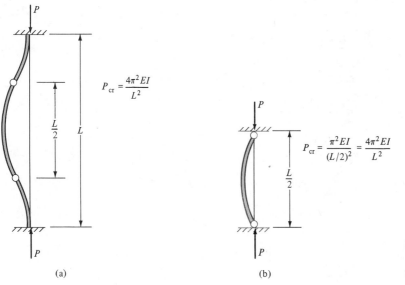

FIGURE 12.4
Equivalent length to simulate a fixed-fixed column.

Equation (h) is plotted in Fig. 12.4*a*. The slope of the column is zero at $x = 0$, $L/2$, and L. The curvature is zero at $x = L/4$ and $3L/4$. These relations can be verified by differentiating Eq. (h) with $k = 2\pi/L$.

$$\frac{dv}{dx} = \frac{M_A}{P} \frac{2\pi}{L} \sin \frac{2\pi x}{L} \qquad \textbf{(l)}$$

$$\frac{d^2v}{dx^2} = \frac{4M_A \pi^2}{PL^2} \cos \frac{2\pi x}{L} \qquad \textbf{(m)}$$

Equation (m) gives d^2v/dx^2, which is proportional to bending moment, as zero at the quarter points along the column. Hence, the midsection of the column is an equivalent pinned-pinned column, as shown in Fig. 12.4*b*. Finally, if an equivalent length of $L/2$ is substituted into the Euler formula, the result will be the same as Eq. (k).

$$P_{cr} = \frac{\pi^2 EI}{(L/2)^2} = \frac{4\pi^2 EI}{L^2}$$

Hence, an equivalent length to simulate a fixed-fixed column as pinned-pinned is $L/2$.
b. A general solution for the fixed-fixed column follows the previous analysis, except that the buckled shape may not be symmetrical as shown in Fig. 12.3*b*, and it follows that $M_A \neq M_B$. Also, $V_A = -V_B \neq 0$, and the shear at support A must be included in the analysis.

Equation (b) will be replaced with

$$\frac{d^2v}{dx^2} + k^2v = \frac{(M_A + V_Ax)k^2}{P} \tag{n}$$

Following the same procedure, Eq. (d) is replaced with

$$v = A \sin kx + B \cos kx + \frac{V_Ax}{P} + \frac{M_A}{P} \tag{o}$$

and

$$\frac{dv}{dx} = kA \cos kx - kB \sin kx + \frac{V_A}{P} \tag{p}$$

Then

$$v(0) = 0 \quad \text{gives} \quad B = -\frac{M_A}{P} \tag{q}$$

$$v'(0) = 0 \quad \text{gives} \quad A = -\frac{V_A}{kP} \tag{r}$$

Substituting Eqs. (q) and (r) into (o) and (p) and using the boundary conditions $v(L) = 0$ and $v'(L) = 0$ gives the following results.

$$V_A(kL - \sin kL) + M_Ak(1 - \cos kL) = 0 \tag{s}$$

$$V_A(1 - \cos kL) + M_Ak \sin kL = 0 \tag{t}$$

Equations (s) and (t) are written in matrix form (see Appendix D) as

$$\begin{bmatrix} kL - \sin kL & k(1 - \cos kL) \\ 1 - \cos kL & k \sin kL \end{bmatrix} \begin{Bmatrix} V_A \\ M_A \end{Bmatrix} = 0 \tag{u}$$

The condition for a solution is that $V_A = M_A = 0$ or that the determinant of their coefficients be zero. The determinant of Eq. (u) reduces to

$$2(\cos kL - 1) + kL \sin kL = 0 \tag{v}$$

It can be shown that the smallest root of Eq. (v) is given by $kL = 2\pi$ and the first buckling load is defined by Eq. (k). The next smallest root of Eq. (v) is given by

$$kL = 8.986$$

Solving for P gives

$$P = \frac{(8.986)^2 EI}{L^2} = \frac{8.182\pi^2 EI}{L^2} \tag{w}$$

The buckled shape corresponding to Eq. (w) is shown in Fig. 12.3d and would occur if the column was supported at $x = L/2$. ∎

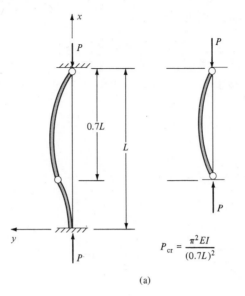

$$P_{cr} = \frac{\pi^2 EI}{(0.7L)^2}$$

(a)

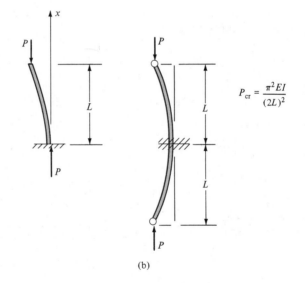

$$P_{cr} = \frac{\pi^2 EI}{(2L)^2}$$

(b)

FIGURE 12.5
(a) Fixed-pinned column;
(b) fixed-free column.

The foregoing example illustrates the idea of an effective or equivalent length. The equivalent length allows the engineer to simulate any long column using the Euler formula and corresponding equivalent length. Very few actual columns can be assumed to have end restraints that satisfy the classical boundary conditions. A restrained or fixed-end column usually undergoes some degree of joint rotation. In practice the boundary condition for the previous example would more than likely be $v(0) = v(L) = 0$, and some specified slope at $x = 0$ and $x = L$. Such boundary conditions lead to complicated transcendental equations for computing the buckling load.

An equivalent length can be obtained by locating the inflection points for a deflected column. Inflection points occur where the curvature changes and represent points of zero moment. The equivalent length is the distance between adjacent points of zero moment.

Two additional cases that can be solved exactly are shown in Fig. 12.5. The fixed-pinned column has an equivalent length of approximately $0.7L$. A fixed-free column is idealized as shown in Fig. 12.5b and has an equivalent length of $2L$.

Long columns are not very difficult to analyze or design. Their failure is attributed to buckling, and usually the corresponding axial stress is well below the allowable stress for the material. In the next section the conditions for buckling to occur will be discussed, along with the limitations of the Euler formula.

12.3 LIMITATIONS OF THE EULER FORMULA

Column buckling corresponds to a flexural failure of the column. The analysis of the previous section was based upon a bending analysis. The column was assumed to deform due to flexural action. It is difficult to establish exactly what causes the initial column buckling. It is assumed that the axial force is applied through the centroid of the section and does not cause an initial bending.

It is conceivable that a slight eccentricity is included by the applied load. The possibility exists that the column was not initially straight; there may be a slight crookedness in the member. There may be an initial imperfection in the cross section of the material. The boundary conditions may not be perfect due to fabrication or loading conditions and may cause some eccentricity of loading. In any case, long columns can be accurately analyzed using the Euler column theory.

As the length of a long column is decreased, eventually some limit must be reached when the column does not qualify as a long column. The physical behavior of the column introduces an additional parameter that enables the engineer to categorize a column as a long column or intermediate column. The limiting length of a column must occur when the material becomes overstressed before the column buckles. The limiting stress is usually the proportional limit where stress and strain are no longer proportional. The proportional limit is the design criterion, and a safety factor is always used that prevents the material from actually reaching such a high and possibly dangerous stress level.

The Euler formula, Eq. (12.3), is dependent upon the material property E, a cross-sectional property I, and the length L. Dividing by cross-sectional area A introduces an axial stress term. Since P is assumed to be axial—that is, applied through the centroid of the cross section—the stress is an average stress distributed uniformly over the cross section.

$$\sigma = \frac{P_{cr}}{A} = \frac{\pi^2 EI}{L^2 A} = \frac{\pi^2 Er^2}{L^2} = \frac{\pi^2 E}{(L/r)^2} \qquad (12.5)$$

where $r^2 = I/A$, the radius of gyration. The quantity (L/r) in Eq. (12.5) is called the *slenderness ratio* and is a convenient way to combine the parameters A, I, and L. Any specified limiting stress is a material parameter, as is E for a given material; hence L/r is the critical design parameter.

Columns are designed in a variety of cross-sectional configurations. They could have a solid cross section, such as a wooden post. Rolled structural shapes are often used as columns. Additionally, a column may be built up using several different structural shapes. The second moment of the area to be used in computing the slenderness ratio always corresponds to the least second moment of the area computed with respect to a principal centroid axis of the cross section. Hence, the term "least" or "lowest" L/r ratio is sometimes used.

A column is determined to be a long column as long as the right-hand side of Eq. (12.5) is less than the proportional limit of the column material. Since L/r is in the denominator, a limiting L/r ratio can be computed for any material, given E and the yield stress.

EXAMPLE 12.2

Compute the limiting value of L/r for a cast aluminum column that has a yield stress of 60 MPa and modulus of elasticity of 70 GPa.

Solution:
Equation (12.5) is written as

$$\left(\frac{L}{r}\right)^2 = \frac{\pi^2 E}{\sigma} = \frac{\pi^2 (70)(10^9) \text{ N/m}^2}{(60)(10^6) \text{ N/m}^2} = 11.5(10^3)$$

$$\frac{L}{r} = 107$$

This result is interpreted that any column with $L/r > 107$ will buckle before the stress reaches 60 MPa. The question concerning the case for L/r less than the limiting value will be discussed in the next section. ∎

To illustrate the relation between material properties and the Euler formula consider the following example.

EXAMPLE 12.3

Construct a plot of L/r versus normal stress, P/A, using the Euler formula for cast aluminum, 0.2 percent carbon hot-rolled steel, copper, and southern pine. Assume the following material properties:

Material	Modulus of Elasticity	Yield Stress
Cast aluminum	$10(10^6)$ psi	$8.7(10^3)$ psi
0.2% carbon, hot-rolled steel	$30(10^6)$ psi	$36(10^3)$ psi
Copper	$17.5(10^6)$ psi	$40(10^3)$ psi
Southern pine	$1.8(10^6)$ psi	$1.2(10^3)$ psi

Indicate the limiting value of L/r using the yield stress as an upper bound for stress.

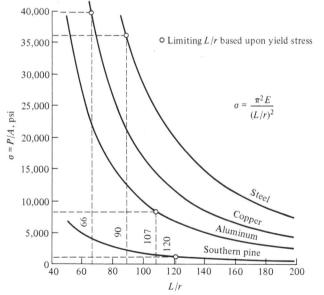

FIGURE 12.6
Euler curve for a column with pinned ends.

Solution:
The results are shown in Fig. 12.6. The Euler curves are similar for all materials. The range of validity for the Euler formula corresponds to any L/r value greater than the limiting value of L/r.

Figure 12.6 represents the distinction between a long column and an intermediate column. An L/r value to the right of the vertical dashed line indicates a long column. An L/r value to the left of the dashed line should qualify the column as intermediate. Note that the Euler curve gives rapidly increasing values of σ as L/r approaches zero. For a given cross section and length, any column can be qualified as either a long column or an intermediate column. Note that Fig. 12.6 does not include a safety factor. Figure 12.6 is incomplete for actually designing a column. The design and analysis for an L/r ratio that falls into the intermediate range will be dealt with using the so-called empirical column formulas. ∎

Columns can be designed using the Euler formula as a single design criterion. In that case, the engineer will always design a long column. A long column is not always the most efficient design. As an example consider the following situation.

EXAMPLE 12.4

A square steel column 20 ft long is to be fabricated using four equal leg angles that are laced together as shown in Fig. 12.7. Note, the lacing is assumed not to add to the load-carrying capability of the structure; it merely holds the angles in place. The angles are L3x3x1/4. Determine the spacing of the angles to ensure that the column qualifies as a long column. Assume a factor of safety of 1.8 and yield stress for the steel as 36 ksi. Compute the maximum axial load that the column should carry. Assume pinned ends.

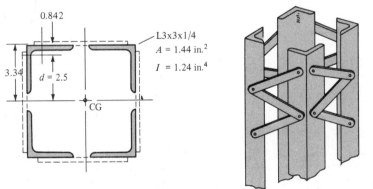

FIGURE 12.7

Solution:
Using Eq. (12.5) gives an equation that can be solved for L/r.

$$\sigma = \frac{\pi^2 E}{(L/r)^2}$$

$$\frac{L}{r} = \left[\frac{\pi^2 (30)(10^6)}{(36)(10^3)} \right]^{1/2} = 90$$

The criterion is $L/r \geq 90$.

$$r = \left(\frac{I}{A} \right)^{1/2} = \frac{(20 \text{ ft})(12 \text{ in./ft})}{90} = \frac{240}{90}$$

$$I = \left(\frac{240}{90} \right)^2 A$$

The properties of an L3x3x1/4 are obtained from the tables and noted in Fig. 12.7.
Substituting into the equation above

$$4(1.24 + 1.44 d^2) \text{ in.}^4 = \left(\frac{240 \text{ in.}}{90} \right)^2 (4)(1.44) \text{ in.}^2$$

$$d = 2.5 \text{ in.}$$

$$I = 4 \left[1.24 + 1.44 (2.5)^2 \right] = 40.96 \text{ in.}^4$$

Use Eq. (12.3) and increase the critical load by a factor of 1.8.

$$P_{cr} = \frac{\pi^2 E I}{1.8 L^2} = \frac{\pi^2 30 (10^6)(40.96)}{1.8 (240)^2} = 117,000 \text{ lb}$$

Computing the stress will show that $\sigma = P_{cr}/A = 117,000/(4)(1.44) = 20.307$ psi, slightly greater than the allowable because of round-off error, but well below the yield stress. The important fact is that the column will buckle. If the spacing d is increased, the column should not buckle, and the engineer is not able to complete the design for lack of a design criterion. ∎

12.4 INTERMEDIATE COLUMNS AND COLUMN DESIGN

The criteria of the previous examples are usually unsatisfactory for efficient column design and analysis. A limiting stress criterion works well for short columns. A short column, for most materials used in structural applications, corresponds to a slenderness ratio that is fairly low. The plot of Fig. 12.6 indicates for steel and aluminum that the division between an intermediate column and a long column corresponds to an L/r ratio that is dependent upon E. The plot of Fig. 12.6 does not include a safety factor. Intermediate columns lie in the range of L/r ratios between short and long. The actual value of the limiting L/r ratio must be computed using an empirical criterion.

The examples of the previous section are intended to illustrate the significance of the Euler equation and the slenderness ratio and their use for defining a long column. The examples are not necessarily to be used as a guide for column design.

Column design is dictated by building codes. Local building codes usually do not specify how to design a column but refer to some accepted design standard. Three common building materials—steel, aluminum, and wood—will be discussed here. Intermediate column formulas are empirical in nature and have developed throughout the years. There are two basic problems to be studied: (1) Given a column, compute the allowable load for that column. (2) Given the axial load, choose a column that meets the design criterion. In either of these problems the column may turn out to be short, intermediate, or long. The ultimate goal of the design engineer is to specify the most economical column (lightest weight) and be confident that it will perform satisfactorily.

Design Formulas for Structural Steel

The American Institute of Steel Construction (AISC) represents the interests of engineers, steel fabricators, and the manufacturers of steel by publishing the *AISC Steel Construction Manual*. The latest edition of the manual gives two design formulas for steel columns, one for long columns and a second for intermediate columns. A short-column formula is not specified as such.

The division between long and intermediate columns is given by a critical L/r ratio denoted by C_c.

$$\frac{L}{r} = C_c = \left(\frac{2\pi^2 E}{\sigma_{yP}}\right)^{1/2} = \left[\frac{2\pi^2(29)(10^6)}{\sigma_{yP}}\right]^{1/2}$$

$$C_c = \frac{23{,}900}{\sqrt{\sigma_{yP}}} \tag{12.6}$$

The AISC recommends $E = 29(10^6)$ psi for steel. If $L/r \geq C_c$, the long-column criterion is in order and is given by

$$\sigma_{\text{allow}} = \frac{12}{23}\frac{\pi^2 E}{(kL/r)^2} = \frac{149(10^6)\ \text{psi}}{(kL/r)^2} \tag{12.7}$$

The k that is included in the L/r ratio is a correction for an equivalent column length as discussed in the preceding section. The factor (12/23) is the factor of safety and is approximately equivalent to dividing by 1.92. Equation (12.7) is the Euler equation—that is, Eq. (12.5) modified by a factor of safety. An L/r ratio less than C_c requires the use of the empirical equation

$$\sigma_{\text{allow}} = \left[1 - \frac{(kL/r)^2}{2C_c^2} \right] \frac{\sigma_{yP}}{\text{F.S.}} \tag{12.8}$$

where

$$\text{F.S.} = \frac{5}{3} + \frac{3}{8} \frac{(kL/r)}{C_c} - \frac{(kL/r)^3}{8C_c^3} \tag{12.9}$$

The factor of safety, F.S., varies according to the L/r ratio, and σ_{allow} varies according to the yield stress, σ_{yP}, for the steel.

A plot of Eqs. (12.7) and (12.8) for several different steels is shown in Fig. 12.8. The intersection of the two curves is the value of $L/r = C_c$ given by Eq. (12.6). Equation (12.8), for low-strength steel, is practically a straight-line extension of the Euler curve for $0 \le L/r \le C_c$. For higher-strength steels the curve is more pronounced. As the slenderness ratio approaches zero, the allowable stress is $\sigma_{yP}/1.67$. The use of Eqs. (12.6)–(12.9) will be illustrated later in this section.

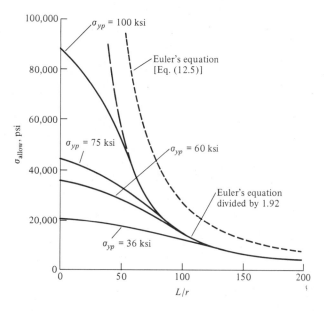

FIGURE 12.8
AISC column formulas:
Eqs. (12.7) and (12.8).

Design Formulas for Aluminum Alloys

Design formulas for two different aluminum alloys, 2014-T6 and 6016-T6, are plotted in Fig. 12.9. The three definite ranges for L/r ratios are discernible.

The design formulas are taken from Section 1, *Specifications for Aluminum Structures*, published by The Aluminum Association.

The column formulas are given below with some change in the nomenclature. It is understood that L is an equivalent column length. For L/r less than S_1,

$$\sigma_{\text{allow}} = \frac{\sigma_{yP}}{k_c n_y} \tag{12.10}$$

$$S_1 = \frac{B_c - n_u \sigma_{\text{allow}}}{D_c} \tag{12.11}$$

For $S_1 \le L/r \le S_2$,

$$\sigma_{\text{allow}} = \frac{B_c - D_c(L/r)}{n_u} \tag{12.12}$$

For L/r greater than S_2,

$$\sigma_{\text{allow}} = \frac{\pi^2 E}{\left[n_u(L/r)^2\right]} \tag{12.13}$$

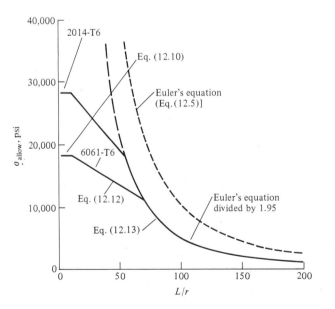

TABLE 12.1 Design Parameters for Aluminum Alloy Columns

	σ_{yP}, ksi	B_c, ksi	D_c, ksi	S_2	n_u*	n_y*	S_1	k_c
2014-T6	53	61.1	0.458	55	1.95	1.65	11	1.12
6061-T6	35	39.4	0.236	68	1.95	1.65	10	1.12

* Factor of safety for buildings.

$$S_2 = \frac{0.41 B_c}{D_c} \tag{12.14}$$

Additionally, the constants in these equations are given by

$$E = 10.9(10^3) \text{ ksi}$$

$$B_c = \sigma_{yP} \left\{ 1 + \left[\frac{\sigma_{yP}}{2.25(10^3)} \right]^{1/2} \right\}$$

$$D_c = \frac{B_c}{10} \left(\frac{B_c}{E} \right)^{1/2}$$

The factors of safety are n_u and n_y. These design parameters are given in Table 12.1.

Equation (12.10) is the zero-slope portion of the curve of Fig. 12.9. Equation (12.12) is a straight line and is the intermediate-column design curve of Fig. 12.9. The Euler curve divided by the factor of safety 1.95 is the long-column portion of Fig. 12.9.

Design Formulas for Wood Columns

The Timber Construction Manual published by the American Institute of Timber Construction is the source of information for designing wood columns. For pin-end conditions or square-end conditions

$$\sigma_{\text{allow}} = \frac{0.30E}{(L/d)^2} \qquad L/d < 50 \tag{12.15}$$

where L is the unsupported length of the column, d is the smallest side dimension, and E is the modulus of elasticity.

The allowable stress for a round column may not exceed the stress for a square column of the same cross-sectional area or the stress as determined by

$$\sigma_{\text{allow}} = \frac{\pi^2 E}{2.727(L/r)^2} = \frac{3.619E}{(L/r)^2} \tag{12.16}$$

TABLE 12.2 Properties of Some Common Wood Used for Construction*

	σ_{max}, psi	E, psi
Douglas Fir		
Select	1400	$1.76(10^6)$
Construction	1200	$1.76(10^6)$
Hemlock		
Select	1200	$1.54(10^6)$
Construction	1100	$1.54(10^6)$
Southern Pine		
Select	2250	$1.76(10^6)$
No. 1	1300	$1.76(10^6)$
No. 2	900	$1.76(10^6)$

* *Source: Timber Construction Manual*, American Institute of Timber Construction.

σ_{allow} in Eqs. (12.15) and (12.16) cannot exceed σ_{max} as given in Table 12.2. Wood-column formulas are modified according to service conditions and loading duration. The allowable stress may be increased for short time loads or decreased for long time loads. The normal duration of loading factor is 1.

The following examples should serve to illustrate the column design formulas.

EXAMPLE 12.5

Compute the load-carrying capacity of the column described in Example 12.4. Assume the spacing of the angles is $d = 4$ in. as shown in Fig. 12.10. Assume $\sigma_{yP} = 36,000$ psi.

Solution:
Compute C_c using Eq. (12.6).

$$C_c = \frac{23,900}{(36,000)^{1/2}} = 126$$

$$I_c = 4[1.24 + 1.44(4^2)] = 97.12 \text{ in.}^4$$

$$\frac{L}{r} = \frac{(20)(12)}{(97.12/5.76)^{1/2}} = 58$$

Since $58 < 126$, the intermediate column formula should be used. The F.S. is computed using Eq. (12.9).

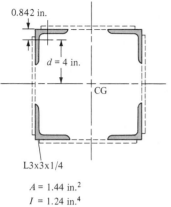

L3x3x1/4

$A = 1.44$ in.^{2}
$I = 1.24$ in.^{4}
$I_{CG} = 4[1.24 + 1.44(4)^2] = 97.12$ in.^{4} **FIGURE 12.10**

$$\text{F.S.} = \frac{5}{3} + \frac{3}{8}\left(\frac{58}{126}\right) - \frac{1}{8}\left(\frac{58}{126}\right)^3 = 1.83$$

$$\sigma_{allow} = \left[1 - \frac{(58/126)^2}{2}\right]\frac{36,000}{1.83} = 17,590 \text{ psi}$$

$$P_{allow} = A\sigma_{allow} = (5.76)(17,590) = 101,320 \text{ lb}$$ ∎

EXAMPLE 12.6

A wide-flange column must support a concentric axial load of 300,000 lb. The width dimensions of the column cannot exceed 13 in. The effective length of the column is $kL = 0.85L$, where L is 14 ft. Use AISC formulas and choose a wide-flange member, assuming $\sigma_{yP} = 42,000$ psi.

Solution:
The design is a trial-and-error procedure that amounts to guessing the proper wide-flange member, then using the formulas to verify the accuracy of the guess. However, as the design iteration proceeds, each guess is guided by the previous iteration. C_c is constant for a given σ_{yP}.

$$C_c = \frac{23,900}{(42,000)^{1/2}} = 117$$

Rather than arbitrarily choosing a wide-flange member, assume a radius of gyration based upon the values given in the Appendix. Assume $r = 2.00$ as an initial guess.

1. Compute the actual L/r.

$$\frac{L}{r} = \frac{(0.85)(14)(12)}{2} = \frac{142}{2} = 71 < 117$$

2. Compute σ_{allow}.

$$F.S. = \frac{5}{3} + \frac{3}{8}\left(\frac{71}{117}\right) - \frac{1}{8}\left(\frac{7}{117}\right)^3 = 1.87$$

$$\sigma_{allow} = \left[1 - \frac{(71/117)^2}{2}\right]\frac{42,000}{1.87} = 18,320 \text{ psi}$$

3. Compute the required area based upon σ_{allow}.

$$A = \frac{P}{\sigma_{allow}} = \frac{300,000}{18,320} = 16.37 \text{ in.}^2$$

From Appendix C it turns out that a W12x58 has an area of 17 in.^{2} and is less than 13 in. in width. However, the least r is 2.51 compared to the assumed r of 2. Hence, the analysis process should be repeated.

1. Compute L/r as follows:

$$\frac{L}{r} = \frac{142}{2.51} = 57 < 117$$

2. Use Eq. (12.8) again, $57/117 = 0.487$.

$$F.S. = \frac{5}{3} + \frac{3(0.487)}{8} - \frac{(0.487)^3}{8} = 1.83$$

$$\sigma_{allow} = \left[1 - \frac{(0.487)^2}{2}\right]\frac{42,000}{1.83} = 20,230 \text{ psi}$$

3. The required area is as follows:

$$A = \frac{300,000}{20,230} = 14.83 \text{ in.}^2$$

The previous choice is still the best section, W12x58. Note that Appendix C is not a complete set of available rolled sections. The AISC manual could provide a more efficient section.

One last computation should be carried out. Check the actual load the column will carry. σ_{allow} for a W12x58 has been computed as $\sigma_{allow} = 20,230$ psi. The actual area is 17.0 in.^{2}. Then

$$P_{allow} = (17 \text{ in.}^2)(20,230 \text{ psi}) = 343,900 \text{ lb}$$

The column is overdesigned considerably. The conscientious design engineer might attempt to find a 10-in. wide-flange member that would satisfy the criterion.   ∎

EXAMPLE 12.7

An aluminum strut is to be 6061-T6 alloy and must carry a concentric axial force of 95,000 lb. The strut is to be 4 ft long and has pinned ends.

Solution:

Aluminum rolled sections have the same shape properties as steel sections. It is permissible to use Appendix C for this example. To begin the design process, assume a long column with minimum allowable L/r ratio of 68 (see Table 12.1). Equation (12.13) gives

$$\sigma_{allow} = \frac{\pi^2(10.9)(10^3)}{1.95(68)^2} = 11.94 \text{ ksi}$$

The required area is $A = 95 \text{ k}/11.94 \text{ ksi} = 7.96 \text{ in.}^2$. Select a W8x28 with $A = 8.25 \text{ in.}^2$ and $r = 1.62$ in.

1. Compute

$$\frac{L}{r} = \frac{(4 \text{ ft})(12 \text{ in./ft})}{1.62} = 29.6$$

   Check Table 12.1: $10 < 29.6 < 68$.

2. Use Eq. (12.12):

$$\sigma_{allow} = \frac{39.4 - (0.236)(29.6)}{1.95} = 16.62 \text{ ksi}$$

3. Check required area:

$$A = 95 \text{ k}/16.62 \text{ ksi} = 5.72 \text{ in.}^2$$

Select a W6x20 with $A = 5.87$ in. and $r = 1.50$ in.

1. Compute

$$\frac{L}{r} = \frac{48}{1.5} = 32$$

   Check Table 12.1: $10 < 32 < 68$.

2. Use Eq. (12.12):

$$\sigma_{allow} = \frac{39.4 - (0.236)(32)}{1.95} = 16.33 \text{ ksi}$$

3. Check required area:

$$A = \frac{95 \text{ k}}{16.33 \text{ ksi}} = 5.81 \text{ in.}^2$$

The W6x20 will be satisfactory. A final computation is in order. Check the allowable load.

$$P_{allow} = A\sigma_{allow} = (5.87 \text{ in.}^2)(16.33 \text{ ksi}) = 95.86 \text{ k}$$

In this case it could be anticipated that P_{allow} would be very near the required load. This is not always the case, but P_{allow} must be equal or greater than the given axial load. ∎

Any method for beginning the design process is acceptable. Two methods have been illustrated in the last two examples. In Example 12.7 a representative value of r was assumed. In Example 12.8 a long column was assumed with the least allowable L/r.

12.5 THE SECANT FORMULA

The secant formula is classical and historically significant. It represents an early attempt directed at solving the problem of a column with initial curvature or approximating the effect of an eccentric axial load. The secant formula allows for some column resistance to bending and like the intermediate-column formula, requires a trial-and-error solution.

Consider the column of Fig. 12.11. The moment at any section x is $-Pv$, therefore, the resulting differential equation is the same as the concentric axially loaded column.

$$\frac{d^2v}{dx^2} + \frac{Pv}{EI} = 0$$

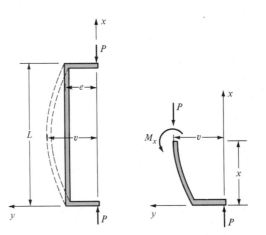

FIGURE 12.11

$$\frac{d^2v}{dx^2} + k^2v = 0 \tag{12.17}$$

$$k^2 = \frac{P}{EI} \tag{a}$$

The solution is the same as Eq. (e) of Section 12.2.

$$v = A \sin kx + B \cos kx \tag{b}$$

The boundary conditions at $x = 0$ and $x = L$ are *not* $v = 0$ but $v = e$. Substituting $x = 0$ into Eq. (b) gives

$$e = B$$

Substituting $x = L$ gives

$$e = A \sin kL + e \cos kL$$

$$A = \frac{e}{\sin kL}(1 - \cos kL) \tag{c}$$

The double-angle formulas, $\sin kL = 2 \sin(kL/2) \cos(kL/2)$ and $2 \sin^2(kL/2) = 1 - \cos kL$ will convert Eq. (c) into a more useful form.

$$v = e\left[\frac{\sin(kL/2)}{\cos(kL/2)}\right] \sin kx + e \cos kx \tag{d}$$

It can be verified that the maximum deflection and maximum bending moment occur at the midpoint of the column. The second derivative of Eq. (d) evaluated at $x = L/2$ gives the maximum moment. Note that the moment is negative and agrees with the coordinates of Fig. 12.11.

$$\frac{d^2v}{dx^2} = -k^2e\left[\frac{\sin(kL/2)}{\cos(kL/2)}\right]\sin(kL/2) - k^2e \cos(kL/2)$$

or, reducing and substituting $k^2 = P/EI$,

$$EI\frac{d^2v}{dx^2} = M = -\frac{Pe}{\cos[(P/EI)^{1/2}(L/2)]} = -Pe \sec[(P/EI)^{1/2}(L/2)]$$

The short-column formula of Chapter 8 is used to compute the stress. The second moment of the area is not the least, but must correspond to the axis about which bending occurs. Then $r^2 = I/A$ and r corresponds to the same axis as I.

$$\sigma = \frac{P}{A} \pm \frac{MC}{I}$$

$$\sigma_{max} = \frac{P}{A}\left(1 + \frac{ec}{r^2} \sec[(P/4EA)^{1/2}(L/r)]\right) \tag{12.18}$$

In the application $\sigma_{\max}$ can be assumed to be the yield stress and a factor of a safety applied to P and a trial-and-error analysis performed.

The secant formula can be compared to the intermediate-column formulas and will give similar results.

12.6 COLUMNS WITH ECCENTRIC LOADING

The secant formula is applicable for columns with eccentric loading. The term Pe of Eq. (12.18) can always represent a bending moment even if the bending moment is not caused directly by the load P. The secant formula is theoretically correct, but due to its nonlinear character it is difficult to use. Design formulas, sometimes referred to as interaction formulas, have developed over the years.

The AISC manual gives specific design formulas for steel. A complete discussion of column interaction formulas is better left to a course in machine design or design of steel structures. The concept is based upon the combined stress problems of Chapter 8. A column can be subjected to an axial force and a bending moment about each axis. The column of Fig. 12.12 would be subject to the stress distribution.

$$\sigma = \frac{-P}{A} \pm \frac{M_x(h/2)}{I_x} \pm \frac{M_y(b/2)}{I_y} \tag{12.19}$$

The maximum stress should occur at the location where all three components are additive. A maximum stress design theory specifies that σ cannot exceed the allowable working stress and is written as

$$\frac{f_a}{F_a} + \frac{f_{bx}}{F_{bx}} + \frac{f_{by}}{F_{by}} \leq 1.0 \tag{12.20}$$

Equation (12.20) is an AISC design formula applicable if $f_a/F_a \leq 0.15$. The notation and terminology is taken from the AISC manual.

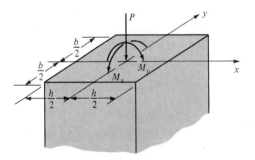

FIGURE 12.12

f_a = Computed axial stress

f_{bx} = Computed bending stress about the x axis

f_{by} = Computed bending stress about the y axis

F_a = Allowable compressive axial stress

F_{bx}, F_{by} = Allowable compressive bending stresses

Equation (12.20) becomes more complicated for $f_a/F_a > 0.15$.

Examples will not be included. The actual design process specified for steel, aluminum, or wood is explained in the various manuals and accompanied by simplifying design formulas, curves, and charts.

12.7 SUMMARY

The design and analysis of columns is divided into three categories:

1. *Long columns.* The design of long columns is based upon the buckling phenomenon and is governed by the Euler column formula.

$$P_{\text{cr}} = \frac{\pi^2 EI}{L^2} \qquad (12.3)$$

2. *Intermediate columns.* Intermediate-column analysis is a combination of theoretical results and experimental results. The empirical column formulas of Section 12.4 are developed according to the material properties and the behavior of a column in the laboratory.

3. *Short columns.* A short column can be designed according to the elementary formula for axially loaded members.

The factor of safety is especially important in column design. Its presence can be seen in all standard column design formulas.

Design formulas for beams with eccentric loading are very specialized. The secant formula can be used for beams subject to combined axial load and bending moment, but due to its nonlinear nature it requires a tedious trial-and-error solution. Eccentric-column formulas are usually based upon limiting the combined bending and axial stresses to be within some predetermined tolerance.

For Further Study

Aluminum Construction Manual, 4th ed., The Aluminum Association, Inc., Washington, D.C., 1982.

Bleich, H. H., *Buckling Strength of Metal Structures*, McGraw-Hill, New York, 1952.

Chajes, A., *Principles of Structural Stability Theory*, Prentice-Hall, Englewood Cliffs, NJ, 1974.

Manual of Steel Construction, 8th ed., American Institute of Steel Construction, Chicago, 1980.

Timber Construction Manual, 3rd ed., American Institute of Timber Construction. Wiley, New York, 1985.

Timoshenko, S. P., and J. M. Gere, *Theory of Elastic Stability*, McGraw-Hill, New York, 1961.

12.8 PROBLEMS

12.1 Derive the expression for the critical buckling load for a fixed-pinned long column in Fig. P12.1. Determine the effective length that should be used to modify the Euler formula to be valid for a fixed-pinned column.

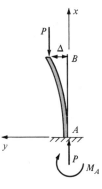

FIGURE P12.2

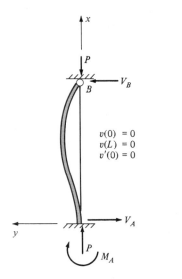

$$v(0) = 0$$
$$v(L) = 0$$
$$v'(0) = 0$$

FIGURE P12.1

12.2 Derive the expression for the critical buckling load for a fixed-free long column of Fig. P12.2. Note that the column is determinate and M_A can be evaluated using statics. Compute the effective length to be used to modify the Euler formula to be valid for a fixed-free column.

12.3 A W6x20 steel structure member is to be used as a column fixed at one end and pinned at the other. Use the Euler criterion of Fig. 12.5 and compute the limiting length of the column. Asssume $\sigma_{yP} = 36$ ksi and $E = 30(10^3)$ ksi.

12.4 Two steel channels, C8x18.75, are connected back to back and used as a column. Assume fixed-free end conditions and compute the limiting length of the column assuming the column must meet the Euler buckling criterion. Assume $\sigma_{yP} = 60$ ksi. Note, r may not be an additive quantity for this column.

12.5 Assume the channels of Problem 12.4 are laced together as shown in Fig. P12.5 and repeat the analysis. Assume that the lacing does not carry load.

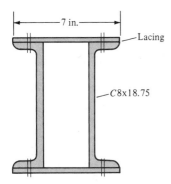

FIGURE P12.5

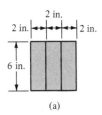

(a)

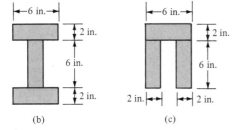

(b)   (c)

FIGURE P12.6

12.7 A round aluminum rod must support an axial load of 30 kN. The rod has pinned ends and is 3.25 m in length. Compute the diameter of the rod using the Euler buckling criterion. $E = 70$ GPa and $\sigma_{yP} = 230$ MPa.

12.8 Select the lightest wide-flange section for a column with fixed ends and 28 ft in length. The axial load is 250,000 lb. Assume steel with yield stress of 42 ksi and $E = 30(10^3)$ ksi. Use the Euler formula as the design criterion with a factor of safety of 1.65.

12.9 A steel column 22 ft in length is pinned at one end and fixed at the other. Select the most economical wide-flange section for an axial load of 150 k. Assume that the Euler formula controls the design with a factor of safety of 2.5. Assume $\sigma_{yP} = 42$ ksi and $E = 30(10^3)$ ksi.

12.10 Four aluminum angles, L4x4x1/4, are arranged as shown in Fig. P12.10. The connecting plates are 1/4 in. thick and are assumed to act with the angles when carrying an axial load. Compute the minimum length at which Euler's formula remains valid for pinned-end conditions. Let $E = 10(10^3)$ ksi and the yield stress is 12 ksi. Properties of steel sections given in Appendix C can be used for aluminum sections.

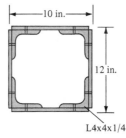

FIGURE P12.10

12.6 The three wood members of Fig. P12.6 are each 2 in. by 6 in. actual size and are to be used as a built-up column 11 ft long. Assume pinned ends and use the Euler formula, Eq. (12.5), as a single design criterion and compute the critical load for each possible configuration of Fig. P12.6. Assume $E = 1.5(10^6)$ psi and a factor of safety of 2.0. If the yield stress of the material is $1.0(10^3)$ psi, determine whether the column will be overstressed.

12.11 A steel support member is rectangular, 15 mm by 40 mm, and is to carry a concentric axial load. The length is 750 mm. Assume $E = 200$ GPa and a yield stress of 220 MPa. Compute the allowable axial load using Euler's formula assuming **a.** a pinned column, **b.** a pinned-fixed column, **c.** a fixed-fixed column.

Problems 12.12 through 12.44 should be solved using the column design formulas of Section 12.4.

12.12 A W8x40 steel member is pinned at both ends and 14 ft long. Use AISC formulas and compute the allowable load for the column. Assume $\sigma_{yP} = 36$ ksi.

12.13 A W16x100 steel member of Fig. P12.13 is pinned-fixed and 22 ft long. Cover plates 12 by 1 in. are attached as shown in the figure. Compute the maximum axial force that can be carried by the column according to AISC specifications. Assume $\sigma_{yP} = 60$ ksi.

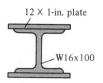

12 X 1-in. plate

W16x100

FIGURE P12.13

12.14 A W8x15 steel member is pinned at both ends and is 12 ft long. Assume $\sigma_{yP} = 42$ ksi and compute the allowable load for the column using AISC formulas.

12.15 Two steel channels are connected together to form a built-up wide-flange member. The channels are C7x9.8 with $\sigma_{yP} = 36$ ksi. Assume a fixed-free column 13 ft long and compute the maximum compressive axial force that can be applied using AISC specifications.

12.16 Assume that the channels of Problem 12.15 are built up to form a square column. The flanges are turned toward the center of the column and laced together. Assume $\sigma_{yP} = 36$ ksi and a fixed-free column 13 ft long. Compute the allowable axial load for the column.

12.17 A square aluminum strut 36 in. long is to be designed as a short column with pinned ends. Compute the dimensions to ensure that short-column equations can be used assuming 2014-T6 aluminum alloy.

12.18 A 6061-T6 aluminum alloy W6x15 section has pinned ends and is 16.5 ft long. Using the properties given in Appendix C, compute the load-carrying capacity of the column.

12.19 Four 2014-T6 aluminum alloy angles are built up as shown in Fig. P12.19. Compute the length of a pinned-end column that could qualify as a long column using this section. If the column length is specified as 9 ft, compute the allowable load.

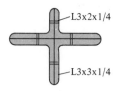

L3x2x1/4

L3x3x1/4

FIGURE P12.19

12.20 A wide-flange member and two channels of 2014-T6 aluminum alloy are built up as shown in Fig. P12.20. Assume a pinned-fixed column 15 ft long and compute the allowable axial load for the column. Use the properties given in Appendix C.

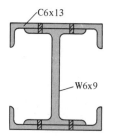

C6x13

W6x9

FIGURE P12.20

12.21 A No. 1 grade southern pine wood column, 8 by 10 in. actual dimensions, is 10 ft long with pinned-end conditions. Compute the axial load the column will support. See Table 12.2. $E = 1.76(10^6)$ psi; $\sigma_{max} = 1300$ psi.

12.22 A select grade southern pine column 6 by 6 in. nominal dimensions is 12 ft long with pinned-end conditions. Compute the allowable axial load for the column. *Note:* For nominal dimensions subtract 1/2 in. from each dimension. $E = 1.76(10^6)$ psi; $\sigma_{max} = 2250$ psi.

12.23 A round column 10 in. in diameter is construction grade hemlock. The column is 20 ft long with pinned ends. Compute the allowable load assuming 10 in. is the nominal dimension. *Note:* Check the square column formula also.

12.24 A round wood column with actual dimensions of 100 mm in diameter is to carry a load of 3.5 kN. Compute the allowable length of a pinned-end column; assuming **a.** Douglas fir, select grade; **b.** hemlock, select grade; **c.** southern pine, select grade. Multiply by 6894 Pa/psi to convert psi to Pa.

12.25 A 150- by 200-mm wood column is 4.2 m in length, with pinned ends. Assume actual dimensions and compare the loading capacities of construction grade Douglas fir and hemlock. Multiply by 6894 Pa/psi to convert psi to Pa.

12.26 Construct a design chart for Eq. (12.15) with L/d versus σ_{allow} for varying values of E. Use log-log graph paper and the data of Table 12.2.

12.27 Select the lightest-weight steel wide-flange member to carry a concentric axial force of 220 k. The column is pinned-end and 15 ft long. Assume $\sigma_{yP} = 50$ ksi.

12.28 Select an economical steel wide-flange column to carry a concentric axial force of 850 k. The column is fixed-pinned with a total length of 9 ft. Assume $\sigma_{yP} = 36$ ksi. Limit your choice to a W12, W14, or W16 column.

12.29 A steel beam 32 ft long is to be used in a special application and can be assumed to be pinned at both ends. Select the most economical wide-flange section to carry an axial force of 95 k. Assume $\sigma_{yP} = 100$ ksi.

12.30 Select the lightest-weight steel wide-flange member to carry a concentric axial force of 250 k. The column has an equivalent length of 0.9 L, where L is 18 ft. Use AISC formulas assuming $\sigma_{yP} = 50$ ksi. Limit your choice to a W10 or W12 column.

12.31 A pinned-end column 16 ft long is to carry an axial load of 275 k. Select a wide-flange member using AISC formulas and compare to a design based upon aluminum alloy formulas. Assume $\sigma_{yP} = 50$ ksi for steel and assume 2014-T6 aluminum alloy. Use the properties of Appendix C for aluminum members.

12.32 A 6061-T6 aluminum alloy column has an equivalent length of 18 ft and must carry an axial load of 350 k. Select a wide-flange member using the section properties of Appendix C.

12.33 The truss of Fig. P12.33 is to be fabricated using 2014-T6 square aluminum alloy rod. Compute the dimensions of member AC to the nearest 1/8 in.

12.34 A 2014-T6 aluminum alloy column has an equivalent length of 8 ft and must carry a concentric axial load of 125 k. Select a wide-flange member using the section properties of Appendix C.

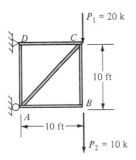

FIGURE P12.33

12.35 Repeat Problem 12.34, assume fixed ends and select the column to ensure that it qualifies as a short column, L/r less than S_1.

12.36 A 6061-T6 aluminum alloy compression member is 18 in. long and can be assumed to have pinned ends. Assume a square cross section and compute the smallest allowable dimension for the structure assuming a concentric axial load of 5000 lb.

12.37 A 2014-T6 aluminum alloy column has an equivalent length of 4 ft and must carry a concentric axial load of 275 k. Select a wide-flange column using the section properties of Appendix C.

12.38 An 2014-T6 aluminum machine element is 0.25 m long and is assumed to have pinned-end conditions. Assume a circular rod and compute the diameter required to carry a concentric axial load of 9 kN. Multiply by 6894 Pa/psi to convert psi to Pa.

12.39 Compute the diameter of a round column 8 ft long with pinned-end conditions that must carry a concentric axial load of 70 k. Assume Douglas fir select grade.

12.40 A square column with equivalent length of 11.5 ft and pinned-end conditions must carry a concentric axial load of 14 k. Compute the required area and corresponding allowable stress. Assume southern pine, grade No. 2.

12.41 A rectangular column is to built up using 4- by 4-in. members. Assume the dimensions are actual, construction grade hemlock and the equivalent length is 14 ft. The column is to carry a concentric axial load of 30,000 lb. Compute the minimum number of 4- by 4-in. members required to carry the load.

12.42 A rectangular column 16 ft long must have a cross section with a width-to-depth ratio of 2 to 3. Assume pinned-end conditions and select grade Douglas fir. Compute the area required to carry a concentric axial load of 35 k.

12.43 Compute the diameter of a round column with an equivalent length of 20 ft and pinned-end conditions.

The concentric axial load is 200 k. Compare the required diameter assuming select, No. 1, and No. 2 southern pine.

12.44 Compute the dimensions of a square column with pinned-end conditions assuming an equivalent length of 11 ft and construction grade hemlock. The column is to carry a concentric axial load of 16,000 lb.

CHAPTER

ENERGY METHODS IN MECHANICS OF MATERIALS

13.1 INTRODUCTION

Energy methods are based upon the concept that for any system in equilibrium there is energy balance between all possible forms of energy for the system. Energy methods, as a method of examining the behavior of structural systems, should be considered to be equally as important as methods based upon the solution of the differential equations that are derived from Newton's laws.

Newton's laws in more general terms are axioms referred to as the *conservation of linear and angular momentum*. A similar axiomatic law is the *conservation of energy* or the *balance of energy*. The conservation of energy is more commonly referred to as the first law of thermodynamics. The idea that several separate topics in mechanics—rigid-body dynamics, particle dynamics, and fluid mechanics—share a common method of analysis was discussed in Chapter 3, Section 3.8. This was referred to loosely as the balance of energy. The terminology associated with energy methods has not developed as rigorously as that of the topics of previous chapters. For instance, the unit load method of Chapter 10 is an interpretation of the principle of virtual work. The unit load method is sometimes called the method of virtual work.

Hence, the terminology can be misleading. Castigliano's theorems will be discussed later in this chapter and although there are two theorems, they are sometimes referenced as Castigliano's theorem without distinguishing between the two. In addition, Castigliano's second theorem is sometimes called the dummy-load method and can become confused with the unit load method. No attempt is made herein to standardize terminology.

The intent of Section 3.8 of Chapter 3 was to introduce the concept of external work and internal strain energy in a most direct manner. In this chapter there will be some repetition of Chapter 3 in order to complete the presentation. However, it is worthwhile to review the derivation of work and strain energy as given in Chapter 3.

Virtual work is a most significant energy concept. The unit load method, as an application of virtual work, was thoroughly discussed in Chapter 10. The reason for including the unit load method in Chapter 10 was to emphasize that it is equally as important as differential equation methods and to introduce a method of analysis that could be extended to three-dimensional indeterminate structures in Chapter 11. As an energy method, that material could properly be included in this chapter.

The idea of potential energy is introduced and the principle of stationary potential energy is briefly discussed. Castigliano's first theorem is presented as a potential energy method and the relationship to the stiffness method of Chapter 11 is demonstrated.

Castigliano's second theorem for linear elastic materials can be an effective tool for structural analysis. The theorem is derived and examples illustrate its use for both two- and three-dimensional structures. An important special case of Castigliano's second theorem is the principle of least work. The principle can be used effectively for the analysis of statically indeterminate structures.

13.2 CONSERVATION OF ENERGY—WORK AND ENERGY

Methods of analysis that are derivable from the axiom of conservation of energy are more difficult conceptually than those that might be obtained from conservation of momentum. For instance, the fundamental concept of statics is the sum of the forces must be balanced, which is the most elementary application of the conservation of linear momentum. The concept is conceptually simplified because a free-body diagram can be drawn that demonstrates the ideas. The forces conceived to represent the conservation of linear momentum are vectors that can be imagined and drawn in space. The quantities that go together to form the equation for the conservation of energy are scalars and deny a neat graphical representation. Imagine trying to draw the quantity *temperature*. It turns out that many of the scalar quantities are derived from vector components.

The equations representing the balance of energy to be used in mechanics of materials will be assumed to have their origin related to the conservation of energy. It must be conceded that similar equations can be derived using the conservation of momentum. For instance, in dynamics the important methods of analysis referred to as work and energy are developed directly from Newton's law. The impetus is mainly philosophical and the final result will be referred to as the *balance of mechanical energy*.

The first law of thermodynamics will be written as the time rate of change of the kinetic energy plus the internal energy is equal to the sum of the rate of work plus all other energies supplied to, or removed from, the system per unit time. Such energies may include thermal energy, chemical energy, or electromagnetic energy. For the thermomechanical system, whereby only thermal and mechanical energies are considered, the first law can be thought of as the balance between the time rate of change in kinetic energy plus internal energy with the rate of change in work plus the rate of change of total heat. The first law of thermodynamics, as described, may be written

$$\Delta T + \Delta U = \Delta W + \Delta Q \qquad (13.1)$$

Several assumptions are in order.[1] The system is assumed to be adiabatic; hence no heat is gained external to the system, and ΔQ is zero. The elastic structural system will be assumed to be at rest, and all loading and deformation processes are assumed to occur slowly. Therefore, velocities are zero and the kinetic energy, ΔT, can be neglected. The balance of the energy equation is

$$\Delta U = \Delta W \qquad (13.2)$$

and is as valid as the first law of thermodynamics.

Equation (13.2) may be compared to Eq. (3.20)

$$U = W$$

in which the total internal energy is equated to the external work. If the change in energies in Eq. (13.2) is imagined to start from a zero

---

[1] The use of the first law of thermodynamics is based on the treatise by H. L. Langhaar, *Energy Methods in Applied Mechanics*, Wiley, New York, 1962.

equilibrium state, Eqs. (3.20) and (13.2) have the same meaning. In the remaining text the Δ will be omitted.

The internal energy is referred to as the strain energy, and W is merely called the work done by externally applied forces. The conservation of energy for linear elastic systems may be stated as: The change of strain energy is balanced by a corresponding change in the work done by externally applied forces or loads. The questions to be answered are: What is the form of the strain energy, and what is the form of the external work for specific problems?

The discussion in Chapter 3 centered about a general expression for the balance of energy and applications for axially loaded members. A special application of Eq. (13.2) was developed in Section 10.8 of Chapter 10 and centered about the principles of virtual work. It is certainly appropriate to review that material at this time, especially if it was omitted as part of Chapter 10.

It was shown in Chapter 3 that the strain energy for a rod of length L loaded with an axial force is

$$U = \frac{1}{2} \int_0^L \frac{P^2 \, dx}{AE} \tag{13.3}$$

and the corresponding external work for an axial deformation of u is

$$W = \frac{Pu}{2} \tag{13.4}$$

The following examples deal with the balance of energy as it is formulated for the basic problems of torsional deformation of circular rods, deformation due to bending, and deformation due to transverse shear. Deformation due to transverse shear was omitted in Chapter 10.

EXAMPLE 13.1

Develop an expression for the strain energy for a circular rod subject to an external torque and show that the balance of energy is satisfied.

Solution:
A circular bar is similar to a torsional spring with spring constant JG/L. The equation relating torque and angular rotation is

$$T = \frac{JG\phi}{L} \tag{a}$$

The external work is

$$W = \int_0^\phi T \, d\phi = \int_0^\phi \frac{JG\phi}{L} \, d\phi = \frac{JG\phi^2}{2L} \tag{b}$$

Substituting for ϕ from Eq. (a),

$$W = \frac{T\phi}{2} \tag{c}$$

The strain energy is computed as one-half the shear stress times the shear strain.

$$U = \frac{1}{2} \int_v \tau\lambda \, dv = \frac{1}{2} \int_v \frac{\tau^2}{2G} \, dv \tag{d}$$

Recall from Chapter 4 that $\lambda = \tau/G$, and

$$\tau = \frac{Tr}{J} \tag{e}$$

$$J = \int_A r^2 \, dA \tag{f}$$

Substituting into Eq. (d)

$$U = \frac{1}{2} \int_0^L \int_A \frac{T^2 r^2}{J^2 G} \, dA \, dx = \frac{1}{2} \int_0^L \frac{T^2}{GJ} \, dx = \frac{T^2 L}{2GJ} \tag{13.5}$$

Equation (13.5) is the expression for strain energy. Substituting Eq. (a) gives an expression that is equivalent to Eq. (c).

$$U = \frac{T\phi}{2} = W$$

and verifies the balance of energy.   ■

EXAMPLE 13.2

Develop an expression for the strain energy due to bending. Show that the conservation of energy is satisfied for the beam of Fig. 13.1.

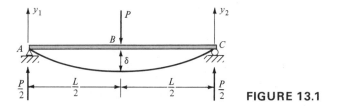

FIGURE 13.1

Solution:

The strain energy due to bending, neglecting shear strain energy, is

$$U = \frac{1}{2} \int_v \sigma_x \varepsilon_x \, dV = \frac{1}{2} \int_v \frac{\sigma_x^2}{E} \, dV \tag{a}$$

Substituting $\varepsilon_x = \sigma_x/E$ and

$$\sigma_x = \frac{My}{I} \tag{b}$$

$$I = \int_A y^2 \, dA \tag{c}$$

gives

$$U = \frac{1}{2} \int_0^L \int_A \frac{M^2 y^2}{EI^2} \, dA \, dx = \frac{1}{2} \int_0^L \frac{M^2}{EI} \, dx \tag{13.6}$$

Equation (13.6) is the general expression for the strain energy due to bending. The strain energy for the beam of Fig. 13.1 is obtained by substituting the moment equation into Eq. (13.6).

$$M_{AB} = \frac{Px}{2}$$

Assuming an origin at the right end

$$M_{CB} = \frac{Px}{2}$$

$$U = \frac{1}{2EI} \int_0^{L/2} \frac{P^2 x^2}{4} \, dx + \frac{1}{2EI} \int_0^{L/2} \frac{P^2 x^2}{4} = \frac{P^2 L^3}{96EI} \tag{d}$$

The external work done by the force P as it moves through the deflection δ is

$$W = \frac{P\delta}{2} \tag{e}$$

δ can be evaluated using the superposition tables.

$$\delta = \frac{PL^3}{48EI} \tag{f}$$

$$W = \frac{P^2 L^3}{96EI} \tag{g}$$

Comparing Eqs. (d) and (g), it is confirmed that $U = W$.

Equation (13.2) can be used to compute the deflection at the point where a concentrated load is applied to a beam. ∎

EXAMPLE 13.3

Use the balance-of-energy equation to compute the deflection at the point where the concentrated force is applied for the beam of Fig. 13.2.

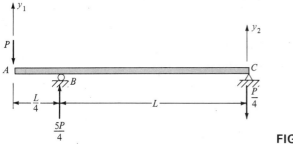

FIGURE 13.2

Solution:
Reactions are computed as shown in Fig. 13.2. Moment equations are written for each segment of the beam using the left and right ends, respectively, as the origin.

$$M_{AB} = -Px$$

$$M_{CB} = \frac{-Px}{4}$$

Using Eq. (13.6), the strain energy is evaluated as

$$U = \frac{1}{2} \int_0^{L/4} \frac{P^2 x^2}{EI}\, dx + \frac{1}{2} \int_0^L \frac{P^2 x^2}{16EI}\, dx = \frac{P^2 L^3}{384EI} + \frac{P^2 L^3}{96EI}$$

$$U = \frac{5P^2 L^3}{384EI}$$

The work done by the force P as it moves through a deformation δ is

$$W = \frac{P\delta}{2}$$

By Eq. (13.2),

$$\frac{P\delta}{2} = \frac{5P^2 L^3}{384EI}$$

$$\delta = \frac{5PL^3}{192EI}$$

The use of Eq. (13.2) as illustrated by Example 13.3 is limited to problems in which the deflection is to be computed at the point where the concentrated force is applied. The unit load method presented in Chapter 10 is obviously a more general interpretation of an energy method.

EXAMPLE 13.4

Develop an expression for the strain energy due to transverse shear. Compute the deflection caused by shear at the free end of a fixed-free beam subject to the loading shown in Fig. 13.3. Compare the deflection due to shear with the deflection caused by bending.

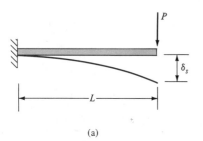

(a)

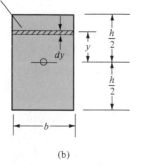

(b)   **FIGURE 13.3**

Solution:

The general expression for strain energy due to transverse shear appears in the same form as strain energy for torsional shear.

$$U = \frac{1}{2} \int_v \tau \lambda \, dV = \frac{1}{2} \int_v \frac{\tau^2}{G} \, dV \qquad \textbf{(a)}$$

The difference is the interpretation of τ.

The variation of transverse shear stress over the cross section of a beam is discussed in Section 7.5 of Chapter 7. Problem 7.29 represents the derivation of the expression defining the shear stress variation. Let the cross section be defined as shown in Fig. 13.3b. Note that Fig. 13.3b is different from Fig. P7.29. In this example a general expression for τ is desirable rather than an expression for the shear stress at a particular location. The shear stress is given by

$$\tau = \frac{VQ}{It} = \frac{V}{It} \int_A y\,dA = \frac{V}{Ib} \int_y^{h/2} yb\,dy$$

$$= \frac{V}{I} \frac{y^2}{2}\bigg|_y^{h/2} = \frac{V}{2I}\left[\left(\frac{h}{2}\right)^2 - y^2\right] \tag{b}$$

Equation (a) gives an expression for the strain energy for constant shear, V. Substitute Eq. (b) into Eq. (a)

$$U = \frac{1}{2G} \int_v \frac{V^2}{4I^2}\left[\left(\frac{h}{2}\right)^2 - y^2\right]^2 dV$$

$$= \frac{V^2}{8I^2G} \int_{-h/2}^{h/2} \int_0^L \left[\left(\frac{h}{2}\right)^2 - y^2\right]^2 b\,dx\,dy = \frac{V^2Lbh^5}{240GI^2}$$

where dV, the volume element, has been replaced by $b\,dx\,dy$. For a rectangular cross section,

$$I = \frac{bh^3}{12}$$

$$U = \frac{V^2Lbh^5}{240G(b^2h^6/144)} = \frac{3}{5}\frac{V^2L}{bhG} = \frac{3V^2L}{5AG} \tag{c}$$

where bh is the area of the rectangular cross section.

The fixed-free beam of Fig. 13.3a has constant shear equal to P. The work done by the force P is $P\delta_s/2$, where δ_s is the shear deflection due to the shear loading.

$$\frac{P\delta_s}{2} = \frac{3}{5}\frac{PL}{AG}$$

$$\delta_s = \frac{6}{5}\frac{PL}{AG} \tag{d}$$

The factor 6/5 is referred to as a shape factor or shear coefficient and varies depending upon the shape of the cross section. The deflection due to bending for the beam of Fig. 13.3a is

$$\delta_b = \frac{PL^3}{3EI} \tag{e}$$

For a material such as steel, where $E = 30(10^3)$ ksi and $G = 12(10^3)$ ksi, the results of Eqs. (d) and (e) can be compared. Let $G = E/2.5$, $A = bh$ and $I = bh^3/12$. The total deflection of the beam is

$$\delta_T = \delta_b + \delta_s = \frac{12PL^3}{3Ebh^3} + \frac{6}{5}\frac{PL}{bhE/2.5} = \frac{PL^3}{3EI}\left(1 + 0.75\frac{h^2}{L^2}\right)$$

For a very short beam where h approaches the value of L, the shear deformation is significant. For most structural systems shear deformation is negligible. ■

13.3 POTENTIAL ENERGY

Potential energy is defined in most dynamics textbooks as the capacity of a system to perform work as a result of the relative positions of the various components of the system. If a spring-type structure is compressed, the material of the spring is stressed and deformed, and energy is stored in the spring. It can be visualized that the elastic potential energy of the spring is energy stored in the spring and is equal to the work required for an external force to compress the spring.

Potential energy is related to relative position. It can be imagined that the deformed position of a structure is the reference configuration from which the potential is measured. Internally stored energy (strain energy) will potentially be the energy required to return the structure to its undeformed position. This is in keeping with the concept in dynamics of referencing a force or body to a datum line. As the structure moves from the loaded position to the unloaded position, the external force does negative work; hence, external forces can be thought of as having potential energy.

The total potential energy is the combination of the strain energy and the negative potential energy of the external forces, or

$$\Pi = U - W_P \tag{13.7}$$

In previous engineering mechanics courses (dynamics), the potential energy was combined with the kinetic energy and the change in energy from one reference state to another was set equal to zero. If Eq. (13.7) was used in this manner, it would actually reduce to the statement of conservation of energy of the previous section. The concept of potential energy will be used to develop the *principle of stationary potential energy*, from which the condition for the principle of *minimum* potential energy can be posed.

13.4 PRINCIPLE OF STATIONARY POTENTIAL ENERGY

The principle of stationary potential energy may be thought of as a condition for equilibrium. The stationary value may be neutral, mini-

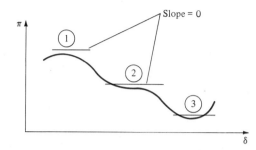

FIGURE 13.4

mum or maximum. Conceptually, let the potential Π be a function of the displacement at some point in the structure. As deformation occurs, the relation between Π and the displacement δ might occur as in Fig. 13.4. The change in Π with respect to δ, the slope of the curve of Fig. 13.4, could be zero at any one of three points. That is, $\partial\Pi/\partial\delta = 0$ indicates (1) a maximum, (2) a neutral value, or (3) a minimum. If the structure is unstable, the stationary potential energy may be maximum or neutral. If the structure is in stable equilibrium, the potential energy is a minimum. Hence, the *principle of minimum potential energy* is formally written as

$$\frac{\partial\Pi}{\partial\delta} = 0 \tag{13.8}$$

The potential energy methods are essentially displacement methods. To analyze a structure using potential energy methods the strain energy must be written in terms of unknown joint displacements. All loads on the structure must correspond to equivalent joint loadings. Direct use of potential energy methods is more applicable to axially loaded structures. The next section deals with a potential energy theorem; in particular, Castigliano's first theorem.

13.5 CASTIGLIANO'S FIRST THEOREM

Alberto Castigliano was an Italian engineer and mathematician who in 1879 published his theorems on structural analysis. These theorems are of such historical significance that they bear his name today.

Castigliano's first theorem is a potential energy theorem and the similarity to Eq. (13.8) will be obvious. Consider the structure of Fig. 13.5. A truss-type structure is used, but a beam or any general type of structure could give the same result. Assume that the loads are applied slowly and simultaneously. It is essential for this discussion that the relation between load and deformation remain elastic. As the loads are applied, the deformations, δ, occur. The potential energy of the system can be written using Eq. (13.7).

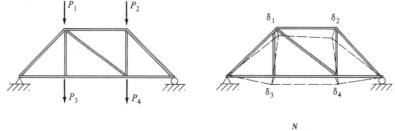

FIGURE 13.5

$$\Pi = U - P_1\delta_1 - P_2\delta_2 - P_3\delta_3 - P_4\delta_4 = U - \sum_{i=1}^{N} P_i\delta_i \qquad \textbf{(a)}$$

It must be kept in mind that U is a function of the deflections, δ_i. Imagine that one of the deflections, δ_1, is modified by a small increment. There will be a slight change in the potential energy. According to Eq. (13.8), the change in potential energy with respect to the deflection δ_1 should be a minimum.

$$\frac{\partial\Pi}{\partial\delta_1} = \frac{\partial U}{\partial\delta_1} - P_1 = 0$$

or

$$\frac{\partial U}{\partial\delta_1} = P_1 \qquad \textbf{(b)}$$

Similarly,

$$\frac{\partial U}{\partial\delta_2} = P_2 \qquad \frac{\partial U}{\partial\delta_3} = P_3 \qquad \frac{\partial U}{\partial\delta_4} = P_4 \qquad \textbf{(c)}$$

Obviously, this process could be extended to any number of external loads and their corresponding deflections. A general expression is

$$\frac{\partial U}{\partial\delta_i} = P_i \qquad \textbf{(13.9)}$$

Equation (13.9) is the mathematical statement of Castigliano's first theorem.

The use of Eq. (13.9) is dependent upon being able to write the strain energy in terms of displacements, either joint translations or rotations. The strain energy expressions—for axial force, Eq. (13.3); torsional shear, Eq. (13.5); bending, Eq. (13.6); and transverse shear, Eq. (c) of Example 13.4—are formulated in terms of the applied load rather than displacements. An expression that relates strain energy directly to strain—and hence displacement—is more applicable. Castigliano's first theorem can be used for structures composed of materials with non-linear stress-strain properties. The use of the strain energy expressions enumerated above limits the applications to linear stress–strain properties since all of those equations are based upon $\sigma = E\varepsilon$ or $\tau = G\gamma$.

The next section of this chapter deals with complementary energy and its relation to strain energy. A general relation between differential strain energy and differential strain will be developed in that section. The following examples will be based upon the strain energy equations previously developed. It must be kept in mind that the examples are to illustrate the use of Castigliano's first theorem and that other methods of analysis could be considered more efficient.

EXAMPLE 13.5

The truss of Fig. 13.6 has a single vertical force P applied at joint B. Assume a linear relation between stress and strain and compute the displacement of joint B and the axial force in each member using Castigliano's first theorem. Assume AE is the same for both members.

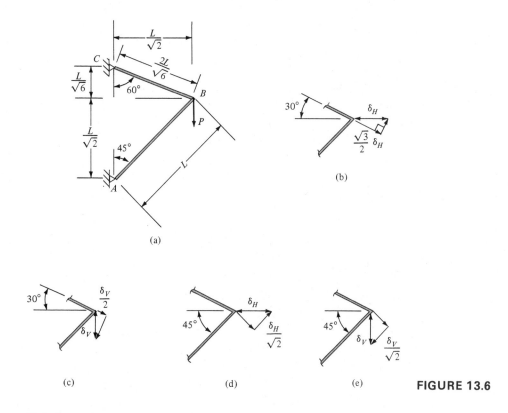

FIGURE 13.6

Solution:
The unknown deflection of joint B is sketched in Fig. 13.6. The horizontal component is δ_H and the vertical is δ_V. The deformation of member BC is shown in Figs. 13.6b and c and is an elongation for the assumed horizontal movement of joint B.

$$\delta_{BC} = \frac{\sqrt{3}\delta_H}{2} + \frac{\delta_V}{2} \tag{a}$$

The elongation of member AB is shown in Figs. 13.6d and e. δ_H would cause elongation and δ_V would cause shortening of the member.

$$\delta_{AB} = \frac{\delta_H}{\sqrt{2}} + \frac{\delta_V}{\sqrt{2}} \tag{b}$$

Equation (13.3) is the only equation that is thus far available for writing the strain energy due to axial force. Prior to using Eq. (13.3), the deflections δ_{AB} and δ_{BC} must be written in terms of axial force. Since the relation between force and deformation is linear, the standard formula $u = PL/AE$ can be used.

$$P_{AB} = \frac{\delta_{AB}AE}{L} = \frac{AE(\delta_H - \delta_V)}{\sqrt{2}L} \tag{c}$$

$$P_{BC} = \frac{\delta_{BC}AE}{(2L/\sqrt{6})} = \frac{\sqrt{6}AE(\sqrt{3}\delta_H + \delta_V)}{4L} \tag{d}$$

Substituting into Eq. (13.3) gives the expression for the strain energy.

$$U = \frac{1}{2}\int_0^L \frac{AE}{2L^2}(\delta_H - \delta_V)^2\,dx + \frac{1}{2}\int_0^{2L/\sqrt{6}} \frac{6AE}{16L^2}(\sqrt{3}\delta_H + \delta_V)^2\,dx \tag{e}$$

Let an x axis represent the axial coordinate for each member. Integrating and rearranging Eq. (e) gives the following.

$$U = \frac{AE}{4L}\left[(\delta_H^2 - 2\delta_H\delta_V + \delta_V^2) + \frac{3}{2\sqrt{6}}(3\delta_H^2 + 2\sqrt{3}\delta_H\delta_V + \delta_V^2)\right] \tag{f}$$

With the strain energy written as a function of the deformations, Eq. (13.9) can be used to generate two equations in terms of the unknown deflections.

$$\frac{\partial U}{\partial \delta_H} = \frac{AE}{4L}\left[(2\delta_H - 2\delta_V) + \frac{3}{2\sqrt{6}}(6\delta_H + 2\sqrt{3}\delta_V)\right] = 0 \tag{g}$$

Equation (g) is set equal to zero since the actual horizontal force at joint B is zero.

$$\frac{\partial U}{\partial \delta_V} = \frac{AE}{4L}\left[(-2\delta_H + 2\delta_V) + \frac{3}{2\sqrt{6}}(2\sqrt{3}\delta_H + 2\delta_V)\right] = P \tag{h}$$

where P is the actual vertical force at joint B. Equations (g) and (h) are solved for δ_V and δ_H.

$$\delta_V = \frac{1.241PL}{AE} \tag{i}$$

$$\delta_H = \frac{-0.027PL}{AE} \tag{j}$$

Substituting Eqs. (i) and (j) into Eqs. (c) and (d) gives the axial force in members AB and BC, respectively.

$$P_{AB} = \frac{AE}{2L}(-0.027 - 1.24)\frac{PL}{AE} = -0.897P$$

$$P_{BC} = \frac{6AE}{4L}[\sqrt{3}(-0.027 + 1.24]\frac{PL}{AE} = 0.731P$$

Member AB is in compression, as indicated by the negative sign. The results can be verified using statics for the determinate structure. ∎

> This illustrative example obviously is not practical for such an elementary problem, but it is applicable for an indeterminate truss with little additional work. However, Castigliano's first theorem has been illustrated. The method is a displacement formulation with definite similarities to the displacement (stiffness) method of Chapter 11. The solution is formulated as a set of simultaneous equations with displacement as the unknowns.

EXAMPLE 13.6

Use Castigliano's first theorem to solve Example 11.15.

Solution:
The beam of Example 11.15 is shown in Fig. 13.7a. The beam is statically indeterminate to the first degree. Prior to formulating the solution, it is worthwhile to derive an expression for the strain energy in a beam subject to joint rotations at both ends. A fixed-fixed beam subject to joint rotations and corresponding end moments is shown in Fig. 13.8. Appendix F can be used to write the expression for each end moment in terms of the joint rotations θ_A and θ_B.

$$M_A = \frac{4EI\theta_A}{L} + \frac{2EI\theta_B}{L} \tag{a}$$

$$M_B = \frac{2EI\theta_A}{L} + \frac{4EI\theta_B}{L} \tag{b}$$

The strain energy for a beam, such as that of Fig. 13.8, subject to end moments and joint rotations can be written as

$$U = \frac{1}{2}M_A\theta_A + \frac{1}{2}M_B\theta_B \tag{c}$$

Substituting Eqs. (a) and (b) into (c) gives the strain energy in terms of joint rotations.

$$U = \frac{2EI(\theta_A^2 + \theta_A\theta_B + \theta_B^2)}{L} \tag{d}$$

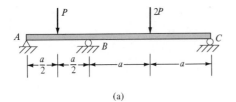

(a)

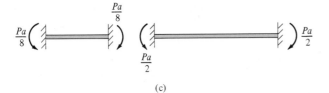

(b)

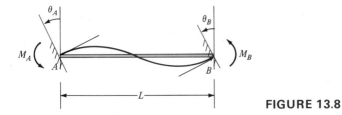

(c)

FIGURE 13.7

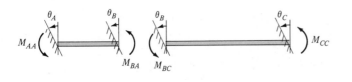

FIGURE 13.8

The beam of Fig. 13.7 can be divided into two separate beams and the beam of Fig. 13.8 used as a model to write the strain energy in terms of θ_A, θ_B, and θ_C. The moments at each end of the beams of Fig. 13.7b are written according to Eqs. (a) and (b) using Fig. 11.8 as a model.

$$M_{AA} = \frac{4EI\theta_A}{a} + \frac{2EI\theta_B}{a} \qquad \text{(e)}$$

$$M_{BA} = \frac{2EI\theta_A}{a} + \frac{4EI\theta_B}{a} \qquad \text{(f)}$$

$$M_{BC} = \frac{2EI\theta_B}{a} + \frac{EI\theta_C}{a} \tag{g}$$

$$M_{CC} = \frac{EI\theta_B}{a} + \frac{2EI\theta_C}{a} \tag{h}$$

The strain energy is

$$U = \frac{1}{2}M_{AA}\theta_A + \frac{1}{2}M_{BA}\theta_B + \frac{1}{2}M_{BC}\theta_B + \frac{1}{2}M_{CC}\theta_C \tag{i}$$

Substitute Eqs. (e)–(h) into Eq. (i).

$$U = \frac{1}{2}\left(\frac{4EI\theta_A^2}{a} + \frac{4EI\theta_A\theta_B}{a} + \frac{6EI\theta_B^2}{a} + \frac{2EI\theta_C\theta_B}{a} + \frac{2EI\theta_C^2}{a}\right) \tag{j}$$

The fixed-end moments caused by the concentrated loads are computed using Appendix F, as illustrated in Example 11.15, and are shown in Fig. 13.7c as moments acting on the joints. The interpretation of Eq. (13.9) is

$$\frac{\partial U}{\partial \theta_i} = M_i \tag{13.10}$$

where M_i is the moment caused by externally applied loads and applied at the joint where θ_i occurs. Counterclockwise moments and rotations are assumed positive.

$$\frac{\partial U}{\partial \theta_A} = \frac{1}{2}\left(\frac{8EI\theta_A}{a} + \frac{4EI\theta_B}{a}\right) = \frac{-Pa}{8} \tag{k}$$

$$\frac{\partial U}{\partial \theta_B} = \frac{1}{2}\left(\frac{4EI\theta_A}{a} + \frac{12EI\theta_B}{a} + \frac{2EI\theta_C}{a}\right) = \frac{Pa}{8} - \frac{Pa}{2} \tag{l}$$

$$\frac{\partial U}{\partial \theta_C} = \frac{1}{2}\left(\frac{2EI\theta_B}{a} + \frac{4EI\theta_C}{a}\right) = \frac{Pa}{2} \tag{m}$$

Equations (k)–(m) are identical to Eqs. (a)–(c) of Example 11.15. It follows that Castigliano's first theorem yields the same results as the stiffness method. The problem is completed identically to Example 11.15 and will not be completed here. ∎

Castigliano's first theorem can be applied to beams and frames where joint translations and rotations occur simultaneously.

13.6 STRAIN ENERGY VERSUS COMPLEMENTARY ENERGY AND CASTIGLIANO'S SECOND THEOREM

Thus far, all discussion of energy methods—strain energy and external work in Chapter 3, virtual work and the unit load method of Chapter 10 and the preceding sections of this chapter—has assumed linear elastic load deflection relations. This is not out of the ordinary since most materials that are studied using the methods of mechanics of materials are, in fact, assumed to be linearly elastic. A nonlinear elastic material was defined in Chapter 3.

Nonlinear behavior may occur due to material nonlinearities or geometric nonlinearities. Material nonlinear behavior is a result of an actual nonlinear relation between stress and strain. A geometric nonlinearity is a result of deforming the structure beyond the point where assumption of small deformation gradients is applicable. For example, to derive the equations for beam deflections, the assumption of $\sin \theta \simeq \theta$ imposes a definite limitation. Chapter 12, on column analysis, gives examples of problems that can become geometrically nonlinear. The discussion in this section is limited to assumed material nonlinearities.

Figure 13.9 represents an assumed nonlinear stress-strain relation. For simplicity imagine that a bar is loaded in tension or compression and Fig. 13.9 is the result. Let σ_1 and ε_1 correspond to some level of applied stress and strain. It was established in Chapter 3 that the area between the curve and ε axis is equal to the strain energy per unit volume. The simplifying assumption of a linear stress strain curve allows the strain energy to be evaluated as $\int_v \sigma\varepsilon \, dv/2$. The nonlinear curve requires a different formulation. The area above the curve, between the curve and the σ axis, is called the *complementary energy per unit volume*. Figure 13.9 is a graphical representation of the relation between strain energy and complementary energy.

Due to the nonlinear stress-strain relation, a different method other than that of Chapter 3 must be used to evaluate the strain energy. Let the curve of Fig. 13.9 be plotted as deformation versus load, as shown in Fig. 13.10. A differential element $d\delta$ is introduced as shown. The area under the curve is the strain energy per unit volume.

$$U = \int P \, d\delta$$

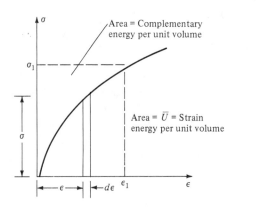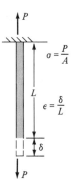

FIGURE 13.9
Strain energy and complementary energy.

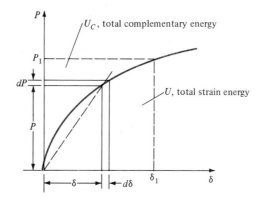

FIGURE 13.10

Differentiating both sides of this equation gives

$$dU = P \, d\delta \qquad \text{or} \qquad \frac{dU}{d\delta} = P \tag{13.11}$$

Imagine that the structure of Fig. 13.5 produces the load deflection relation, such that U is made up of contributions from several loads. The total derivative of Eq. (13.11) is written as a partial derivative with respect to a specific displacement.

$$\frac{\partial U}{\partial \delta_i} = P_i \tag{13.12}$$

It turns out that Eq. (13.12) is identical to Eq. (13.9); thus, it is equivalent to Castigliano's first theorem. Another form of Eq. (13.12) can be obtained using Fig. 13.9. Let $\bar{U}$ be the strain energy per unit volume; then

$$\bar{U} = \int \sigma \, d\varepsilon \tag{13.13}$$

The area U_c, referred to as the complementary energy, can be evaluated using Fig. 13.10.

$$U_c = \int \delta \, dP$$

Differentiating this equation gives

$$dU_c = \delta \, dP \qquad \text{or} \qquad \frac{dU_c}{dP} = \delta \tag{13.14}$$

The same argument as before suggests that Eq. (13.14) should be written

$$\frac{\partial U_c}{\partial P_i} = \delta_i \tag{13.15}$$

Equation (13.15) is the mathematical statement of Castigliano's second theorem. Castigliano's second theorem becomes a powerful energy method for structures with *linear* stress-strain properties. The dashed line of Fig. 13.10 represents a linear stress-strain curve, and obviously $\bar{U} = U = U_c$. Then Castigliano's second theorem for linear elastic materials is

$$\frac{\partial U}{\partial P_i} = \delta_i \tag{13.16}$$

Equation (13.16) is desirable from an analysis viewpoint because U can be evaluated, for the basic structures that have been studied, directly in terms of the loads, P_i.

The following examples will illustrate applications to trusses and beams.

EXAMPLE 13.7

Compute the vertical deflection of point B for the truss of Fig. 13.6.

Solution:
The force in each member of the truss can be computed using statics; however, the forces were computed in Example 13.5 as

$$P_{AB} = -0.897P = F_{AB}$$

$$P_{BC} = 0.731P = F_{BC}$$

Equation (13.16) must be in terms of the strain energy for an axial load. Let the axial force be F and the applied load be P (F is a function of P).

$$\frac{\partial U}{\partial P} = \frac{\partial}{\partial P}\left[\frac{1}{2}\int_0^L \frac{F^2}{AE}\,dx\right] = \int_0^L \frac{F}{AE}\frac{\partial F}{\partial P}\,dx = \delta \tag{13.17}$$

The deflection δ is at the point where P is applied and in the direction of P. For this example,

$$\delta_v = \int_0^L \frac{(-0.897P)}{AE}(-0.897)\,dx + \int_0^{2L/\sqrt{6}} \frac{(0.731P)}{AE}(0.731)\,dx$$

$$= \frac{1.241PL}{AE}$$

and agrees with Example 13.5. ∎

In the previous example the deflection was computed at the point of applied load and in the direction of the applied load. To compute a deflection at a point where there is no load, a fictitious force can be applied. After computing the partial derivative with respect to the fictitious load, the load is set equal to zero. Consider the following example.

EXAMPLE 13.8

Compute the horizontal deflection at point B for the truss of Fig. 13.6.

Solution:
A fictitious force Q is applied at B in the horizontal direction, as shown in Fig. 13.11a. A free body of joint B yields the following equations.

$$Q = H_{BC} + H_{AB}$$
$$V_{BC} = V_{AB}$$
$$H_{AB} = V_{AB}$$
$$H_{BC} = \sqrt{3}V_{BC} \tag{a}$$

The results are shown in Fig. 13.11b. The results for Q are added to the previous results for the actual P. The force in each member is

$$F_{AB} = -0.897P + 0.518Q$$
$$F_{BC} = 0.731P + 0.731Q \tag{b}$$

The deflection δ_H is given by

$$\delta_H = \frac{\partial U}{\partial Q} = \frac{\partial}{\partial Q}\left[\frac{1}{2}\int_0^L \frac{F^2}{AE}\,dx\right] = \int_0^L \frac{F}{AE}\frac{\partial F}{\partial Q}\,dx \tag{c}$$

where differentiation is with respect to the load at B in the direction of the desired deflection. The load Q is fictitious and will be removed after the partial derivative is evaluated. Since the order of differentiation and integration can be interchanged, it will expedite the analysis to compute $\partial F/\partial Q$. Set Q equal to zero and then integrate.

$$\frac{\partial F_{AB}}{\partial Q} = 0.518$$

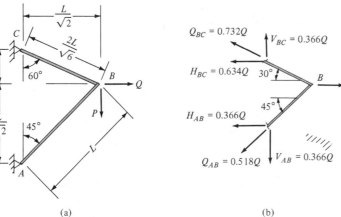

(a)　　　　　(b)　　　　**FIGURE 13.11**

$$\frac{\partial F_{BC}}{\partial Q} = 0.731 \qquad \qquad \textbf{(d)}$$

Let $Q = 0$.

$$\delta_H = \int_0^L \frac{(-0.897P)(0.518)}{AE} \, dx + \int_0^{2L/\sqrt{6}} \frac{(0.731P)(0.731)}{AE} \, dx \qquad \textbf{(e)}$$

$$= -0.465P + 0.436P = -0.028P$$

and agrees with Eq. (j) of Example 13.5.   ◼

Statically determinate beam problems using Castigliano's second theorem are straightforward to analyze for deflections. The strain energy due to bending is given by Eq. (13.6), and the bending moment in terms of applied loads is easily formulated. Castigliano's second theorem for linear elastic structures, Eq. (13.16), subject to a bending load P_i at the point where the deflection is to be computed is as follows:

$$v_i = \frac{\partial U}{\partial P_i} = \frac{\partial}{\partial P_i}\left[\frac{1}{2}\int_0^L \frac{M^2}{EI} \, dx\right] = \int_0^L \frac{M}{EI}\frac{\partial M}{\partial P_i} \, dx \qquad \textbf{(13.18)}$$

where the integration is over the total length of the structure. The use of Eq. (13.18) is illustrated by the following example.

EXAMPLE 13.9

Compute the deflection at the free end of the fixed-free beam of Fig. 13.12. Assume that EI is constant for the beam.

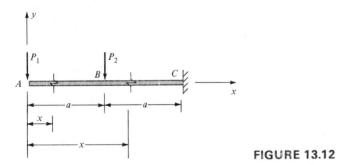

FIGURE 13.12

Solution:
Moment equations are written for beam segments AB and BC using the free end as the origin.

$$M_{AB} = -P_1 x \tag{a}$$

$$M_{BC} = -P_1 x - P_2(x - a) \tag{b}$$

Since P_1 is applied at the free end, Eq. (13.18) is evaluated using the load P_1.

$$\frac{\partial M_{AB}}{\partial P_1} = -x \tag{c}$$

$$\frac{\partial M_{BC}}{\partial P_1} = -x \tag{d}$$

$$v = \int_0^a \frac{P_1 x^2 \, dx}{EI} + \int_a^{2a} \left[P_1 x^2 + P_2(x^2 - ax) \right] \frac{dx}{EI} \tag{e}$$

$$= \frac{P_1 a^3}{3EI} + \frac{7 P_1 a^3}{3EI} + \frac{5}{6} \frac{P_2 a^3}{EI}$$

$$= \frac{8}{3} \frac{P_1 a^3}{EI} + \frac{5}{6} \frac{P_2 a^3}{EI} \qquad\blacksquare$$

EXAMPLE 13.10

Compute the deflection at the midpoint of the structure shown in Fig. 13.13. Assume EI to be constant.

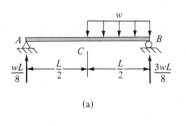

(a)

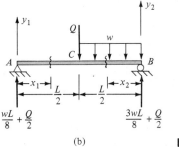

(b)   **FIGURE 13.13**

Solution:
The reactions are computed as illustrated in Fig. 13.13a. A fictitious load Q is applied at C, and the reactions are modified accordingly. Moment equations are written using A for the origin of segment AC and B for the origin of segment BC.

$$M_{AC} = \frac{wLx}{8} + \frac{Qx}{2}$$

$$\frac{\partial M_{AC}}{\partial Q} = \frac{x}{2}$$

$$M_{BC} = \frac{3wLx}{8} + \frac{Qx}{2} - \frac{wx^2}{2}$$

$$\frac{\partial M_{BC}}{\partial Q} = \frac{x}{2}$$

Let Q equal zero and substitute into Eq. (13.18).

$$v_c = \frac{1}{EI} \int_0^{L/2} \frac{wLx^2}{16}\,dx + \frac{1}{EI} \int_0^{L/2} \left(\frac{3}{16} wLx^2 - \frac{wx^3}{4} \right) dx$$

$$= \frac{1}{EI}\left[\frac{wLx^3}{48} \right]_0^{L/2} + \frac{1}{EI}\left[\frac{3wLx^3}{48} - \frac{wx^4}{16} \right]_0^{L/2}$$

$$= \frac{5wL^4}{768EI}$$ ∎

Consider an additional beam analysis problem: the computation of the slope at a given point. The statement of Castigliano's second theorem to relate strain energy to beam rotation would be

$$\theta_i = \frac{\partial U}{\partial C_i} = \frac{1}{2} \int_0^L \frac{\partial M^2}{\partial C_i} \frac{dx}{EI} = \int_0^L \frac{M}{EI} \frac{\partial M}{\partial C_i}\,dx \qquad \textbf{(13.19)}$$

where C_i is a bending couple applied at the point where the rotation is to be computed. The following example should illustrate the use of Eq. (13.19).

EXAMPLE 13.11

Compute the slope at the free end of the fixed-free beam of Fig. 13.14. Assume that EI is constant.

Solution:
A fictitious couple is applied at point B since an actual couple does not exist at that point. The coordinate system is assumed at point B and the moment equation written for the entire beam.

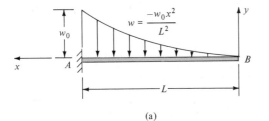

(a)

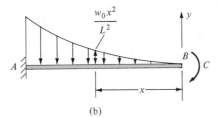

(b)   **FIGURE 13.14**

$$M = -w_0\left(\frac{x^2}{L^2}\right)\left(\frac{x}{3}\right)\left(\frac{x}{4}\right) - C$$

$$\frac{\partial M}{\partial C} = -1$$

Let C equal zero and substitute into Eq. (13.19).

$$\theta_B = \int_0^L \frac{w_0 x^4}{12EIL^2}(1)\,dx = \frac{w_0 x^5}{60EIL^2}\bigg|_0^L = \frac{w_0 L^3}{60EI} \qquad \blacksquare$$

Three-dimensional structures can be analyzed using Castigliano's second theorem. Equation (13.6) should be reformulated, indicating the scalar product between the stress term and the strain term. Equation (13.6) can be visualized as the scalar product of one-half the moment times the slope. Recalling that $d\theta = M\,dx/EI$ for bending moment

$$dU = \left(\frac{1}{2}\right)\mathbf{M}\cdot d\theta = \left(\frac{1}{2}\right)\mathbf{M}\cdot\frac{\mathbf{M}\,dx}{EI}$$

$$U = \frac{1}{2}\int_0^L \mathbf{M}\cdot\frac{\mathbf{M}\,dx}{EI} \qquad\qquad\qquad\qquad \textbf{(13.20)}$$

The moment vector is composed of three components as shown in Fig. 13.15. The axial vector is a torque and is placed along the x axis. The rotation vector $d\theta_x$ is the axial rotation defined in Chapter 4; $d\phi = d\theta_x = M_x\,dx/JG$.

$$\mathbf{M} = M_x\mathbf{i} + M_y\mathbf{j} + M_z\mathbf{k} \qquad\qquad\qquad\qquad \textbf{(a)}$$

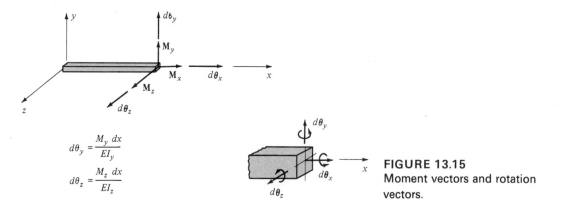

$$d\theta_y = \frac{M_y \, dx}{EI_y}$$

$$d\theta_z = \frac{M_z \, dx}{EI_z}$$

FIGURE 13.15
Moment vectors and rotation vectors.

$$d\theta = \frac{M_x}{JG} dx \, \mathbf{i} + \frac{M_y \, dx}{EI_y} \mathbf{j} + \frac{M_z \, dx}{EI_z} \mathbf{k} \qquad \textbf{(b)}$$

Equation (13.20) becomes

$$U = \frac{1}{2} \int_0^L (M_x \mathbf{i} + M_y \mathbf{j} + M_z \mathbf{k}) \cdot \left(\frac{M_x}{JG} \mathbf{i} + \frac{M_y}{EI_y} \mathbf{j} + \frac{M_z}{EI_z} \mathbf{k} \right) dx$$

$$U = \frac{1}{2} \int_0^L \left(\frac{M_x^2}{JG} + \frac{M_y^2}{EI_y} + \frac{M_z^2}{EI_z} \right) dx \qquad \textbf{(13.21)}$$

Define three forces, $P\mathbf{i}$, $Q\mathbf{j}$, and $R\mathbf{k}$ as forces, either fictitious or actual, acting at a point where deflections due to moments are to be computed.

$$\mathbf{F} = P\mathbf{i} + Q\mathbf{j} + R\mathbf{k} \qquad \textbf{(c)}$$

A statement, equivalent to Eq. (13.18), is to take the partial derivative of the strain energy U with respect to F. The derivative of Eq. (13.21) with respect to F can be written to give δ, the deflection vector.

$$\delta = \frac{\partial U}{\partial \mathbf{F}} = \frac{1}{2} \int_0^L \frac{\partial}{\partial \mathbf{F}} \left(\frac{M_x^2}{JG} + \frac{M_y^2}{EI_y} + \frac{M_z^2}{EI_z} \right) dx \qquad \textbf{(d)}$$

The term $\partial / \partial \mathbf{F}$ can be interpreted as an operator.

$$\frac{\partial}{\partial \mathbf{F}} = \left(\frac{\partial}{\partial P} \right) \mathbf{i} + \left(\frac{\partial}{\partial Q} \right) \mathbf{j} + \left(\frac{\partial}{\partial R} \right) \mathbf{k} \qquad \textbf{(e)}$$

and

$$\delta = u\mathbf{i} + v\mathbf{j} + w\mathbf{k}$$

where u, v, and w are the beam deflections in the x, y and z directions, respectively. Equation (d) represents three equations of the form

$$u\mathbf{i} = \frac{\partial U}{\partial P} \mathbf{i} = \frac{1}{2} \int_0^L \frac{\partial}{\partial P} \left(\frac{M_x^2}{JG} + \frac{M_y^2}{EI_y} + \frac{M_z^2}{EI_z} \right) \mathbf{i} \, dx$$

$$u\mathbf{i} = \int_0^L \left(\frac{M_x}{JG} \frac{\partial M_x}{\partial P} + \frac{M_y}{EI_y} \frac{\partial M_y}{\partial P} + \frac{M_z}{EI_z} \frac{\partial M_z}{\partial P} \right) \mathbf{i}\, dx$$

$$v\mathbf{j} = \int_0^L \left(\frac{M_x}{JG} \frac{\partial M_x}{\partial Q} + \frac{M_y}{EI_y} \frac{\partial M_y}{\partial Q} + \frac{M_z}{EI_z} \frac{\partial M_z}{\partial Q} \right) \mathbf{j}\, dx \qquad \textbf{(13.22)}$$

$$w\mathbf{k} = \int_0^L \left(\frac{M_x}{JG} \frac{\partial M_x}{\partial R} + \frac{M_y}{EI_y} \frac{\partial M_y}{\partial R} + \frac{M_z}{EI_z} \frac{\partial M_z}{\partial R} \right) \mathbf{k}\, dx$$

Equations (13.22) are valid for deflections due to bending moments and axial torque. If axial forces and shears are to be included, use equations of the form of Eq. (13.3) for axial forces and Eq. (c) of Example 13.4 for shear forces. Equations to compute beam slopes due to bending and rotation due to torque would require equations of the form of Eq. (13.5) for torque and Eq. (13.19) for bending moments.

EXAMPLE 13.12

Compute the horizontal and vertical deflections due to bending at the center of the simply supported beam of Fig. 13.16a. Assume that E, I_y, and I_z are constant.

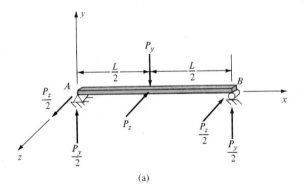

(a)

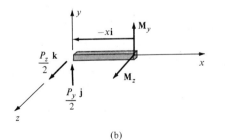

(b)

FIGURE 13.16

Solution:

This example is somewhat elementary but will illustrate the use of Eq. (13.22). The reactions are computed and shown in Fig. 13.16a. A free-body section is shown in Fig. 13.16b. Assume positive vector moments acting at the cut section. The moment equations are computed by summing moments about the cut section.

$$M_x \mathbf{i} + M_y \mathbf{j} + M_z \mathbf{k} + (-xi) \times \left[\left(\frac{P_y}{2} \right) \mathbf{j} + \left(\frac{P_z}{2} \right) \mathbf{k} \right] = 0$$

$$M_x \mathbf{i} = 0 \qquad M_y \mathbf{j} = -\left(\frac{P_z}{2} \right) x \mathbf{j} \qquad M_z \mathbf{k} = \left(\frac{P_y}{2} \right) x \mathbf{k}$$

Identify the moment equations with Eq. (13.22).

$$Q \Rightarrow P_y \qquad R \Rightarrow P_z \qquad P = 0$$

Equation (13.22) gives $u = 0$ and gives v and w as follows:

$$v = 2 \int_0^{L/2} \frac{M_z}{EI_z} \frac{\partial M_z}{\partial P_y} \, dx \tag{a}$$

$$w = 2 \int_0^{L/2} \frac{M_y}{EI_y} \frac{\partial M_y}{\partial P_z} \, dx \tag{b}$$

All other terms drop out. The integrals are multiplied by 2 in order to account for the beam segment BC, which is identical to segment AB for this problem. Compute the indicated partial derivatives.

$$\frac{\partial M_z}{\partial P_y} = \frac{x}{2} \qquad \frac{\partial M_y}{\partial P_z} = \frac{-x}{2}$$

Substitute into Eqs. (a) and (b).

$$v = 2 \int_0^{L/2} \frac{P_y}{2} \frac{x^2}{2EI_y} \, dx = \frac{P_y L^3}{48 EI_z}$$

$$w = 2 \int_0^{L/2} \frac{P_z}{2} \frac{x^2}{2EI_y} \, dx = \frac{P_z L^3}{48 EI_y}$$

The deflections are in the direction of the applied loads. ∎

The next example will illustrate Eq. (13.22) for a more general problem.

EXAMPLE 13.13

Compute the horizontal and vertical deflections of point D for the structure of Fig. 13.17. The structure is made of circular pipe with $G = E/2.5$. Neglect shear and axial deformation.

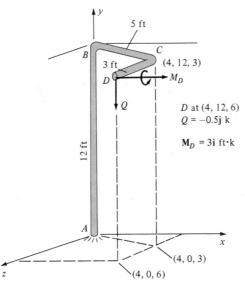

5 ft

B

3 ft

C

D (4, 12, 3)

M_D

Q

12 ft

D at $(4, 12, 6)$

$Q = -0.5\mathbf{j}$ k

$\mathbf{M}_D = 3\mathbf{i}$ ft·k

A

$(4, 0, 3)$

$(4, 0, 6)$

z

x

(a)

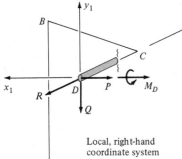

Local, right-hand coordinate system at D

$$\mathbf{r} = (0, 0, 0) - (0, 0, z_1) = -z_1\mathbf{k}$$

$$\mathbf{F} = -P\mathbf{i} - Q\mathbf{j} - R\mathbf{k}$$

Free body of CD in local coordinate system

(b)

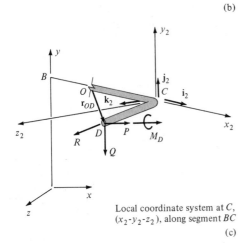

Local coordinate system at C, $(x_2\text{-}y_2\text{-}z_2)$, along segment BC

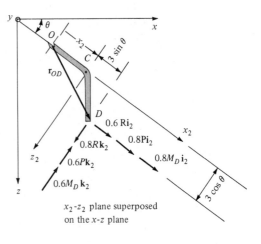

$x_2\text{-}z_2$ plane superposed on the $x\text{-}z$ plane

(c)

FIGURE 13.17

609

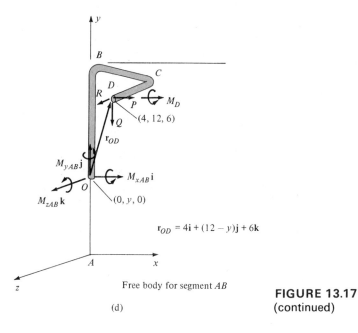

Free body for segment AB

(d)

FIGURE 13.17
(continued)

Solution:
The solution will be obtained by writing moment equations along each segment of the structure. Local coordinates will be used for each segment rather than using the coordinate system located at point A. This means that the integration must be related to each local coordinate system for each segment. To compute the vertical deflection at D, the force Q should not be evaluated until after the partial derivatives have been computed. Fictitious forces P and R should be located at D and will be assumed positive in the coordinate system of Fig. 13.17a. A free body of segment CD is shown in Fig. 13.17b. Assume positive moments at the cut section and sum moments at the cut section.

$$\Sigma \mathbf{M}_{CD} = 0$$

$$M_{xCD}\mathbf{i} + M_{yCD}\mathbf{j} + M_{zCD}\mathbf{k} + (-z_1\mathbf{k}) \times (-P\mathbf{i} - Q\mathbf{j} - R\mathbf{k}) - M_D\mathbf{i} = 0$$

$$M_{xCD}\mathbf{i} = (Qz_1 + M_D)\mathbf{i} \qquad M_{yCD}\mathbf{j} = -Pz_1\mathbf{j} \qquad M_{zCD}\mathbf{k} = 0 \tag{a}$$

$$\frac{\partial M_{xCD}}{\partial P} = 0 \qquad \frac{\partial M_{yCD}}{\partial P} = -z_1 \qquad \frac{\partial M_{zCD}}{\partial P} = 0$$

Let $P = R = 0$. The actual loading Q and M_D do not cause point O to deflect in the x direction for segment CD; hence, $u\mathbf{i} = 0$.

$$\frac{\partial M_{xCD}}{\partial Q} = z_1 \qquad \frac{\partial M_{yCD}}{\partial Q} = 0 \qquad \frac{\partial M_{zCD}}{\partial Q} = 0$$

Let $P = R = 0$; then

$$v\mathbf{j} = \int_0^3 \frac{M_{xCD}}{EI_x} \frac{\partial M_{xCD}}{\partial Q} dz_1 \mathbf{j} = \int_0^3 \left(\frac{Qz_1^2 + M_D z_1}{EI_x} \right) dz_1 \mathbf{j}$$

$$v\mathbf{j} = \frac{9}{EI_x} \left(Q + \frac{M_D}{2} \right) \mathbf{j} \tag{b}$$

Equation (b) is the contribution to the deflection in the y direction obtained from segment CD. Since R does not appear in the moment equations, $\mathbf{wk} = 0$ for segment CD.

Segment BC is shown in Fig. 13.17c. Member BC does not lie in the x-y-z coordinate system. A local coordinate system x_2-y_2-z_2 is used, in which x_2 corresponds to the axis of member BC. The integration will be along x_2. The moment M_D and the forces P and R must be rotated into the x_2-y_2-z_2 coordinate system in order to compute the moment equations. The differential is still with respect to P, Q, and R. A plane view of the coordinate transformation is shown in Fig. 13.17c. The free-body cut is denoted by point O. The position vector is measured from O to D. The moment M_D and forces P and R must be transformed into components in the x_2-y_2-z_2 system. The angle θ of Fig. 13.17c is used to compute the transformation matrix (see the appendix to Chapter 8).

$$\sin \theta = \frac{3}{5} = 0.6 \qquad \cos \theta = \frac{4}{5} = 0.8$$

The length CD is 3 ft along the positive z direction. The distance CD in x_2 and z_2 components is

$$\begin{Bmatrix} CD_{x_2} \\ CD_{z_2} \end{Bmatrix} = \begin{bmatrix} \cos \theta & \sin \theta \\ -\sin \theta & \cos \theta \end{bmatrix} \begin{Bmatrix} 0 \\ 3 \end{Bmatrix} = \begin{Bmatrix} 3 \sin \theta \\ 3 \cos \theta \end{Bmatrix} = \begin{Bmatrix} 1.8 \\ 2.4 \end{Bmatrix}$$

$$\mathbf{CD} = 1.8\mathbf{i}_2 + 2.4\mathbf{k}_2$$

$$\mathbf{r}_{OD} = (x_2 + 1.8)\mathbf{i}_2 + 2.4\mathbf{k}_2$$

$$\begin{Bmatrix} P_{x_2} \\ P_{z_2} \end{Bmatrix} = \begin{bmatrix} \cos \theta & \sin \theta \\ -\sin \theta & \cos \theta \end{bmatrix} \begin{Bmatrix} P \\ 0 \end{Bmatrix} = \begin{Bmatrix} P \cos \theta \\ -P \sin \theta \end{Bmatrix}$$

$$\mathbf{P} = 0.8P\mathbf{i}_2 - 0.6P\mathbf{k}_2$$

$$\begin{Bmatrix} R_{x_2} \\ R_{z_2} \end{Bmatrix} = \begin{bmatrix} \cos \theta & \sin \theta \\ -\sin \theta & \cos \theta \end{bmatrix} \begin{Bmatrix} 0 \\ R \end{Bmatrix} = \begin{Bmatrix} R \sin \theta \\ R \cos \theta \end{Bmatrix}$$

$$\mathbf{R} = 0.6R\mathbf{i}_2 + 0.8R\mathbf{k}_2$$

and

$$\mathbf{M}_D = 0.8M_D\mathbf{i}_2 - 0.6M_D\mathbf{k}_2$$

$$\mathbf{Q} = -Q\mathbf{j}_2$$

Assume a positive moment vector at O; use the free body OCD, and construct moment equations by summing moments about point O. Note that

M_{x_2} is a torque

M_{y_2} is bending about the y_2 axis

M_{z_2} is bending about the z_2 axis

$$M_{x_2}\mathbf{i}_2 + M_{y_2}\mathbf{j}_2 + M_{z_2}\mathbf{k}_2 + 0.8M_D\mathbf{i}_2 - 0.6M_D\mathbf{k}_2$$

$$+ \begin{vmatrix} \mathbf{i}_2 & \mathbf{j}_2 & \mathbf{k}_2 \\ x_2 + 1.8 & 0 & 2.4 \\ 0.8P + 0.6R & -Q & -0.6P + 0.8R \end{vmatrix} = 0$$

$$M_{x_2}\mathbf{i}_2 = (-0.8M_D - 2.4Q)\mathbf{i}_2 \tag{c}$$

$$\begin{aligned} M_{y_2}\mathbf{j}_2 &= [-2.4(0.8P + 0.6R) + (x_2 + 1.8)(-0.6P + 0.8R)]\mathbf{j}_2 \\ &= [-P(3 + 0.6x_2) + 0.8Rx_2]\mathbf{j}_2 \end{aligned} \tag{d}$$

$$M_{z_2}\mathbf{k}_2 = [0.6M_D + (x_2 + 1.8)Q]\mathbf{k}_2 \tag{e}$$

Equations (13.22) require that partial derivatives be computed with respect to P, R, and Q for each of Eqs. (c), (d), and (e). To compute u (displacement in the x direction),

$$\frac{\partial M_{x_2}}{\partial P} = 0 \qquad \frac{\partial M_{y_2}}{\partial P} = -(3 + 0.6x_2) \qquad \frac{\partial M_{z_2}}{\partial P} = 0 \tag{f}$$

To compute v (displacement in the y direction),

$$\frac{\partial M_{x_2}}{\partial Q} = -2.4 \qquad \frac{\partial M_{y_2}}{\partial Q} = 0 \qquad \frac{\partial M_{z_2}}{\partial Q} = x_2 + 1.8 \tag{g}$$

To compute w (displacement in the z direction),

$$\frac{\partial M_{x_2}}{\partial R} = 0 \qquad \frac{\partial M_{y_2}}{\partial R} = 0.8x_2 \qquad \frac{\partial M_{z_2}}{\partial R} = 0 \tag{h}$$

Let $P = R = 0$ in Eqs. (c), (d), and (e).

$$u = 0$$

$$\begin{aligned} v &= \int_0^5 (0.8M_D + 2.4Q)\frac{(2.4)}{JG}\,dx + \int_0^5 \frac{[0.6M_D(x_2 + 1.8) + Q(x_2 + 1.8)^2]}{EI_{z_2}}\,dx_2 \\ &= \frac{9.6M_D + 28.8Q}{JG} + \frac{12.9M_D + 102.9Q}{EI_{z_2}} \end{aligned} \tag{i}$$

$$w = 0$$

Finally, segment AB can be evaluated using the free body of Fig. 13.17d. Fictitious forces P and R are applied at D. The moment at O is computed in terms of the structure coordinate system, where y is the axial coordinate. (*Note:* The moment equation is written using an equivalent system as in Chapter 8.)

$$M_{xAB}\mathbf{i} + M_{yAB}\mathbf{j} + M_{zAB}\mathbf{k} = M_D\mathbf{i} + \begin{vmatrix} \mathbf{i} & \mathbf{j} & \mathbf{k} \\ 4 & (12 - y) & 6 \\ P & -Q & R \end{vmatrix}$$

$$M_{xAB}\mathbf{i} = [M_D + 6Q + R(12 - y)]\mathbf{i}$$

$$M_{yAB}\mathbf{j} = -(4R - 6P)\mathbf{j}$$

$$M_{zAB}\mathbf{k} = -[4Q + (12 - y)P]\mathbf{k}$$

M_{xAB} is bending about the x axis

M_{yAB} is torque

M_{zAB} is bending about the z axis

$$\frac{\partial M_{xAB}}{\partial P} = 0 \qquad \frac{\partial M_{yAB}}{\partial P} = 6 \qquad \frac{\partial M_{zAB}}{\partial P} = -12 + y$$

$$\frac{\partial M_{xAB}}{\partial Q} = 6 \qquad \frac{\partial M_{yAB}}{\partial Q} = 0 \qquad \frac{\partial M_{zAB}}{\partial Q} = -4$$

$$\frac{\partial M_{xAB}}{\partial R} = (12 - y) \qquad \frac{\partial M_{yAB}}{\partial R} = -4 \qquad \frac{\partial M_{zAB}}{\partial R} = 0$$

Let $P = R = 0$ and use the format given by Eq. (13.22).

$$u\mathbf{i} = \int_0^{12} \frac{4Q}{EI_z}(12 - y)\,dy\,\mathbf{i}$$

$$v\mathbf{j} = \int_0^{12} \frac{6(M_D + 6Q)}{EI_x}\,dy\mathbf{j} + \int_0^{12} \frac{4Q4}{EI_z}\,dy\mathbf{j}$$

$$w\mathbf{k} = \int_0^{12} \frac{(M_D + 6Q)}{EI_x}(12 - y)\,dy\,\mathbf{k}$$

$$u = \frac{288Q}{EI_z} \tag{j}$$

$$v = \frac{72(M_D + Q)}{EI_x} + \frac{192Q}{EI_z} \tag{k}$$

$$w = (72M_D + 432Q)EI_x \tag{l}$$

The total deflection is obtained by adding the results for the three segments. The problem is simplified by specifying that each segment of the structure has the same geometry. All second moments of the area are equal, $J = 2I$, $G = E/2.5$, and u is given by Eq. (j).

$$u = \frac{288Q}{EI} \qquad \text{(deflection of point } D \text{ in the } x \text{ direction)}$$ **(m)**

v is given by the sum of Eqs. (b), (i), and (k):

$$v = \frac{375.9Q + 89.4M_D}{EI} + \frac{9.6M_D + 28.8Q}{JG} \qquad \begin{array}{l}\text{(the deflection of point } D \text{ in}\\ \text{the } y \text{ direction)}\end{array}$$

w is given by Eq. (l), as follows:

$$w = \frac{72M_D + 432Q}{EI} \qquad \text{(deflection of point } D \text{ in the } z \text{ direction)}$$

Values of M_D, Q, E, and I can be substituted to give a final answer.   ■

13.7   THE PRINCIPLE OF LEAST WORK

The principle of least work is a special case of Castigliano's second theorem; that is,

$$\frac{\partial U}{\partial P_i} = \delta_i = 0$$ **(13.23)**

Equation (13.23) is the special case of Eq. (13.16) when the deflection δ_i at the point where P_i is applied is zero. The equation is most useful for statically indeterminate structures. The name "least work" may be a misnomer. Perhaps Eq. (13.23) should just be referred to as Castigliano's second theorem. It is, however, claimed that the name "least work" is used to prevent the misconception that Eq. (13.23) is a minimum-energy principle. The principle of minimum potential energy, discussed in a previous section, includes both strain energy and external work in order to express a minimum principle. The least work concept is merely an application of Castigliano's second theorem when the deflection can be specified as zero or a predetermined value.

EXAMPLE 13.14

Solve the statically indeterminate beam problem of Example 11.8 using the principle of least work.

Solution:
The beam of Fig. 11.10 is shown in Fig. 13.18. The problem is to compute the reactions for the beam of Fig. 13.18a; then assume support B settles an amount δ and compute the reactions as shown in Fig. 13.18b. Since the deflection is zero at A, B, and C for the beam of Fig. 13.18a, Eq. (13.23) can be applied to either reaction.

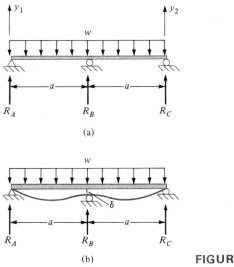

(a)

(b)   **FIGURE 13.18**

Choose the reaction that leads to the most efficient formulation of the problem. For the beam of Fig. 13.18a choose R_A. The moment equation is obtained using two beam segments.

$$M_{AB} = R_A x - \frac{wx^2}{2} \tag{a}$$

$$M_{CB} = R_C x - \frac{wx^2}{2} = R_A x - \frac{wx^2}{2} \tag{b}$$

Unknown reactions must be written in terms of R_A because the partial derivative is with respect to R_A. Summing moments about B gives $R_C = R_A$.

$$\frac{\partial M_{AB}}{\partial R_A} = x \qquad \frac{\partial M_{CB}}{\partial R_A} = x$$

Equation (13.23) is written, using Eq. (13.18) as a model, as follows:

$$\frac{\partial U}{\partial R_A} = 0 = \frac{2}{EI} \int_0^a \left(R_A x^2 - \frac{wx^3}{2} \right) dx$$

$$\frac{R_A a^3}{3} - \frac{wa^4}{8} = 0 \qquad \text{or} \qquad R_A = \frac{3}{8} wa$$

The result agrees with Example 11.8. The remaining reactions can be obtained from statics (see Example 11.8).

The beam of Fig. 13.18b is analyzed by choosing to evaluate Eq. (13.23) at point B. The deflection is δ rather than zero. Hence

$$\frac{\partial U}{\partial R_B} = v_B = -\delta \qquad \text{(negative because } \delta \text{ and } R_B \text{ are in opposite directions)}$$

R_A is written in terms of R_B using statics

$$R_A = wa - \frac{R_B}{2}$$

and is substituted into Eqs. (a) and (b)

$$M_{AB} = wax - \frac{R_B x}{2} - \frac{wx^2}{2} = M_{BC}$$

$$\frac{\partial M_{AB}}{\partial R_B} = -\frac{x}{2} = \frac{\partial M_{BC}}{\partial R_B}$$

$$\frac{2}{EI} \int_0^a \left(-\frac{wax^2}{2} + \frac{R_B x^2}{4} + \frac{wx^3}{4} \right) dx = -\delta$$

$$-\frac{wa^4}{6} + \frac{R_B a^3}{12} + \frac{wa^4}{16} = \frac{-\delta EI}{2}$$

$$R_B = \frac{5wa}{4} - \frac{6EI\delta}{a^3}$$

Again, the result agrees with Example 11.8.

EXAMPLE 13.15

Compute the reactions for the beam of Fig. 13.19. Write the least work equations in terms of R_A and M_A.

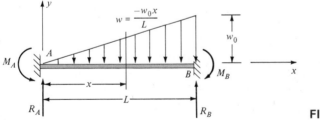

FIGURE 13.19

Solution:
The beam is statically indeterminate to the second degree. Two equations will be required to compute the reactions R_A and M_A. The same moment equations can be used for both M_A and R_A.

$$M = R_A x - M_A - \frac{w_0 x^3}{6L}$$

$$\frac{\partial M}{\partial R_A} = x$$

$$\frac{\partial M}{\partial M_A} = -1$$

$$\frac{\partial U}{\partial R_A} = 0 = \int_0^L \left(\frac{R_A x^2 - M_A x - w_0 x^4}{6L} \right) dx$$

$$\frac{R_A L^3}{3} - \frac{M_A L^2}{2} - \frac{w_0 L^4}{30} = 0 \qquad\qquad \text{(a)}$$

$$\frac{\partial U}{\partial M_A} = 0 = \int_0^L \left(\frac{-R_A x + M_A + w_0 x^3}{6L} \right) dx$$

$$\frac{R_A L^2}{2} - M_A L - \frac{w_0 L^3}{24} = 0 \qquad\qquad \text{(b)}$$

Solving Eqs. (a) and (b) gives

$$R_A = \frac{3w_0 L}{20} \qquad M_A = \frac{w_0 L^2}{30}$$

The equations of statics give the reactions at B.

$$R_B = \frac{7w_0 L}{20} \qquad M_B = \frac{w_0 L^2}{20} \qquad\qquad \blacksquare$$

The complex structures of Chapter 11 can be analyzed using the principle of least work.

EXAMPLE 13.16

Use the principle of least work to compute the reactions for the structure of Fig. 13.20.

Solution:
Choose a support that remains fixed in position. The reaction at B will be used to formulate Eq. (13.23).

$$M_x = R_B x - \frac{wx^2}{2}$$

There will be a term similar to Eq. (13.17) for the rod.

$$\frac{\partial U}{\partial R_B} = \frac{\partial}{\partial R_B} \left[\frac{1}{2} \int_0^a \frac{R_B^2}{AE} dx + \frac{1}{2} \int_0^L \frac{M^2 dx}{EI} \right] = 0$$

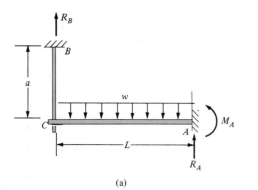

(a)

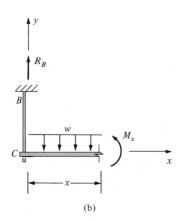

(b)

FIGURE 13.20

$$\int_0^a \frac{R_B}{AE} \frac{\partial R_B}{\partial R_B} dx + \int_0^L \frac{M}{EI} \frac{\partial M}{\partial R_B} dx = 0$$

$$\frac{R_B a}{AE} + \int_0^L \frac{1}{EI} \left(R_B x - \frac{wx^2}{2} \right) x dx = 0$$

$$\frac{R_B a}{AE} + \frac{R_B L^3}{3EI} - \frac{wL^4}{8EI} = 0$$

$$R_B = \left(\frac{wL^4}{8EI} \right) \bigg/ \left(\frac{a}{AE} + \frac{L^2}{3EI} \right)$$

M_A and R_A can be computed using the equations of statics.

As an additional example, consider a statically indeterminate shaft subject to a torque.

EXAMPLE 13.17

Use the principle of least work to compute the torsional reactions for the shaft of Fig. 13.21.

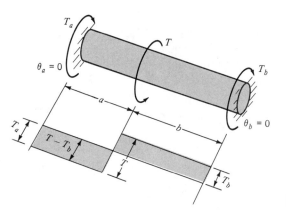

FIGURE 13.21

Solution:

Compute the change in strain energy with respect to point b. The torque must be written in terms of T_b along each section of the shaft as shown in the torque diagram above.

$$U = \frac{1}{2} \int_0^a \frac{T_a^2}{JG} \, dx + \frac{1}{2} \int_0^b \frac{T_b^2}{JG} \, dx$$

$$T_a = T - T_b$$

$$\frac{\partial}{\partial T_b} (T - T_b) = -1$$

$$\frac{\partial U}{\partial T_b} = 0 = \int_0^a \frac{(T - T_b)}{JG} (-1) \, dx + \int_0^b \frac{T_b}{JG} \, dx$$

$$(T_b - T)a + T_b b = 0$$

$$T_b(a + b) = Ta$$

$$T_b = \frac{Ta}{L}$$

$$T_a = T - T_b = \frac{Tb}{L}$$

■

As a final example, consider a complex beam problem with temperature dependence. Example 11.19 will be used to illustrate a method for analyzing a structure subject to a temperature differential.

EXAMPLE 13.18

Steel fixed-free beams are connected at their free ends by a rod as shown in Fig. 13.22. Beam AB is an S4x9.5, 10 ft long; and beam CD is an S5x10, 8 ft long. The rod is magnesium, originally 12 ft in length with an area of 4 in.^{2}. Compute the axial force in the rod if the temperature of the rod decreases 60°F. Given $\alpha = 14.5(10^{-6})$°F and E for magnesium = $6.5(10^6)$ psi.

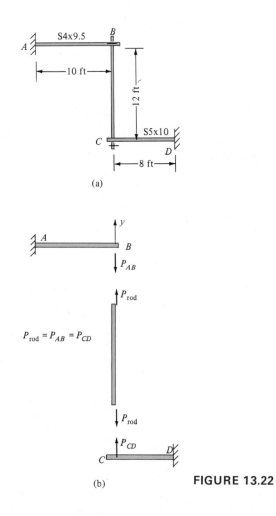

(a)

(b)

FIGURE 13.22

Solution:

A free body of the structure is shown in Fig. 13.22b. The deformation of the rod due to the temperature decrease is given by $\delta_T = \alpha \Delta T L$. Since this deformation is known, Eq. (13.23) can be formulated in terms of axial force in the rod. The partial derivative of the strain energy is with respect to the axial force in the rod and may be set equal to the deformation of the rod. Note that $P_{\text{rod}} = P_{AB} = P_{CD} = P$.

$$M_{AB} = -P_{AB}x \qquad \frac{\partial M_{AB}}{\partial P} = -x$$

$$M_{CD} = P_{CD}x \qquad \frac{\partial M_{CD}}{\partial P} = x$$

Axial force in the rod is P_{rod}; $\partial P_{\text{rod}}/\partial P = 1$.

$$\delta_T = 14.5(10^{-6})(60)(144) = \int_0^{120} \frac{Px^2\, dx}{30(10^6)(6.79)}$$

$$+ \int_0^{96} \frac{Px^2\, dx}{30(10^6)(12.3)} + \int_0^{144} \frac{P\, dx}{6.5(10^6)(4)}$$

This equation is identical to that of Example 11.19 and can be solved for P.

$$0.125 = [2.83(10^{-3}) + 0.79(10^{-3}) + 0.0005(10^{-3})]P$$

$$P = 34.5 \text{ lb} \qquad\qquad \blacksquare$$

13.8 SUMMARY

Numerous topics, which are given the general title energy methods, have been discussed in this chapter. An important concept is the idea of potential energy, the potential that a system has to do work. Potential energy, for the purposes of this chapter, was written in the simple form

$$\Pi = U - W_P \qquad\qquad (13.7)$$

The principle of stationary potential energy follows directly from Eq. (13.7) and was written

$$\frac{\partial \Pi}{\partial \delta} = 0 \qquad\qquad (13.8)$$

Potential energy methods are limited in structural analysis. However, their application in more advanced study is pertinent to the study of mechanics.

Castigliano's first theorem was shown to be a potential energy method. Also, the use of Castigliano's first theorem proved to be somewhat awkward compared to Castigliano's second theorem.

Castigliano's second theorem and the theorem of least work prove to be the most applicable methods of the chapter for structural analysis. However, while the methods can be applied to both determinate and indeterminate systems, they were limited to structural materials with linear stress-strain properties. Once again the use of vector analysis was demonstrated as an aid in organizing the analysis of a three-dimensional structure.

13.9 APPENDIX: CASTIGLIANO'S SECOND THEOREM FOR THREE-DIMENSIONAL STRUCTURES

In this appendix a general form of Eq. (13.22) will be derived. The derivation is based upon vector analysis and is intended as a mathematical supplement for this chapter.

Define a moment vector, $\mathbf{M}$, which is the moment expression written at a cut section.

$$\mathbf{M} = M_x\mathbf{i} + M_y\mathbf{j} + M_z\mathbf{k} \tag{A13.1}$$

Define a force vector, $\mathbf{F}$, which is the force applied at the point where deflections are to be computed. $\mathbf{F}$ may represent real or fictitious forces.

$$\mathbf{F} = P\mathbf{i} + Q\mathbf{j} + R\mathbf{k} \tag{A13.2}$$

Define a vector operator, $\mathbf{L}$.

$$\mathbf{L} = \frac{\partial}{\partial P}\mathbf{i} + \frac{\partial}{\partial Q}\mathbf{j} + \frac{\partial}{\partial R}\mathbf{k} \tag{A13.3}$$

Define a displacement vector, $\boldsymbol{\delta}$, which represents the displacements at the point where $\mathbf{F}$ is applied.

$$\boldsymbol{\delta} = u\mathbf{i} + v\mathbf{j} + w\mathbf{k} \tag{A13.4}$$

Castigliano's second theorem for a linear elastic structure to compute deflections due to bending loads is obtained by combining Eqs. (13.16) and (13.6). Recall that U is a scalar quantity. Equation (13.16) is written using Eqs. (A13.4) and (A13.3).

$$\boldsymbol{\delta} = \mathbf{L}U \tag{A13.5}$$

Equation (13.6) in vector form must be formulated as the scalar product between moment and corresponding beam rotations. Define a vector of beam rotations.

$$d\boldsymbol{\theta} = \frac{M_x}{JG}\,dl\,\mathbf{i} + \frac{M_y}{EI_y}\,dl\,\mathbf{j} + \frac{M_z}{EI_z}\,dl\,\mathbf{k} \tag{A13.6}$$

The strain energy due to bending moments and torque, recognizing that the internal strain energy is equal to external work, for a linear elastic structure is

$$U = \frac{1}{2} \int \mathbf{M} \cdot d\boldsymbol{\theta} \tag{A13.7}$$

Assume that beam properties are constant and independent of the axial coordinate l. Let

$$d\boldsymbol{\theta} = \frac{\mathbf{M} \, dl}{\beta} \tag{A13.8}$$

where β represents the beam properties corresponding to each component of $\mathbf{M}$. Equation (A13.8) is substituted into Eq. (A13.7) and the result substituted into Eq. (A13.5).

$$\boldsymbol{\delta} = \mathbf{L}\left(\frac{1}{2} \int \mathbf{M} \cdot \frac{\mathbf{M}}{\beta} \, dl\right) \tag{A13.9}$$

$$\boldsymbol{\delta} = \frac{1}{2} \int \left[(\mathbf{LM}) \cdot \frac{\mathbf{M}}{\beta} + \mathbf{M} \cdot \frac{\mathbf{LM}}{\beta}\right] dl \tag{A13.10}$$

The vector operation $(\mathbf{LM})$ is not normally covered in an elementary course in vector analysis. However, $\mathbf{LM}$ can be interpreted as follows:

$$\mathbf{LM} = \left(\frac{\partial}{\partial P} \mathbf{i} + \frac{\partial}{\partial Q} \mathbf{j} + \frac{\partial}{\partial R} \mathbf{k}\right)(M_x \mathbf{i} + M_y \mathbf{j} + M_z \mathbf{k})$$

$$\mathbf{LM} = \frac{\partial M_x}{\partial P} \mathbf{ii} + \frac{\partial M_y}{\partial P} \mathbf{ij} + \frac{\partial M_z}{\partial P} \mathbf{ik} + \frac{\partial M_x}{\partial Q} \mathbf{ji} + \frac{\partial M_y}{\partial Q} \mathbf{jj}$$

$$+ \frac{\partial M_x}{\partial Q} \mathbf{jk} + \frac{\partial M_x}{\partial R} \mathbf{ki} + \frac{\partial M_y}{\partial R} \mathbf{kj} + \frac{\partial M_z}{\partial R} \mathbf{kk} \tag{A13.11}$$

The quantities $\mathbf{ii}$, $\mathbf{ij}$, and so on, are referred to as unit dyads. The total quantity $\mathbf{LM}$ is called a dyadic or second-order tensor. The term $(\mathbf{LM}) \cdot \mathbf{M}/\beta$ is the product of the dyadic and a vector. The product is formed by multiplying Eq. (A13.11) with Eq. (A13.5) and taking the scalar product of $\mathbf{M}/\beta$ with each term of $\mathbf{LM}$.

$$(\mathbf{LM}) \cdot \frac{\mathbf{M}}{\beta} = (\mathbf{LM}) \cdot \frac{M_x}{JG} \, dl\,\mathbf{i} + (\mathbf{LM}) \cdot \frac{M_y}{EI_y} \, dl\,\mathbf{j} + (\mathbf{LM}) \cdot \frac{M_z}{EI_z} \, dl\,\mathbf{k} \tag{A13.12}$$

The first term of the multiplication would be

$$\frac{\partial M_x}{\partial P} \mathbf{ii} \cdot \frac{M_x}{JG} \, dl\,\mathbf{i} = \frac{\partial M_x}{\partial P} \frac{M_x}{JG} \, dl\,\mathbf{i}(\mathbf{i} \cdot \mathbf{i}) = \frac{\partial M_x}{\partial P} \frac{M_x}{JG} \, dl\,\mathbf{i} \quad (\text{since } \mathbf{i} \cdot \mathbf{i} = 1)$$

The second term would be

$$\frac{\partial M_y}{\partial P} \frac{M_x}{JG} dl\,\mathbf{i}(\mathbf{j} \cdot \mathbf{i}) = 0 \qquad (\text{since } \mathbf{j} \cdot \mathbf{i} = 0)$$

The third term would result in zero. However, the fourth term of Eq. (A13.11) times the first term of Eq. (A13.12) would be

$$\frac{\partial M_x}{\partial Q} \frac{M_x}{JG} dl\,\mathbf{j}(\mathbf{i} \cdot \mathbf{i}) = \frac{\partial M_x}{\partial Q} \frac{M_x}{JG} dl\,\mathbf{j}$$

Following through the multiplication term by term will give the final result. Prior to writing the result note that

$$(\mathbf{LM}) \cdot \frac{\mathbf{M}}{\beta} = \mathbf{M} \cdot \frac{(\mathbf{LM})}{\beta}$$

and can be verified by expanding both terms. Equation (A13.10) becomes

$$
\begin{aligned}
u\mathbf{i} + v\mathbf{j} + w\mathbf{k} = \int \Bigg[& \left(\frac{\partial M_x}{\partial P} \frac{M_x}{JG} + \frac{\partial M_y}{\partial P} \frac{M_y}{EI_y} + \frac{\partial M_z}{\partial P} \frac{M_z}{EI_z} \right)\mathbf{i} \\
& + \left(\frac{\partial M_x}{\partial Q} \frac{M_x}{JG} + \frac{\partial M_y}{\partial Q} \frac{M_y}{EI_y} + \frac{\partial M_z}{\partial Q} \frac{M_z}{EI_z} \right)\mathbf{j} \\
& + \left(\frac{\partial M_x}{\partial R} \frac{M_x}{JG} + \frac{\partial M_y}{\partial R} \frac{M_y}{EI_y} + \frac{\partial M_z}{\partial R} \frac{M_z}{EI_z} \right)\mathbf{k} \Bigg] dl
\end{aligned}
$$

$$\text{(A13.13)}$$

Equation (A13.13) is identical to Eq. (13.21) with the exception that the axial coordinate has been called l.

For Further Study

Laible, J. P., *Structural Analysis*, Holt, Rinehart and Winston, New York, 1985.

Langhaar, H. L., *Energy Methods in Applied Mechanics*, Wiley, New York, 1962.

Reddy, J. N., *Energy and Variational Methods in Applied Mechanics*, Wiley, New York, 1984.

Tauchert, T. R., *Energy Principles in Structural Mechanics*, McGraw-Hill, New York, 1974.

Washizu, K., *Variational Methods in Elasticity and Plasticity*, 3rd ed. Pergamon, New York, 1982.

13.10 PROBLEMS

13.1–13.5 Use the method of Section 13.2 to compute the deflection at the point of application of the concentrated force in Figs. P13.1–P13.5.

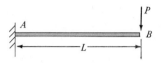

FIGURE P13.1

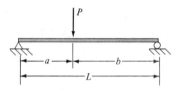

FIGURE P13.2

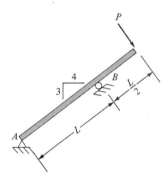

FIGURE P13.3

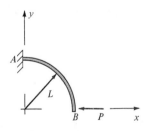

FIGURE P13.4

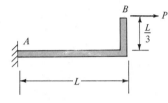

FIGURE P13.5

For Problems 13.6–13.11 use the method of Section 13.2 to compute the specified deformation.

13.6 Compute the rotation at A for the beam of Fig. P13.6.

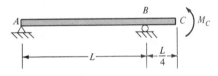

FIGURE P13.6

13.7 Compute the rotation at point C for the beam of Fig. P13.7.

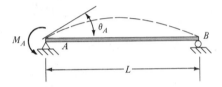

FIGURE P13.7

13.8 Compute the rotation at B for the structure of Fig. P13.8.

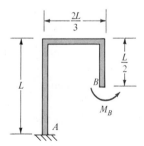

FIGURE P13.8

13.9 Compute the rotation of the circular shaft at the free end for the structure of Fig. P13.9.

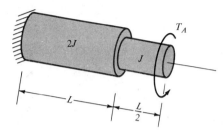

FIGURE P13.9

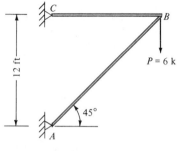

FIGURE P13.12

13.10 Compute the deflection at joint *B* for the truss of Fig. P13.10.

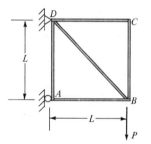

FIGURE P13.10

13.11 Compute the deflection at joint *B* for the truss of Fig. P13.11.

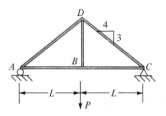

FIGURE P13.11

Castigliano's first theorem (Section 13.5) should be used to solve Problems 13.12–13.19.

13.12 Compute the displacement of joint *B* and the axial force in each member of Fig. P13.12. Assume the truss members are steel with a cross-sectional area of 2 in.².

13.13 Compute the displacement of joint *C* and the axial force in each member of Fig. P13.13.

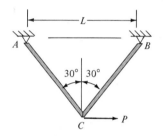

FIGURE P13.13

13.14 Compute the displacement at *B* and the axial force in each member of Fig. P13.14.

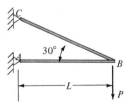

FIGURE P13.14

13.15 Compute the reactions for the structure of Fig. P13.15.

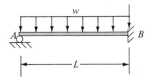

FIGURE P13.15

13.16 Compute the reactions for the structure of Fig. P13.16.

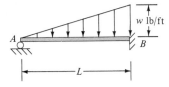

FIGURE P13.16

13.17 Compute the reactions for the structure of Fig. P13.17.

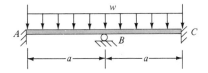

FIGURE P13.17

13.18 Compute the reactions for the structure of Fig. P13.18.

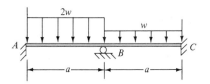

FIGURE P13.18

13.19 Compute the reactions for the structure of Fig. P13.19.

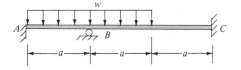

FIGURE P13.19

Use Castigliano's second theorem (Section 13.6) for Problems 13.20–13.48.

13.20 Compute the deflection at the free end of the beam of Fig. P13.20.

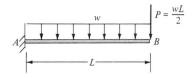

FIGURE P13.20

13.21 Compute the deflection at the free end of the beam of Fig. P13.21. Assume $EI = 40(10^6) \text{ N·m}^2$.

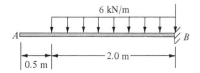

FIGURE P13.21

13.22 Compute the deflection at the one-third points of the beam of Fig. P13.22. Assume that EI is constant.

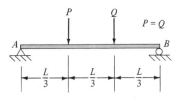

FIGURE P13.22

13.23 Compute the deflection at the center of the simple span of Fig. P13.23. Assume that E is constant.

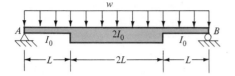

FIGURE P13.23

13.24 Compute the deflection at the free end of the fixed-free beam of Fig. P13.24.

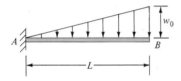

FIGURE P13.24

13.25 Compute the deflection at C for the beam of Fig. P13.25. Assume that EI is constant.

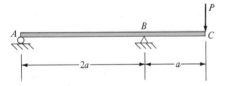

FIGURE P13.25

13.26 Compute the deflection at point B for the beam of Fig. P13.26.

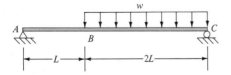

FIGURE P13.26

13.27 Compute the deflection at the center of the beam of Fig. P13.27.

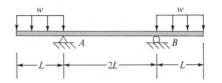

FIGURE P13.27

13.28 Compute the deflection at the free end of the beam of Fig. P13.28. Let $E = 6(10^6)$ psi and $I_0 = 200$ in.^{4}.

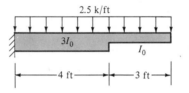

FIGURE P13.28

13.29 Compute the rotation at the supports for the beam of Fig. P13.29.

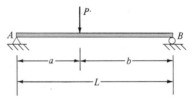

FIGURE P13.29

13.30 Compute the rotation at point B for the beam of Fig. P13.20.

13.31 Compute the rotation at point A for the beam of Fig. P13.21.

13.32 Compute the rotation at point A for the beam of Fig. P13.22.

13.33 Compute the rotation at point B for the beam of Fig. P13.23.

13.34 Compute the rotation at the free end for the beam of Fig. P13.24.

13.35 Compute the rotation midway between the supports for the beam of Fig. P13.25.

13.36 Compute the rotation at midspan for the beam of Fig. P13.26.

13.37 Compute the rotation of the free end for the beam of Fig. P13.28.

13.38 Compute the horizontal and vertical deflection at point B for the truss of Fig. P13.12.

13.39 Compute the horizontal and vertical deflection at point C for the truss of Fig. P13.13.

13.40 Compute the horizontal and vertical deflection for point B for the truss of Fig. P13.10.

13.41 Compute the horizontal deflection of joint A for the truss of Fig. P13.11.

13.42 Compute the deflection at midspan for the beam of Fig. P13.42.

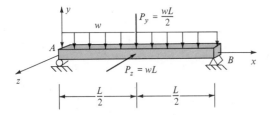

FIGURE P13.42

13.43 Compute the deflection at the free end of the beam of Fig. P13.43. Assume steel and an S4x9.5 member. Include axial deformation.

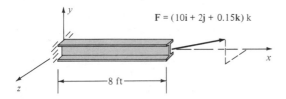

FIGURE P13.43

13.44 A circular shaft is fabricated as shown in Fig. P13.44. Assume $G = E/2.5$ and compute the deflection at point C in terms of I and E caused by the couple applied at C. The beam lies in the x-z plane.

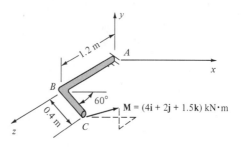

FIGURE P13.44

13.45 Compute the horizontal and vertical deflection and rotation of point B for the structure of Fig. P13.8.

13.46 Compute the deflection at the free end, point B, of the steel structure of Fig. P13.46.

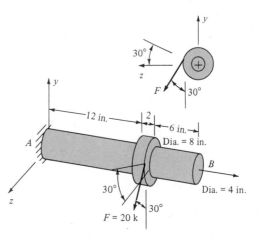

FIGURE P13.46

13.47 Compute the deflections at point C in all three directions of Fig. P13.47. The structure is steel and of circular cross section.

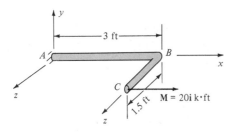

FIGURE P13.47

13.48 Compute the deflection at point C for the structure of Fig. P13.48. The structure is of circular cross section and $G = E/2.5$.

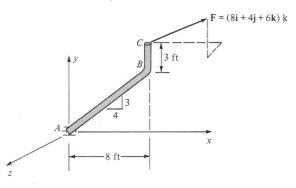

FIGURE P13.48

Use the principle of least work (Section 13.7) to solve Problems 13.49–13.80.

13.49 Compute the reactions for the beam of Fig. P13.49. Use the reaction at B to evaluate the partial derivative.

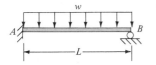

FIGURE P13.49

13.50 Compute the reaction for the beam of Fig. P13.49. Use the moment at A to evaluate the partial derivative.

13.51 Compute the reactions for the structure of Fig. P13.51.

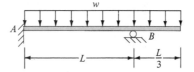

FIGURE P13.51

13.52 Compute the reactions for the beam of Fig. P13.52. Assume $E = 30(10^6)$ psi and $I = 300$ in.^{4}.

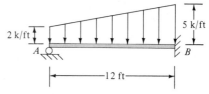

FIGURE P13.52

13.53 Compute the reactions for the structure of Fig. P13.53.

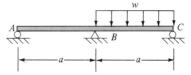

FIGURE P13.53

13.54 Assume that the support B in Problem 13.53 settles an amount δ and compute the reactions in terms of δ.

13.55 The fixed-fixed beam of Fig. P13.55 settles an amount δ. Compute the reactions.

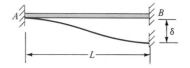

FIGURE P13.55

13.56 The fixed support of Fig. P13.56 rotates an amount θ. Compute the reactions in terms of θ.

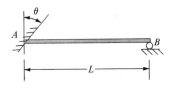

FIGURE P13.56

13.57 Compute the reactions for the fixed-fixed beam of Fig. P13.57.

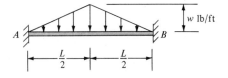

FIGURE P13.57

13.58 Compute the reactions at A and B for the structure of Fig. P13.58.

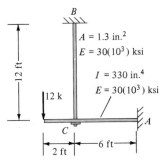

FIGURE P13.58

13.59 Compute the reactions for the structure of Fig. P13.59.

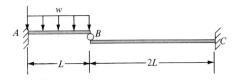

FIGURE P13.59

13.60 Compute the reactions for the structure of Fig. P13.60. The strain energy for the spring is $(1/2)k\delta^2$, where δ is the deflection of the spring.

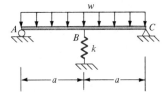

FIGURE P13.60

13.61 Compute the reactions for the beam of Fig. P13.61. Assume EI to be the same for both beams. Let A be the area of the rod. E for the rod is the same as for the beams.

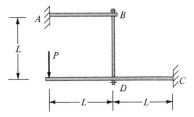

FIGURE P13.61

13.62 Compute the reactions for the beam of Fig. P13.62 in terms of P, A, E, and L.

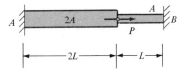

FIGURE P13.62

13.63 Compute the reactions for the beam of Fig. P13.63 in terms of P, L, A, and E.

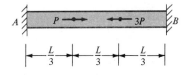

FIGURE P13.63

13.64 The axial force for the beam of Fig. P13.64 varies linearly from A to B; $P = P_0 x/L$. Compute the reactions in terms of P_0, L, A, and E.

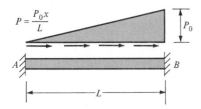

$$P = \frac{P_0 x}{L}$$

FIGURE P13.64

13.65 The applied axial force is zero between A and C and is given by $P = P_0 \sin(\pi x/L)$ between C and B of Fig. P13.65. Compute the reactions at A and B in terms of P, L, A, and E.

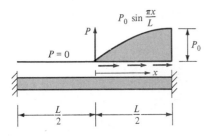

$$P_0 \sin \frac{\pi x}{L}$$

$$P = 0$$

FIGURE P13.65

13.66 The stepped shaft of Fig. P13.66 is fixed at both ends. Compute the reactions.

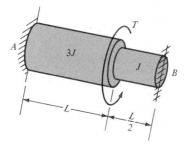

FIGURE P13.66

13.67 The steel shaft of Fig. P13.67 has torques of $T_1 = 5$ kN·m and $T_2 = 8$ kN·m. The diameter of the shaft is 45 mm. Compute the reactions at A and B.

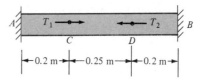

$$\vdash\!\!-0.2\,\text{m}\!-\!\!\vdash\!\!-0.25\,\text{m}\!-\!\!\vdash\!\!-0.2\,\text{m}\!-\!\vdash$$

FIGURE P13.67

13.68 The steel shaft of Fig. P13.68 is fixed at both ends. Assume the pulley thickness can be neglected and compute the reactions at A and B. Assume $r = 2$ in. and $L = 20$ in., and obtain the solution in terms of P_y.

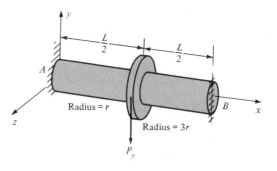

Radius = r

Radius = $3r$

P_y

FIGURE P13.68

13.69 A circular shaft is subjected to the torque distribution of Fig. P13.69. Compute the reactions in terms of T_0, L, J, and G.

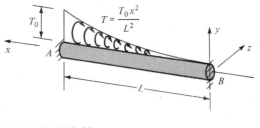

$$T = \frac{T_0 x^2}{L^2}$$

FIGURE P13.69

13.70 Assume that the force P of Problem 13.58 is zero and compute the reactions if the temperature of the rod decreases 75°F. $\alpha = 6.5(10^{-6})/°F$.

13.71 Assume the force P of Problem 13.58 is 12 k as shown in the figure and the temperature of the rod decreases 75°F. Compute the reactions.

13.72 Compute the magnitude of the force P of Fig. P13.58 to produce zero deformation of the rod if the temperature of the rod drops 15°F.

13.73 The bar of Fig. P13.73 is aluminum, 10 mm by 30 mm in cross section. Rods A and C are aluminum, and rod B is steel. Assume the rods to be supported such that they can withstand compressive forces. Compute the axial stress in each rod and the maximum bending moment in the bar for a 50°C temperature decrease in rod B.

13.76 Assume P equals zero and point A is prevented from moving vertically for the structure of Fig. P13.75. Assume all other conditions of Problem 13.75 and compute the axial force in the wire and the reaction at A.

13.77 The structure of Fig. P13.77 is made of circular steel pipe, fixed at A and prevented from moving vertically at B. Moments M_x, M_y, and M_z are applied at point C. Compute the reactions at A and B.

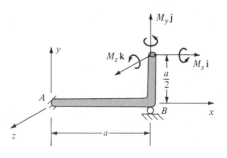

FIGURE P13.77

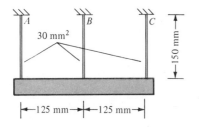

FIGURE P13.73

13.74 Repeat Problem 13.73 if the 50°C temperature drop is in rod A only.

13.75 The beams of Fig. P13.75 are steel with a cross section of 10 mm by 3 mm. A copper wire connects points D and E. The wire is 1 mm in diameter. Compute the axial force in the wire if the temperature of the wire drops 30°C.

13.78 Compute the reactions for the structure of Fig. P13.78. Neglect the strain energy due to shear. $G = E/2.5$.

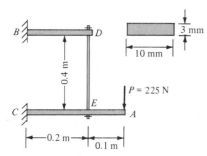

FIGURE P13.75

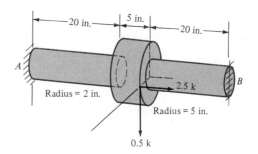

FIGURE P13.78

13.79 Compute the reactions for the frame of Fig. P13.79.

13.80 Compute the reactions for the structure of Fig. P13.80.

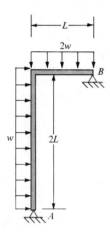

FIGURE P13.79

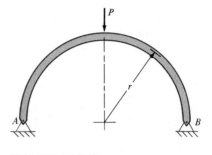

FIGURE P13.80

CHAPTER

INELASTIC MATERIAL BEHAVIOR AND FAILURE THEORIES

14.1 INTRODUCTION

In this chapter an attempt is made to introduce the reader to the complicated world of nonlinear material behavior. Inelastic material behavior is idealized as a bilinear relation between stress and strain. The idealization is not unreasonable when the actual behavior of common structural materials is studied. Steel, in some cases, is quite accurately assumed to be perfectly plastic, meaning that the second segment of the bilinear curve is horizontal or of zero slope. Aluminum and several other common ductile materials can be modeled as bilinear.

Brittle materials are not discussed in this chapter. Brittle materials are subject to catastrophic fracture and failure. The state of the art of knowledge concerning the analysis of brittle materials would, in the author's opinion, justify not including an inference that the methods of this chapter be applied to brittle materials.

The ductile materials have an inherent ability to deform and remain intact well beyond the level of stress that will be considered as yield stress in this chapter. Design and analysis based upon inelastic behavior requires the use of a factor of safety. The analysis of a structure using inelastic criteria does not mean that the design engineer has any intention of letting the material be stressed near the elastic limit or yield strength of the material.

The first few sections of this chapter deal with stress analysis situations where only one normal or one shear stress component occurs. A problem of combined normal stresses, where the stresses act in the same direction, is illustrated later.

A general biaxial-stress state requires the use of a yield theory. Two of the most popular failure theories are presented but are limited to two-dimensional stress states. The maximum shear stress theory, sometimes referred to as the Tresca failure theory, is a principal stress theory. The derivation is straightforward but care must be exercised in using the theory. The use of the theory for biaxial stress states is illustrated in Section 14.6. The maximum distortion energy theory is often referred to as the von Mises failure theory. The derivation is more involved but the use of the failure criterion is straightforward. The von Mises theory, as presented here, is also a principal stress theory. In other words, principal stresses must be computed before either failure theory can be used.

14.2 BEHAVIOR OF AXIALLY LOADED STRUCTURES

The discussion of plastic behavior of Section 3.2, Chapter 3, should be reviewed. As stated, plasticity is a special case of inelastic material behavior. Plasticity effects in axially loaded members are readily apparent in the laboratory. The stress-strain curve of Fig. 3.3 is often idealized as perfectly plastic for strain levels of 0.005 or less. Many materials do not have such a well-defined yield plateau as low-carbon steel (see Fig. 3.3.). In this chapter stress-strain curves will be assumed to be bilinear, as shown in Fig. 3.6. The analysis of structures for the inelastic range is similar to the analysis of previous sections. The computation of axial stress is independent of material properties.

$$\sigma = \frac{P}{A} \tag{14.1}$$

However, there are limitations on the use of the equation. If the stress computed using Eq. (14.1) exceeds the maximum stress the material will withstand, the analysis is obviously wrong. For plastic analysis, the use of Eq. (14.1) must be coupled with the actual material behavior. Consider the following elementary example.

EXAMPLE 14.1

The statically indeterminate axially loaded structure of Fig. 14.1a is made of two materials having the stress-strain properties shown in Fig. 14.1b. Assume that the bar DE is rigid and compute the following.

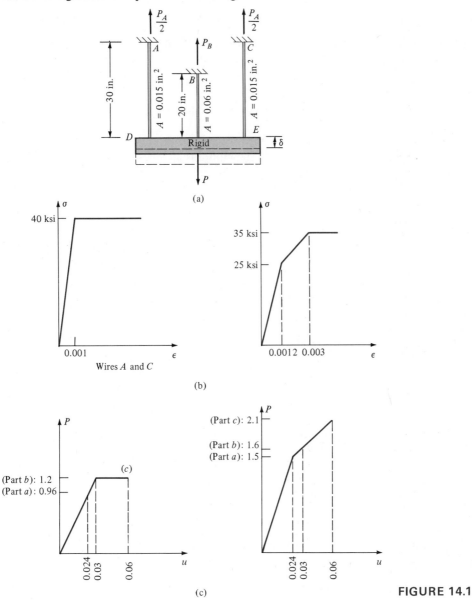

FIGURE 14.1

a. The force P required to cause first yielding of either wire A or B.
b. The force P for the maximum stress in one of the wires. Which wire yields first?
c. The maximum force P for both wires to yield.
d. The deformation corresponding to part *b* above.
e. Construct a load versus deformation diagram for both wires showing all of the results computed above.

Solution:
Wires A and C are a perfectly plastic material, as shown in Fig. 14.1*b*. The maximum stress that can be carried by the wires A and C is 40 ksi. Young's modulus is computed as

$$E_A = E_C = \frac{40 \text{ ksi}}{0.001} = 40(10^3) \text{ ksi} \tag{a}$$

The property of wire B is approximated with a bilinear stress-strain curve. First yielding occurs at a stress level of 25 ksi, at which time the modulus changes and the maximum yield level is 35 ksi.

$$E_{B1} = \frac{25 \text{ ksi}}{0.0012} = 20.8(10^3) \text{ ksi} \tag{b}$$

$$E_{B2} = \frac{35 \text{ ksi} - 25 \text{ ksi}}{0.003 - 0.0012} = 5.55(10^3) \text{ ksi} \tag{c}$$

Wires A and C are assumed to act together with a total area of 0.03 in.^{2}. The equilibrium equation is independent of the material behavior.

$$P_A + P_B = P \tag{d}$$

where P_A is the total force in both wires A and C.
a. The axial stress equation is independent of material behavior, $\sigma = P/A$. There are two cases to be considered for first yielding:

$$P_A = (40 \text{ ksi})(0.03 \text{ in.}^2) = 1.2 \text{ k} \tag{e}$$

$$P_B = (25 \text{ ksi})(0.06 \text{ in.}^2) = 1.5 \text{ k} \tag{f}$$

Now, compute the axial-force distribution for each wire and determine which yield load is reached first. The deformation for a wire is $u = PL/AE$. The deformation condition is

$$u_A = u_B$$

$$\frac{P_A(30 \text{ in.})}{2(0.015 \text{ in.})(40)(10^3) \text{ ksi}} = \frac{P_B(20 \text{ in.})}{(0.06)(20.8)(10^3) \text{ ksi}}$$

$$P_A = 0.64P_B \quad \text{or} \quad P_B = 1.56P_A \tag{g}$$

Substituting Eq. (g) into Eq. (d) gives

$$P_A = 0.391P \quad \text{and} \quad P_B = 0.609P \tag{h}$$

From Eq. (e),

$$P = \frac{1.2}{0.391} = 3.07 \text{ k}$$

or, from Eq. (f),

$$P = \frac{1.5}{0.609} = 2.46 \text{ k}$$

This result indicates that a force $P = 2.46$ k will cause wire B to reach a stress of 25 ksi; using Eq. (h) as a check gives

$$P_A = (0.391)(2.46) = 0.96 \text{ k} \qquad P_B = (0.609)(2.46) = 1.50 \text{ k}$$

$$\sigma_A = \frac{0.96}{0.03} = 32 \text{ ksi} \qquad \sigma_B = \frac{1.5}{0.06} = 25 \text{ ksi}$$

b. The maximum stress condition occurs when $\sigma_A = 40$ ksi or $\sigma_B = 35$ ksi. Equation (d) is still valid. An increase in deformation, beyond the deformation produced by $P = 2.46$ k, is

$$u_A = u_B$$

$$\frac{P_A(30)}{(0.03)(40)(10^3)} = \frac{P_B(20)}{(0.06)(5.55)(10^3)}$$

$$P_A = 2.4P_B \qquad P_B = 0.416P_A$$

$$P_A = 0.706P_i \qquad P_B = 0.294P_i \tag{i}$$

where P_i is the increase in P computed in part *a*. The force in wire A can increase by

$$P_{Ai} = 1.20 - 0.96 = 0.24 \text{ k}$$

whereas wire B can have a maximum force of

$$P_B = (35 \text{ ksi})(0.06 \text{ in.}) = 2.1 \text{ k}$$

$$P_{Bi} = 2.1 - 1.5 = 0.6 \text{ k}$$

Using Eq. (i) gives the two cases:

$$(1) \quad P_i = \frac{0.24}{0.706} = 0.34$$

$$(2) \quad P_i = \frac{0.6}{0.294} = 2.04$$

An increase of 0.34 k in P will cause wire A to yield at 40 ksi.

$$P = 2.46 + 0.34 = 2.80 \text{ k}$$

$$P_A = 0.96 + (0.706)(0.34) = 1.20 \text{ k}$$

$$P_B = 1.50 + (0.294)(0.34) = 1.60 \text{ k}$$

c. The maximum force P corresponds to the case when $\sigma_A = 40$ ksi and $\sigma_B = 35$ ksi. Wire A will not carry more stress than computed for the force of part **b**. However, as P is increased, wire A will deform while wire B carries the force until wire B reaches a yield condition.

$$P_{B\,\text{max}} = (35)(0.06) = 2.1 \text{ k}$$

$$P_{Bi} = 2.1 - 1.6 = 0.5 \text{ k}$$

The maximum force P is $2.8 \text{ k} + 0.5 \text{ k} = 3.3 \text{ k}$. Wire A carries 1.20 k and wire B carries 2.1 k.

d. The deformation in part **b** corresponds to $P_A = 1.20$ k or $P_B = 1.60$ k. Either force should give the same result.

$$u_A = \frac{(1.2)(30)}{(0.03)(40)(10^3)} = 0.03 \text{ in.}$$

$$u_B = \frac{(1.5)(20)}{(0.06)(20.8)(10^3)} + \frac{(0.1)(20)}{(0.06)(5.55)(10^3)} = 0.024 + 0.006 = 0.03 \text{ in.}$$

and agrees with u_A.

e. Figure 14.1c illustrates the result of computing deformations for parts **a**, **b**, and **c**.

$$u_A = \frac{(0.96)(30)}{(0.03)(40)(10^3)} = 0.024 \text{ in.} = u_B \qquad \text{(for part } \boldsymbol{a})$$

The deformation for part **b** was computed in part **d**. Finally,

$$u_B = \frac{(1.5)(20)}{(0.06)(20.8)(10^3)} + \frac{(0.6)(20)}{(0.06)(5.55)(10^3)} = 0.060 \text{ in.} \qquad \text{(for part } \boldsymbol{c})$$

Note, wire A deforms from 0.03 in. to 0.06 in. with no increase in axial force. ∎

14.3 INELASTIC BEHAVIOR OF BEAMS

Beams exhibiting inelastic behavior react differently than axially loaded members. This is due primarily to the effect of the strain distribution. The flexure formula as developed in Chapter 6 is a function of the coordinate of the cross section.

$$\sigma = \frac{My}{I} \tag{14.2}$$

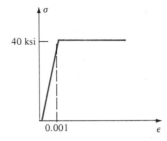

FIGURE 14.2

Perfectly plastic stress-strain relation.

The strain varies in a similar way, linearly with the y coordinate. In this analysis it will be assumed that the material behaves the same in tension and compression. To illustrate inelastic stress distribution, assume the perfectly plastic material of Fig. 14.2.

Consider the beam section of Fig. 14.3 (next page), which can be assumed to be rectangular in cross section. It must be kept in mind that strain is a geometrical quantity. The moment shown on the section of Fig. 14.3a produces the strain distribution shown in Fig. 14.3a, where it is assumed that the strain has reached a value of 0.001. The strain distribution is linear in the coordinate normal to the neutral axis. The maximum strain occurs at the farthest distance from the neutral axis. The strain distribution is given as

$$\varepsilon = \frac{\varepsilon_{max}y}{c} \tag{14.3}$$

The relation between stress and strain is linear for Figs. 14.3a and b; that is, $\sigma = E\varepsilon$. According to Fig. 14.2, the corresponding stress is $\sigma = 40$ ksi. An increase in bending moment M, shown in Fig. 14.3c, produces an increase in strain. The strain distribution is shown in Fig. 14.3c. The stress at the top of the section cannot increase beyond 40 psi. However, the material between the top of the section and the neutral axis of the section can support additional stress. The stress distribution of Fig. 14.3d illustrates the stress distribution. The material begins to yield and reaches a plastic state that is referred to as partially plastic.

The cross section will continue to resist additional bending moment until the fully plastic condition of Fig. 14.3f occurs. The strain distribution continues to be linear. The applied moment is called the plastic moment. Any increase in moment beyond that of Fig. 14.3e cannot be resisted by the beam. If the moment is increased the section will deform until it is completely bent out of shape or ruptures.

Consider the following example that illustrates the computations associated with the foregoing discussion.

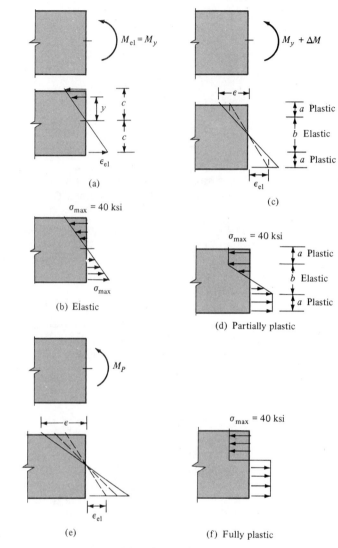

FIGURE 14.3
Plastic behavior for a beam.

EXAMPLE 14.2

A beam of rectangular cross section, 150 mm by 200 mm, may be characterized by the stress-strain diagram of Fig. 14.4a.
a. Compute the maximum moment for elastic behavior, the yield moment.
b. Compute the fully plastic moment.

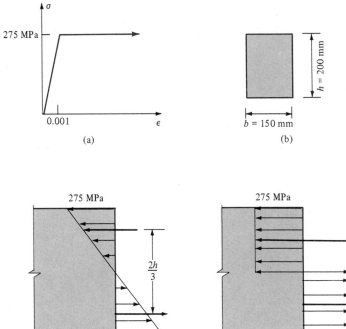

(a)

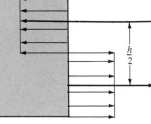

(b)

(c)

(d)

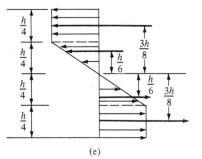

(e)

FIGURE 14.4

c. Compute the moment when one-half the cross section is fully plastic.
d. Compute the shape factor.

Solution:
a. The stress distribution corresponding to the maximum elastic moment is shown
in Fig. 14.4c. The moment can be computed using the flexure formula.

$$M_y = \frac{\sigma_y I}{c} = \frac{(275\text{ MPa})(bh^3/12)}{(h/2)} = \frac{275bh^2}{6}$$

Equilibrium of moments about the neutral axis using Fig. 14.4c gives the same result.

$$M_y = 275b \left(\frac{h}{2}\right)\left(\frac{1}{2}\right)\left(\frac{2h}{3}\right) = \frac{275bh^2}{6}$$

b. The fully plastic moment is obtained by summing moments about the neutral axis of Fig. 14.4d.

$$M_P = 275b \left(\frac{h}{2}\right)\left(\frac{h}{2}\right) = \frac{275bh^2}{4}$$

c. The stress distribution for one-half the cross section to be plastic is shown in Fig. 14.4e. Again, the forces are summed about the neutral axis.

$$M = 275b \left(\frac{h}{4}\right)\left(\frac{3h}{8}\right)(2) + 275b \left(\frac{h}{4}\right)\left(\frac{1}{2}\right)\left(\frac{h}{6}\right)(2) = 275\left(\frac{11}{48}\right)bh^2$$

Substituting $b = 0.15\ m$ and $h = 0.2\ m$,

$M_y = 275\ \text{kN·m}$ (for part **a**)

$M_P = 412.5\ \text{kN·m}$ (for part **b**)

$M = 378\ \text{kN·m}$ (for part **c**)

d. The shape factor is defined as the ratio of the fully plastic moment to the yield moment. For a rectangular cross section, using the results of parts **a** and **b**,

$$\text{Shape factor} = f = \frac{M_P}{M_y} = \frac{275bh^2/4}{275bh^2/6} = 1.5 \qquad\blacksquare$$

In elastic analysis the neutral axis coincides with the centroid axis. For a fully plastic moment the neutral axis coincides with the centroid axis for sections that are symmetrical with respect to the bending axis. For a section, such as a T-shape, the neutral axis is obtained by balancing the tensile force of the stress distribution with the compressive force of the stress distribution.

EXAMPLE 14.3

For the plastic moment compute the location of the plastic neutral axis for the cross section in Fig. 14.5. Assume a perfectly plastic material.

Solution:
The forces associated with the fully plastic stress distribution are shown in Fig. 14.5b. Define y as the distance to the neutral axis from the base of the section.

$$F_1 + F_2 = F_3$$

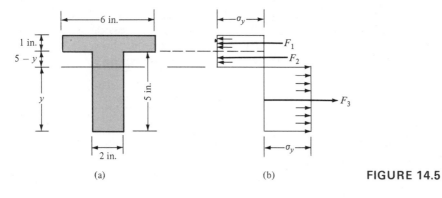

(a)   (b)   **FIGURE 14.5**

$$\sigma_y(6)(1) + \sigma_y(2)(5 - y) = \sigma_y(2)(y)$$

$$16 = 4y \quad \text{or} \quad y = 4 \text{ in.}$$

It is obvious that the plastic neutral axis divides the area equally above and below the neutral axis. ■

14.4 INELASTIC BEHAVIOR OF TORSION MEMBERS

The process of inelastic deformation of a circular shaft subject to torque is similar to the bending deformation of the previous section. A review of Chapter 4 illustrates that the torsional shear stresses vary linearly from the center of the shaft to the outside surface. A diagram similar to Fig. 14.3 can be imagined for the shear stress distribution. The elastic shear stress distribution is elastic and linear until the stress at the outside surface reaches the shear stress yield point for the material. This stress distribution is shown in Fig. 14.6a. As deformation continues, the stress at the outside radius remains constant and penetrates into the section as shown in Fig. 14.6b. For perfectly plastic material be-

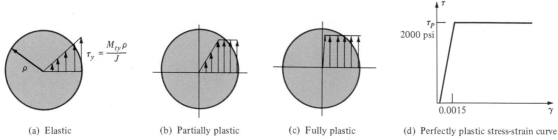

(a) Elastic   (b) Partially plastic   (c) Fully plastic   (d) Perfectly plastic stress-strain curve

FIGURE 14.6
Plastic behavior of a torsion member.

havior as depicted in Fig. 14.6d, the final stress distribution will be fully plastic as shown in Fig. 14.6c. The fully plastic stress distribution will not penetrate completely to the axis of the member; however, the error caused by assuming a rectangular stress block will be very small.

EXAMPLE 14.4

A circular shaft 4 in. in diameter is made of a material having the shear stress-strain properties of Fig. 14.6d. Compute the yield torque and the fully plastic torque.

Solution:
The yield torque is computed using Eq. (4.11).

$$M_{ty} = \frac{\tau_y J}{\rho} = \frac{(20{,}000 \text{ psi})}{2 \text{ in.}} \frac{\pi (2 \text{ in.})^4}{2} = 25.13(10^4) \text{ in.} \cdot \text{lb}$$

The fully plastic torque is obtained by considering the section of Fig. 14.7. A differential force is represented by the fully plastic stress times a differential ring element.

$$dF = \tau_P \, dA = \tau_P 2\pi r \, dr$$

$$dM_{tP} = 2\pi \tau_P r^2 \, dr$$

$$M_{tP} = 2\pi \tau_P \int_0^\rho r^2 \, dr = \frac{2\pi \tau_P \rho^3}{3} \tag{14.4}$$

Substituting into Eq. (14.4) gives the fully plastic torque.

$$M_{tP} = \frac{2\pi (20{,}000)(2)^3}{3} = 33.51(10^4) \cdot \text{in.} \cdot \text{lb} \qquad \blacksquare$$

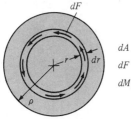

$$dA = 2\pi r \, dr$$
$$dF = \tau_P \, dA$$
$$dM_{tP} = r \, dF$$

FIGURE 14.7

EXAMPLE 14.5

A shaft 150 mm in diameter is subject to a torsional moment of 115 kN·m. Assume a perfectly plastic material with $\tau_y = 145$ MPa and compute the percentage of the cross section that exhibits plastic behavior. Sketch the result.

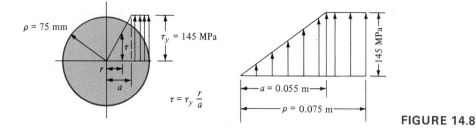

FIGURE 14.8

Solution:

A formula will be developed based on Fig. 14.8a. Let the distance from the center of the shaft to the point where the plastic stress has penetrated be a. The shear stress at a location between $r = 0$ and $r = a$ is $\tau = \tau_y r / a$. Write the moment as

$$dM_t = r\, dF = r\tau\, dA \qquad \text{(a)}$$

Using Fig. 14.8 as a guide will give a relation between M_t, τ_y, and a.

$$M_t = \int_0^a (r)\left(\frac{\tau_y r}{a}\right)(2\pi r\, dr) + \int_a^\rho (r)(\tau_y)(2\pi r\, dr) \qquad \text{(b)}$$

$$M_t = \left(\frac{2\pi \tau_y}{a}\right)\left[\frac{r^4}{4}\right]_0^a + (2\pi \tau_y)\left[\frac{r^3}{3}\right]_a^\rho \qquad \text{(c)}$$

$$M_t = \frac{\tau_y \pi (4\rho^3 - a^3)}{6} \qquad \text{(d)}$$

Substitute into Eq. (d) and solve for a.

$$115(10^3)\, \text{N} \cdot \text{m} = \frac{145(10^6)\, \text{N/m}^2\ \pi[(4)(0.075\ \text{m})^3 - a^3]}{6}$$

$$a^3 = 1.716(10^{-4})\, \text{m}^3$$

$$a = 0.0555\ \text{m} = 55.5\ \text{mm}$$

The total area of the cross section is

$$A = \pi(0.075)^2 = 1.767(10^{-2})\, \text{m}^2$$

That portion of the cross section that has become plastic is

$$A_P = \pi[(0.075)^2 - (0.0555)^2] = 8.0(10^{-3})\, \text{m}^2$$

The percentage of the cross section that is plastic is

$$\frac{A_P}{A} = \frac{8.0(10^{-3})}{1.767(10^{-2})} = 45.3\ \text{percent} \qquad \blacksquare$$

14.5   INELASTIC BEHAVIOR AND COMBINED STRESSES

The previous examples of inelastic behavior have been limited to situations when the normal stress or shear stress is attributable to one external effect. In this section the question is posed concerning combined normal stresses due to an axial force and a bending load. This problem is still not a completely general situation. The stresses remain normal; shear stresses will not be considered. There are two separate effects and the idea is to examine the interaction of the two normal stresses.

EXAMPLE 14.6

A rectangular beam, 3 by 10 in., is subject to both a tensile axial force and a negative bending moment, as shown in Fig. 14.9a. Assume that perfectly plastic material property of Fig. 14.2 for both tension and compression.

a. Compute the allowable axial force associated with the maximum elastic stress distribution if $M = 60$ ft·k.

b. Compute the allowable axial force associated with a fully plastic stress distribution if $M = 200$ ft·k.

Solution:

a. The maximum elastic stress distribution occurs when the yield stress develops at either the top or bottom of the section. For the beam loading specified for this problem, the yield stress will be compressive, as shown in Fig. 14.9b. When combined bending and axial stress reaches σ_y, the stress is given by

$$\sigma_y = \frac{P}{A} + \frac{Mc}{I} \tag{a}$$

The magnitude of P/A or Mc/I is dependent upon the relative magnitude of P and M. If $M = 0$, $\sigma_y = P_y/A$, or if $P = 0$, $\sigma_y = M_y c/I$. If each term of Eq. (a) is divided by σ_y, the result is a nondimensional relation defining a relation between P and M.

$$1 = \frac{P}{P_y} + \frac{M}{M_y} \tag{b}$$

Equation (b) is shown in Fig. 14.9c and is referred to as an interaction curve. For $M = 60$ k·ft, the bending stress is

$$\sigma_b = \frac{(60)(12)(5)}{(3)(10^3)/12} = 14.4 \text{ ksi}$$

$$\frac{\sigma_b}{\sigma_y} = \frac{14.4}{40} = 0.36$$

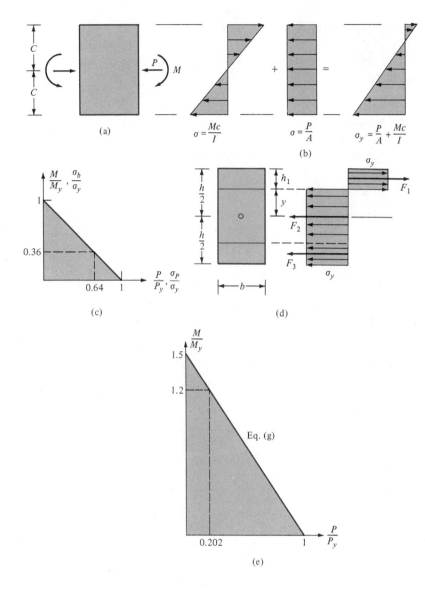

FIGURE 14.9

$$\frac{\sigma_P}{\sigma_y} = 1 - 0.36 = 0.64$$

The computation is illustrated by the dashed line of Fig. 14.9c.

$$\sigma_P = 0.64\sigma_y = 25.6 \text{ ksi}$$

$$P = \sigma_P A = (25.6)(3)(10) = 768 \text{ k (maximum allowable axial force)}$$

b. A formula similar to Eq. (b) will be developed for a fully plastic section. An idealized fully plastic section is shown in Fig. 14.9*d*. The combined axial compression and plastic bending stress distribution is defined by three forces. The compressive force F_1 acts on an area bh_1 and is given by

$$F_1 = \sigma_y bh_1 \tag{c}$$

F_1 is balanced by F_3. Hence, F_1 and F_3 combine to resist the applied bending moment. The remaining stress distribution is a force F_2 acting through the centroid of the section. F_2 produces no moment and can be assumed to resist the action of the axial force.

$$F_2 = \sigma_y(2yb) = P \qquad \text{or} \qquad y = \frac{P}{2\sigma_y b} \tag{d}$$

Equation (c) is equivalent to

$$F_1 = F_3 = \sigma_y b\left[\left(\frac{h}{2}\right) - y\right]$$

$$M = 2F_1\left[y + \left(\frac{h_1}{2}\right)\right] = \sigma_y b\left[\left(\frac{h^2}{4}\right) - y^2\right] \tag{e}$$

Comparing the first term of Eq. (e) to part *b* of Example 14.2, it can be seen that $\sigma_y bh^2/4$ is the fully plastic moment, M_P.

$$M = M_P - \sigma_y by^2$$

Substituting Eq. (d)

$$M = M_P - \frac{P^2}{4b\sigma_y} \tag{f}$$

For a rectangular cross section $M_P = 3M_y/2$ (Example 14.2), also $M_P = \sigma_y bh^2/4$. Divide Eq. (f) by M_P.

$$M/M_P = 1 - \frac{P^2}{(4b^2h^2\sigma_y^2)/4}$$

or

$$1 = \frac{2M}{3M_y} + \left(\frac{P}{P_y}\right)^2 \tag{g}$$

$$M_y = \frac{(40\ \text{ksi}/5\ \text{in.})(3\ \text{in.})(10\ \text{in.})^3}{12} = 2000\ \text{in.·k} = 167\ \text{ft·k}$$

$$P_y = (40\ \text{ksi})(3\ \text{in.})(10\ \text{in.}) = 1200\ \text{k}$$

$$1 = \frac{2}{3}\left(\frac{200}{167}\right) + \frac{P^2}{(1200)^2}$$

$$P^2 = (1200)^2(1 - 0.798)$$

$$P = 539 \text{ k}$$

An interaction curve is shown in Fig. 14.9e for Eq. (g). The solution for $(P/P_y)^2 = 0.202$ corresponding to $M/M_y = (1.5)(0.798)$ is shown dashed in Fig. 14.9e. ■

Interaction curves that combine normal stresses on different planes or combined normal and shear stresses are not so straightforward as Example 14.6. The next topic is an introduction to failure theories that can be thought of as an extension of Example 14.6 to more complicated stress situations.

14.6 FAILURE THEORIES

Failure of a structure or any part of a structure occurs when the stresses exceed some predetermined limit. When combined stress states are present, the interaction of inelastic stress components becomes very complicated analytically. Failure theories have been postulated for the purpose of identifying stress situations that could cause material failure. In this section, two of the most popular failure theories will be discussed. Other theories will be mentioned, primarily because of their historical significance. A theory proposed by H. Tresca in 1864, referred to as the *maximum shear stress theory*, has been shown to be valid for numerous applications in stress analysis. In 1913 R. von Mises proposed his version of the *maximum distortion energy theory*, which has survived as a popular failure theory. These two failure theories, sometimes referred to as a yield criterion, are usually called the Tresca theory and von Mises theory. Several other scientists and engineers have contributed to the development of these and other less popular theories.

Maximum Shear Stress Theory

The maximum shear stress theory is sometimes called the Coulomb theory and assumes that yielding occurs when the maximum shear stress reaches the value of the maximum shear stress corresponding to yielding in a simple tension test. Yielding due to normal tensile stress corresponds to the axially loaded problem of Section 14.2. The associated shear stress is determined by considering the discussion of Chapter 9 on combined stresses and transformation of stresses. The stress state of Fig. 14.10 corresponds to the theory in the most elementary way. The shear stress associated with the element is, by Eq. (9.16),

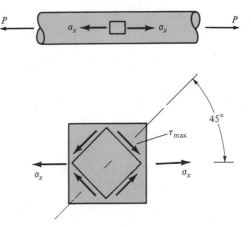

FIGURE 14.10

$$\tau_{max} = \left[\left(\frac{\sigma_x - \sigma_y}{2} \right)^2 + \tau_{xy}^2 \right]^{1/2} = \frac{\sigma_x}{2}$$

If σ_x is the yield stress, the failure theory is

$$\tau_{max} = \frac{\sigma_{yield}}{2} \tag{a}$$

Equation (a) merely states that the yield stress in pure shear is one-half the yield stress in simple tension.

Consider a state of biaxial tension as shown in Fig. 14.11. Note the shear stress is zero; hence, according to Eq. (9.13), σ_x and σ_y are the principal stresses. The maximum shear stress becomes

$$\tau_{max} = \frac{\sigma_x - \sigma_y}{2} = \frac{\sigma_1 - \sigma_2}{2} = \frac{\sigma_{yield}}{2} \tag{b}$$

The maximum shear stress theory in terms of the principal stresses and σ_{yield} for uniaxial tension is written as

$$\sigma_1 - \sigma_2 = \sigma_{yield} \tag{14.5}$$

for biaxial stress states. The principal stresses σ_1 and σ_2 are arbitrarily identified as σ_1 being larger than σ_2.

A general biaxial stress state is shown in Fig. 14.11*b*. In order to use Eq. (14.5) the principal stresses must be computed; hence, the Tresca failure theory is a principal stress failure theory.

Sometimes there is difficulty in visualizing the use of Eq. (14.5). Throughout this textbook the emphasis has been on two-dimensional stress states. In both Chapters 8 and 9 the effect of a third dimension was basically neglected. This procedure has not presented any serious difficulties until now. Consider the element of Fig. 14.11*c*, where a third normal stress, $\sigma_z = 0$, is included. The zero normal stress in the third

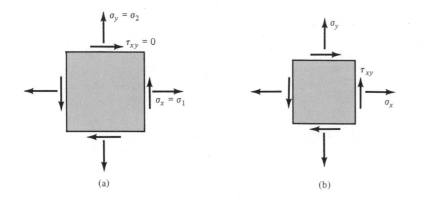

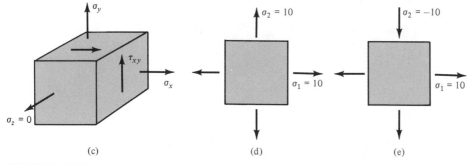

FIGURE 14.11

dimension always exists for a biaxial state of stress. To illustrate its effect, consider the stress states of Figs. 14.11d and e.

Application of Eq. (14.5) to Fig. 14.11d gives

$$\sigma_1 - \sigma_2 = 10 - 10 = 0 \quad \text{and is always less than } \sigma_{\text{yield}}$$

and actually says that since σ_1 and σ_2 remain positive, no matter how large they become, a failure condition may never occur. This result physically should not exist. The stresses of Fig. 14.11e give $\sigma_1 - \sigma_2 = 10 + 10 = 20$ and as σ_1 and σ_2 increase in magnitude but remain opposite in sign, a yield condition will eventually occur.

The situation of Fig. 14.11d is accounted for by considering the third normal stress, $\sigma_z = 0$. For a biaxial stress state when σ_1 and σ_2 are of the same sign, $\sigma_z = 0$ is substituted for the second principal stress. The largest numerical value of either σ_1 or σ_2 is combined with $\sigma_z = 0$.

Maximum Distortion Energy Theory

The energy of distortion is best illustrated by considering a three-dimensional stress element such as that introduced in Chapter 8. The theory assumes that yielding begins or failure occurs when the distor-

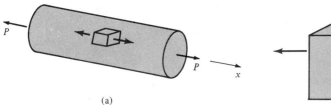

(a)

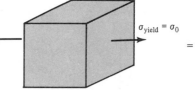

$\sigma_{\text{yield}} = \sigma_0$

=

(b)

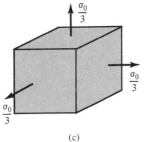

$\dfrac{\sigma_0}{3}$

$\dfrac{\sigma_0}{3}$

$\dfrac{\sigma_0}{3}$

(c)

+

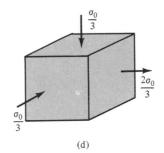

$\dfrac{\sigma_0}{3}$

$\dfrac{2\sigma_0}{3}$

$\dfrac{\sigma_0}{3}$

(d)

FIGURE 14.12

tion part of the strain energy reaches the value of the distortion energy corresponding to failure in uniaxial tension. The failure theory will be derived for a general biaxial stress state.

The strain energy can be divided into two parts: a hydrostatic part and a distortion part. Consider the uniaxial stress case of Fig. 14.12. A three-dimensional element is shown in uniaxial tension. Let the stress be the yield stress, σ_{yield}. A hydrostatic state of stress occurs when the normal stresses acting on each face of the element are equal, either tension or compression. The hydrostatic stress case is shown in Fig. 14.12c where $\sigma_{\text{yield}}/3$ acts on each face. The stress situation shown in Fig. 14.12d is the distortion part of the stress and when added to the hydrostatic case gives the original uniaxial stress case. The hydrostatic case is merely the average of the normal stress acting on each face. In this case the stresses of Fig. 14.12 are principal stresses.

The strain energy per unit volume for an element subject to the action of principal stresses as shown in Fig. 14.13 is written using the definition of the previous chapter.

$$U = \frac{1}{2}\sigma_x \varepsilon_x + \frac{1}{2}\sigma_y \varepsilon_y + \frac{1}{2}\sigma_z \varepsilon_z \qquad (14.6)$$

For a three-dimensional stress case the strains in terms of stress are given by referring to Eqs. (3.3)–(3.5)

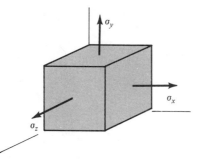

FIGURE 14.13

$$\varepsilon_x = \frac{\sigma_x - v(\sigma_y + \sigma_z)}{E} \qquad (3.3)$$

$$\varepsilon_y = \frac{\sigma_y - v(\sigma_x + \sigma_z)}{E} \qquad (3.4)$$

$$\varepsilon_z = \frac{\sigma_z - v(\sigma_x + \sigma_y)}{E} \qquad (3.5)$$

Identifying the stress components with Fig. 14.12d gives

$$\sigma_x = \frac{2\sigma_0}{3} \qquad \sigma_y = -\frac{\sigma_0}{3} \qquad \sigma_z = \frac{-\sigma_0}{3}$$

The strain energy of distortion is obtained by substituting into Eq. (14.6) as follows:

$$U = \frac{1}{2E}\left(\frac{2}{3}\sigma_0\right)\left[\frac{2}{3}\sigma_0 - v\left(-\frac{\sigma_0}{3} - \frac{\sigma_0}{3}\right)\right] + \frac{1}{2E}\left(-\frac{\sigma_0}{3}\right)$$

$$\times \left[-\frac{\sigma_0}{3} - v\left(\frac{2}{3}\sigma_0 - \frac{\sigma_0}{3}\right)\right] + \frac{1}{2E}\left(-\frac{\sigma_0}{3}\right)\left[-\frac{\sigma_0}{3} - v\left(-\frac{\sigma_0}{3} + \frac{2}{3}\sigma_0\right)\right]$$

$$U = \frac{(1 + v)\sigma_0^2}{3E} \qquad (14.7)$$

Equation (14.7) is the strain energy of distortion for a uniaxial stress state. If σ_0 is the yield stress in simple tension or compression, then Eq. (14.7) represents the yielding parameter for the von Mises theory.

The strain energy of distortion for a general biaxial principal stress state is obtained using a similar process. The element of Fig. 14.14a is divided into hydrostatic and distortion states of stress. Note that the hydrostatic state is the average of the principal stresses. The state of stress in Fig. 14.14c is the stress state that must be added to Fig. 14.14b to obtain Fig. 14.14a.

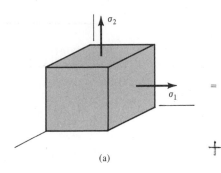

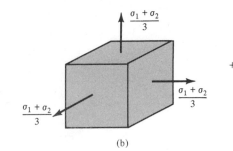

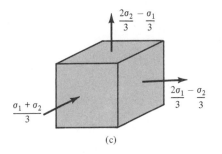

FIGURE 14.14

(a)

(b)

(c)

Substitute the stresses of Fig. 14.14c into Eqs. (3.3)–(3.5) and then into Eq. (14.6). The result, after considerable simplification, is

$$U = \frac{1 + v}{3E} (\sigma_1^2 + \sigma_2^2 - \sigma_1 \sigma_2) \qquad (14.8)$$

The failure theory is that U given by Eq. (14.8) cannot exceed U given by Eq. (14.7) or

$$\sigma_1^2 + \sigma_2^2 - \sigma_1 \sigma_2 = \sigma_0^2 \qquad (14.9)$$

Equation (14.9) is also a principal stress failure theory. In order to use either failure theory the principal stresses must be computed first. It should be pointed out that the strain energy of distortion can be written in terms of a general stress state. However, the theory required to simplify the general failure theory is beyond the scope of this text.

Equation (14.9) is applicable for a biaxial stress state without considering the effects of $\sigma_z = 0$ as discussed for Eq. (14.5). The effects of a distortion and hydrostatic pressure were separated in the derivation.

Several examples will illustrate the use of the failure theories.

EXAMPLE 14.7

The stresses in a machine member are computed to be

$$\boldsymbol{\sigma} = \begin{pmatrix} \sigma_{xx} & \tau_{xy} \\ \tau_{xy} & \sigma_{yy} \end{pmatrix} = \begin{pmatrix} 120 & 60 \\ 60 & 80 \end{pmatrix} \text{MPa}$$

Investigate material failure using the Tresca failure theory and the von Mises failure theory if the yield stress in uniaxial tension is 150 MPa.

Solution:

The principal stressess must be computed prior to investigating material failure:

$$\sigma_{1,2} = \frac{\sigma_x + \sigma_y}{2} \pm \left[\left(\frac{\sigma_x - \sigma_y}{2} \right)^2 + \tau_{xy}^2 \right]^{1/2}$$

$$\sigma_{1,2} = \frac{120 + 80}{2} \pm \left[\left(\frac{120 - 80}{2} \right)^2 + (60)^2 \right]^{1/2} = 100 \pm 63.2$$

$$\sigma_1 = 163.2 \text{ MPa} \qquad \sigma_2 = 36.8 \text{ MPa}$$

Equation (14.5) gives the result for the Tresca failure theory. Note σ_1 and σ_2 are of the same sign; therefore, let $\sigma_2 = \sigma_z = 0$ in Eq. (14.5).

$$(163.2 - 0) \text{ MPa} = 163.2 \text{ MPa} > 150 \text{ MPa}$$

Since the result, 163.2 MPa, is greater than the yield stress in uniaxial tension of 150 MPa, the stress state will cause failure. See the discussion following Eq. (14.5) in the text.

Equation (14.9) gives the result for the von Mises failure theory.

$$(163.2)^2 + (36.8)^2 - (163.2)(36.8) = 21{,}983$$

$$\sigma_0^2 = (150)^2 = 22{,}500$$

Since 21,983 is less than 22,500, the material does satisfy the failure criterion in this case.   ■

EXAMPLE 14.8

A spherical pressure vessel 4 m in diameter is subject to an internal pressure of 250 kPa. Compute the wall thickness according to the Tresca and von Mises failure theories. Assume $\sigma_{\text{yield}} = 220$ MPa and a factor of safety of 1.8.

Solution:

A review of Chapter 8 will show that the normal stresses are given by $\sigma = pr/2t$. The normal stresses σ_x and σ_y are equal for an element in any orientation, and hence they are principal stresses.

$$\sigma_x = \sigma_y = \sigma_1 = \sigma_2 = \frac{(250)(10^3)(2) \text{ Pa·m}}{2t \text{ m}} = \frac{250}{t} \text{ kPa}$$

The Tresca theory, Eq. (14.5) with $\sigma_2 = 0$, since both principal stresses are positive, gives

$$\frac{250(10^3)}{t} - 0 = \frac{220(10^6)}{1.8}$$

$$t = 2.05(10^{-3}) \text{ m} = 2.1 \text{ mm}$$

The von Mises theory, with $\sigma_1 = \sigma_2$, gives

$$\left[\frac{500(10^3)}{2t}\right]^2 + \left[\frac{500(10^3)}{2t}\right]^2 - \left[\frac{500(10^3)}{2t} \cdot \frac{500(10^3)}{2t}\right] = \left[\frac{220(10^6)}{1.8}\right]^2$$

$$t = 2.05(10^{-3}) \text{ m} = 2.1 \text{ mm}$$

It turns out that both theories give the same result for this spherical pressure vessel. ∎

Maximum Principal Stress Theory

The maximum principal stress theory is sometimes referred to as the maximum normal stress theory and was originally proposed by Rankine. Essentially, the theory predicts that failure or inelastic action occurs when the maximum principal stress at a point reaches the yield stress of the material as determined from the uniaxial tension or compression test. The theory has serious limitations because any effect of shear stress is neglected.

Maximum Normal Strain Theory

St. Venant is given credit for first suggesting the maximum normal strain theory. The theory postulates that inelastic material behavior occurs when the maximum strain at a point reaches a value equal to that which occurs when inelastic behavior begins in the same material when subjected to a uniaxial test. Again, the failure theory has limited application in engineering.

Total Energy Theory

The total energy theory was first proposed by Beltrami and suggests that inelastic behavior first occurs at a point in a material body when the energy per unit volume absorbed is equal to the energy absorbed per unit volume when the material is subjected to a uniaxial test and reaches a yield condition. The total energy theory is not frequently used in engineering analysis.

14.7 SUMMARY

Inelastic material behavior encompasses an extremely large and complicated volume of analytical and experimental data. In this chapter, at best, a brief introduction has been accomplished.

The one-dimensional problems of axially loaded structures allow the correlation of theoretical and experimental results. Other elementary problems can be reproduced in the laboratory; however, a vast majority of real engineering applications of theoretical results cannot be verified experimentally. Failure theories, based upon elementary analyses, are postulates of material behavior for more complicated stress analysis situations. The references that follow this summary offer a beginning point for additional study.

In the introduction to this chapter it was stated that failure theories for brittle materials would not be discussed. The section on fracture mechanics in the next chapter should logically follow this chapter and give some insight to the behavior of a brittle material.

For Further Study

Boresi, A. P. and O. M. Sidebottom, *Advanced Mechanics of Materials*, 4th ed., Wiley, New York, 1985.

Chakrabarty, J., *Theory of Plasticity*, McGraw-Hill, New York, 1987.

Mendelson, A., *Plasticity: Theory and Applications*, Macmillan, New York, 1968.

14.8 PROBLEMS

14.1 A rod is subjected to a uniaxial tension. Assume the properties of Fig. P14.1 and compare the elongation of the rod as the material becomes perfectly plastic. In Fig. P14.1a assume $\sigma_1 = 35$ ksi and $\varepsilon_1 = 0.0016$. In Fig. P14.1b assume $\sigma_1 = 45$ ksi, $\varepsilon_1 = 0.0005$, $\sigma_2 = 60$ ksi, and $\varepsilon_2 = 0.002$. Compute your result in terms of P, L, and A.

14.2 Assume that both materials of Fig. P14.2 are perfectly plastic, with material properties as shown in Fig. P14.1a. Assume that both areas are 4.0 in.^{2}. For member A, $\sigma_1 = 20$ ksi and $\varepsilon_1 = 0.001$; for member B, $\sigma_1 = 40$ ksi and $\varepsilon_1 = 0.0012$.
a. Compute the load, P, for first yield to begin.
b. Compute the load that corresponds to yielding for both materials to yield.

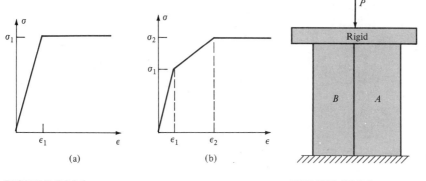

(a)   (b)

FIGURE P14.1

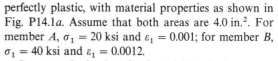

FIGURE P14.2

14.3 Repeat Problem 14.2 assuming that material A is bilinear, as shown in Fig. P14.1b. Assume $\sigma_1 = 30$ ksi, $\varepsilon_1 = 0.001$, $\sigma_2 = 45$ ksi, and $\varepsilon_2 = 0.003$. Material B remains the same as given in Problem 14.2.

14.4 Use the materials properties of Fig. 14.1a. In Fig. P14.4, wire A is perfectly plastic with $\sigma_1 = 200$ MPa and $E_A = 220$ GPa. Wire B is perfectly plastic with $\sigma_1 = 135$ MPa and $E_B = 180$ GPa. Compute the maximum force P that the structure will support for one wire to yield. Compute the maximum force P for both wires reaching yield conditions and the corresponding deformation of each wire.

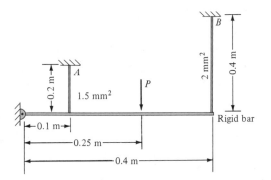

FIGURE P14.4

14.5 Repeat Problem 14.4 assuming that wire B is bilinear with properties as shown in Fig. P14.1b with $\sigma_1 = 150$ MPa, $E_1 = 180$ GPa, $\sigma_2 = 225$ MPa, and $E_2 = 50$ GPa. Assume all other properties to be the same.

14.6 The truss of Fig. P14.6 is made of bilinear material such as Fig. P14.1b. All truss members have an area of 2 in.^{2}. Compute the maximum value of P, assuming a design criterion that none of the members can exceed an allowable stress of σ_2. Let $\sigma_1 = 40$ ksi, $\sigma_2 = 55$ ksi, $\varepsilon_1 = 0.001$, and $\varepsilon_2 = 0.0035$.

14.7 A beam of rectangular cross section, 3 in. by 10 in., is characterized by the stress-strain diagram of Fig. P14.1a with $\sigma_1 = 32$ ksi and $E = 30(10^3)$ ksi.
a. Compute the maximum moment for elastic behavior.
b. Compute the fully plastic moment.

14.8 Assume that the beam of Problem 14.7 is made of a material with stress-strain properties of Fig. P14.1b. Compute the moment that can be carried by the section when the stress at the top of the section reaches σ_2. Assume $\sigma_1 = 30$ ksi, $\sigma_2 = 40$ ksi, $\varepsilon_1 = 0.0015$, and $\varepsilon_2 = 0.004$.

14.9 A T-beam, shown in Fig. P14.9, is made of a perfectly plastic material such as Fig. P14.1a.
a. Compute the maximum moment for elastic behavior.
b. Compute the fully plastic moment.
c. Compute the shape factor for the section.
Let $\sigma_1 = 350$ MPa and $\varepsilon_1 = 0.0012$.

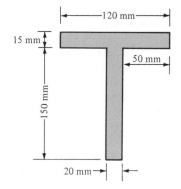

FIGURE P14.9

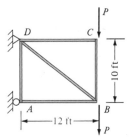

FIGURE P14.6

14.10 A beam shown in Fig. P14.10 has the material properties of Fig. P14.1a with $\sigma_1 = 42$ ksi and $E = 28(10^3)$ ksi. Compute the maximum elastic moment and the fully plastic moment. Compute the shape factor for the section.

14.11 A solid circular shaft 60 mm in diameter has the stress-strain relation of Fig. P14.1a. Compute the yield torque and the fully plastic torque. Let the yield stress be 85 MPa.

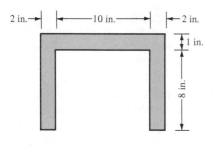

FIGURE P14.10

14.12 A solid circular shaft is made of a perfectly plastic material with $\sigma_y = 15,000$ psi. The radius of the shaft is 1.5 in.
a. Compute the yield torque.
b. Compute the fully plastic torque.
c. Assume an applied torque of 7 ft·k and compute the percentage of the cross section that is plastic.

14.13 A hollow circular shaft is made of a perfectly plastic material with $\sigma_y = 110$ MPa. The outside and inside radii are 130 mm and 100 mm, respectively. Compute the initial yield torque and the fully plastic torque.

14.14 A hollow circular shaft 4 in. in diameter is subject to a torque of 7500 in.·lb.
a. Compute the wall thickness based upon developing the maximum elastic torque that could act on the cross section assuming a perfectly plastic material with $\sigma_y = 12,000$ psi.
b. Compute the wall thickness assuming that the section is fully plastic.
c. Compute the wall thickness assuming a fully plastic section and a factor of safety of 2.

14.15 A rectangular beam, 3 in. by 10 in., is subject to a compressive force of 95 k. Assume a perfectly plastic material with $\sigma_y = 40$ ksi.
a. Compute the bending moment that could be applied to the beam assuming that the maximum elastic stress distribution occurs.
b. Compute the bending moment that can be applied assuming a fully plastic stress distribution.

14.16 A square beam 150 mm by 150 mm is subject to a positive moment of 90 kN · m and a compressive axial force. Assume a perfectly plastic material with $\sigma_y = 225$ MPa.
a. Compute the maximum allowable axial force assuming the maximum elastic stress distribution.

b. Compute the axial force assuming a fully plastic stress distribution.

A state of stress is given for Problems 14.17–14.22. Assuming a yield stress of 50 ksi in uniaxial tension or compression, determine whether the state of stress satisfies the yield criterion of Section 14.6.

14.17 $\sigma = \begin{pmatrix} 28 & 0 \\ 0 & -22 \end{pmatrix}$ ksi

14.18 $\sigma = \begin{pmatrix} -18 & 5 \\ 5 & 30 \end{pmatrix}$ ksi

14.19 $\sigma = \begin{pmatrix} 5 & 45 \\ 45 & -10 \end{pmatrix}$ ksi

14.20 $\sigma = \begin{pmatrix} 30 & -15 \\ -15 & 20 \end{pmatrix}$ ksi

14.21 $\sigma = \begin{pmatrix} 20 & 40 \\ 40 & -30 \end{pmatrix}$ ksi

14.22 $\sigma = \begin{pmatrix} 45 & 30 \\ 30 & 20 \end{pmatrix}$ ksi

14.23 A circular solid shaft 2 in. in diameter is subject to an axial force of 50 k. Compute the allowable torque that could be applied according to each of the failure theories. Let $\sigma_{yield} = 32$ ksi and assume a factor of safety of 1.5.

14.24 A 4-in.-diameter standard section of pipe is subject to a bending moment of 45 in.·k and a torque of 15 in.·k. For a factor of safety of 2, determine whether the pipe is satisfactory. Compare both failure theories and assume a yield stress of 32 ksi.

14.25 A section of pipe is subject to an axial compressive force of 60 k, a positive bending moment of 6.2 ft·k, and a torque of 5.5 ft·k. Using the strain energy of distortion theory and a factor of safety of 2.5, select a standard pipe to satisfy the design. Assume $\sigma_{yield} = 36$ ksi.

14.26 A cylindrical pressure vessel has a radius of 12 ft and is subject to an internal pressure of 220 psi. Compute the wall thickness of the vessel, assuming a factor of safety of 1.8. Compare the maximum shear theory with distortion energy theory for a yield stress of 60 ksi.

14.27 A cylindrical pressure vessel is 60 in. in diameter with a wall thickness of 0.25 in. Compute the allowable internal pressure using the distortion energy failure theory.

14.28 A spherical pressure vessel is made of a material with σ_{yield} of 60 ksi. The diameter of the vessel is 6 ft and the wall thickness is 0.375 in. Assume a factor of safety of 2 and compute the allowable internal pressure using both failure theories.

14.29 Begin with the stresses of Fig. 14.14c, substitute in Eqs. (3.3)–(3.5), then into Eq. (14.6) and derive Eq. (14.8).

14.30 A short beam is subjected to a shear force of $V = 20$ k and a negative bending moment of $M = 2$ ft·k. For a material with yield strength of 22 ksi compute the dimensions for a square cross section using both failure theories.

14.31 Repeat Problem 14.30 assuming that an axial tensile force of 120 k is applied to the beam in addition to the shear and moment.

CHAPTER

STRESS CONCENTRATION

15.1 INTRODUCTION

Stress concentration is significant subject and earned its place in a mechanics of materials textbook many years ago. The topic has been treated here almost entirely as an empirical study. The mathematics and advanced theory for solving the simplest problem would require many pages in this book and such an approach would not serve the purpose of the text. Hence, stress concentration has been treated in an elementary way with selected illustrative design charts. Many problems in stress concentration analysis have been solved experimentally. The experimental methods are, for the most part, unique and sophisticated and many times require advanced study to understand the experiment and analyze the data.

15.2 STRESS CONCENTRATION

The basic concept of stress concentration was introduced in Chapter 3, Section 3.5. Stress concentrations occur at locations in structures where there is an abrupt change in geometry such as holes, notches, corners, or where concentrated forces are applied.

In this section design equations, charts, or curves will be presented and examples are given to illustrate their use. No attempt will be made to derive or defend the design methods. Computations are based on stress concentration factors, k, which are defined as the ratios of the stress caused by the stress concentration to the nominal stress. Nominal stress may be thought of as the stress in a member with no imperfections in its geometry. The discussion in this section is limited to structures that have been studied in previous chapters. References at the end of this chapter contain additional design data.

Tension Members

Two cases will be discussed for tension members. The classical case for an infinite plate in tension with a circular hole in the center, as shown

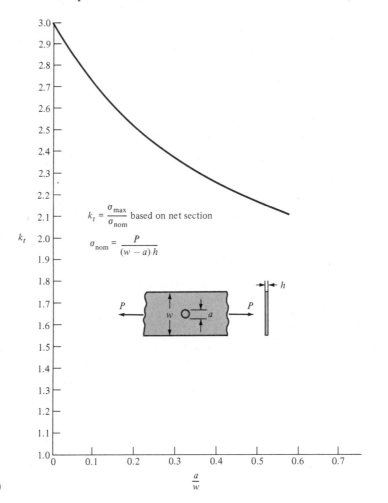

$$k_t = \frac{\sigma_{max}}{\sigma_{nom}} \text{ based on net section}$$

$$\sigma_{nom} = \frac{P}{(w - a)\,h}$$

FIGURE 15.1
Stress concentration factor for axial loading of a finite-width plate with a tranverse hole. (*Source:* R. E. Peterson, *Stress Concentration Design Factors*, Wiley, 1953. *Original source:* R. C. J. Howland, "On the Stress in the Neighborhood of a Circular Hole in a Strip under Tension," *Phil. Trans. Roy. Soc.* (*London*) *A*, vol. 229, p. 67, 1929–1930.)

FIGURE 15.2
Stress concentration factor for the tension case of a flat bar with a shoulder fillet. (*Source:* R. E. Peterson, *Stress Concentration Design Factors*, Wiley, 1953. *Original sources:* M..M. Frocht, "Factors of Stress Concentration Photoelastically Determined," *Trans, ASME*, vol. 57, p. A-67, 1935. M. M. Frocht and D. Landsberg, "Factors of Stress Concentration in Bars with Deep Sharp Grooves and Fillets in Tension," *Proc. SESA* vol. 8, p. 149, 1951. M. M. Leven and J. B. Hartman, "Factors of Stress Concentration for Flat Bars and Shafts with Centrally Enlarged Sections," *Proc. SESA*, vol. 9, p. 53, 1951.

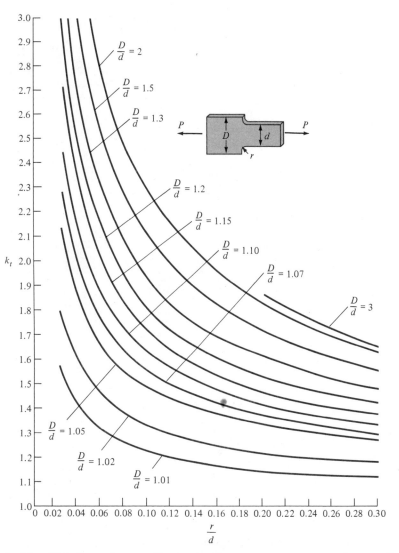

in Fig. 15.1, is used as an illustration in numerous texts on theory of elasticity. It turns out that even though the curve is for a hole specified to be in the center of the plate, the stress concentration factor is fairly accurate if the hole is displaced from the center of the plate. However, the hole should be at least a distance equal to the radius of the hole away from the edge of the plate.

Figure 15.2 is for a thin plate or flat bar with shoulder fillets. The stress concentration factor is a function of the width of the bar and the radius of the fillet.

The stress concentration is given by the elementary formula for axially loaded members modified by the stress concentration factor.

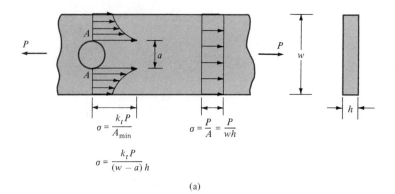

$$\sigma = \frac{k_t P}{A_{\min}}$$

$$\sigma = \frac{P}{A} = \frac{P}{wh}$$

$$\sigma = \frac{k_t P}{(w-a)\,h}$$

(a)

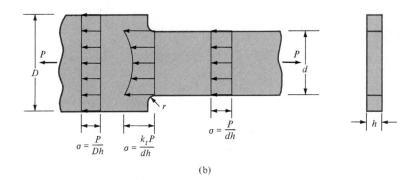

$$\sigma = \frac{P}{Dh} \qquad \sigma = \frac{k_t P}{dh} \qquad \sigma = \frac{P}{dh}$$

FIGURE 15.3

(b)

$$\sigma = \frac{k_t P}{A_{\min}} \tag{15.1}$$

The stress distribution is shown, approximately, in Fig. 15.3. The stress for a hole in a plate is quite high at the edge of the hole. At a distance away from the hole, approximately equal to the diameter of the hole, the effect of the stress concentration is negligible. However, a design is based upon the stress at the edge of the hole. The stress for the fillet of Fig. 15.3b is based upon the correction factor being applied to the nominal stress using the area (dh). Consider the following illustrative examples.

EXAMPLE 15.1

A flat plate 400 mm in width and 10 mm thick has a hole 30 mm in diameter located at the center of the plate. Compute the stress concentration factor at the edge of the hole.

Solution:
The stress concentration factor is for point A of Fig. 15.3a. The geometry is defined by the following parameters.

$$w = 400 \text{ mm} \qquad h = 10 \text{ mm} \qquad a = 30 \text{ mm}$$

$$\frac{a}{w} = \frac{30}{400} = 0.075$$

The curve of Fig. 15.1 gives $k_t = 2.78$.

$$\sigma_A = \frac{k_t P}{(w - a)h}$$ ∎

EXAMPLE 15.2

A flat bar with a shoulder fillet is shown in Fig. 15.4. Compute the normal stress at sections a-a and b-b and the stress at point A.

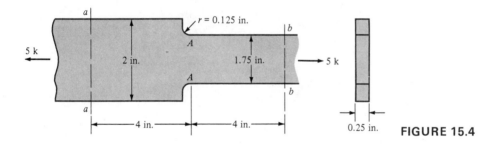

FIGURE 15.4

Solution:
The stress at section a-a, well removed from the effects of the stress concentration, is given by the axial stress formula.

$$\sigma_{a-a} = \frac{5 \text{ k}}{(2)(0.25) \text{ in.}^2} = 10 \text{ ksi}$$

Similarly, at section b-b,

$$\sigma_{b-b} = \frac{5 \text{ k}}{(1.75)(0.25) \text{ in.}^2} = 11.43 \text{ ksi}$$

The stress concentration factor is obtained from Fig. 15.2.

$$\frac{D}{d} = \frac{2}{1.75} = 1.14$$

$$\frac{r}{d} = \frac{0.125}{1.75} = 0.071$$

Enter the chart with $r/d = 0.071$, and go upward to intersect with the D/d curve of 1.14. Extrapolating between the curves and reading k_t at the left gives $k_t = 1.86$. Use Eq. (15.1).

$$\sigma_A = \frac{(1.86)(5 \text{ k})}{(1.75)(0.25)} = 21.26 \text{ ksi}$$

From a design viewpoint the stress at the fillet is a significant parameter. The stress at the fillet is more than twice the stress at section *a-a*.   ■

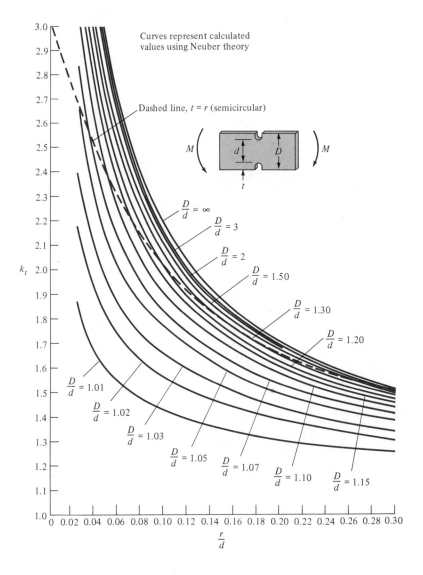

FIGURE 15.5
Stress concentration factor for a notched flat bar in bending. (*Source*: R. E. Peterson, *Stress Concentration Design Factors*, Wiley, 1953.)

Flexural Members

Flexural members are subject to stress concentration in much the same way as tension members. A flat bar subjected to a bending moment will suffer stress concentration at holes, notches, grooves, or fillets, as illustrated in Fig. 15.5. Only one set of design curves is shown here. A flat bar with a circular notch is shown along with a set of curves. The stress concentration factor k_t is used in conjunction with the flexure formula.

$$\sigma = \frac{k_t Mc}{I_{\min}} \tag{15.2}$$

where $I_{\min}$ corresponds to the cross section where the notch occurs.

EXAMPLE 15.3

The flat bar of Fig. 15.6 is subject to a positive moment of 2500 ft·lb. Use Fig. 15.5 to compute the flexural stress at the base of the notch.

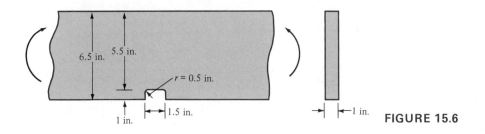

FIGURE 15.6

Solution:
The maximum normal stress at a distance removed from the notch is given by the flexural formula.

$$\sigma = \frac{Mc}{I} = \frac{(2500)(12)(6.5/2)}{(1)(6.5)^3/12} = 4260 \text{ psi}$$

The stress at the base of the notch is given by Eq. (15.2); that is,

$$\sigma = k_t \frac{(2500)(12)(5.5/2)}{(1)(5.5)^3/12} = k_t(5950) \text{ psi}$$

The stress concentration factor is obtained from Fig. 15.5.

$$D = 6.5 \text{ in.} \qquad d = 5.5 \text{ in.} \qquad r = 0.5 \text{ in.}$$

$$\frac{D}{d} = 1.18 \qquad \frac{r}{d} = 0.091$$

Enter the chart at the bottom, $r/d = 0.091$, move upward to intersect with the curve $D/d = 1.18$. Read k_t at the left.

$k_t = 2.06$ $\sigma = (2.06)(5950)$ psi $= 12{,}258$ psi   ■

Torsion Members

Stress concentration may appear in various forms for a shaft subjected to torsional stresses. A design chart for the case of a shaft with a shoulder fillet has been included as Fig. 15.7. The shear stress at the base

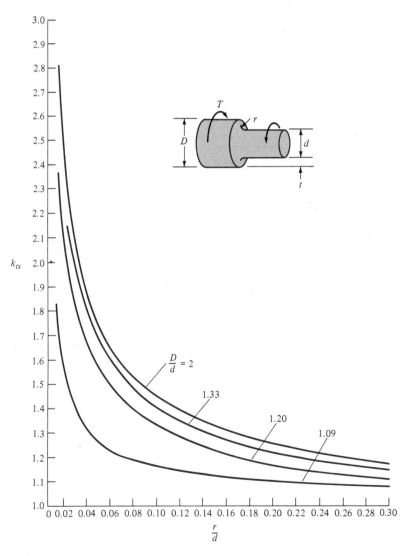

FIGURE 15.7
Stress concentration factor for the torsion of a shaft with a shoulder fillet. (*Source:* R. E. Peterson, *Stress Concentration Design Factors*, Wiley, 1953. *Original source:* L. S. Jacobsen, "Torsional Stress Concentration in Shafts of Circular Cross-Section and Variable Diameter," *Trans. ASME*, vol. 47, p. 619, 1925.)

of the fillet is given by

$$\tau' = \frac{k_{ts}M_t\rho}{J_{min}}$$ (15.3)

The polar second moment of the area corresponds to the smaller section of the shaft of Fig. 15.7.

EXAMPLE 15.4

A circular shaft as shown in Fig. 15.8 is subjected to a torque. The shear stress is to be limited to 85 MPa. Based upon the stress at the base of the fillet compute the maximum allowable torque. $D = 37.5$ mm; $d = 25$ mm; and $r = 5$ mm.

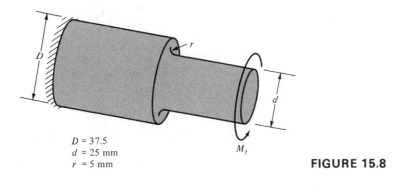

D = 37.5
d = 25 mm
r = 5 mm

M_t

FIGURE 15.8

Solution:

$$\frac{D}{d} = \frac{37.5}{25} = 1.5 \qquad \frac{r}{d} = \frac{5}{25} = 0.2$$

Enter the chart with $r/d = 0.2$, read upward and extrapolate for the value of $D/d = 1.5$. The stress concentration factor is read at the left as $k_{ts} = 1.24$. Equation (15.3) can be solved for allowable torque.

$$M_t = \frac{\tau J_{min}}{k_{ts}\rho} = \frac{[85(10^6)\pi(0.025/2)^4/2]}{(1.24)(0.025/2)} = 210.3 \text{ N·m}$$ ∎

The examples and charts are merely representative of the vast amount of data available for computing stress concentration factors. These charts are included (with the permission of John Wiley and Sons) to serve primarily as an illustration of the topic.

For Further Study

Peterson, R. E., *Stress Concentration Design Factors*, Wiley, New York, 1953.

Roark, R. J., and W. C. Young, *Formulas for Stress and Strain*, McGraw-Hill, New York, 1975.

Savin, G. N., *Stress Concentration Around Holes*, Pergamon, New York, 1961.

Seely, F. B., and T. O. Smith, *Advanced Mechanics of Materials*, Wiley, New York, 1952.

15.3 PROBLEMS

For Problems 15.1 through 15.4 use Fig. P15.1 and compute the stress concentration factor for a circular hole in a plate. Compare the stress at a point A to the stress for a solid plate.

15.1 $w = 3$ in.; $a = 1$ in.; $h = 0.25$ in.; $P = 8$ k.

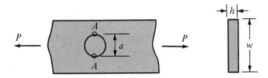

FIGURE P15.1

15.2 $w = 400$ mm; $a = 200$ mm; $h = 20$ mm; $P = 10$ kN.

15.3 $w = 1000$ mm; $a = 50$ mm; $h = 20$ mm; $P = 60$ kN.

15.4 $w = 265$ mm; $a = 40$ mm; $h = 30$ mm; $P = 40$ kN.

For Problems 15.5–15.7 use Fig. P15.5 and compute the stress concentration factor for the shoulder fillet. Compare the stress at point A to the nominal stress at a point removed from the shoulder.

15.5 $D = 4$ in.; $d = 2$ in.; $r = 0.1$ in.; $h = 0.1$ in.; $P = 2$ k.

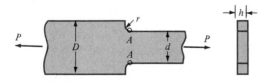

FIGURE P15.5

15.6 $D = 75$ mm; $d = 70$ mm; $r = 2.5$ mm; $h = 10$ mm; $P = 9.5$ kN.

15.7 $D = 50$ mm; $d = 25$ mm; $r = 5$ mm; $h = 12$ mm; $P = 6.3$ kN.

15.8 Compare the stress concentration factor at points A and B for the flat bar of Fig. P15.8. Compute the stresses at points A and B.

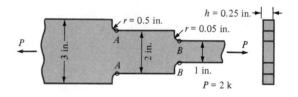

FIGURE P15.8

For Problems 15.9–15.12 use Fig. P15.9 and compute the stress concentration factor caused by the notch in the flat bar. Compare the bending stress at the notch to the bending stress at a section removed from the notch.

15.9 $D = 1.5$ in.; $d = 1.25$ in.; $r = 0.1$ in.; $h = 0.25$ in.; $M = 125$ ft·lb.

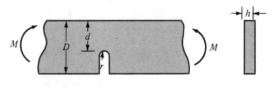

FIGURE P15.9

15.10 $D = 125$ mm; $d = 100$ mm; $r = 10$ mm; $h = 25$ mm; $M = 300$ N·m.

15.11 $D = 300$ mm; $d = 280$ mm; $r = 6$ mm; $h = 20$ mm; $M = 525$ N·m.

15.12 $D = 300$ mm; $d = 175$ mm; $r = 15$ mm; $h = 20$ mm; $M = 525$ N·m.

For Problems 15.13–15.15 use Fig. P15.13 and compute the stress concentration factor at the fillet. Compare the shear stress at the fillet to the shear stress at locations removed from the fillet.

15.13 $D = 125$ mm; $d = 100$ mm; $r = 10$ mm; $M_t = 450$ N·m.

15.14 $D = 2$ in.; $d = 1.5$ in.; $r = 0.15$ in.; $M_t = 125$ ft·lb.

15.15 $D = 2$ in.; $d = 1.0$ in.; $r = 0.25$ in.; $M_t = 125$ ft·lb.

15.16 Compute the maximum moment that can be applied to the shaft of Fig. P15.16. Assume an allowable shear stress of 8000 psi.

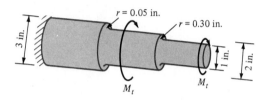

FIGURE P15.16

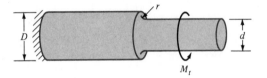

FIGURE P15.13

APPENDIX

PROPERTIES OF SOME PLANE AREAS

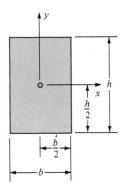

Rectangle

$A = bh$

$I_x = \dfrac{bh^3}{12}$

$I_y = \dfrac{hb^3}{12}$

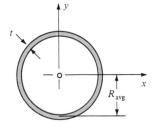

Thin tube

$A = 2\pi R_{avg}\, t$

$I_x = I_y \simeq \pi R_{avg}^3\, t$

$J \simeq 2\pi R_{avg}^3\, t$

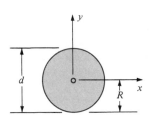

Circle

$A = \pi R^2 = \dfrac{\pi d^4}{4}$

$I_x = I_y = \dfrac{\pi R^4}{4} = \dfrac{\pi d^4}{64}$

$J = I_x + I_y = \dfrac{\pi R^4}{2} = \dfrac{\pi d^4}{32}$

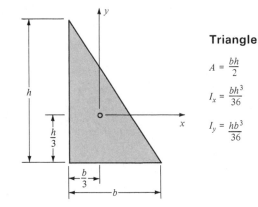

Triangle

$A = \dfrac{bh}{2}$

$I_x = \dfrac{bh^3}{36}$

$I_y = \dfrac{hb^3}{36}$

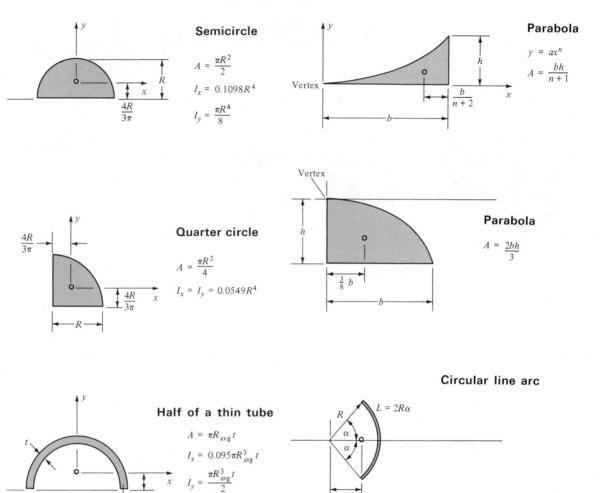

Semicircle

$$A = \frac{\pi R^2}{2}$$

$$I_x = 0.1098 R^4$$

$$I_y = \frac{\pi R^4}{8}$$

Parabola

$$y = ax^n$$

$$A = \frac{bh}{n+1}$$

Quarter circle

$$A = \frac{\pi R^2}{4}$$

$$I_x = I_y = 0.0549 R^4$$

Parabola

$$A = \frac{2bh}{3}$$

Half of a thin tube

$$A = \pi R_{avg} t$$

$$I_x = 0.095 \pi R_{avg}^3 t$$

$$I_y = \frac{\pi R_{avg}^3 t}{2}$$

Circular line arc

$$L = 2R\alpha$$

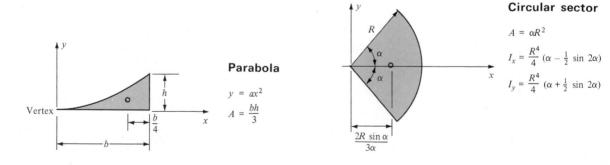

Parabola

$$y = ax^2$$

$$A = \frac{bh}{3}$$

Circular sector

$$A = \alpha R^2$$

$$I_x = \frac{R^4}{4}\left(\alpha - \tfrac{1}{2}\sin 2\alpha\right)$$

$$I_y = \frac{R^4}{4}\left(\alpha + \tfrac{1}{2}\sin 2\alpha\right)$$

APPENDIX

PHYSICAL PROPERTIES OF SOME MATERIALS

Properties of Several Metals

Material	Density		Coefficient of Thermal Expansion ($°F^{-1}$)	Yield Strength		Ultimate Strength		Modulus of Elasticity		Shear Modulus	
	lb/in.^{3}	kN/m^3		ksi	MPa	ksi	MPa	ksi	GPa	ksi	GPa
Aluminum alloy	0.096	26	12.5×10^{-6}	30	205	45	310	10,000	70	4000	28
Brass, rolled	0.300	82	11.0×10^{-6}	25	172	60	415	14,000	96	5000	35
Bronze	0.300	82	10.5×10^{-6}	20	140	35	240	12,000	83	—	—
Cast iron, gray	0.260	70	6.0×10^{-6}	6	40	20	140	15,000	105	6000	42
Copper, drawn	0.322	88	10.0×10^{-6}	38	260	55	380	18,000	120	6000	42
Steel											
Low-carbon	0.283	77	6.5×10^{-6}	40	280	60	415	30,000	200	12,000	83
High-strength	0.283	77	6.5×10^{-6}	100	700	160	1120	30,000	200	12,000	83

Properties of Several Types of Wood

Material	Density		Allowable Strengths										Modulus of Elasticity	
			Bending		Shear		Compression Parallel		Compression Perpendicular					
	lb/ft^3	kN/m^3	psi	MPa	psi	MPa	psi	MPa	psi	MPa			ksi	GPa
Douglas fir	30	4.7	1250	8.7	90	0.62	1000	7.0	400	2.8			1600	11
White pine	25	4.0	1100	7.6	65	0.45	800	5.5	250	1.7			1200	8.3
Redwood	23	3.7	1500	10.3	100	0.70	1200	8.3	400	2.8			1400	9.7

Properties of Some Other Materials

Material	Density		Coefficient of Thermal Expansion ($°F^{-1}$)	Yield Strength		Ultimate Strength		Modulus of Elasticity	
	lb/ft^3	kN/m^3		ksi	MPa	ksi	MPa	ksi	GPa
Plastics									
Polystyrene	55	8.9	38×10^{-6}	7	48	—	—	450	3.1
Acrylic	75	11.8	40×10^{-6}	8	55	—	—	420	2.9
Phenolic	95	15.0	20×10^{-6}	7.5	52	—	—	1000	6.9
Glass									
Soda lime	154	24.3	5.0×10^{-6}	5–10	35–70	—	—	10,000	70
Fused silicate	137	21.6	2.0×10^{-6}	5–10	35–70	—	—	12,000	83
Concrete	150	23.5	5.5×10^{-6}	—	—	4	28	3500	25

APPENDIX

PROPERTIES OF SELECTED STRUCTURAL SHAPES

Properties of W Shapes

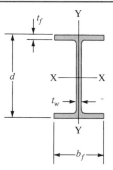

Designation	Area A, in.^{2}	Depth d, in.	Web Thickness t_w, in.	Flange Width b_f, in.	Flange Thickness t_f, in.	Axis X-X I, in.^{4}	Axis X-X S, in.^{3}	Axis X-X r, in.	Axis Y-Y I, in.^{4}	Axis Y-Y S, in.^{3}	Axis Y-Y r, in.	Torsional Constant J, in.^{4}
W 24x146	42.9	27.38	0.605	13.965	0.975	4580	371	10.3	391	60.5	3.01	13.4
x94	27.7	24.31	0.515	9.065	0.875	2700	222	9.87	109	24.0	1.98	5.26
W 21x122	35.9	21.68	0.600	12.390	0.960	2960	273	9.09	305	49.2	2.92	8.98
x93	27.3	21.62	0.580	8.420	0.930	2070	192	8.70	92.9	22.1	1.84	6.03
x50	14.7	20.83	0.380	6.530	0.535	984	94.5	8.18	24.9	7.64	1.30	1.14
W 18x106	31.1	18.73	0.590	11.200	0.940	1910	204	7.84	220	39.4	2.66	7.48
x65	19.1	18.35	0.450	7.590	0.750	1070	117	7.49	54.8	14.4	1.69	2.73
x40	11.8	17.90	0.315	6.015	0.525	612	68.4	7.21	19.1	6.35	1.27	0.81

Properties of W Shapes (Continued)

Designation	Area A, in.^{2}	Depth d, in.	Web Thickness t_w, in.	Flange Width b_f, in.	Flange Thickness t_f, in.	Axis X-X I, in.^{4}	Axis X-X S, in.^{3}	Axis X-X r, in.	Axis Y-Y I, in.^{4}	Axis Y-Y S, in.^{3}	Axis Y-Y r, in.	Torsional Constant J, in.^{4}
W 16x100	29.4	16.97	0.585	10.425	0.985	1490	175	7.10	186	35.7	2.51	7.73
x89	26.2	16.75	0.525	10.365	0.875	1300	155	7.05	163	31.4	2.49	5.45
x67	19.7	16.33	0.395	10.235	0.665	954	117	6.96	119	23.2	2.46	2.39
x45	13.3	16.13	0.345	7.035	0.565	586	72.7	6.65	32.8	9.34	1.57	1.11
x26	7.68	15.69	0.250	5.500	0.345	301	38.4	6.26	9.59	3.49	1.12	0.26
W 14x176	51.8	15.22	0.830	15.650	1.310	2140	281	6.43	838	107	4.02	26.5
x145	42.7	14.78	0.680	15.500	1.090	1710	232	6.33	677	87.3	3.98	15.2
x109	32.0	14.32	0.525	14.605	0.860	1240	173	6.22	447	61.2	3.73	7.12
x74	21.8	14.17	0.450	10.070	0.785	796	112	6.04	134	26.6	2.48	3.88
x48	14.1	13.79	0.340	8.030	0.595	485	70.3	5.85	51.4	12.8	1.91	1.46
x34	10.0	13.98	0.285	6.745	0.455	340	48.6	5.83	23.3	6.91	1.53	0.57
x26	7.69	13.91	0.255	5.025	0.420	245	35.3	5.65	8.91	3.54	1.08	0.36
W 12x152	44.7	13.71	0.870	12.480	1.400	1430	209	5.66	454	72.8	3.19	25.8
x96	28.2	12.71	0.550	12.160	0.900	833	131	5.44	270	44.4	3.09	6.86
x58	17.0	12.19	0.360	10.010	0.640	475	78.0	5.28	107	21.4	2.51	2.10
x50	14.7	12.19	0.370	8.080	0.640	394	64.7	5.18	56.3	13.9	1.96	1.78
x35	10.3	12.50	0.300	6.560	0.520	285	45.6	5.25	24.5	7.47	1.54	0.74
x19	5.57	12.16	0.235	4.005	0.350	130	21.3	4.82	3.76	1.88	0.822	0.18
W10x77	22.6	10.60	0.530	10.190	0.870	455	85.9	4.49	154	30.1	2.60	5.11
x60	17.6	10.22	0.420	10.080	0.680	341	66.7	4.39	116	23.0	2.57	2.48
x49	14.4	9.98	0.340	10.000	0.560	272	54.6	4.35	93.4	18.7	2.54	1.39
x45	13.3	10.10	0.350	8.020	0.620	248	49.1	4.32	53.4	13.3	2.01	1.51
x39	11.5	9.92	0.315	7.985	0.530	209	42.1	4.27	45.0	11.3	1.98	0.98
x22	6.49	10.17	0.240	5.750	0.360	118	23.2	4.27	11.4	3.97	1.33	0.24
W 8x58	17.1	8.75	0.510	8.220	0.810	228	52.0	3.65	75.1	18.3	2.10	3.34
x40	11.7	8.25	0.360	8.070	0.560	146	35.5	3.53	49.1	12.2	2.04	1.12
x31	9.13	8.00	0.285	7.995	0.435	110	27.5	3.47	37.1	9.27	2.02	0.54
x28	8.25	8.06	0.285	6.535	0.465	98.0	24.3	3.45	21.7	6.63	1.62	0.54
x18	5.26	8.14	0.230	5.250	0.330	61.9	15.2	3.43	7.97	3.04	1.23	0.17
x15	4.44	8.11	0.245	4.015	0.315	48.0	11.8	3.29	3.41	1.70	0.876	0.14
W 6x25	7.34	6.38	0.320	6.080	0.455	53.4	16.7	2.70	17.1	5.61	1.52	0.46
x20	5.87	6.20	0.260	6.020	0.365	41.4	13.4	2.66	13.3	4.41	1.50	0.24
x15	4.43	5.99	0.230	5.990	0.260	29.1	9.72	2.56	9.32	3.11	1.46	0.10
x9	2.68	5.90	0.170	3.940	0.215	16.4	5.56	2.47	2.19	1.11	0.905	0.04

Properties of S Shapes

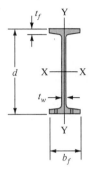

Designation	Area A, in.^{2}	Depth d, in.	Web Thickness t_w, in.	Flange Width b_f, in.	Flange Thickness t_f, in.	Axis X-X I, in.^{4}	Axis X-X S, in.^{3}	Axis X-X r, in.	Axis Y-Y I, in.^{4}	Axis Y-Y S, in.^{3}	Axis Y-Y r, in.	Torsional Constant J, in.^{4}
S 24x100	29.3	24.00	0.745	7.245	0.870	2390	199	9.02	47.7	13.2	1.27	7.58
S 20x96	28.2	20.30	0.800	7.200	0.920	1670	165	7.71	50.2	13.9	1.33	8.39
S 18x70	20.6	18.00	0.711	6.251	0.691	926	103	6.71	24.1	7.72	1.08	4.15
x54.7	16.1	18.00	0.461	6.001	0.691	804	89.4	7.07	20.8	6.94	1.14	2.37
S 15x50	14.7	15.00	0.550	5.640	0.622	486	64.8	5.75	15.7	5.57	1.03	2.12
x42.9	12.6	15.00	0.411	5.501	0.622	447	59.6	5.95	14.4	5.23	1.07	1.54
S 12x50	14.7	12.00	0.687	5.477	0.659	305	50.8	4.55	15.7	5.74	1.03	2.82
S 12x35	10.3	12.00	0.428	5.078	0.544	229	38.2	4.72	9.87	3.89	0.980	1.08
x31.8	9.35	12.00	0.350	5.000	0.544	218	36.4	4.83	9.36	3.74	1.00	0.90
S 10x35	10.3	10.00	0.594	4.944	0.491	147	29.4	3.78	8.36	3.38	0.901	1.29
x25.4	7.46	10.00	0.311	4.661	0.491	124	24.7	4.07	6.79	2.91	0.954	0.60
S 8x23	6.77	8.00	0.441	4.171	0.426	64.9	16.2	3.10	4.31	2.07	0.798	0.55
x18.4	5.41	8.00	0.271	4.001	0.426	57.6	14.4	3.26	3.73	1.86	0.831	0.34
S 6x17.25	5.07	6.00	0.465	3.565	0.359	26.3	8.77	2.28	2.31	1.30	0.675	0.37
x12.5	3.67	6.00	0.232	3.332	0.359	22.1	7.37	2.45	1.82	1.09	0.705	0.17
S 4x9.5	2.79	4.00	0.326	2.796	0.293	6.79	3.39	1.56	0.903	0.646	0.569	0.12

Properties of Angles—Equal Legs and Unequal Legs

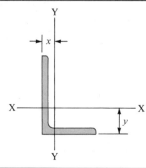

Size and Thickness, in.	Weight per Foot lb	Area, in	Axis X-X				Axis Y-Y			
			I, in.^{4}	S, in.^{3}	r, in.	y, in.	I, in.^{4}	S, in.^{3}	r, in.	x, in.
L 8x8x1	51.0	15.0	89.0	15.8	2.44	2.37	89.0	15.8	2.44	2.37
$\frac{1}{2}$	26.4	7.75	48.6	8.36	2.50	2.19	48.6	8.36	2.50	2.19
L 8x6x1	44.2	13.0	80.8	15.1	2.49	2.65	38.8	8.92	1.73	1.65
$\frac{1}{2}$	23.0	6.75	44.3	8.02	2.56	2.47	21.7	4.79	1.79	1.47
L 8x4x1	37.4	11.0	69.6	14.1	2.52	3.05	11.6	3.94	1.03	1.05
$\frac{1}{2}$	19.6	5.75	38.5	7.49	2.59	2.86	6.74	2.15	1.08	0.859
L 6x6x1	37.4	11.0	35.5	8.57	1.80	1.86	35.5	8.57	1.80	1.86
$\frac{3}{4}$	28.7	8.44	28.2	6.66	1.83	1.78	28.2	6.66	1.83	1.78
$\frac{1}{2}$	19.6	5.75	19.9	4.61	1.86	1.68	19.9	4.61	1.86	1.68
L 6x4x$\frac{3}{4}$	23.6	6.94	24.5	6.25	1.88	2.08	8.68	2.97	1.12	1.08
$\frac{1}{2}$	16.2	4.75	17.4	4.33	1.91	1.99	6.27	2.08	1.15	0.987
L 5x5x$\frac{3}{4}$	23.6	6.94	15.7	4.53	1.51	1.52	15.7	4.53	1.51	1.52
$\frac{1}{2}$	16.2	4.75	11.3	3.16	1.54	1.43	11.3	3.16	1.54	1.43
L 5x3x$\frac{1}{2}$	12.8	3.75	9.45	2.91	1.59	1.75	2.58	1.15	0.829	0.750
$\frac{1}{4}$	6.6	1.94	5.11	1.53	1.62	1.66	1.44	0.614	0.861	0.657
L 4x4x$\frac{3}{4}$	18.5	5.44	7.67	2.81	1.19	1.27	7.67	2.81	1.19	1.27
$\frac{1}{2}$	12.8	3.75	5.56	1.97	1.22	1.18	5.56	1.97	1.22	1.18
$\frac{1}{4}$	6.6	1.94	3.04	1.05	1.25	1.09	3.04	1.05	1.25	1.09
L 4x3x$\frac{1}{2}$	11.1	3.25	5.05	1.89	1.25	1.33	2.42	1.12	0.864	0.827
$\frac{1}{4}$	5.8	1.69	2.77	1.00	1.28	1.24	1.36	0.599	0.896	0.736
L 3x3x$\frac{1}{2}$	9.4	2.75	2.22	1.07	0.898	0.932	2.22	1.07	0.898	0.932
$\frac{1}{4}$	4.9	1.44	1.24	0.577	0.930	0.842	1.24	0.577	0.930	0.842
L 3x2x$\frac{3}{8}$	5.9	1.73	1.53	0.781	0.940	1.04	0.543	0.371	0.559	0.539
$\frac{1}{4}$	4.1	1.19	1.09	0.542	0.957	0.993	0.392	0.260	0.574	0.493
L 2x2x$\frac{3}{8}$	4.7	1.36	0.479	0.351	0.594	0.636	0.479	0.351	0.594	0.636
$\frac{1}{4}$	3.19	0.938	0.348	0.247	0.609	0.592	0.348	0.247	0.609	0.592

Properties of American Standard Channels

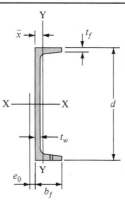

Designation	Area A, in.^{2}	Depth d, in.	Web Thickness t_w, in.	Flange Width b_f, in.	Flange Average Thickness t_f, in.	$\bar{x}$, in.	Shear Center Location e_0, in.	Axis X-X I, in.^{4}	Axis X-X S, in.^{3}	Axis X-X r, in.	Axis Y-Y I, in.^{4}	Axis Y-Y S, in.^{3}	Axis Y-Y r, in.
C 15x50	14.7	15.00	0.716	3.716	0.650	0.798	0.583	404	53.8	5.24	11.0	3.78	0.867
x33.9	9.96	15.00	0.400	3.400	0.650	0.787	0.896	315	42.0	5.62	8.13	3.11	0.904
C 12x30	8.82	12.00	0.510	3.170	0.501	0.674	0.618	162	27.0	4.29	5.14	2.06	0.763
x20.7	6.09	12.00	0.282	2.942	0.501	0.698	0.870	129	21.5	4.61	3.88	1.73	0.799
C 10x30	8.82	10.00	0.673	3.033	0.436	0.649	0.369	103	20.7	3.42	3.94	1.65	0.669
x25	7.35	10.00	0.526	2.886	0.436	0.617	0.494	91.2	18.2	3.52	3.36	1.48	0.676
x15.3	4.49	10.00	0.240	2.600	0.436	0.634	0.796	67.4	13.5	3.87	2.28	1.16	0.713
C 9x20	5.88	9.00	0.448	2.648	0.413	0.583	0.515	60.9	13.5	3.22	2.42	1.17	0.642
x13.4	3.94	9.00	0.233	2.433	0.413	0.601	0.743	47.9	10.6	3.48	1.76	0.962	0.669
C 8x18.75	5.51	8.00	0.487	2.527	0.390	0.565	0.431	44.0	11.0	2.82	1.98	1.01	0.599
x11.5	3.38	8.00	0.220	2.260	0.390	0.571	0.697	32.6	8.14	3.11	1.32	0.781	0.625
C 7x14.75	4.33	7.00	0.419	2.299	0.366	0.532	0.441	27.2	7.78	2.51	1.38	0.779	0.564
x 9.8	2.87	7.00	0.210	2.090	0.366	0.540	0.647	21.3	6.08	2.72	0.968	0.625	0.581
C 6x13	3.83	6.00	0.437	2.157	0.343	0.514	0.380	17.4	5.80	2.13	1.05	0.642	0.525
x 8.2	2.40	6.00	0.200	1.920	0.343	0.511	0.599	13.1	4.38	2.34	0.693	0.492	0.537
C 4x 7.25	2.13	4.00	0.321	1.721	0.296	0.459	0.386	4.59	2.29	1.47	0.433	0.343	0.450
x 5.4	1.59	4.00	0.184	1.584	0.296	0.457	0.502	3.85	1.93	1.56	0.319	0.283	0.449

Properties of Pipe

	Dimensions			Weight per Foot lb Plain Ends	Properties			
Nominal Diameter, in.	Outside Diameter, in.	Inside Diameter, in.	Wall Thickness, in.		$A,$ in.^{2}	$I,$ in.^{4}	$S,$ in.^{3}	$r,$ in.
Standard Weight								
1	1.315	1.049	.133	1.68	.494	.087	.133	.421
$1\frac{1}{2}$	1.900	1.610	.145	2.72	.799	.310	.326	6.23
2	2.375	2.067	.154	3.65	1.07	.666	.561	.787
3	3.500	3.068	.216	7.58	2.23	3.02	1.72	1.16
4	4.500	4.026	.237	10.79	3.17	7.23	3.21	1.51
5	5.563	5.047	.258	14.62	4.30	15.2	5.45	1.88
8	8.625	7.981	.322	28.55	8.40	72.5	16.8	2.94
Extra Strong								
1	1.315	9.57	.179	2.17	.639	.106	.161	.407
$1\frac{1}{2}$	1.900	1.500	.200	3.63	1.07	.391	.412	.605
2	2.375	1.939	.218	5.02	1.48	.868	.731	.766
3	3.500	2.900	.300	10.25	3.02	3.89	2.23	1.14
4	4.500	3.826	.337	14.98	4.41	9.61	4.27	1.48
5	5.563	4.813	.375	20.78	6.11	20.7	7.43	1.84
8	8.625	7.625	.500	43.39	12.8	106	24.5	2.88
Double-Extra Strong								
2	2.375	1.503	.436	9.03	2.66	1.31	1.10	.703
3	3.500	2.300	.600	18.58	5.47	5.99	3.42	1.05
4	4.500	3.152	.674	27.54	8.10	15.3	6.79	1.37
5	5.563	4.063	.750	38.55	11.3	33.6	12.1	1.72
8	8.625	6.875	.875	72.42	21.3	162	37.6	2.76

APPENDIX

ELEMENTARY MATRIX ALGEBRA

Matrix algebra, much like vector analysis, is a mathematical tool. A matrix is used to represent an array of quantities. The array is written as a single entity, a single letter. In this brief introduction the matrix is used to represent a system of linear algebraic equations.

Consider a set of two equations in two unknowns.

$$3x + 2y = 6$$
$$2x + 4y = 3$$
(D.1)

Define three matrix quantities as the three arrays

$$\begin{bmatrix} 3 & 2 \\ 2 & 4 \end{bmatrix} \begin{Bmatrix} x \\ y \end{Bmatrix} = \begin{Bmatrix} 6 \\ 3 \end{Bmatrix}$$
(D.2)

The first is a square matrix, the other two are referred to as column matrices. These particular matrices are used to represent a set of simultaneous equations. The matrices may be defined as follows:

$$[A] = \begin{bmatrix} 3 & 2 \\ 2 & 4 \end{bmatrix} \qquad \{x\} = \begin{Bmatrix} x \\ y \end{Bmatrix} \qquad \{b\} = \begin{Bmatrix} 6 \\ 3 \end{Bmatrix}$$

Equation (D.2) can be written in shorthand notation as

$$[A]\{x\} = \{b\} \tag{D.3}$$

Matrix multiplication is the process of multiplying matrix $\{x\}$ by matrix $[A]$. $\{x\}$ is said to be *premultiplied* by $[A]$ or $[A]$ is *postmultiplied* by $\{x\}$.

The following example will illustrate the process of matrix multiplication. Define the following matrices.

$$[A] = \begin{bmatrix} A_{11} & A_{12} \\ A_{21} & A_{22} \end{bmatrix} \quad [B] = \begin{bmatrix} B_{11} & B_{12} \\ B_{21} & B_{22} \end{bmatrix} \tag{D.4}$$

$$[A][B] = \begin{bmatrix} A_{11} & A_{12} \\ A_{21} & A_{21} \end{bmatrix}\begin{bmatrix} B_{11} & B_{12} \\ B_{21} & B_{22} \end{bmatrix}$$

$$= \begin{bmatrix} A_{11}B_{11} + A_{12}B_{21} & A_{11}A_{12} + A_{12}B_{22} \\ A_{21}B_{11} + A_{22}B_{21} & A_{21}B_{12} + A_{22}B_{22} \end{bmatrix} \tag{D.5}$$

It follows the matrix multiplication is not commutative; that is,

$$[A][B] \neq [B][A] \tag{D.6}$$

The associative law for matrix multiplication holds.

$$[A]([B][C]) = ([A][B])[C] \tag{D.7}$$

Matrix addition is the process of adding two matrices term by term.

$$\begin{bmatrix} A_{11} & A_{12} \\ A_{21} & A_{22} \end{bmatrix} + \begin{bmatrix} B_{11}B_{12} \\ B_{21}B_{22} \end{bmatrix} = \begin{bmatrix} A_{11} + B_{11} & A_{12} + B_{12} \\ A_{21} + B_{21} & A_{22} + B_{22} \end{bmatrix} \tag{D.8}$$

Matrix manipulations are distributive.

$$[A]([B] + [C]) = [A][B] + [A][C] \tag{D.9}$$

The matrices used in this text are either square or column. A square matrix with all off-diagonal terms zero is called a *diagonal* matrix.

$$\begin{bmatrix} A_{11} & 0 \\ 0 & A_{22} \end{bmatrix}$$

A *symmetric* matrix has corresponding off-diagonal terms equal, or $A_{ij} = A_{ji}$.

$$\begin{bmatrix} A_{11} & A_{12} & A_{13} \\ A_{12} & A_{22} & A_{23} \\ A_{13} & A_{23} & A_{33} \end{bmatrix} \quad \begin{matrix} A_{12} = A_{21} \\ A_{13} = A_{31} \\ A_{23} = A_{32} \end{matrix}$$

A *skew symmetric* matrix has the corresponding off-diagonal terms equal but opposite in sign, $A_{ij} = -A_{ji}$.

APPENDIX

SUPERPOSITION TABLES

Case E1

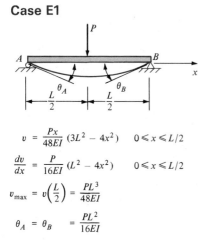

$$v = \frac{Px}{48EI}(3L^2 - 4x^2) \qquad 0 \leq x \leq L/2$$

$$\frac{dv}{dx} = \frac{P}{16EI}(L^2 - 4x^2) \qquad 0 \leq x \leq L/2$$

$$v_{max} = v\left(\frac{L}{2}\right) = \frac{PL^3}{48EI}$$

$$\theta_A = \theta_B = \frac{PL^2}{16EI}$$

Case E2

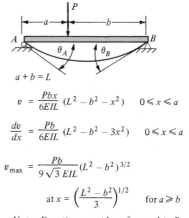

$$a + b = L$$

$$v = \frac{Pbx}{6EIL}(L^2 - b^2 - x^2) \qquad 0 \leq x \leq a$$

$$\frac{dv}{dx} = \frac{Pb}{6EIL}(L^2 - b^2 - 3x^2) \qquad 0 \leq x \leq a$$

$$v_{max} = \frac{Pb}{9\sqrt{3}\,EIL}(L^2 - b^2)^{3/2}$$

$$at\ x = \left(\frac{L^2 - b^2}{3}\right)^{1/2} \qquad for\ a \geq b$$

Note: Equations must be referenced to B as the origin, for $a \leq x \leq L$.

Case E3

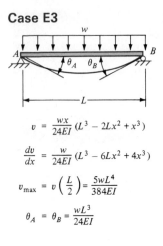

$$v = \frac{wx}{24EI} (L^3 - 2Lx^2 + x^3)$$

$$\frac{dv}{dx} = \frac{w}{24EI} (L^3 - 6Lx^2 + 4x^3)$$

$$v_{max} = v\left(\frac{L}{2}\right) = \frac{5wL^4}{384EI}$$

$$\theta_A = \theta_B = \frac{wL^3}{24EI}$$

Case E5

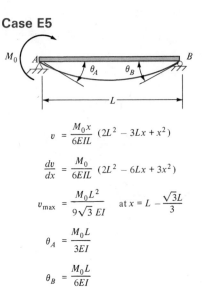

$$v = \frac{M_0 x}{6EIL} (2L^2 - 3Lx + x^2)$$

$$\frac{dv}{dx} = \frac{M_0}{6EIL} (2L^2 - 6Lx + 3x^2)$$

$$v_{max} = \frac{M_0 L^2}{9\sqrt{3}\, EI} \quad \text{at } x = L - \frac{\sqrt{3}L}{3}$$

$$\theta_A = \frac{M_0 L}{3EI}$$

$$\theta_B = \frac{M_0 L}{6EI}$$

Case E4

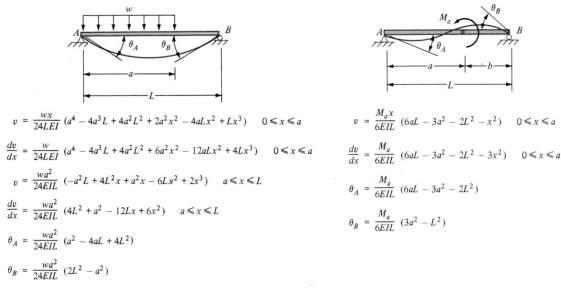

$$v = \frac{wx}{24LEI} (a^4 - 4a^3L + 4a^2L^2 + 2a^2x^2 - 4aLx^2 + Lx^3) \quad 0 \leqslant x \leqslant a$$

$$\frac{dv}{dx} = \frac{w}{24LEI} (a^4 - 4a^3L + 4a^2L^2 + 6a^2x^2 - 12aLx^2 + 4Lx^3) \quad 0 \leqslant x \leqslant a$$

$$v = \frac{wa^2}{24EIL} (-a^2L + 4L^2x + a^2x - 6Lx^2 + 2x^3) \quad a \leqslant x \leqslant L$$

$$\frac{dv}{dx} = \frac{wa^2}{24EIL} (4L^2 + a^2 - 12Lx + 6x^2) \quad a \leqslant x \leqslant L$$

$$\theta_A = \frac{wa^2}{24EIL} (a^2 - 4aL + 4L^2)$$

$$\theta_B = \frac{wa^2}{24EIL} (2L^2 - a^2)$$

Case E6

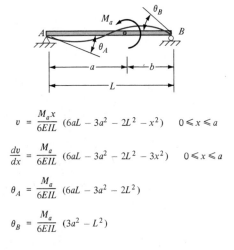

$$v = \frac{M_a x}{6EIL} (6aL - 3a^2 - 2L^2 - x^2) \quad 0 \leqslant x \leqslant a$$

$$\frac{dv}{dx} = \frac{M_a}{6EIL} (6aL - 3a^2 - 2L^2 - 3x^2) \quad 0 \leqslant x \leqslant a$$

$$\theta_A = \frac{M_a}{6EIL} (6aL - 3a^2 - 2L^2)$$

$$\theta_B = \frac{M_a}{6EIL} (3a^2 - L^2)$$

Case E7

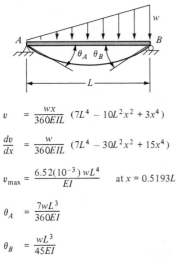

$$v = \frac{wx}{360EIL} (7L^4 - 10L^2x^2 + 3x^4)$$

$$\frac{dv}{dx} = \frac{w}{360EIL} (7L^4 - 30L^2x^2 + 15x^4)$$

$$v_{max} = \frac{6.52(10^{-3}) wL^4}{EI} \quad \text{at } x = 0.5193L$$

$$\theta_A = \frac{7wL^3}{360EI}$$

$$\theta_B = \frac{wL^3}{45EI}$$

Case E9

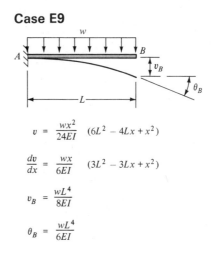

$$v = \frac{wx^2}{24EI} (6L^2 - 4Lx + x^2)$$

$$\frac{dv}{dx} = \frac{wx}{6EI} (3L^2 - 3Lx + x^2)$$

$$v_B = \frac{wL^4}{8EI}$$

$$\theta_B = \frac{wL^4}{6EI}$$

Case E8

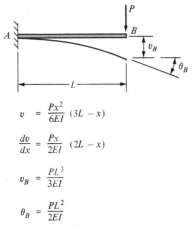

$$v = \frac{Px^2}{6EI} (3L - x)$$

$$\frac{dv}{dx} = \frac{Px}{2EI} (2L - x)$$

$$v_B = \frac{PL^3}{3EI}$$

$$\theta_B = \frac{PL^2}{2EI}$$

Case E10

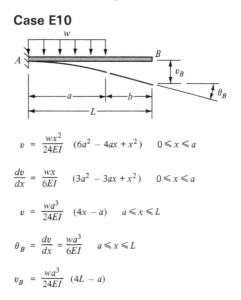

$$v = \frac{wx^2}{24EI} (6a^2 - 4ax + x^2) \quad 0 \leqslant x \leqslant a$$

$$\frac{dv}{dx} = \frac{wx}{6EI} (3a^2 - 3ax + x^2) \quad 0 \leqslant x \leqslant a$$

$$v = \frac{wa^3}{24EI} (4x - a) \quad a \leqslant x \leqslant L$$

$$\theta_B = \frac{dv}{dx} = \frac{wa^3}{6EI} \quad a \leqslant x \leqslant L$$

$$v_B = \frac{wa^3}{24EI} (4L - a)$$

Case E11

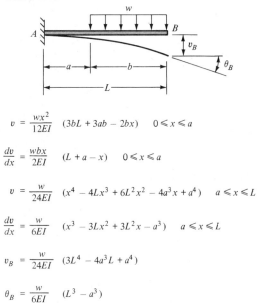

$$v = \frac{wx^2}{12EI} \; (3bL + 3ab - 2bx) \quad 0 \leqslant x \leqslant a$$

$$\frac{dv}{dx} = \frac{wbx}{2EI} \quad (L + a - x) \quad 0 \leqslant x \leqslant a$$

$$v = \frac{w}{24EI} \; (x^4 - 4Lx^3 + 6L^2x^2 - 4a^3x + a^4) \quad a \leqslant x \leqslant L$$

$$\frac{dv}{dx} = \frac{w}{6EI} \; (x^3 - 3Lx^2 + 3L^2x - a^3) \quad a \leqslant x \leqslant L$$

$$v_B = \frac{w}{24EI} \; (3L^4 - 4a^3L + a^4)$$

$$\theta_B = \frac{w}{6EI} \quad (L^3 - a^3)$$

Case E12

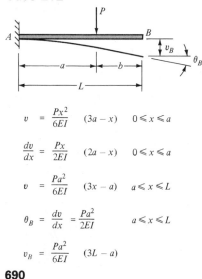

$$v = \frac{Px^2}{6EI} \quad (3a - x) \quad 0 \leqslant x \leqslant a$$

$$\frac{dv}{dx} = \frac{Px}{2EI} \quad (2a - x) \quad 0 \leqslant x \leqslant a$$

$$v = \frac{Pa^2}{6EI} \quad (3x - a) \quad a \leqslant x \leqslant L$$

$$\theta_B = \frac{dv}{dx} = \frac{Pa^2}{2EI} \quad a \leqslant x \leqslant L$$

$$v_B = \frac{Pa^2}{6EI} \quad (3L - a)$$

690

Case E13

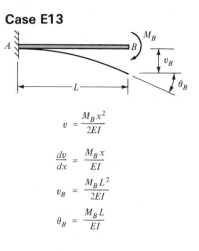

$$v = \frac{M_Bx^2}{2EI}$$

$$\frac{dv}{dx} = \frac{M_Bx}{EI}$$

$$v_B = \frac{M_BL^2}{2EI}$$

$$\theta_B = \frac{M_BL}{EI}$$

Case E14

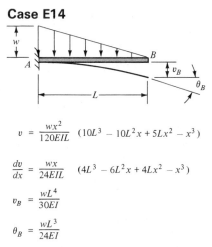

$$v = \frac{wx^2}{120EIL} \; (10L^3 - 10L^2x + 5Lx^2 - x^3)$$

$$\frac{dv}{dx} = \frac{wx}{24EIL} \quad (4L^3 - 6L^2x + 4Lx^2 - x^3)$$

$$v_B = \frac{wL^4}{30EI}$$

$$\theta_B = \frac{wL^3}{24EI}$$

Case E15

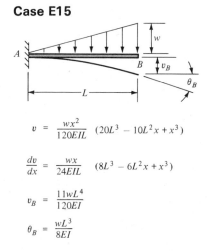

$$v = \frac{wx^2}{120EIL} \; (20L^3 - 10L^2x + x^3)$$

$$\frac{dv}{dx} = \frac{wx}{24EIL} \quad (8L^3 - 6L^2x + x^3)$$

$$v_B = \frac{11wL^4}{120EI}$$

$$\theta_B = \frac{wL^3}{8EI}$$

FIXED-FIXED BEAM FORMULAS

Case F1

$$M_A = M_B = \frac{PL}{8}$$

$$R_A = R_B = \frac{P}{2}$$

Case F2

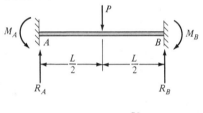

$$M_A = \frac{Pab^2}{L^2}$$

$$M_B = \frac{Pba^2}{L^2}$$

$$R_A = \frac{Pb^2}{L^3}\ (3a + b)$$

$$R_B = \frac{Pa^2}{L^3}\ (a + 3b)$$

Case F3

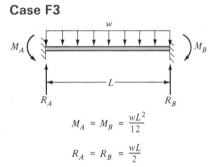

$$M_A = M_B = \frac{wL^2}{12}$$

$$R_A = R_B = \frac{wL}{2}$$

Case F4

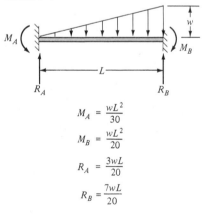

$$M_A = \frac{wL^2}{30}$$

$$M_B = \frac{wL^2}{20}$$

$$R_A = \frac{3wL}{20}$$

$$R_B = \frac{7wL}{20}$$

Case F5

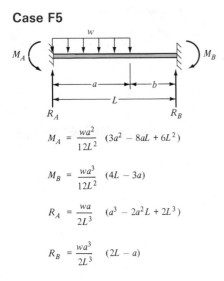

$$M_A = \frac{wa^2}{12L^2} \ (3a^2 - 8aL + 6L^2)$$

$$M_B = \frac{wa^3}{12L^2} \ (4L - 3a)$$

$$R_A = \frac{wa}{2L^3} \ (a^3 - 2a^2L + 2L^3)$$

$$R_B = \frac{wa^3}{2L^3} \ (2L - a)$$

Case F6

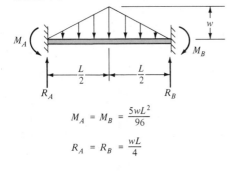

$$M_A = M_B = \frac{5wL^2}{96}$$

$$R_A = R_B = \frac{wL}{4}$$

Case F7

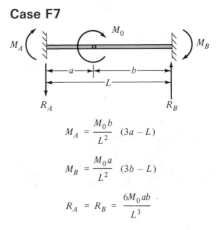

$$M_A = \frac{M_0 b}{L^2} \ (3a - L)$$

$$M_B = \frac{M_0 a}{L^2} \ (3b - L)$$

$$R_A = R_B = \frac{6M_0 ab}{L^3}$$

Case F8

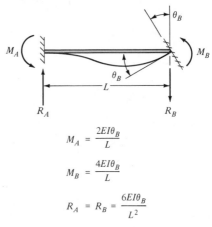

$$M_A = \frac{2EI\theta_B}{L}$$

$$M_B = \frac{4EI\theta_B}{L}$$

$$R_A = R_B = \frac{6EI\theta_B}{L^2}$$

Case F9

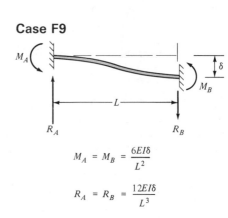

$$M_A = M_B = \frac{6EI\delta}{L^2}$$

$$R_A = R_B = \frac{12EI\delta}{L^3}$$

APPENDIX

COMPUTER APPLICATIONS

G.1   INTRODUCTION

The four computer programs in this appendix are intended to introduce the reader to the use of the computer for solving a problem in mechanics of materials. The programs were written with the idea that any student, no matter what level of expertise he or she might have with programming, could be challenged. The programs are written in BASIC and are intended to be implemented on any personal computer that has the BASIC capability.

G.2   PROGRAMS

Vector Product

The vector product program will perform the computation for a vector product. The input is defined at the beginning of the program and does not need any discussion. This program is quite elementary and is intended to be implemented by the novice programmer.

Scalar Product

The scalar product program will perform the computation for the scalar product of a unit vector and a general vector. Again, the program is elementary and the discussion and figure at the beginning of the listing is sufficient to understand the input of data and execution of the program.

Suggested Problem Assignment: Combine Programs 1 and 2 to perform the computation for a triple scalar product. The vector operation is $A \cdot (B \times C)$.

Stress Transformation

The stress transformation program uses the matrix multiplication that is illustrated in Chapter 9. The input for the program is defined in the remarks at the beginning of the program. This program illustrates the application of a matrix multiplication. The theory discussed in Chapter 9 is limited to a two-dimensional stress transformation. The computer program, however, is a three-dimensional transformation with an application that is two-dimensional. The transformation matrix about the z axis for the figure shown in the program listing is

$$\begin{bmatrix} \cos \theta & \sin \theta & 0 \\ -\sin \theta & \cos \theta & 0 \\ 0 & 0 & 1 \end{bmatrix}$$

and is the three-dimensional counterpart of the transformation matrix that is derived in Chapter 8. A transformation about the y axis is

$$\begin{bmatrix} \cos \theta & \sin \theta & 0 \\ 0 & 1 & 0 \\ -\sin \theta & 0 & \cos \theta \end{bmatrix}$$

and about the x axis is

$$\begin{bmatrix} 1 & 0 & 0 \\ 0 & \cos \theta & \sin \theta \\ 0 & -\sin \theta & \cos \theta \end{bmatrix}$$

The issue of three-dimensional stress transformation was not discussed in Chapter 9, but the extension of the matrix theory to three dimensions is straightforward.

Suggested Problem Assignment

1. Modify the stress transformation program so that the matrix multiplications are in a subroutine.

2. Modify the stress transformation program to allow for additional values of θ to be input.

3. Write a program to compute the principal stresses using the principal stress equations of Chapter 9.

4. Modify the stress transformation program to compute the transformation about any given axis. For instance, a rotation of 30° about the z axis superimposed on a rotation of 45° about the y axis.

Truss Analysis

The truss analysis program is considerably more involved than the previous programs. However, a brief study of the listing will show that the majority of the program is dedicated to inputting the data. The program solves the equations that are obtained by using the method of joints to analyze a statically determinate truss. The program is not limited to truss analysis, but will solve any set of simultaneous equations.

This program will solve for the forces in each truss member. It then requests input data that will be used to compute the displacement of a specified point. After that computation it will request information for computing the displacement of still another joint. The analyst can exit from the program in between any of these separate computation routines. If program statements 1230 through 2010 are omitted, a general program to solve a set of simultaneous equations results.

The program illustrates the use of subroutines. The matrix multiplication subroutine is not general since it specifically multiplies a column matrix by a square matrix. The equations are solved using an elementary method of matrix inversion. The choice of matrix inversion gives an efficient approach for incorporating the unit load method for computing a joint displacement. The inversion routine requires that all diagonal terms are nonzero. A more versatile method could be used, but this routine will be easier for the beginning programmer to follow. Admittedly, the input routine is more tedious. Hence, the program allows the analyst several opportunities to modify the input data without starting over. The input routine for the matrices will continue to request input for each matrix until the last element has been input. These ideas will be illustrated with an example.

The truss of Fig. G.1 has four joints and two equations of equilibrium can be written at each joint. The eight unknowns are the five member forces and three reactions. The equations of equilibrium are written assuming each member is in tension and using the x-y coordinates shown in the figure. The number to the right of each equation represents the order that they appear in the matrix.

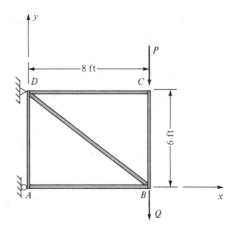

FIGURE G.1

Joint A

$$F_{AD} = 0 \tag{4}$$

$$F_{AB} + H_A = 0 \tag{8}$$

Joint B

$$F_{BC} + (3/5) F_{BD} - Q = 0 \tag{5}$$

$$-F_{AB} - (4/5) F_{BD} = 0 \tag{1}$$

Joint C

$$-F_{BC} - P = 0 \tag{2}$$

$$-F_{CD} = 0 \tag{3}$$

Joint D

$$V_D - F_{AD} - (3/5) F_{BD} = 0 \tag{6}$$

$$H_D + F_{CD} + (4/5) F_{BD} = 0 \tag{7}$$

The eight equations are written in matrix format in the order of the numbers that appear on the right of the equations. This ordering assures that there are no zeros on the diagonal of the matrix.

$$
\begin{bmatrix}
-1 & 0 & 0 & 0 & -.8 & 0 & 0 & 0 \\
0 & -1 & 0 & 0 & 0 & 0 & 0 & 0 \\
0 & 0 & -1 & 0 & 0 & 0 & 0 & 0 \\
0 & 0 & 0 & 1 & 0 & 0 & 0 & 0 \\
0 & 1 & 0 & 0 & .6 & 0 & 0 & 0 \\
0 & 0 & 0 & -1 & -.6 & 1 & 0 & 0 \\
0 & 0 & 1 & 0 & .8 & 0 & 1 & 0 \\
1 & 0 & 0 & 0 & 0 & 0 & 0 & 1
\end{bmatrix}
\begin{bmatrix}
F_{AB} \\
F_{BC} \\
F_{CD} \\
F_{AD} \\
F_{BD} \\
V_D \\
H_D \\
H_A
\end{bmatrix}
=
\begin{bmatrix}
0 \\
P \\
0 \\
0 \\
Q \\
0 \\
0 \\
0
\end{bmatrix}
$$

The computer input corresponding to the matrix equation would be as follows:

8			number of equations
1,	1,	−1	I, J, E(I, J) matrix input
1,	5,	−.8	
2,	2,	−1	
3,	3,	−1	
4,	4,	1	
5,	2,	1	
5,	5,	.6	
6,	4,	−1	
6,	5,	−.6	
6,	6,	1	
7,	3,	1	
7,	5,	.8	
7,	7,	1	
8,	1,	1	
8,	8,	1	
2,	P		I, S(I) input data
5,	Q		
8,	0		

For P = 100 and Q = 200 the results will be computed as

1	−400
2	−100
3	0
4	0
5	500
6	300
7	−400
8	400

These results are interpreted according to the order of the column matrix of unknowns and may be checked by hand to determine if the program is working correctly.

The program will request data to compute a joint displacement. Assume that the vertical displacement of joint *B* is required. The input response would be

5,   1

The 5 corresponds to the fifth equation in the matrix, the sum of the forces at joint B in the vertical direction. The program will request length, area, and modulus of elasticity data for each member. The truss equations that correspond to a reaction should be assigned a length of zero. The area and modulus of elasticity for a reaction should be assigned a value of 1 to avoid a division by zero. The deflection is computed using Eq. (10.16) of Chapter 10.

$$1 \cdot \Delta = \sum_{n=1}^{N} \frac{\mathbf{p} \cdot \mathbf{P}L}{AE}$$

Assume that all truss members have an area of 2 in.^{2} and modulus of $10(10^6)$ psi. The input data will be:

Length, in.	Area, in.^{2}	Modulus of Elasticity, psi
96	2	10000000
72	2	10000000
96	2	10000000
72	2	10000000
120	2	10000000
0	1	1
0	1	1
0	1	1

The program will return the result

THE DEFLECTION IS   .00756

Suggested Problem Assignment

1. Solve Example 10.9 using the Truss Analysis program.

2. Modify the program to output the truss parameters, length, area and modulus of elasticity with an option to change their values.

3. Replace the matrix inversion subroutine with a pivotal routine that checks for a zero on the diagonal. A pivotal routine will rearrange the equations and continue the solution if a zero occurs on the diagonal. See the text by Robert W. Hornbeck, *Numerical Methods*, Quantum Publishers, New York, 1975, p. 294, for a pivotal inversion program.

```
10 REM                     VECTOR PRODUCT
20 REM *
30 REM     PROGRAM TO COMPUTE THE CROSS PRODUCT OF TWO VECTORS.
40 REM     BOTH VECTORS ARE INPUT FROM THE TERMINAL IN
50 REM     COMPONENT FORM. THE COMPUTATION IS A ^ B AS SHOWN BELOW.
60 REM *
70 REM              |  i      j      k |
80 REM              | Ax     Ay     Az |
90 REM              | Bx     By     Bz |
100 REM *
110 REM    RESPOND TO THE PROMPTS FOR INPUT.
120 REM *
130 WRITE "INPUT Ax"
140 INPUT X
150 WRITE "INPUT Ay"
160 INPUT Y
170 WRITE "INPUT Az"
180 INPUT Z
190 WRITE "INPUT Bx"
200 INPUT A
210 WRITE "INPUT By"
220 INPUT B
230 WRITE "INPUT Bz"
240 INPUT C
250 U=Y*C-Z*B
260 V=Z*A-X*C
270 W=X*B-Y*A
280 PRINT
290 PRINT
300 PRINT "YOUR INPUT DATA IS"
310 PRINT
320 PRINT "VECTOR A"
330 PRINT "Ax=", X
340 PRINT "Ay=",Y
350 PRINT "Az=",Z
360 PRINT
370 PRINT "VECTOR B"
380 PRINT "Bx=",A
390 PRINT "By=",B
400 PRINT "Bz=",C
410 PRINT
420 PRINT "THE VECTOR PRODUCT IS"
430 PRINT
440 PRINT U,"i"
450 PRINT V,"j"
460 PRINT W,"k"
470 END
```

```
10 REM                    SCALAR PRODUCT
20 REM *
30 REM     PROGRAM TO COMPUTE THE SCALAR PRODUCT OF A FORCE VECTOR
40 REM     AND A UNIT VECTOR. THE DIRECTION OF THE UNIT VECTOR IS
50 REM     DEFINED BY COORDINATES OF TWO POINTS, A AND B. THE
60 REM     FORCE VECTOR MUST BE GIVEN IN COMPONENT FORM.
70 REM     SEE THE FIGURE AND RESPOND TO THE PROMPTS GIVEN BY
80 REM     THE PROGRAM.
90 REM *
```

```
100 REM *
110 REM                    | Y                    * B
120 REM                    |       / F      *
130 REM                    |      /       *
140 REM                    |     /      *
150 REM                    |    /     *
160 REM                    |   /    *
170 REM                    |/_____ X
180 REM              *     A
190 REM          *
200 REM      *
210 REM      Z
220 REM *
230 REM *
240 WRITE "INPUT X COORDINATE OF POINT A"
250 INPUT A
260 WRITE "INPUT Y COORDINAT OF POINT A"
270 INPUT B
280 WRITE "INPUT Z COORDINATE OF POINT A"
290 INPUT C
300 WRITE "INPUT X COORDINATE OF POINT B"
310 INPUT D
320 WRITE "INPUT Y COORDINATE OF POINT B"
330 INPUT E
340 WRITE "INPUT Z COORDINATE OF POINT B"
350 INPUT F
360 WRITE "INPUT X COMPONENT OF FORCE VECTOR"
370 INPUT P
380 WRITE "INPUT Y COMPONENT OF FORCE VECTOR"
390 INPUT Q
    WRITE "INPUT Z COMPONENT OF FORCE VECTOR"
410 INPUT R
420 PRINT  "YOUR INPUT DATA IS"
430 PRINT
440 PRINT  "COORDINATES OF POINT A"
450 PRINT  "(",A,B,C,")"
460 PRINT  "COORDINATES OF POINT B"
470 PRINT  "(",D,E,F,")"
480 PRINT
490 PRINT  "COMPONENTS OF FORCE VECTOR"
500 PRINT
510 PRINT  P,"i"
520 PRINT  Q,"j"
530 PRINT  R,"k"
540 X=D-A
550 Y=E-B
560 Z=F-C
570 S=X*X+Y*Y+Z*Z
580 S=SQR(S)
590 X=X/S
600 Y=Y/S
610 Z=Z/S
620 T=X*P+Y*Q+Z*R
630 WRITE
640 PRINT      "THE UNIT VECTOR IS"
650 WRITE
660 PRINT X, "i"
670 PRINT Y,"j"
680 PRINT Z,"k"
690 WRITE
700 PRINT "THE SCALAR PRODUCT IS"
710 WRITE
720 PRINT  T
730 END
```

```
10 REM                    STRESS TRANSFORMATION
20 REM *
30 REM       GIVEN AN ANGLE THETA IN DEGREES, A TRANS-
40 REM       FORMATION MATRIX IS COMPUTED AND THE TRANSFORMED
50 REM       STRESSES ARE COMPUTED USING A MATRIX MULTIPLICATION
60 REM *
70 REM *
80 REM              T
90 REM    [S]´ = [T]  [S]  [T]
100 REM *
110 REM *
120 REM                           ^ Y
130 REM                           |
140 REM                           |        Z IS NORMAL  TO THE PAGE
150 REM                           |
160 REM            *************
170 REM            *           *
180 REM            *           *        X
190 REM            *           * -------->
200 REM            *           *
210 REM            *           *
220 REM            *************
230 REM *
240 REM    THE TRANSFORMATION IS ABOUT THE Z AXIS WITH THETA
250 REM    MEASURED POSITIVE IN THE FIGURE. THE TRANSFORMATION
260 REM    MATRIX IS THREE-DIMENSIONAL, BUT IN THIS PROGRAM
270 REM    IS LIMITED TO TWO-DIMENSIONS.
280 REM *
290 DIM  T(3,3), S(3,3), R(3,3), A(3,3),B(3,3)
300 WRITE "INPUT THE VALUE OF THETA IN DEGREES"
310 INPUT C
320 P=3.141593
330 B=C*P/180
340 T(1,1)=COS(B)
350 T(1,2)=SIN(B)
360 T(2,1)=-T(1,2)
370 T(2,2)=T(1,1)
380 T(3,3)=1
390 WRITE "INPUT SIGMA X"
400 INPUT S(1,1)
410 WRITE "INPUT SIGMA Y"
420 INPUT S(2,2)
430 WRITE "INPUT SHEAR STRESS"
440 INPUT S(1,2)
450 S(2,1)=S(1,2)
460 REM *
470 REM     TRANSPOSE OF T(I,J) IS B(I,J)
480 FOR I = 1 TO 3
490 FOR J = 1 TO 3
500 B(J,I)=T(I,J)
510 NEXT J
520 NEXT I
530 REM *
540 REM  MATRIX MULTIPLICATION    [A]=[B]*[S]
550 FOR I = 1 TO 3
560 FOR J = 1 TO 3
570 FOR K = 1 TO 3
580 A(I,J)=A(I,J)+B(I,K)*S(K,J)
590 NEXT K
600 NEXT J
610 NEXT I
620 REM *
630 REM  MATRIX MULTIPLICATION   [R]=[A]*[T]
640 FOR I = 1 TO 3
```

```
650 FOR J = 1 TO 3
660 FOR K = 1 TO 3
670 R(I,J)=R(I,J)+A(I,K)*T(K,J)
680 NEXT K
690 NEXT J
700 NEXT I
710 REM *
720 REM     FINAL RESULTS ARE R(I,J)
730 REM     THE RESULTS WILL BE WRITTEN IN MATRIX FORMAT
740 PRINT
750 PRINT
760 PRINT "THE RESULTS IN MATRIX FORMAT ARE;"
770 PRINT
780 PRINT "STRESS IN X-Y COORDINATES"
790 PRINT
800 FOR I = 1 TO 3
810 PRINT S(I,1),S(I,2),S(I,3)
820 NEXT I
830 PRINT
840 PRINT "FOR A ROTATION OF", C, "DEGREES"
850 PRINT "OR",B,"RADIANS"
860 PRINT "THE STRESSES ARE AS FOLLOWS IN MATRIX FORMAT"
870 PRINT
880 FOR I =1 TO 3
890 PRINT R(I,1),R(I,2),R(I,3)
900 NEXT I
910 PRINT
920 PRINT "THE TRANSFORMATION MATRIX IS"
930 PRINT
940 FOR I = 1 TO 3
950 PRINT T(I,1),T(I,2),T(I,3)
960 NEXT I
970 END
```

```
10 REM              TRUSS ANALYSIS
20 REM *
30 REM     THIS PROGRAM USES A MATRIX INVERSION AND
40 REM     MATRIX MULTIPLICATION FOR TRUSS ANALYSIS.
50 REM     THE FORCE IN EACH TRUSS MEMBER AND REACTIONS
60 REM     ARE COMPUTED. THERE IS AN OPTION TO COMPUTE
70 REM     JOINT DEFLECTIONS USING THE UNIT LOAD METHOD.
80 REM *
90 REM     THE JOINT EQUATIONS ARE WRITTEN AS
100 REM *
110 REM          [E] {X} = {S}
120 REM *
130 REM     WHERE [E] IS THE COEFFICIENT MATRIX OF JOINT
140 REM     EQUATIONS. {S} IS A COLUMN MATRIX OF JOINT
150 REM     LOADS. {X} REPRESENTS A COLUMN MATRIX OF UNKNOWN
160 REM     MEMBER FORCES.
170 REM     THE SOLUTION IS OBTAINED AS
180 REM *
190 REM                    -1
200 REM          {X} = [E]    {S}
210 REM *
220 DIM E(20,20), F(20), S(20), X(20), U(20), D(20)
230 DIM P(20),Q(20),R(20,20)
240 WRITE  "INPUT THE TOTAL NUMBER OF EQUATIONS"
```

```
250 INPUT N
260 FOR I = 1 TO N
270 S(I)=0
280 FOR J = 1 TO N
290 E(I,J)=0
300 NEXT J
310 NEXT I
320 REM        INPUT THE NONZERO VALUES OF [E]
330 REM        INPUT ROW NUMBER, COLUMN NUMBER AND [E] VALUE.
340 WRITE   "INPUT [E] COEFFICIENTS IN THE FORMAT OF ROW
350 WRITE   "NUMBER, COLUMN NUMBER AND [E] VALUE"
360 WRITE
370 WRITE "INPUT   I,   J,   E(I,J)"
380 INPUT    I,   J,   E(I,J)
390 IF I = N THEN IF J = N GOTO 410
400 GOTO 370
410 WRITE
420 WRITE "INPUT {S} IN FORMAT OF ROW NUMBER AND {S} VALUE"
430 WRITE
440 WRITE   "INPUT I,   S(I)
450 INPUT I,S(I)
460 IF I = N GOTO 470 ELSE 440
470 WRITE   "WOULD YOU LIKE TO VIEW THE INPUT DATA ?"
480 WRITE   "E(I,J) DATA FIRST"
490 WRITE "1 FOR YES, 0 FOR NO"
500 INPUT L
510 IF L = 0 GOTO 670
520 FOR J = 1 TO N
530 FOR K = 1 TO N
540 IF E(J,K) = 0 GOTO 560
550 WRITE J,K,E(J,K)
560 NEXT K
570 NEXT J
580 WRITE
590 WRITE "WOULD YOU LIKE TO MODIFY THE E(I,J) DATA?"
600 WRITE "1 FOR YES, 0 FOR NO"
610 INPUT L
620 IF L = 0 GOTO 680
630 WRITE "INPUT I, J, E(I,J), OR 0,0,0 FOR NO MORE INPUT"
640 INPUT I,J,E(I,J)
650 IF I = 0 GOTO 670
660 GOTO 640
670 REM *
680 REM *
690 WRITE
700 WRITE "WOULD YOU LIKE TO VIEW THE S(I) DATA?"
710 WRITE "1 FOR YES, 0 FOR NO"
720 INPUT L
730 IF L = 0 GOTO 870
740 FOR J = 1 TO N
750 WRITE J,S(J)
760 NEXT J
770 REM *
780 WRITE "DO YOU WANT TO MODIFY THE S(I) DATA ?"
790 WRITE "1 FOR YES, 0 FOR NO"
800 INPUT L
810 IF L = 0 GOTO 870
820 WRITE " INPUT I, S(I) OR 0,0 FOR NO MORE INPUT DATA"
830 INPUT I,S(I)
840 IF I = 0 GOTO 860
850 GOTO 830
860 GOTO 470
```

```
870 REM *
880 REM   INVERT THE [E] MATRIX
890 FOR I = 1 TO N
900 FOR J = 1 TO N
910 R(I,J)=E(I,J)
920 NEXT J
930 NEXT I
940 GOSUB 2030
950 FOR I = 1 TO N
960 P(I)=0
970 Q(I)=S(I)
980 NEXT I
990 REM   MATRIX MULTIPLICATION
1000 GOSUB 2340
1010 FOR J = 1 TO N
1020 X(J)=P(J)
1030 NEXT J
1040 REM   OUTPUT THE RESULTS FOR FORCE
1050 PRINT "FORCE IN EACH MEMBER"
1060 FOR I = 1 TO N
1070 PRINT I,X(I)
1080 NEXT I
1090 REM *
1100 WRITE
1110 WRITE "WOULD YOU LIKE TO CHECK YOUR ORIGINAL INPUT"
1120 WRITE "DATA AND POSSIBLY TRY AGAIN"
1130 WRITE "1 FOR YES, 0 FOR NO"
1140 INPUT L
1150 IF L = 0 GOTO 1230
1160 FOR J = 1 TO N
1170 S(J)=Q(J)
1180 FOR I = 1 TO N
1190 E(J,I)=R(J,I)
1200 NEXT I
1210 NEXT J
1220 GOTO 470
1230 REM       TRUSS DEFLECTION BY THE UNIT LOAD METHOD
1240 WRITE
1250 WRITE
1260 WRITE "DO YOU WANT TO COMPUTE A JOINT DISPLACEMENT?"
1270 WRITE "1 FOR YES, 0 FOR NO"
1280 INPUT L
1290 IF L = 0 GOTO 2410
1300 FOR J = 1 TO N
1310 P(J)=0
1320 Q(J)=0
1330 NEXT J
1340 WRITE "INPUT AN [S] MATRIX"
1350 WRITE "PUT A 1 IN THE ROW CORRESPONDING TO JOINT"
1360 WRITE     "AND DIRECTION OF THE DISPLACEMENT"
1370 WRITE   "INPUT I, S(I)"
1380 INPUT I, Q(I)
1390 GOSUB 2340
1400 FOR J = 1 TO N
1410 U(J)=P(J)
1420 NEXT J
1430 WRITE
1440 WRITE "YOU MUST INPUT THE LENGTH OF EACH MEMBER"
1450 WRITE "IN THE ORDER THAT WAS USED IN THE ORIGINAL"
1460 WRITE "[S] MATRIX. A REACTION HAS ZERO LENGTH."
1470 WRITE "AREA AND MODULUS OF ELASTICITY MUST BE INPUT AS"
1480 WRITE "1 FOR REACTIONS TO AVOID A DIVISION BY ZERO."
```

```
1490 FOR J = 1 TO N
1500 T(J)=0
1510 G(J)=0
1520 D(J)=0
1530 NEXT J
1540 WRITE "CHECK THAT YOUR UNITS ARE CONSISTENT"
1550 WRITE "INPUT I, LENGHT(I)"
1560 WRITE
1570 INPUT I, T(I)
1580 IF I = N GOTO 1590 ELSE 1550
1590 WRITE
1600 WRITE "INPUT THE AREA OF EACH MEMBER"
1610 WRITE "INPUT I, AREA(I)"
1620 INPUT I, D(I)
1630 IF I = N GOTO 1640 ELSE 1610
1640 REM *
1650 WRITE "INPUT THE MODULUS OF ELASTICITY FOR EACH MEMBER"
1660 WRITE "INPUT I, MOD. OF ELAST.(I)"
1670 INPUT I,G(I)
1680 IF I = N GOTO 1690 ELSE 1660
1690 REM *
1700 REM    COMPUTE DEFLECTION
1710 W=0
1720 FOR J = 1 TO N
1730 W=W+X(J)*U(J)*T(J)/(G(J)*D(J))
1740 NEXT J
1750 PRINT
1760 PRINT "THE DEFLECTION IS ", W
1770 WRITE
1780 WRITE "DO YOU WANT TO COMPUTE ANOTHER JOINT DISPLACEMENT"
1790 WRITE "1 FOR YES, 0 FOR NO"
1800 INPUT L
1810 IF L = 0 GOTO 2400
1820 FOR J = 1 TO N
1830 P(J)=0
1840 Q(J)=0
1850 NEXT J
1860 WRITE "INPUT A [S] MATRIX CORRESPONDING TO THE JOINT"
1870 WRITE "PLACE A 1 AT THE JOINT LOCATION AND IN THE"
1880 WRITE "DIRECTION OF THE DESIRED DISPLACEMENT"
1890 WRITE "INPUT I, S(I)
1900 INPUT I, Q(I)
1910 GOSUB 2340
1920 FOR J = 1 TO N
1930 U(J)=P(J)
1940 NEXT J
1950 W=0
1960 FOR J = 1 TO N
1970 W=W+X(J)*U(J)*T(J)/(G(J)*D(J))
1980 NEXT J
1990 PRINT
2000 PRINT "THE DEFLECTION IS",W
2010 GOTO 1770
2020 REM    SUBROUTINE TO INVERT A MATRIX
2030 FOR K = 1 TO N
2040 FOR I = 1 TO N
2050 F(I)=0
2060 NEXT I
2070 F(K) = 1
2080 FOR J = 1 TO N
2090 IF J = K GOTO 2180
```

```
2100 IF E(K,K) = 0 GOTO 2280
2110 B=E(K,K)
2120 C=E(J,K)
2130 IF C = 0 GOTO 2180
2140 FOR I = 1 TO N
2150 E(J,I)=E(J,I)-E(K,I)*C/B
2160 NEXT I
2170 F(J)=F(J)-F(K)*C/B
2180 NEXT J
2190 FOR J = 1 TO N
2200 E(K,J)=E(K,J)/B
2210 NEXT J
2220 F(K)=F(K)/B
2230 FOR J = 1 TO N
2240 E(J,K)=F(J)
2250 NEXT J
2260 NEXT K
2270 GOTO 2300
2280 PRINT "ZERO ON DIAGONAL, ROW",K
2290 GOTO 1100
2300 RETURN
2310 REM *
2320 REM *
2330 REM     SUBROUTINE FOR MATRIX MULTIPLICATION
2340 FOR I = 1 TO N
2350 FOR K = 1 TO N
2360 P(I)=P(I)+E(I,K)*Q(K)
2370 NEXT K
2380 NEXT I
2390 RETURN
2400 REM *
2410 END
```

APPENDIX H

ANSWERS TO SELECTED PROBLEMS

CHAPTER 2

2.1 100 MPa; 14,500 psi   **2.2** 6366 psi; 43.89 MPa   **2.3** 10.7 mm

2.4 $\sigma_{AB} = 0$; $\sigma_{BC} = -576$ MPa; $\sigma_{CD} = -120$ MPa; $\sigma_{AD} = 0$; $\sigma_{AC} = 679.2$ MPa

2.5 545 psi comp. (left); 545 psi ten. (right)   **2.6** From left, -80 MPa; 120 MPa; 80 MPa

2.7 A, 10.42; B, -281.25; C, -11.25 MPa   **2.8** $w_1 = 15.4$ in.; $w_2 = 7.06$ in.; $w_3 = 7.85$ in.

2.9 A, 703.125 MPa; B, 1171.88 MPa   **2.10** 21.88 MPa

2.11 (a) 0.437, use 7/16 in.; (b) (bearing), $t_1 = 0.229$, $A = 10.3$, $t_2 = 0.457$, $A = 20.6$; (normal), $t_1 = 0.074$, $t_2 = 0.148$; (c) for $t_2 = 0.427$, $h = 0.274$; for $t_2 = 0.148$, $h = 0.846$

2.12 (a) 16,500; (b) 26,500; (c) 33,750; (d) 60,000   **2.13** (a) 4.4 m^2; (b) 0.00323 m^2

2.14 (a) 31.8 mm; (b) 40 mm, 20 mm;   **2.15** (a) 0.0687 m, 0.0687 m, 0.0442 m; (b) 1850 mm^2

2.16 (a) A, 46.74; B, 23.37; C, 83.3 ksi; (b) 10.65 ksi; (c) 183.6 ksi

2.17 (a) 186 ksi; (b) 256 ksi; (c) 3.5 in.^{2} **2.18** (a) 0.5 ksi; (b) 1.128 in.; (c) 0.448 in.

2.19 8.47 ksi; 3.60 ksi   **2.20** $\theta = 30°$; $\sigma = 0.25P/A$; $\tau = 0.433P/A$ **2.21** 0.002; 198.95 GPa

2.22 0.0025 in.; 4.25(10^{-4}) in.; 5.67(10^{-4}) in.   **2.23** 0.248; 70.45 GPa   **2.24** 0.75 mm; 0.0134 mm

2.25 $(1 - 2v)\varepsilon$ **2.26** (a) 6366 psi; (b) 0.636(10^{-3}) in.; (c) 0.382(10^{-3}) in.; (d) 0.382(10^{-4}) in.; (e) $-4.8(10^{-3})$ in.; (f) 1.194(10^{-3}) in.^{2} **2.27** 9.0(10^{-4})   **2.28** $\gamma L^2/2E$ **2.29** 681.35(10^{-6}) in.

2.30 -3.56 ksi; 14.22 ksi; 16.0 ksi; $-7.12(10^{-4})$ in.; 4.977(10^{-3}) in.; 10.31(10^{-3}) in.

2.31 0.0248 in.   **2.32** $-0.6(10^{-4})$; 1.2(10^{-4}); 2.2(10^{-4})   **2.33** 0.000142; 0.01236; 0.01248 m

2.34 1.095 in.^{2}; 0.667 in.^{2}; 0.746 in.^{2}; 0.00402 in.; -0.00318 in.; 0.00352 in.   **2.35** $2.22P_0L/AE$

2.36 $-0.02P_0L/AE$ **2.37** $P_0[Lx - (x^3/3L)]/2AE$ **2.38** $P_0[\pi x + L\sin(\pi x/L)]/AE\pi^2$

CHAPTER 3

3.1 (a) 12.4(10^6) psi; (b) 35,400 psi; (c) 49,515 psi; (d) 50.75 psi; (e) 49,510 psi

3.2 (a) 29.1(10^3) ksi; (b) 40 ksi; (c) 44 ksi; (d) 0.028 ksi; (e) 48.5 ksi; (f) 45 ksi

3.3 (a) 14.86(10^6) psi; (b) 27,700 psi   **3.4** 6000; 0.0   **3.5** 0.0042; 0.0013   **3.6** 4.167

3.7 700 lb   **3.8** (a) 17.5, 8.8 ksi; (b) 18.3, 8 ksi   **3.9** 25.7 ksi   **3.10** 0.0172 in., 19,850 psi

3.11 0.067 in.   **3.12** 100.6°F   **3.13** 67°F   **3.14** 75°F   **3.15** 0.05 mm   **3.16** $P^2L/2AE$

3.17 $P^2L(1 + k)/2AE\,k$ **3.18** 0.886t   **3.19** 111 MPa; 186 MPa

CHAPTER 4

4.1 60, 75, 50 N·m   **4.2** $-12.5i$ kN·m   **4.3** 18.38 ft·k   **4.4** 14,260 psi   **4.5** 5.42, 6.40 MPa
4.6 0.024 in.   **4.7** 17.28, 36.13, 31.42 ksi   **4.8** 612, 816, 408, 4150 MPa
4.9 16.4, 8.5, 16.0 ksi   **4.10** 122.7 N·m   **4.11** 4.13 in.   **4.12** 34, 47, 0 ksi   **4.13** 0.856 rad
4.14 0.00103 rad   **4.15** 11.14, 4.77, 90.54 ksi; 0.0093, 0.0143, 0.2155
4.16 140, 1121 MPa; 0.0538, 0.4848   **4.17** 13.69, 8.60, 11.80, 14.35 MPa; -0.0048, -0.0018, 0.0051, 0.0132 rad.
4.18 10,470 in·lb   **4.19** 14.15, 5.97 ksi; $(10.2, 13.9, 24.1)(10^{-3})$ rad.
4.20 6.19 mm   **4.21** 3.70 in.   **4.23** 14.72 ksi   **4.24** 0.0723 rad   **4.25** 2.06 in.
4.26 46.7 mm, 0.025 rad, 0.039 rad   **4.27** 0.26 in.   **4.28** 0.00575 rad
4.29 $275F(10^3)$ Pa, $55.08F(10^6)/G$, 4ϕ **4.30** 7.55, 60 N·m; 75.1 kPa, 0.597 MPa; $9(10^{-5})$ rad
4.31 $T(x = 0.5) = 1.25$ kN·m; $\phi(x = 0) = 0.0233(10^{-3})$ rad.
4.32 $T(x = 12) = 24$ in.·lb c.c.w.; $T(x = 36) = 60$ in.·lb c.w.; $\phi(x = 0) = 1.95(10^{-6})$; $\phi(x = 12) = 2.93(10^{-6})$; $\phi(x = 24) = 3.17(10^{-6})$; $\phi(x = 18.86) = 3.49(10^{-6})$
4.33 $T(x = 32) = 40$, $T(x = 48) = 80$ in.·lb; $\phi(x = 0) = 5.52(10^{-6})$, $\phi(x = 32) = 3.82(10^{-6})$
4.34 $T(x = 0.5) = 10/\pi$, $T(x = 1.0) = 20/\pi$; $\phi(x = 0) = 64.8(10^{-6})$
4.35 $\phi = 2TL(R_2^3 - R_1^3)/[3G\pi(R_1R_2)^3(R_2 - R_1)]$ **4.36** 501 psi; 0.00236 rad
4.37 $b = 0.294$ m; $0.156(10^{-3})$ **4.38** 30 mm   **4.39** 116.3 psi; 0.00018 rad   **4.40** 1809 psi
4.41 86.5 MPa   **4.42** 635 psi   **4.43** 108 MPa   **4.44** 15.4 mm   **4.45** 12,600   **4.46** 7.52 hp
4.47 52.8, 23.5 MPa; 0.0756 rad   **4.48** 1811, 5432 psi; 0.04 rad   **4.49** 0.496 in.
4.50 AB, 0.478 in.; CD, 0.322 in.   **4.51** 0.0381 hp   **4.52** 0.75 in.

CHAPTER 5

5.1 (a) determinate; (d) determinate; (f) first degree indeterminate
5.2 (a) $A_y = 7wL/27$; (b) $A_y = 90$ N; (c) 468.75, 93.75 N; (d) $A_x = wL$, $B_y = 5wL/6$; (e) $C_y = 94.285$ N
5.3 $M(x = 12) = -120$ ft·k   **5.4** $M = -x^2 + 40x - 375$
5.5 $(0 < x < 7)$, $M_x = 31x - 3x^2/2 - 223.5$; $(7 < x < 15)$, $M_x = 10x - 150$
5.6 $V(x = 10) = -24$ k; $M(x = 6) = -72$ ft·k   **5.7** $(6 < x < 12)$, $M_x = -15(x - 6) - x^3/12$ **5.8** $M_x = M_0$
5.9 $(3 < x < 6)$, $M_x = -40x - 200$ **5.10** $M_{max} = 9wL^2/128$ at $x = 3L/8$
5.11 $(2.5 < x < 4)$, $V_x = -5$, $M_x = -5x + 50$; $M_{max} = 38.28$ N·m   **5.12** $M_{max} = 16.786$ N·m at $x = 2.5$ m.
5.13 $M_{max} = 172.5$ N·m at $x = 5$ m; $(5 < x < 8)$, $M_x = -57.5x + 460$ **5.14** $M_{max} = -200$ N·m
5.15 $(6 < x < 16)$, $M_x = -36(x - 3) + 30.8(x - 6)$; $M_{max} = -160$
5.16 $(L/3 < x < 4L/3)$, $M_x = -3M_0x/L + 2M_0$
5.17 $(0 < x < 3)$, $V_x = 3.875 - 5x^2/6$; $(3 < x < 4)$, $M_x = 3.875x - 7.5(x - 2) - (x - 3)^3/2$
5.18 $M_{max} = 0.06415\,wL^2$ at $x = 0.5774\,L$ **5.19** $M_{max} = 5.67$ ft·k at $x = 3.78$ ft
5.20 $(1.5 < x < 3)$, $M_x = -10x^2 + 45x - 45$
5.21 $M(x = 0.5) = -0.685$; $(1.0 < x < 1.4)$, $V_x = 3.43 - 30(x - 10)$; $M_x = 3.43(x - 0.7) - 30(x - 1)^2/2$
5.22 $M(x = 4) = -16$; $M(x = 13) = 8$; $M(x = 20) = -23.33$
5.23 $V(x = 3) = -21.354$ and 17.5 kN; $M(x = 3) = -9.378$ kN·m
5.24 $(8 < x < 14)$, $M_x = -228.67 + 24.5x - 3.5x^3/84$ **5.26** $M_{max} = 130.67$ ft·k at $x = 9.33$ ft
5.29 $w = 2$ N/m   **5.30** $V_a = 14.14$ k; $M_a = 80$ ft·k; $H_a = -35.36$ k   **5.31** $V_a = 40$; $M_a = -60$; $H_a = -30$
5.32 $M_a = 18$ kN·m; $V_a = H_a = 0$ **5.33** $V_a = -2$; $M_a = -15$; $H_a = 24$
5.34 $V_a = 10.71$ k; $M_a = 55.86$ k·ft; $H_a = 9.09$ k   **5.35** $V_a = -0.75$ kN; $M_a = 4.5$ kN·m; $H_a = -5.22$ kN
5.36 $M_a = 12.397$ ft·k   **5.37** $V_a = 0$; $H_a = -32$ kN   **5.38** $M_{max} = 31.25$ kN·m
5.39 $M = -48$ ft·k, 8 ft above A **5.40** $M_{max} = 31.25$ at midspan   **5.41** $V_c = 8$; $H_c = -9.33$
5.42 $M(x = 3.5) = -30$ kN·m   **5.43** $H_a = 6$ k, $V_{max} = \pm 6$ k, $M_{max} = -12$ ft·k for the beam
5.44 $M_z(x = 0) = -600$; $M_y(x = 0) = -432$ ft·k   **5.45** $M_{zmax} = 46.4$; $M_{ymax} = 79$ kN·m
5.46 $M_{xmax} = 45$; $M_{ymax} = 112.5$ kN·m   **5.47** $M_y(x = 2) = 20$ ft·k; $V_{yB} = -266.67$ k
5.48 $M_{ymax} = 187.5$; $M_{zmax} = -84.375$ kN·m   **5.49** $M_x(z = 2) = -426.67$ ft·k
5.50 $(0 < x < 3)$, $\mathbf{M} = 100\mathbf{i} - 50(3 - x)\mathbf{j} + 30(3 - x)\mathbf{k}$ **5.51** $\mathbf{M}_{AB} = -60\mathbf{i} + 1.875x\mathbf{j}$

5.52 $(0 < z < 4)$, $\mathbf{M} = 6\mathbf{j}$ kN·m   **5.53** $\mathbf{M} = -PR \cos \theta\mathbf{k}$
5.54 $M'_x = -144 \cos \theta$; $M'_y = -60(1 - \sin \theta) + 96 \cos \theta$; $M'_z = -144(1 - \sin \theta)$

CHAPTER 6
6.1 6.78 in.   **6.2** 64.87 mm   **6.3** 13.30; 16.48 in.   **6.4** 5.48; 8.03 in.
6.5 0.0642; 0.1456 m   **6.6** 314.2 in.^{4} **6.7** 21.274(10^6) mm^4 **6.8** 11,503; 7180.5 in.^{4}
6.9 4406; 2133 in.^{4} **6.10** 2.77(10^{-4}); 6.36(10^{-5}) m^4 **6.11** (a) 5400; (b) 3600 psi
6.12 21.6; 43.2 ksi   **6.13** 44.815 ksi   **6.14** 2.70; 0.90 MPa   **6.15** 0.34 m
6.16 $b = 150$; $h = 200$; $\sigma = 8.367$ MPa   **6.17** 20.8 ksi   **6.18** 96,000 psi   **6.19** 157; 351 MPa
6.20 1774 psi   **6.21** 0.569 m, 0.427 m   **6.22** 0.194 m   **6.23** 7.156 in.
6.24 19.78; 11.63; 10.32; 8.13 ksi   **6.25** 3.817 MPa   **6.26** 6400; 3200; 0; -3200; -6400 lb
6.27 372.9 kN; 52.3 MPa   **6.28** 27 k   **6.29** 13.44; 23.0; 12.8; 7.68 ksi
6.30 (a) 17.31 in.; (b) a and b, 10.18; (c) 14.20; (d) 17.31; (e) 7.2 in.; (c) 16618; 12091 in.^{3}
6.31 $\sigma(x = 0) = 11.2$ ksi; $\sigma(x = 60) = 3.875$ ksi   **6.32** 231.9 in.^{4} **6.33** 2115.6 in.^{4}
6.34 861.78 in.^{4} **6.35** 36.6 ksi   **6.36** S15X42.9   **6.37** W8X58   **6.38** 8.78 k/ft   **6.39** W21X93
6.40 S12X50   **6.41** 0.278 in.   **6.42** 2.77 in. (not economical)   **6.43** 2.44 ft; $t = 0.504$ in.
6.44 810 psi   **6.45** 1.89; 4.24 ft   **6.46** S6X17.25   **6.47** 1.79 ft
6.48 (a) 3240; (b) 3703; (c) 12,960 psi   **6.50** 0.296, 1.853 ksi   **6.51** 5.827; 63.323 kPa
6.52 $\sigma_1 = 0.307$; $\sigma_2 = -0.154$; $\sigma_3 = -0.36$ ksi   **6.53** $\sigma_{al} = 1.96$; $\sigma_{st} = 5.88$ ksi   **6.54** (b) 34.57 lb
6.55 $\sigma_i = 72.7$ MPa   **6.58** $\sigma_i = 13,780$; $\sigma_0 = -10,170$ psi   **6.59** $\sigma_i = 20.07$; $\sigma_0 = 5.86$ MPa
6.60 6.5 N·m

CHAPTER 7
7.1 64.86 kN; 129.72 kN/m   **7.2** 428.9 kN/m   **7.3** 2.57 k; 0.215 k/in.   **7.4** 1.62 in.
7.5 5.89 in.   **7.7** $\bar{y} = 0.775b$ **7.9** $\bar{y} = 200.9$ **7.11** 48.8 mm   **7.12** 1033.3N
7.13 (a) 11.1 in.; (b) 15.76 psi   **7.14** 5.97 in.   **7.15** 9.54 k   **7.17** 67.48 mm
7.18 $S(x = 2) = 6.83$ in.; $S(x = 8) = 6.50$ in.   **7.19** 10.27 kN/m; 10.51 kN/m; 0
7.21 2.53 ksi   **7.22** 8 ksi   **7.23** 11.76 ksi   **7.24** $0.075V$; $0.070V$ psi   **7.25** 4.44 ft
7.26 37.5 psi   **7.27** 0.32 m   **7.28** 11.85 in.   **7.30** 0.375; 0.281 ksi   **7.31** 11.85; 10.61 in.
7.33 1.55 ksi   **7.34** 7.0 ksi   **7.35** 3.37 in.   **7.36** 4.91 in.   **7.37** 0.502 k/in.
7.38 0.740; 1.031 k/in.   **7.39** 145.23 kN   **7.40** 25.43 mm   **7.41** 11.44 mm, left of centroid
7.42 $e_x = 3.98$ in., left of centroid; $e_y = 3.69$ in. below centroid   **7.43** 2.89 in. left of web

CHAPTER 8
8.1 A, 245.3; B, 5.3 MPa   **8.2** A, -10.95; B, -3.75 ksi   **8.3** $b = 5.95$ in.
8.4 A, $\sigma = 4.888$; $\tau = 1.332$ GPa   **8.5** 6.22 in., use 6.5 in.
8.6 $\sigma_A = 81.33$; $\sigma_B = 1.33$; $\tau_B = 1.6$ MPa   **8.7** $\sigma_A = 10.57$; $\sigma_B = -0.307$; $\tau_B = -0.323$ ksi
8.8 $\sigma_A = -21.15$; $\sigma_B = -1.39$; $\tau_B = -1.04$ ksi   **8.9** $\sigma_A = -18.0$; $\tau_A = 0.69(-z)$; $\sigma_B = -33.3$ ksi; $\tau_B = 0$
8.10 $\sigma_A = -10.42$; $\tau_A = 0.938(+z)$; $\sigma_B = -41.67$; $\tau_B = 0.625(+y)$ MPa
8.11 $\sigma_A = 0$; $\tau_A = 4.26(+z)$; $\sigma_B = 9.17$; $\tau_B = 4.67(-y)$ MPa
8.12 $\sigma_A = -99.5$; $\tau_A = 16.82(-z)$; $\sigma_B = -114.0$; $\tau_B = 17.95(+x)$ MPa
8.13 $\sigma_A = 0$; $\tau_A = 6.71(+y)$; $\tau_B = 7.27(+y)$ ksi
8.14 $\sigma_A = -45.1$; $\tau_A = 29.2(-y)$; $\sigma_B = -162.2$; $\tau_B = 30.56(+z)$ MPa
8.15 $\sigma_A = 17.81$; $\tau_A = 1.98(-z)$; $\sigma_B = -13.85$; $\tau_B = 1.88(+y)$ ksi; $\sigma_C = 35.16$; $\tau_C = 1.14(+z)$; $\sigma_D = -28.35$; $\tau_D = 1.52(-y)$ ksi
8.16 $\sigma_A = -73.67$; $\tau_A = 77.59(-x)$; $\sigma_B = -125.08$; $\tau_B = 80.63(+y)$; $\sigma_C = -92.09$; $\tau_C = 15.64(+z)$; $\sigma_D = 65.45$; $\tau_D = 14.71(-y)$; $\sigma_E = 21.31$; $\tau_E = 1.17(-x)$; $\sigma_F = -23.46$; $\tau_F = 0.94(-z)$ MPa
8.17 $\sigma_A = -8814.1$; $\tau_A = 196(+y)$; $\sigma_B = -6484$; $\tau_B = 261(+z)$ psi
8.18 $\sigma_A = 2823$; $\tau_A = 5641(-y)$; $\sigma_B = -3760$; $\tau_B = 5641(+z)$ psi
8.19 97,000; $29,700(-y)$ psi   **8.20** -64.97; 56.63 psi   **8.21** $\sigma_A = -887.47$; $\sigma_B = -612.53$ psf

8.22 $\sigma_A = -62.6$; $\sigma_B = -34.32$ kPa   **8.23** 8786; 4334 psi   **8.24** $\sigma_A = 10.67$; $\sigma_C = -17.33$ MPa
8.25 $\sigma_A = 0.582$; $\sigma_B = -4.01$ ksi   **8.26** $\sigma_A = 1.25$; $\sigma_B = -4.75$ ksi   **8.27** 5.6; 0.64 ksi
8.28 $\sigma_A = 0$; $\tau_A = 52.86(-z)$ MPa   **8.29** $\sigma_D = -21.89$; $\tau_D = 2.72(+z)$; $\sigma_E = -15.43$; $\tau_E = 2.0(-y')$ MPa
8.30 $956(+y')$ kPa   **8.31** -37.8; $18.7(-y')$ ksi   **8.32** 451; $12.4(+z')$ MPa
8.33 $\sigma_D = 101$; $\sigma_E = 6320$ psi   **8.34** $N = -9.6$ kN; $V = 7.2$ kN; $M = 14.4$ kN·m
8.35 $\sigma_C = -14.65$; $\tau_C = 18.43$; $\sigma_D = 436.5$ psi   **8.36** $\tau_C = 0.901$; $\sigma_D = 18.0$ MPa
8.37 $N = 0.985$ k; $V = 0.174$ k; $M = 1.11$ ft·k   **8.38** 52.48; 1.96 MPa
8.39 $\sigma(z = -1 \text{ in.}) = 13.9$; $\tau(+y) = 0.187$ ksi   **8.40** 0.18 in.   **8.41** $\sigma_A = 77.94$; $\tau_B = 48.46$ MPa
8.42 A, -270, 2340; B, 2610, 2340 psi   **8.43** $\sigma_{Ay} = 77.7$; $\sigma_{Ax} = 36$; $\tau_B = 47.7$; $\sigma_{By} = 18$ ksf
8.44 $\sigma_A = 8.639$; $\sigma_B = -4.295$ ksi   **8.45** $\sigma_A = 13{,}750$; $\sigma_B = -7814$ psi   **8.46** 16.88 kN

CHAPTER 9
9.1 -1258; -617; -881 psi   **9.2** -0.165; -1.25; -0.45 ksi   **9.3** 15.96 kN   **9.4** 6; 6; 0 MPa
9.5 0; 0; 5 MPa   **9.6** -9.17; 14.1; -30.8 **9.7** -3.75; 5.75; -1.85 MPa
9.8 (a) 84.7 at $-13.3°$; -4.7 at $76.7°$; 44.7 at $-58.3°$; (b) 42.7; 37.3; -44.6 ksi
9.9 (a) 10; -10; ± 10 MPa; (b) 3.4; -3.4; -9.4 MPa   **9.10** (a) 8; 0; ± 4 ksi; (b) 6; 2; -3.5 ksi
9.11 (a) -0.4 at $118.2°$; -7.6; $\tau = \pm 3.6$ or ± 3.8 MPa; (b) -2.1; -4.6; -3 MPa
9.12 (a) 12.9 at $58.2°$; -4.9; ± 8.94 ksi; (b) 2.98; 5.02; 8.88 ksi
9.13 (a) -4.3; at $19.3°$; -10.7; $\tau = \pm 3.2$ or ± 5.35 MPa; (b) -5.7; -9.3; -2.7 MPa
9.20 $\tau_{\max} = \sigma_2/2$ **9.21** $P = 10.5$ kN   **9.22** 14,070 in.·lb   **9.23** 34.96 k   **9.24** 372.7 N·m
9.25 T in.·lb $= 0.0842\ P$ lb   **9.26** A, 35.6; -8.1; ± 21.85 MPa; B, ± 16.67 **9.27** $b = 9.65$ in.
9.28 12.28; 2.72; ± 4.75 or ± 6.14 ksi; $-60.76°$ **9.29** 3.30; -2.28 ksi   **9.30** $60.77°$
9.31 $55.13°$ and $66.41°$, θ as in Fig. P9.29   **9.32** 26.23 mm   **9.33** $\tau = \sigma/2$ **9.34** 0.79 in.
9.35 9818 lb; $6.8(10^{-5})$ **9.36** 4372; -6984; ± 5678 psi; $\theta_{N1} = 96.2°$ **9.37** 300 psi
9.38 $\varepsilon_x = \varepsilon_0$; $\varepsilon_y = [2(\varepsilon_{60} + \varepsilon_{120}) - \varepsilon_0]/3$ **9.40** $I'_{xy} = bh(h^2 - b^2) \sin\theta\cos\theta/12$
9.41 $I'_{xy} = bh[(7h^2/4) - b^2]\sin\theta\cos\theta/12$ **9.42** $I_x = 167.3$; $I_y = 95.3$; $I_{xy} = -74.1$ in.^{4}
9.43 A, 2.49; B, -35.47; C, 31.80 MPa   **9.44** A, -82.28; B, 81.63; C, -64.23; D, 17.68 ksi

CHAPTER 10
10.1 1.818 m; 0.0864 N·m   **10.2** 1979.17 in.   **10.3** 0.024 in.   **10.4** 0.0225 in.
10.5 $v = (P/6EI)(-x^3 + 3L^2x - 2L^3)$ **10.6** $v = (wx^2/24EI)(-6L^2 + 4Lx - x^2)$ **10.7** $v = M_0 x^2/2EI$
10.8 $v = (w_0/120\ EIL)(-x^5 + 5L^4x - 4L^5)$ **10.9** $v = (w_0 x^2/120EIL)(-20L^3 + 10L^2x - x^3)$
10.10 $v = (w_0/360EIL^2)(-x^6 + 6L^5x - 5L^6)$
10.11 $(0 < x < L)$; $v = (Px/9EI)(x^2 - 5L^2)$; $(L < x < 3L)$; $v = (P/72EI)(-4x^3 + 36Lx^2 - 76L^2x + 12L^3)$
10.12 $(0 < x < L - a)$; $v = (wx^2/24EI)[-6(L - a)^2 + 4x(L - a) - x^2]$; $(L - a < x < L)$;
$v = [w(L - a)^3/24EI](-4x + L - a)$
10.13 $(0 < x_1 < L)$; $v = (P/6EI)(-x^3 + 18L^2x - 26L^3)$; $(0 < x_2 < L)$; $v = (Px^2/2EI)(x - 4L)$
10.14 $v = -(w_0 L^4/EI\pi^4)\sin(\pi x/L)$
10.15 $(0 < x < L)$; $v = (M_0 x/18EIL)(x^2 + 3L^2)$; $(L < x < 3L)$; $v = (M_0/18EIL)(x^3 - 9Lx^2 + 21L^2x - 9L^3)$
10.16 $(0 < x_1 < L)$; $v = (w_0 x/360EIL)(10L^2x^2 - 3x^4 - 7L^4)$; $(0 < x_2 < L)$;
$v = -(w_0 x/360EIL)(10L^2x^2 - 3x^4 - 7L^4)$
10.17 origin at left end; $(0 < x < L)$; $v = (wxL^3/3EI)(x - L)$
origin at left support; $(0 < x < 4L)$; $v = (wLx/48EI)(16L^2 - x^2)$
origin at right end; $(0 < x < L)$; $v = (w/24EI)(-x^4 + 20L^3x - 19L^4)$
10.18 origin at left end; $(0 < x_1 < L)$; $v = (wL^3/3EI)(L - x)$
origin at left support; $(0 < x_2 < 2L)$; $v = (w/24EI)(Lx^3 - x^4 - L^3x)$
origin at right end; $(0 < x_3 < L)$; $v = (wL^3/3EI)(L - x)$
10.19 $(0 < x_1 < L)$; $v = (M_0 x/18EIL)(x^2 + 3L^2)$; $(0 < x_2 < 2L)$; $v = (M_0 x/18EIL)(6L^2 - x^2)$
10.20 $v = (wx/24EI)(2Lx^2 - x^3 - L^3) + (v_1/EIL)(x - L)$ **10.21** $v = (Px/60EI)(-30Lx + 10x^2 + 3L^2)$

10.22 $M = wL^2/8$ **10.23** $v = (PL/2EI_0)(-x^2 + 2Lx - L^2)$

10.24 $(0 < x < 0.25)$; $L = 0.6$ m $v = w(4Lx^3 - 6L^2x^2)/24E_1I$ $(0.25 < x < 0.6)$
$v = w(4Lx^3 - 6L^2x^2 - x^4)/24E_2I + w(0.0144x - 0.0015)/E_2I$ $v(x = 0.6) = -1.478(10^{-3})$ m

10.25 $(0 < x < L)$; $v = w(-2x^4 + 36L^3x - 43L^4)/48EI_0$; $(L < x2L)$; $v = w(-x^4 + 32L^3x - 40L^4)/48EI_0$

10.26 origin at the left end, $(0 < x < a)$; $v = wax^2(x - 3a)/12EI_0 - 7wa^3x/24EI_1$
origin at the left support, $(0 < x < a)$; $v = w(-x^4 + 2ax^3 + 6a^2x^2 - 7a^3x - 7a^4)/24EI_1 - wa^4/6EI_0$
origin at the right end, $(0 < x < a)$; $v = wa\{[x^2(x - 3a)/12I_0] - 7a^2x/24I_1\}/E$
$I_0 = 1.0(10^{-4})$, $I_1 = 3.375(10^{-4})$ origin at left support, $x = 0.25$, $v = -1.377(10^{+6})/E$ m

10.27 $7wL^4/128EI$ **10.28** $1.514(10^{-3})$ m   **10.29** $11wL^4/24EI$ **10.30** $73wa^4/240EI$ **10.31** $PL^3/48EI$

10.32 $u_V = 6.97(10^{-3})$ m; $u_H = 0$ **10.33** $u_E = 10.2Pl/AE$; $u_D = 22.5Pl/AE$ **10.34** $464.86P/AE$ in.

10.35 0.051 in.   **10.36** $u_C = 1.274Fl/AE$; $u_A = 0.5Fl/AE$ **10.37** $u_C = 0.0438$ m; $u_D = 0.0146$ m

10.38 $PL^3/48EI$; $PL^3/16EI$ **10.39** $v_B = wL^4/8EI$; $\theta_C = wL^3/2EI$ **10.40** $v = 11wL^4/120EI$; $\theta = wL^3/8EI$

10.41 $v = 9.806(10^{-3}) wL^4/EI$; $\theta = 1.804(10^{-3}) wL^3/EI$ **10.42** $v = wL^4/384EI$ up; $\theta = wL^3/48EI$

10.43 $v = 5M_0L^2/4EI$; $\theta = 3M_0L/2EI$ **10.44** $v = 41wa^4/24EI$; $\theta = 7wa^3/6EI$

10.45 $v = 17wL^4/16EI_0$; $\theta = 3wL^3/4EI_0$ **10.46** $53wa^4/36EI$ **10.47** $0.0259wL^4/EI_0$

10.48 $v_A = 5.11$ mm; $u_B = 5.71(10^{-3})$ mm   **10.49** $v_A = 1.397$ in.; $v_B = 0.461$ in.   **10.50** $2PL^3/3EI + 4PL/AE$

10.51 $v_C = 0.0216$ m; $u = 5.96(10^{-5})$ m   **10.52** TL/JG **10.53** $TL^2/8JG$ **10.54** $9wa^4/8EI$

10.55 $4Pa^3/EI$ **10.56** $v_H = 9M_0L^2/32EI$ (left); $v_V = M_0L^2/2EI$ (up)   **10.57** $0.196wL^4/EI$

10.58 $v_A = 7wL^4/24EI$; $v_D = wL^4/3EI \rightarrow$ **10.59** 1.52 in.

10.60 $v_x = -36/EI$; $v_y = 75/2EI$; $v_z = (16/JG) - (46/EI)$; ft

10.61 $\theta_x = 4/EI + 10/JG$; $\theta_y = 20/EI + 8/JG$; $\theta_z = 21/EI$

10.62 $v_z = -5.00P/EI - 1.03P/JG$ **10.75** $2wa^3/EI$ **10.76** $PaL^2/8EI$ **10.77** $10wL^4/243EI$

10.78 $v_A = Pa^3/EI$; $v_{max} = 4Pa^3/9\sqrt{3}EI$ at $x = 2a/\sqrt{3}$ from right end

10.79 0.992 in.   **10.80** $M_0L^2/2EI$; M_0L/EI **10.81** $2.2(10^{-4}) wL^4/EI$

10.82 $v_A = 7M_0a^2/6EI$; $v_{max} = 4M_0a^2/9\sqrt{3}EI$ at $x = 2a/\sqrt{3}$ from right end

10.83 $21wL^4/480EI$; $11wL^3/192EI$

10.84 $v_A = wL^4/48EI$; $v_B = 5wL^4/144EI$; $v_{max} = 0.0209wL^4/EI$ at $x = 0.5275L$ from left support.

CHAPTER 11

11.1 $P/5$; $4P/5$ **11.2** 120 k   **11.3** $R_A = 3P/4$; $R_B = P/2$ **11.5** $P_A = 7.467$; $P_B = 6.533$

11.6 $T_A = 5T/9$ **11.7** $T_A = 0.458T$ **11.8** 3.83 in.   **11.9** $T_A = T_1/4$; $T_B = 3T_1/4$

11.10 83.33; 16.67 **11.11** 41.55, 16.90, 41.55 percent of P **11.12** Area $A = (0.65)$(Area B)

11.13 $P_S = 207.1$; $P_M = 67.9$ kN   **11.14** $R_A = 51.38$; $R_B = 3.62$ kN   **11.15** $41,667$ psi

11.16 $26,582$, $10,855$ psi   **11.17** $C = 0.778 P$; $B = 0.519 P$ **11.18** $C = 30.45$; $B = 38.05$; $A = -18.5$ k

11.19 $B = 1.77$; $C = 6.91$; $A = -4.18$ **11.20** $B = 2PK_1/(K_1 + 4K_2)$ **11.21** $B = 2.86$; $C = 5.72$

11.22 $A = wL/10$; $B = 2wL/5$ **11.23** $A = 22.87$; $B = 4.13$ k; $\sigma = 31.33$ ksi

11.24 $A = wa/8$; $B = 99wa/48$; $C = 39wa/48$ **11.25** $A = 3wa/4$; $B = 11wa/4$; $C = -3wa/2$

11.26 $A = -P/4$; $B = 7P/8$; $C = 3P/8$ **11.27** $A = -3P/4$; $B = 7P/4$, $M = PL/2$

11.28 $A = -41wL/80$; $B = 121wL/80$; $M = 43wL^2/240$ **11.29** $A = M_0/L$; $C = -M_0/L$

11.30 $A = 43wL/128$; $B = 21wL/128$; $M = -11wL^2/128$

11.31 $A = 151wa/320$; $B = 9wa/320$; $M = -53wa^2/480$

11.32 $A = 13w_0a/16$; $15w_0a/8$; $5w_0a/16$ **11.33** $A = 2M_0/L$; $B = 0$; $C = -2M_0/L$

11.34 $A = 359wa/128$; $B = 137wa/64$; $C = 263wa/128$ **11.35** $.B = 3wL/8 - 3EI\delta/L^3$

11.36 $B = 15wL/16 - 6EI\delta/L^3$ **11.37** $M_B = wL^2/15 - 3EI\theta/L$ **11.38** $M = 7wL^2/128 - 3EI\theta/L$

11.39 $M = wL^2/12$; $R = wL/2$ **11.41** Case F2   **11.42** Case F4

11.43 $A = 8.17$; $B = 20.97$; $C = 25.09$; $D = 9.76$ **11.44** $A = 4wL/7$; $B = 13wL/28$; $M = -3wL^2/28$

11.45 A; $-9M_0/8L$; $-3M_0/16$ and B; $9M_0/8L$; $-5M_0/16$ **11.47** $A = 23wL/60$; $B = 6wL/5$;
$C = 9wL/20$; $D = -wL/30$ **11.48** $A = 7w_0L/60$; $B = 13w_0L/60$; $M = -w_0L^2/30$

11.49 $A = w_0L/10$; $B = 2w_0L/5$; $M = wL^2/15$ **11.50** $A = 23w_0L/40$; $B = 11w_0L/120$; $M = -19w_0L^2/120$
11.51 A; $R = 0.423w_0L$; $M = -0.0642w_0L^2$ and B; $R = 0.2136w_0L$; $M = -0.0465w_0L^2$
11.52 A; $R = -0.0770w_0L$; $M = 0.0191w_0L^2$ and B; $R = -0.2864w_0L$; $M = 0.0368w_0L$
11.53 A; $R = -w_0L/12$; $M = w_0L^2/60$ and B; $R = -w_0L/12$; $M = w_0L^2/60$ **11.54** Case F8   **11.55** Case F9
11.56 $A = -3EI\theta_A/L$; $M = 3EI\theta_A/L^2$ **11.57** $A = 44wL/81$; $B = 10wL/81$; $M = -8wL^2/81$
11.58 $A = wL/10$; $B = 2wL/5$; $M = -wL^2/15$ **11.59** $A = -B = -3M_0/2L$; $M = M_0/2$
11.60 A; $R = 17wL/16$; $M = -3wL^2/16$ and C; $R = 7wL/16$; $M = -wL^2/16$ and B; $R = 3wL/2$; $M = -wL^2/8$
11.61 A; $R = 9wL/10$; $M = -3wL^2/20$ and B; $R = wL$; $M = -wL^2/12$ and C; $R = wL/10$; $M = wL^2/60$
11.62 A; $R = 5wL/4$; $M = -wL^2/6$ and B; $R = -wL$; $M = wL^2/12$ and C; $R = 3wL/4$; $M = -wL^2/6$
11.63 A; $R = 17wa/48$; $M = -5wa^2/144$ and B; $R = 91wa/64$; $M = -13wa^2/72$ and C; $R = 43wa/192$;
$M = -37wa^2/288$
11.64 A; $R = 13wa/80$; $M = -3wa^2/80$ and B; $R = wa/2$; $M = -5wa^2/120$ and C; $R = 27wa/80$; $M = -11wa^2/240$
11.65 A; $R = 9wa/14$; $M = -75wa^2/448$ and B; $R = 675wa/448$ and C; $R = 157wa/448$
11.66 A; $R = -3wa/16$; $M = wa^2/16$ and B; $R = 11wa/16$
11.67 B; $R = 49wL/40$ and C; $R = 31wL/40$; $M = -13wL^2/120$ **11.68** $R_A = 5wa/6$; $R_B = 3wa/2$; $R_C = -wa/3$
11.69 (a) $R_A = 477.8$ lb; $R_B = 6722.2$; $M_A = -1178$ ft·lb; (b) $R_B = 5877.2$; (c) 0.994 in.
11.70 $R_A = 22.89$; $R_B = 4.11$; $M_A = -39.88$ ft·k
11.71 (a) $R_A = 2502.2$; $R_B = 1497.8$; $M_A = -4017.6$ ft·lb; (b) $R_B = 1449.1$ lb
11.72 $R_A = 18M_0kL^2/(8kL^3 + 81EI)$ **11.73** $R = 1.478$ k (tension); $R_D = 18.52$ k; $M_D = -68.18$ k ft
11.74 108 lb   **11.75** 2.155 kN; 6.76 mm   **11.76** $3.532w$ **11.77** $R_A = -239$; $R_B = 318$; $R_C = -79$ lb
11.78 $6EI/L^3$ **11.79** 30.8 MPa   **11.80** 58.33 MPa   **11.81** 44.42 MPa; 71.07 MPa
11.82 30.88 kN (right)   **11.83** 49.1 kN (comp.)   **11.84** 105.77   **11.85** 5268t; 2199c, psi
11.86 0.327 rad   **11.87** 0.245 rad (tighten); 4172 psi ten.   **11.88** 146.9°F
11.89 $P_{al} = 10{,}999$ lb comp.; $P_{cu} = 11{,}046$ lb comp.; $P_{st} = 47$ lb ten.
11.90 $R_A = R_B = 435.2$; $R_C = 870.4$ lb   **11.91** $R_C = 2R_A = 4.06$ lb; 1828 psi   **11.92** 147.6
11.93 al., 30.98 lb ten.; st., 61.96 lb comp.   **11.94** $R_B = -1235.4$; $R_A = R_C = 617.7$ N
11.95 $R_B = 864.8$; $R_A = R_C = -432.4$ N   **11.96** $R_B = -283.4$ N; $\sigma = 11.65$ MPa
11.97 $R_{0D} = 1.73$ lb; $M = 17.3$ ft·lb   **11.98** 0.0606 N (ten.)   **11.99** 2.06 N; 1.2 mm
11.100 198°F   **11.101** 288 MPa   **11.102** 180°F; 35.1 ksi
11.103 $H_A = -6.42$; $H_B = 1.59$; $H_d = 4.83$ kN   **11.104** $R_A = R_C = 4.71$; $R_B = 6.58$ kN
11.105 $H_A = -14.14$; $H_C = -15.86$; $V_C = -5.86$ k   **11.106** $F_{AE} = 6.49$ kN ten.
11.107 $F_{FB} = 4.12$ kN ten.   **11.108** $F_{BF} = 7.75$ k comp.   **11.109** $R_B = 15wL/32$
11.110 $V_A = 11wL/20$; $H_B = wL/30$ **11.111** $H_D = 100$ k   **11.112** $V_A = 7.11$ k
11.113 $R_A = -0.1875$ kN   **11.114** $R_B = 0.75$ kN; $\sigma = 18.4$ MPa   **11.115** $V_B = 12$ k; $M_{AX} = -24$ ft·k
11.116 $R_B = 13.5$ k   **11.117** 5.49 k ten.   **11.118** $R_C = 9.36$ N; $\mathbf{M}_A = 7.489\mathbf{i} - 14.04\mathbf{k}$

CHAPTER 12
12.3 16.2 ft   **12.4** 28.94 in.   **12.5** 4.83 ft   **12.6** 45,880; 32,287; 59,476 lb   **12.7** 0.237 m
12.8 w10x39   **12.9** w10x39   **12.10** 30.35 ft   **12.11** (a) 39.48 kN; (b) 80.57 kN; (c) not an Euler column
12.12 176.1 k   **12.13** 1353.7 k   **12.14** 24.48 k   **12.15** 44.8 k   **12.16** 110.2 k   **12.17** 11.32 in.
12.18 13,290 lb   **12.19** 58 in.; 27,674 lb   **12.20** 128.6 k   **12.21** 104 k   **12.22** 23.3 k   **12.23** 33.87 k
12.24 (a) 1.685 m; (b) 1.704 m; (c) 1.330 m   **12.25** 139.3 kN; 121 kN   **12.27** w10x45   **12.28** w14x145
12.29 w12x58   **12.30** w10x49   **12.31** st, w10x49; al, w12x96   **12.32** w14x109   **12.33** 125 in.
12.34 w8x28   **12.35** w10x49   **12.36** 0.77 in.   **12.37** w10x39   **12.38** 13.2 mm   **12.39** 5.69 in.
12.40 4.74 in.^{2} **12.41** 4   **12.42** 60.52 in.^{2} **12.43** 13.85; 14.0; 16.82 in.   **12.44** 4.96 in.

CHAPTER 13
13.1 $PL^3/3EI$ **13.2** $Pa^2b^2/3EIL$ **13.3** $5PL^3/24EI$ **13.4** $PL^3\pi/4EI$ **13.5** $10PL^3/81EI$
13.6 $M_AL/3EI$ **13.7** $7M_CL/12EI$ **13.8** $13M_BL/6EI$ **13.9** T_AL/JG **13.10** $3.828PL/AE$

13.11 $243PL/72EA$ **13.12** $\delta_H = PL/AE$; $\delta_V = 3.828PL/AE$ **13.13** $\delta_H = 2PL/AE$; $\delta_V = 0$
13.14 $\delta_H = \sqrt{3}PL/AE$; $\delta_V = 7.62PL/AE$ **13.15** $R_A = 3wL/8$ **13.16** $R_A = wL/10$
13.17 $R_B = wa$; $R_A = R_C = wa/2$; $M_A = M_C = -wa^2/12$ **13.18** see Prob. 11.60   **13.19** see Prob. 11.63
13.20 $7wL^4/24EI$ **13.21** 0.4 mm   **13.22** $5PL^3/162EI$ **13.23** $57wL^4/24EI$
13.24 $11w_0L^4/120EI$ **13.25** Pa^3/EI **13.26** $2wL^4/3EI$ **13.27** $wL^4/4EI$ **13.28** 0.3844 in.
13.29 $\theta_A = Pab(b-a)/3EIL$ **13.30** $5wL^3/12EI$ **13.31** $2(10^{-4})$ rad   **13.32** $PL^2/6EI$
13.33 $7wL^3/4EI_0$ **13.34** $w_0L^3/8EI$ **13.35** $Pa^3/12EI$ **13.36** $0.123wL^3/EI$
13.37 0.0066 rad   **13.41** $4PL/3AE$ **13.42** $v = 3wL^4/128EI_z$; $w = wL^4/48EI_y$
13.43 $u = 0.0115$; $v = 2.89$; $w = 1.63$ in.   **13.44** $(2.4\mathbf{i} + 0.032\mathbf{j} - 0.56\mathbf{k})/EI$
13.45 $u = 11M_BL^2/24EI$; $v = 8M_BL^2/9EI$ **13.46** $(-0.0684\mathbf{j} + 0.0395\mathbf{k})$ in.
13.47 $v = -(22.5/EI + 90/JG)$ **13.48** $(2072\mathbf{i} - 1387\mathbf{j} + 3760.4\mathbf{k})(1728/EI)$
13.49 $R_B = 3wL/8$ **13.50** $M_A = wL^2/8$ **13.51** $R_B = 57wL/72$; $M_A = 7wL^2/72$
13.52 12.6 k; 29.4 k; -64.8 ft·k   **13.53** $-wa/16$; $5wa/8$; $7wa/16$ **13.54** $R_B = 6EI\delta + 5wa/8$
13.55 Case F9   **13.56** $R_B = -R_A = 3EI\theta/L^2$; $M_A = 3EI\theta/L$ **13.57** Case F.6
13.58 $R_B = 13.14$; $M_A = -17.16$ ft·k   **13.59** $R_A = 23wL/24$; $M_A = -11wL^2/24$
13.60 $R_B = 5wa^4k/[4(a^3k + 6EI)]$ **13.61** $R_A = 5APL^2/[2(2AL^2 + 3I)]$
13.62 $R_A = R_B = P/2$ **13.63** $R_A = P/3$; $R_B = 5P/3$ **13.64** $R_A = P_0L/3$; $R_B = P_0L/6$
13.65 $R_A = P_0L(\pi - 2)/\pi^2$; $R_B = 2P_0L/\pi^2$ **13.66** $T_A = 3T/5$; $T_B = 2T/5$ **13.67** $T_A = -1$; $T_B = 4$ kN·m
13.68 $T_A = -3rP_y/2$; $V_A = P_y/2$; $M_A = -P_yL/8$ **13.69** $T_A = -T_0L/4$; $T_B = -T_0L/12$
13.70 $R_B = 4.32$ k   **13.71** $R_B = 17.46$ k   **13.72** 2.53 k
13.73 $\sigma_A = -5.61$ MPa; $\sigma_B = 11.22$ MPa; $\sigma_b = 14.03$ MPa
13.74 $\sigma_A = 5.14$ MPa; $\sigma_B = -10.28$ MPa; $\sigma_b = 12.89$ MPa
13.75 196.3 N   **13.76** 0.331 N; 0.172 N   **13.77** $\mathbf{R}_B = -(3M_z/2a)\mathbf{j}$; $\mathbf{M}_A = -M_x\mathbf{i} - M_y\mathbf{j} + (M_z/2)\mathbf{k}$
13.78 $\mathbf{R}_A = -1.25\mathbf{i} + 0.209\mathbf{j} - 0.402\mathbf{k}$ $\mathbf{M}_A = -1.21\mathbf{i} + 2.98\mathbf{j} - 2.27\mathbf{k}$ $\mathbf{R}_B = -1.25\mathbf{i} + 0.291\mathbf{j} + 0.402\mathbf{k}$
$\mathbf{M}_B = -1.29\mathbf{i} + 2.61\mathbf{j} + 1.67\mathbf{k}$ **13.79** $V_A = 7wL/4$; $H_A = 5wL/8$ **13.80** $V_A = V_B = P/2$; $H_A = H_B = P/\pi$

CHAPTER 14

14.1 (a) $4.57(10^{-5})\,PL/A$; (b) $(10P_{TOTAL} - 8.89P_1)(10^{-5})\,L/A$ **14.2** (a) 213.36 k; (b) 240 k
14.3 (a) $P = 253.3$; $P_A = 120$; $P_B = 133.3$ k; (b) 340 k
14.4 B yields at 481.4 N; $P_{max} = 657.7$ N; $u_A = 1.818(10^{-4})$; $u_B = 7.272(10^{-4})$ m
14.5 B yields at 535.2 N; $P_{max} = 809.8$ N; $u_A = 2.33(10^{-4})$; $u_B = 9.33(10^{-4})$ m
14.6 BD first yields at $P = 25.64$ k; AB first yields at $P = 33.33$ k; BD second yield at $P_{max} = 35.26$ k
14.7 1600; 2400 in.·k   **14.8** 2515.59 in.·k   **14.9** 43.97; 77.18 kN·m; 1.75
14.10 2880.56; 5334 in.·k; 1.85   **14.11** 3.605; 4.807 kN·m   **14.12** 6.625; 8.833 ft·k; 11.64 percent
14.13 30.8; 34.5 kN·m   **14.14** 0.025; 0.025; 0.05 in.   **14.15** 1842; 2981.2 in.·k   **14.16** 1.46 kN; 3.67 MN
14.17 $51 > 50$; $1957 < 2500$ **14.18** no failure; no failure   **14.19** $91.24 > 50$; $6250 > 2500$
14.20 $40.8 < 50$; $1374 < 2500$ **14.21** failure; failure   **14.22** $65 > 50$; $4225 > 2500$
14.23 11.16; 12.88 in.·k   **14.24** $14.76 < 16$; $212.4 < 256$ **14.25** 8 in.   **14.26** 0.95 in.; 0.82 in.
14.27 $0.0096\sigma_0$ **14.28** 0.625 ksi   **14.30** 1.87; 1.65 in.   **14.31** 2.79; 2.47; 2.79; 2.44 in.

CHAPTER 15

15.1 10.67 ksi; $k = 2.33$; 37.28 ksi   **15.2** 1250 kPa; $k = 2.16$; 5400 kPa
15.3 3 MPa; $k = 2.85$; 17.1 MPa   **15.4** 5.03 MPa; $k = 2.6$; 15.41 MPa
15.5 10 ksi; $k = 2.94$; 29.4 ksi   **15.6** 13.57 MPa; $k = 2.04$; 27.68 MPa
15.7 21 MPa; $k = 1.83$; 38.43 MPa   **15.8** $k = 1.63$; 6.52 ksi; $k = 2.94$; 23.52 ksi
15.9 16 ksi; $k = 2.175$; 50.1 ksi   **15.10** 4.61 MPa; $k = 2.03$; 14.62 MPa
15.11 1.75 MPa; $k = 2.91$; 5.85 MPa   **15.12** 1.75 MPa; $k = 2.23$; 11.47 MPa
15.13 1.17 MPa; 2.29 MPa; $k = 1.36$; 3.12 MPa   **15.14** 955 psi; 2264 psi; $k = 1.41$; 3192 psi
15.15 955 psi; 7639 psi; $k = 1.22$; 9320 psi   **15.16** 6042; 1343 in.·lb

INDEX